Acoustical Imaging

Volume 11

Acoustical Imaging

A Continuation Order Plan is available for this series. A continuation order will bring delivery of each new volume immediately upon publication. Volumes are billed only upon actual shipment. For further information please contact the publisher.

Acoustical Imaging

Volume 11

Edited by
John P. Powers

Naval Postgraduate School
Monterey, California

PLENUM PRESS · NEW YORK AND LONDON

The Library of Congress cataloged the first volume of this series as follows:

International Symposium on Acoustical Holography.

Acoustical holography; proceedings. v. 1-
New York, Plenum Press, 1967-

v. illus. (part col.), ports. 24 cm.

Editors: 1967-   A. F. Metherell and L. Larmore (1967 with H. M. A. el-Sum)
Symposiums for 1967-   held at the Douglas Advanced Research Laboratories,
Huntington Beach, Calif.

1. Acoustic holography–Congresses–Collected works. I. Metherell. Alexander A.,
ed. II. Larmore, Lewis, ed. III. el-Sum, Hussein Mohammed Amin, ed. IV. Douglas
Advanced Research Laboratories. v. Title.
QC244.5.I 5   69-12533

ISBN-13: 978-1-4684-1139-3   e-ISBN-13: 978-1-4684-1137-9
DOI: 10.1007/978-1-4684-1137-9

Proceedings of the Eleventh International Symposium on Acoustical
Imaging, held May 4–7, 1981, in Monterey, California

© 1982 Plenum Press, New York
Softcover reprint of the hardcover 1st edition 1982
A Division of Plenum Publishing Corporation
233 Spring Street, New York, N.Y. 10013

PREFACE

This volume contains forty-one papers presented at the Eleventh
International Symposium on Acoustical Imaging held on 4 - 7 May 1981
in Monterey, California. The objective of this conference series is
to bring together workers in diverse areas and applications of
Acoustical Imaging for interaction and exchange of ideas. People
working in other aspects of scalar wave theory and applications
also benefit from this series.

The papers presented here demonstrate continued growth in the
activity of this field. In this conference there was emphasis on
New Techniques, Acoustic Tomography, Tissue Characterization,
Signal Processing, Inversion Techniques, and Transducers and Arrays.

The success and stimulation of the conference and of the
papers presented in this volume is owed, of course to the authors
and participants. Many thanks are due to the authors and their
co-workers for their diligence and enthusiasm in performing their
research, preparing their manuscripts and presenting their results.
The editor would like to express his appreciation to each and every
one of them.

The editor would like to also thank the Program Committee who
evaluated the abstracts and gave their able advice in the organi-
zation and direction of the conference. The program committee
consisted of Mahfuz Ahmed, University of California at Irvine
Medical Center; Pierre Alais, Universite Pierre et Marie Curie,
Paris; C.B. Burchhardt, Hoffman-La Roche and Co., Basle, Switzerland;
B.P. Hildebrand, Spectron Development Labs, Costa Mesa, California;
C.R. Hill, Royal Marsden Hospital, Sutton, England; Larry W. Kessler,
Sonoscan, Inc., Bensenville, Illinois; Alexander F. Metherell, South
Bay Hospital, Redondo Beach, California; Rolf K. Mueller, University
of Minnesota, Minneapolis, Minnesota; George Sackman, Naval Post-
graduate School, Monterey, California; Frederick L. Thurstone, Duke
University, Durham, North Carolina; Robert C. Waag, University of
Rochester Medical Center, Rochester, New York; Glen Wade, University
of California, Santa Barbara; and Keith Y. Wang, University of Miami,

Florida.   The editor wishes to thank the session chairmen: Keith
Wang, Eric Ash, H. Dale Collins, Rolf Mueller, Frederick Thurstone
and George Sackman for their effort and contribution to provide a
smooth-running conference.

    Special thanks are due to Ms. Darlene Kelley for her help and
assistance in providing her highly capable secretarial support.
The help cf Ms. Ruby Kapsalis, Conference Coordinator at the Naval
Postgraduate School, is also gratefully acknowledged.

    The editor would like to especially thank the Office of Naval
Research (Physics Division) for their sponsorship of this meeting
and the Naval Postgraduate School for graciously providing the
meeting facilities.   The Symposium was held in cooperation with
the IEEE Group on Sonics and Ultrasonics.

    The Twelfth International Symposium on Acoustical Imaging
will be held in London, England, on 19 - 22 July 1982 under the
joint chairmanship of Prof. Eric Ash and Dr. Kit Hill.   The
Thirteenth International Symposium will be held in Minneapolis,
Minnesota in the Fall of 1983 and will be organized by Prof. Rolf
Mueller, Prof. Mos Kaveh and Dr. James Greenleaf.

                                       John Powers

CONTENTS

SIGNAL PROCESSING TECHNIQUES

TRANSDUCERS AND ARRAYS

NEW TECHNIQUES

SCANNING ACOUSTIC MICROSCOPY OF SOLID OBJECTS

USING ASPHERIC LENSES

Frank Pino, David A. Sinclair, Eric A. Ash

Department of Electronic and Electrical Engineering
University College London
Torrington Place, London WC1E 7JE, England

1.   INTRODUCTION

Acoustic microscopes are able to image regions within an opaque
material.   This characteristic is of particular value for a range
of NDE problems, including those arising in metal and semiconductor
components.   Solids have a substantially higher velocity and
impedance than water, a fact which implies potentially severe
problems in high resolution imaging.   The planar interface between
the water and the object results in a foreshortening of the focal
distance as well as in severe aberrations.   Both effects can be
seen in the example presented in Figure 1, which shows a ray optics
diagram for the case when the velocity ratio, (relative to water)
for the lens material is 4.3 and for the object material 3.1.   The
outer rays from the lens are totally internally reflected; the
spherical aberration is severe.   Though not apparent from such a
ray diagram there is also a considerable amount of apodization
arising from the lower efficiency of longitudinal wave transmission
for rays with large angles of incidence.   It is apparent that this
situation implies a substantial degradation of the imaging
performance[1].

The problem has been studied by a number of workers[2,3,4].
One can alleviate the difficulties by resorting to a liquid metal
coupling fluid[2,4].   If, after Jipson, one bases the imaging on
the shear waves which are generated at the interface, there is the
additional advantage of a reduction, by a factor of about two, in
the velocity of the waves in the solid.   There are, however, some
practical problems associated with the use of liquid metals.   The
aim of the present study is to examine what can be done with regard
to the design of the lens itself, to reduce the aberration within

1

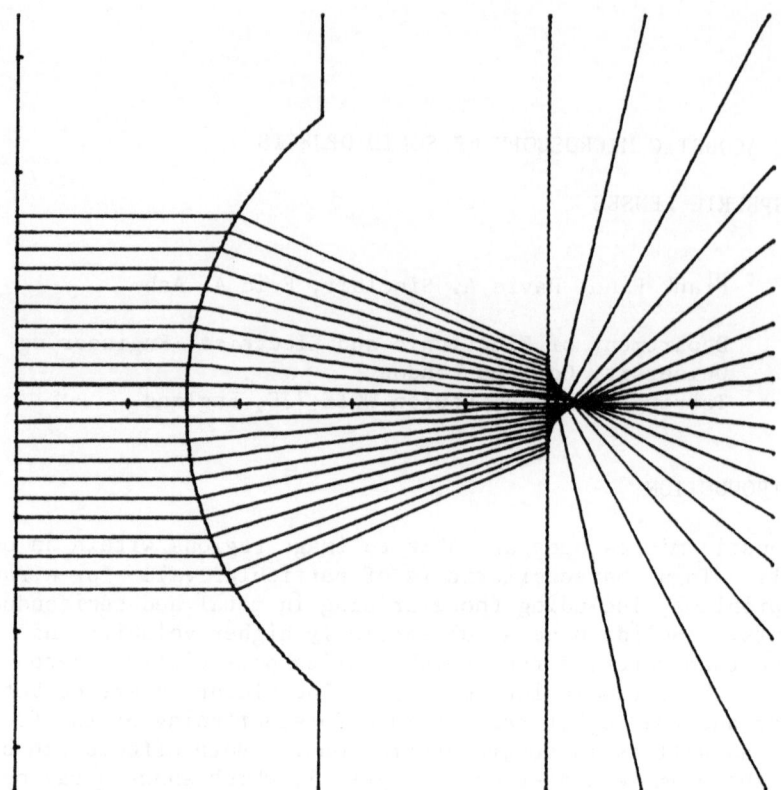

Fig. 1.   Ray diagram for a spherical lens (r_ℓ = 4.3), focussed on
            an internal plane in a solid object (r_o = 3.1).

the object.   Specifically we seek to explore the use of aspheric
lenses, and of spherical lenses adapted to a specific material.

2.   COMPARISON OF LENS GEOMETRIES

    Ray tracing provides a useful insight into various possible
configurations; they are however at best suggestive, and require
confirmation by means of computations which take diffraction into
account.   We have calculated the distribution of the field in the
vicinity of the focal region, using the established techniques of
Fourier optics – though so far in the two-dimensional approximation.
One assumes a wave with a rectangular amplitude plane phase front
incident on the lens, whose only effect is to impose a phase
curvature (i.e. a "thin lens" assumption).   The resulting field
is represented in terms of a set of plane waves, propagated to the
object interface.   Snell's Law provides the new direction of the

individual plane waves, and the amplitudes are calculated from the classic expressions for the transmission of plane waves across an interface[5].   The field distributions are then obtained by summing the plane wave spectrum.

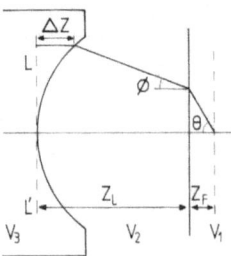

Fig. 2.   Notation used in defining aspheric lens profile.

In most of the cases discussed below, we have restricted the angular aperture of the beam radiated by the lens so as to produce a focal angular aperture of the order of $\pm45^{\circ}$.   This in itself provides a major improvement in the imaging performance relative to the use of a lens designed for focussing in water - the case indicated in Figure 1.   The imaging performance is related to the autocorrelation of the focal distribution; however one can obtain a very good indication of performance by quoting just two figures - the 3dB width, and the sidelobe level.   We will concentrate on these two parameters in assessing various lens structures.   The parameters relating to the imaging arrangement are defined in Figure 2.

The velocity relative to that of water is defined as r_ℓ, r_o for the
lens and object material respectively. The wavelengths in the
liquid and the object will be denoted λ_w and λ_o respectively.

3.   ASPHERIC LENS DESIGN

We will assume that the wave field incident on the back face
of the lens has a constant amplitude over the planar wave front.
This assumption already implies that we cannot expect to find a
lens surface geometry which will give ideal - i.e. diffraction-
limited-focussing in the object.   Nevertheless, one might expect
to approach this condition and further, that the geometrical optics
solution to the problem would provide a reasonable guide.   The aim
is then to find Δz as a function of Θ in Figure 2 so as to produce
a focus in the object plane.   The simplest way of achieving this
requirement is to apply Fermat's principle, by equalising the
transit time from the focus to the plane LL' for all values of Θ.
Direct application of Snell's law then leads to the result

$$\frac{\Delta z}{z_\ell}\left(\frac{r_\ell - \cos\phi}{r_o \cos\phi}\right) = \frac{z_f}{r_o z_\ell}\left(\frac{(1-\cos\phi)}{\cos\Theta}\right) + \left(\frac{1-\cos\phi}{\cos\phi}\right) \qquad (1)$$

where $r_o \sin\phi = \sin\Theta$.

The ray focussing diagram for a lens so designed is shown in
Figure 3 and is, as expected, devoid of aberration.   The general
effect of adopting such an aspheric design is to produce a lens
surface which, by comparison with a normal spherical design, has a
smaller curvature at the periphery than on the axis.   The geometrical change is most marked when $r_\ell \gg r_o$ - a fact which derives
directly from the circumstance that the geometric "error" arising
from the planar object interface is corrected by a geometric change
in the lens, scaled by the factor r_ℓ/r_o.   It is important to
appreciate that the aberrations in a very fast object medium are
not more severe than in a slower medium, provided that the design
is based on the definition of a specified angular aperture within
the solid.

4.   COMPUTED IMAGING FIELDS

An example of the use of an aspheric lens for focussing in a high
r_ℓ/r_o situation is shown in Figure 4; the r_ℓ and r_o figures
correspond to a sapphire lens and a brass object.   The 3dB main
lobe width is $0.67\lambda_o$ where λ_o is the wavelength in the object, and
the sidelobe level is -12dB.   An isometric projection of the wave
amplitude field is shown in Figure 5.   A similar plot for the same
materials, but this time with a focal plane $50\lambda_o$ inside the object
is shown in Figure 6.   The sidelobe level is almost unchanged,
whilst the 3dB width has increased to $0.96\lambda_o$.   The results obtained

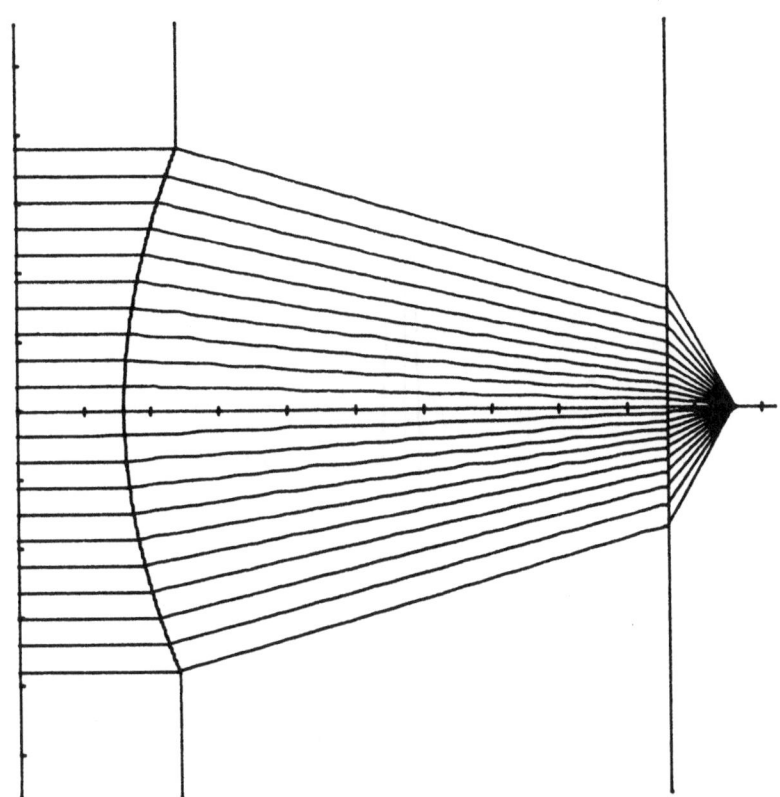

Fig. 3.   Ray diagram for an aspheric lens (r_ℓ = 4.3), focussed on
an internal plane in a solid object (r_o = 3.1).

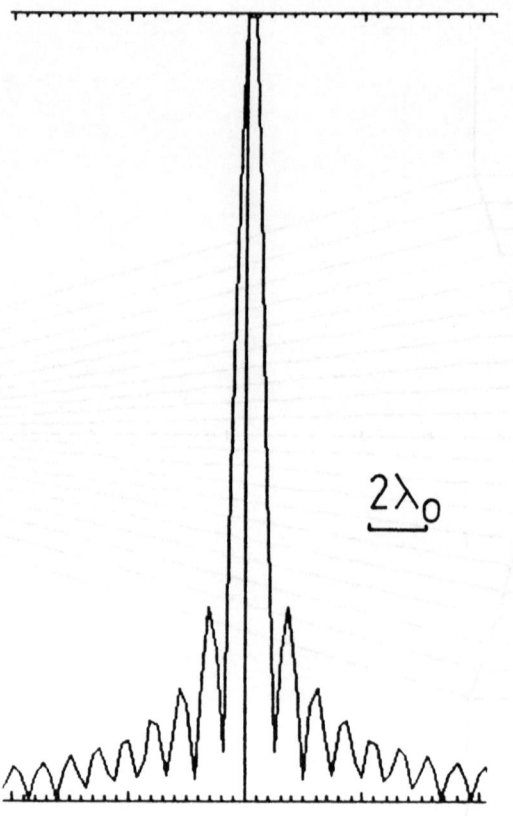

Fig. 4. Focal plane distribution of an aspheric lens (sapphire/ water; $r_\ell = 7.4$) focussed to a depth of $5\lambda_o$ in a brass object ($r_o = 3.1$).

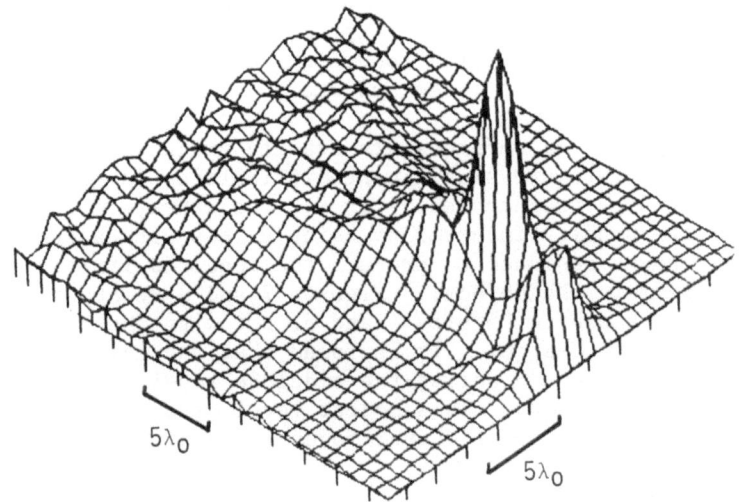

Fig. 5. Focal plane region for lens of Figure 4.

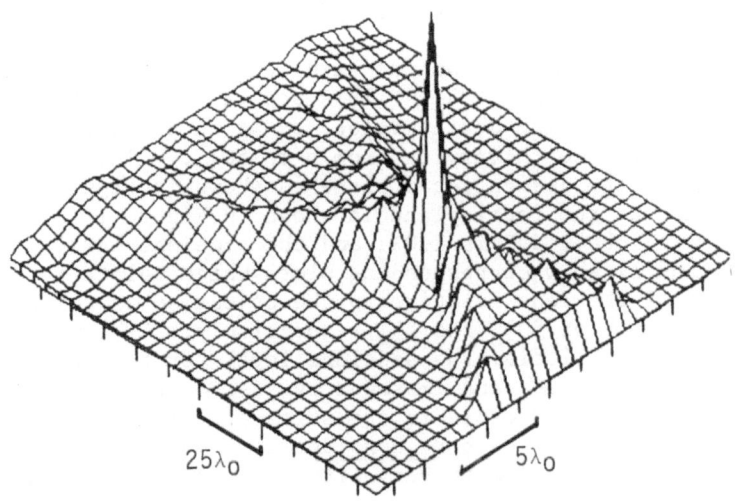

Fig. 6. Focal plane region for an aspheric lens (sapphire/water; $r_\ell = 7.4$) focussed to a depth of $50\lambda_o$ in a brass object ($r_o = 3.1$).

for $r_\ell = r_0 = 3.1$ are very similar, both as regards main lobe width as sidelobe levels. These results using an aspheric lens surface suggest that within a few wavelengths of the surface one can approach diffraction limited focussing, and even at substantial depths, the degradation is not severe.

Whilst aspheric surfaces are readily produced at low frequencies, there are significant difficulties in generating such surfaces at high frequencies. It is, therefore, of interest to discover the penalty incurred by using a lens with a spherical surface, though designed to produce the required limited angular aperture beam in the object. Figures 7 and 8 relate to the same case as that shown in Figures 4 and 5, but with the restriction of a spherical lens design. It is found that the main lobe width is essentially unchanged, whilst the sidelobe level is degraded by about 2.1dB to -9.9dB. This modest level of degradation is maintained in the case of deeper focussing designs.

5. FOCUSSING RANGE

The lenses, whose performance was discussed in the previous section, are designed to produce a focus in a given material, at a specified depth below the surface. In practice, one will wish to be able to focus through a range of depths with a given lens. We have explored the range of focussing for a specific case – that of a lens with $r_\ell = 4.3$, producing a focus within an object, $r_0 = 3.1$, at a depth of $30\lambda_0$ with an angular aperture of $\pm 30^\circ$.

For this design condition the path length in the water, measured on the axis was $213\lambda_w$. This separation was then varied both above and below the design value, so as to produce a focus in the object at a depth varying from 5 to $50\lambda_0$. The field distribution at the design focal plane is shown in Figure 9 while the region around the focal plane is shown in Figure 10. In Figure 9 the 3dB main lobe width is $1.0\lambda_0$ and the sidelobe level -12dB. Figures 11 and 12 show corresponding plots which relate to a displacement of the lens relative to the object so as to produce a focus at a depth of $50\lambda_0$ and $5\lambda_0$ respectively. The main lobe width changes very little; the sidelobe level for the deep focus is degraded by 1dB to 11dB, and improves by just over 1dB to 13.2dB for the focus produced very close to the surface. These results suggest that for the specific case here considered, a very substantial range of focus can be obtained with very little degradation of the focal plane distribution.

6. EXPERIMENTAL MODELLING RESULTS

The experiments were carried out on models, at a frequency of 1.7MHz. They were based on a water tank with two axis mechanical drives, actuated by stepper motors and under computer control.

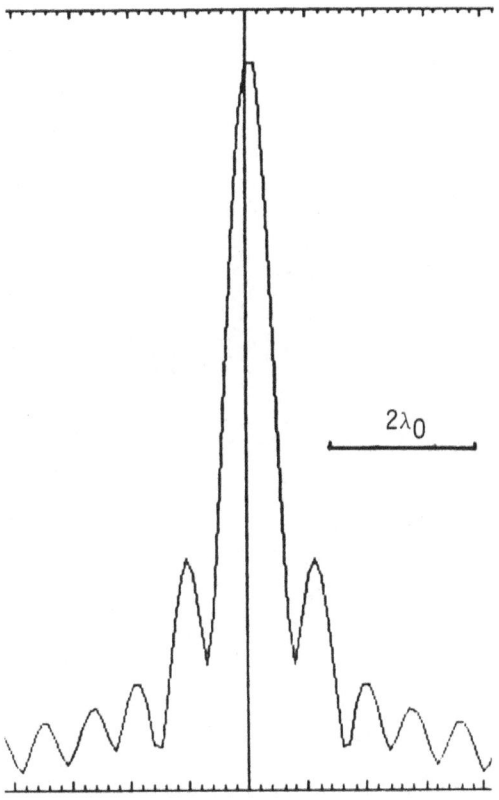

Fig. 7.   Focal plane distribution for spherical lens
(sapphire/water; r_ℓ = 7.4) focussed at a depth of $5\lambda_o$ in
a brass object (r_o = 3.1).

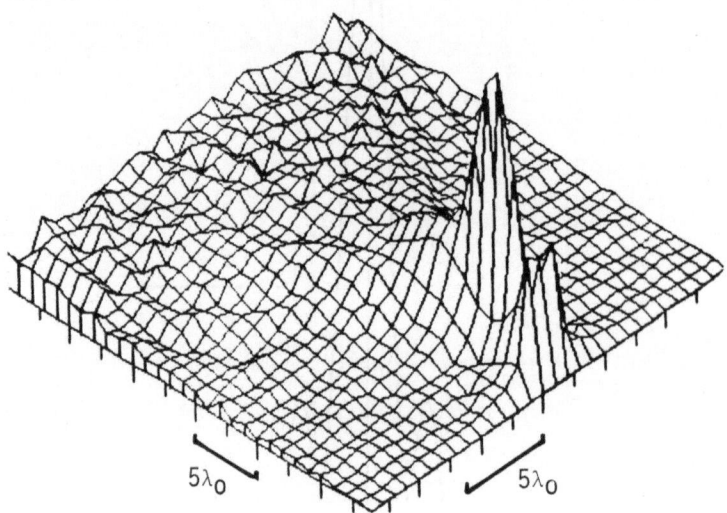

Fig. 8.   Focal plane region for lens of Figure 7.

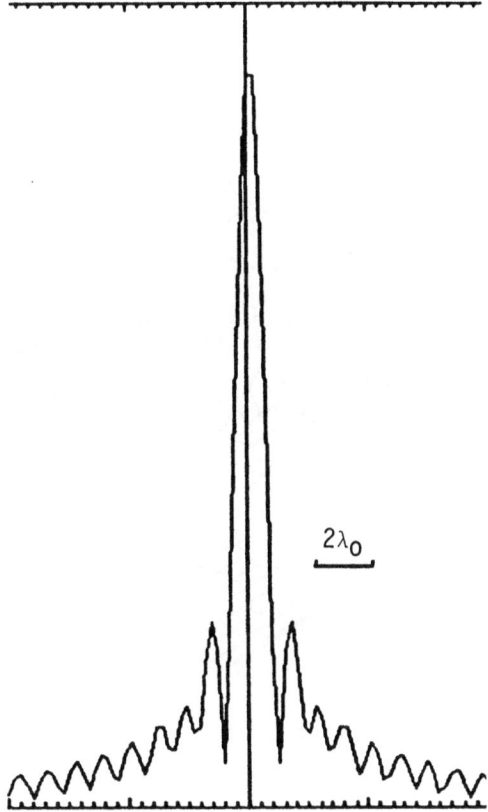

Fig. 9. Focal plane region for a spherical lens
         aluminium/water; r_ℓ = 4.3, focussed at the design depth
         of $30\lambda_o$ in a brass object (r_o = 3.1).

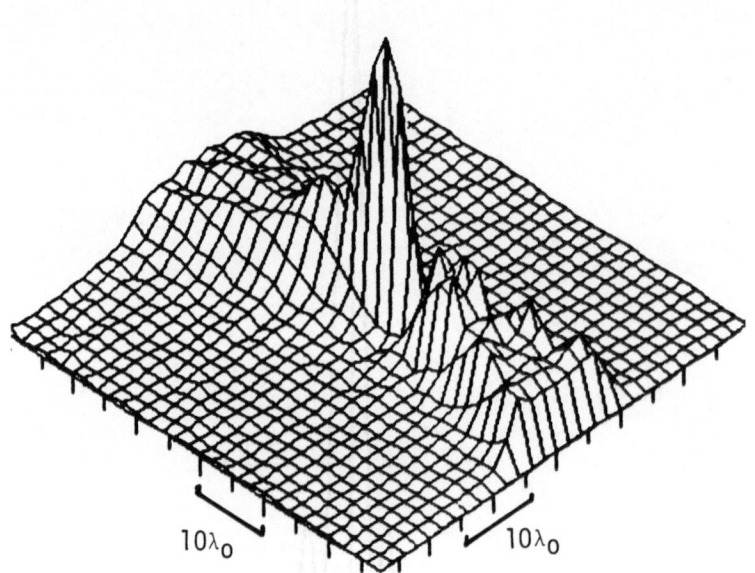

Fig. 10.   Focal plane region for the lens of Figure 9.

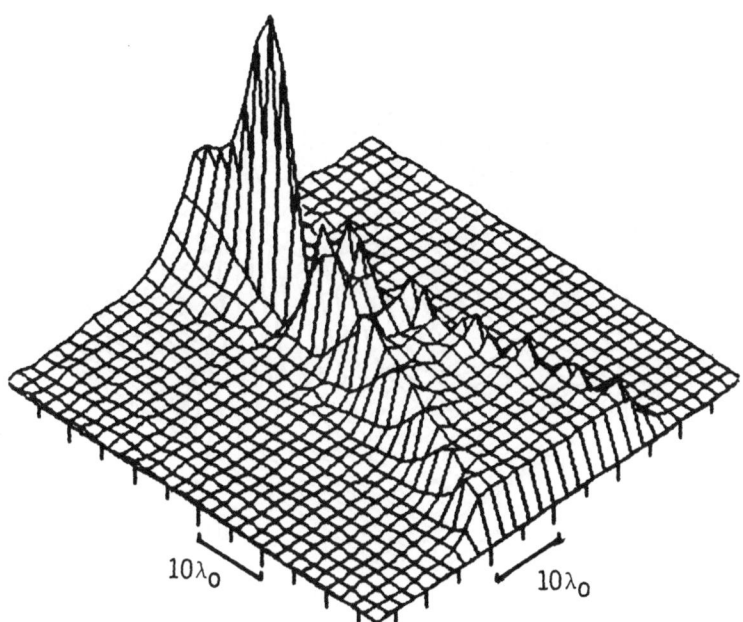

Fig. 11.   Focal plane region for the lens of Figure 9 obtained
when the lens was moved to produce a focus at a depth
of $50\lambda_o$ in the sample.

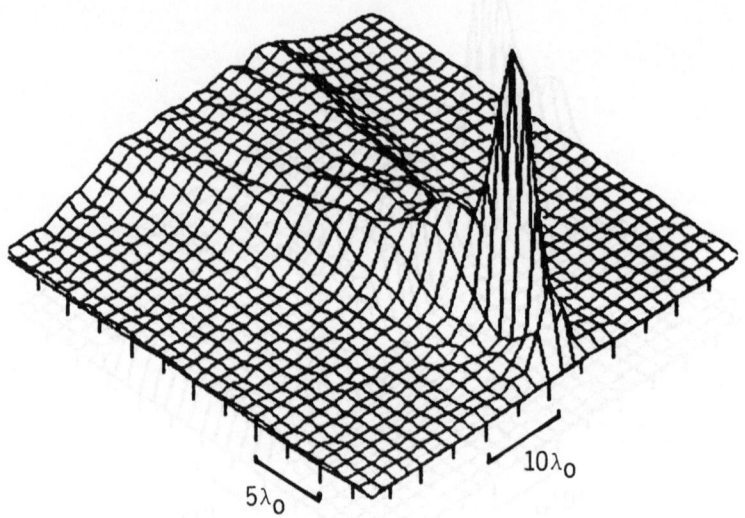

Fig. 12   Focal plane region for the lens of Figure 9 obtained
          when the lens was moved to produce a focus at a depth
          of $5\lambda_o$ in the sample.

Both amplitude and phase signals were digitized and digitally evaluated. The phase was derived by the use of two reference signals in quadrature which were sequentially mixed with the signal to be measured. Amplitude accuracy is estimated at 5%, phase accuracy at 10°.

We performed two types of experiment. The first is indicated in Figure 13. The thickness of the brass object was $1.8\lambda_o$ and the design focal plane was located on the back of the object. This allowed the distribution to be probed by means of a small transducer. The angular aperture in the object was $\pm60^o$.

The second type of experiment, involved the use of a specially prepared object with a series of slots of known width located in its center. The object was illuminated by means of a plane wave transducer, Figure 14. Figures 15a and b show the results obtained with the probing system of Figure 13 for the case of a spherical and an aspherical lens. The measured response is, of course, the convolution of the actual distribution with the probe transducer width – approximately 1 mm. One can deconvolve the measured result to obtain an improved estimate of the width of the main lobe. With this procedure we find the effective width for the spherical lens is $0.81\lambda_o$, and for the aspheric lens $0.45\lambda_o$. These results are within 20% of the values predicted by our computer simulations. The measured sidelobe levels, –10.5dB for the spherical lens and –14dB for the aspheric lens, are again encouragingly close to the predicted values.

Figure 16 shows a comparison between the scanned image obtained by means of the apparatus of Figure 14 and the dimensions of the slots within the object. It is clear that dimensions of the order of λ_o are clearly resolved.

7. CONCLUSIONS

The calculations and preliminary experiments indicate that one can obtain diffraction limited focussing inside solids – even those having high propagation velocities, with little degradation of sidelobe levels. The key issue in the design lies in restriction of the angular aperture of the beam within the object to values of the same order as commonly used in microscopes focussed in water (e.g. $\pm60^o$). The use of aspheric lenses leads to the closest approximation to aberration free focussing. However, the degradation arising from the use of spherical lenses is not, in most cases, very large.

In principle a given lens design is addressed at a particular material and a specified depth of focus. In practice, at least for modest angular apertures ($\pm30^o$) a large range of depths can be imaged, merely by changing the lens-object distance.

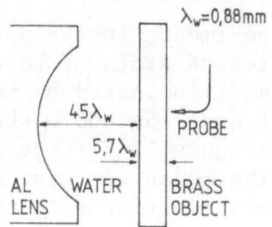

Fig. 13.   Experimental arrangement for focal plane measurement.

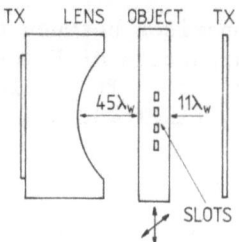

Fig. 14.   Arrangement for transmission resolution test.

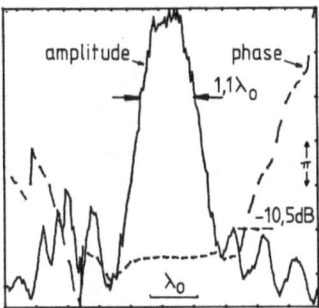

Fig. 15.   Beam probing results for (a) spherical lens

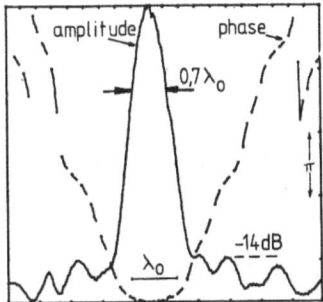

Fig. 15.   Beam probing results for (b) aspheric lens.

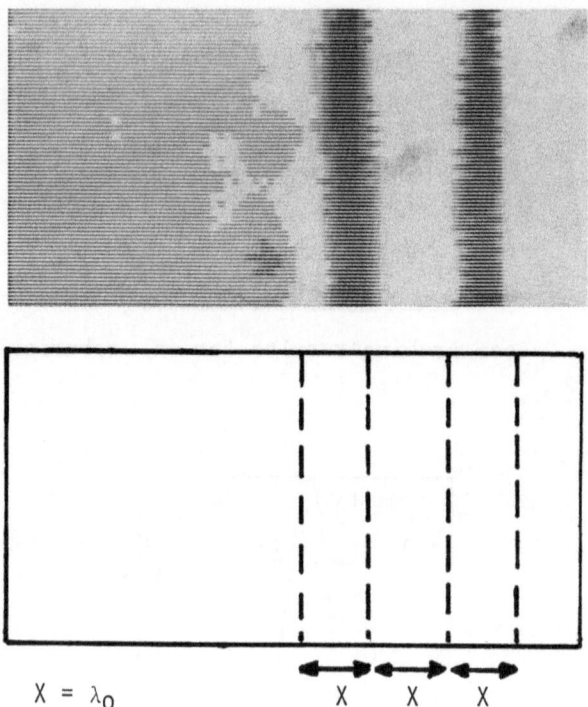

Fig. 16.   Transmission picture of object with buried slots.

Preliminary experimental results are in encouraging agreement with the computed performance.   There is, however, a need to confirm the latter using a full three dimensional analysis.

REFERENCES

1.   D.A. Sinclair and E.A. Ash, Bond Integrity Evaluation using
        Transmission Scanning Acoustic Microscopy, Electr. Lett., 16:
        23 pp 880-882 (1980).
2.   V.B. Jipson, Acoustic Microscopy of Interior Planes, Appl.
        Phys. Lett., 35 pp 385-387 (1979).
3.   B. Nongaillard, J.M. Rouvaen, E. Bridoux, R. Torguet and
        C. Bruneel, Visualisation of Thick Specimens using Reflection
        Acoustic Microscopy, J. Appl. Phys., 50 pp 1245-1249 (1979).
4.   J. Attal, Acoustic Microscopy: Imaging Microelectronic
        Circuits with Liquid Metals, in: "Scanned Image Microscopy,"
        E.A. Ash, ed., Academic Press, London (1980).
5.   L.M. Brekhovskikh, "Waves in Layered Media", Academic Press
        (1980).

ACKNOWLEDGEMENTS

The authors wish to express their gratitude to Mr S. Humphries of Willesden College of Science and Technology for his help in fabricating the aspheric lens used here.   The authors are also grateful to Mr M. Gillett and Mr J. Vansickle of University College London for technical assistance.

A DIGITAL PROCESSING SYSTEM FOR ACOUSTO-OPTIC

VISUALIZATION OF SOUND FIELDS

Henry D. Dardy and Charles F. Gaumond

Acoustics Division
U.S. Naval Research Laboratory
Washington, D.C. 20375

ABSTRACT

The visualization of sound fields provides an important tool to aid in the design of transducers for sonar and medical applications, the study of sound scattering and radiation from objects, or the non-destructive evaluation of materials. The ability to see the sound image through its interaction with light has remained primarily qualitative as the process of constructing a quantitative acoustic image is complex. This work describes how a sensitive schlieren imaging system has been applied to the quantitative presentation and modeling of sound field data under various conditions. This reconstruction process is aided by an interactive digital video processing facility capable of acquiring, correcting, and displaying experimental camera data. Frames can be processed, and if necessary, digitally corrected and enhanced. The added capability afforded by the computational system can be used to mathematically simulate and test under controlled conditions practical models for the sound interaction process. This combined simulation system, together with the experimental imaging system, serves to quantify processes arising from fluctuations, field irregularities, etc. in an absolute fashion. This approach can easily be extended to other than the schlieren imaging as an instructive tool provided one maintains a raster scan format.

1.0   INTRODUCTION

The ability to quantitatively visualize sound as it propagates through a medium presents an interesting challenge to the researcher as well as a useful tool in a number of areas; the design of transducers for sonar and medical applications, the study of scattering and radiation of sound fields from objects, and the non-destructive evaluation of materials are a few representative areas. The need to understand acoustic processes and to image them qualitatively has progressed. The final step of constructing a quantitative acoustic image from passive optical sensing of the acoustic field is complex; one must account for both the acoustic processes involved as well as the optical effects of the readout processes, namely an integrated optic effect, a cumulative buildup of signal in the image plane due to the finite width of the acoustic source.

The need encountered in Navy research to understand in a quantitative fashion the acousto-optic processes completely and to establish them as a useful research tool has led to the development of an interactive, digital processing laboratory facility capable of obtaining digital video images from acoustic experiments and of playing back of these images in a processed and partially animated fashion. Additionally, the same system has been structured so as to allow researchers to model processes and synthesize their images. One can then view a process utilizing pseudo color imagery to display a quantitative spatial view or one can output gray scale images to a hardcopy plotting device.

2.0   THE SCHLIEREN VISUALIZATION PROBLEM

The system has been structured primarily around the schlieren imaging process which offers both a high sensitivity and a wide frequency range for visualization of acoustic fields. The basic components of the schlieren geometry are shown in Figure 1. Briefly, schlieren is a process whereby an acoustic field imposes a spatial phase modulation on a parallel, collimated beam of light traversing orthogonal to the acoustic field of interest. After optical fourier transformation of the output light, one can spatial filter to remove the unmodulated beam in the transform plane of the collection lens; retransformation of the remaining light with a relay lens forms an amplitude image of the sound field. Placing the photosurface of a video camera in the image plane allows one to view in real-time the progression of the sound field. Through synchronization of the sound pulse with the light modulator, one can then resolve the phase of the propagating sound field.

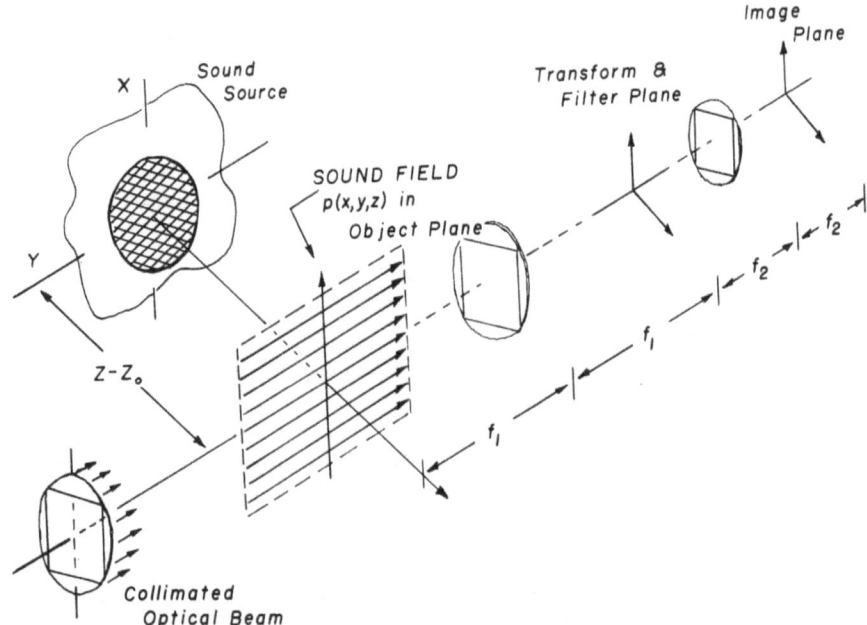

Fig. 1. Diagram of geometry involved in schlieren visualization.

The image formation process of the schlieren system has been treated classically by a number of authors.[1-7] We will only review it briefly here using a Fourier approach.[8] The optical wave fronts are initially plane before traversing the sound field. Using the coordinates shown in Figure 1, we have the light field $E_o(x,z)$ = constant. Assuming Raman-Nath diffraction, the sound field imposes on the optical field a spatial phase modulation which is proportional to the integrated acoustic pressure. We can therefore represent, neglecting time,

$$E_1(x,z) = E_o \exp[i\nu(x,z)] \tag{1}$$

$$\nu(x,z) = (2\pi\kappa/\lambda)\int p(x,y,z)dy \tag{2}$$

where $p(x,y,z)$ is the instantaneous, acoustic pressure distribution, λ is the optical wavelength and κ a piezo-optic coefficient. This optical field is fourier transformed into the filter plane, where the central order, which is the spatial average of the field over the object plane, is blocked.

$$E_2(k_x, k_y) = \iint \exp[i(k_x + k_z z)]E_1(x,z)dxdz$$

$$- \delta(k_x)\delta(k_z)E_2(0,0) \tag{3}$$

The relay lens once again transforms this filtered image to form the intensity pattern which is detected by a video camera tube.

$$E_3(x,z) = E_1(x,z) - E_2(0,0) \tag{4}$$

$$I_3(x,z) = I_0[1 + b^2 - 2b\cos(\nu(x,z))] \tag{5}$$

This resultant image impinging on the video camera is one which spatially describes the sound field of interest propagating in the medium. The system works exceedingly well in the megahertz frequency region, the upper limit being defined by the changeover of Raman-Nath scattering to Bragg scattering[9]; it can likewise work well down to 100 kilohertz and lower. At these lower frequencies, however, a number of system constraints begin to affect image quality, i.e.: finite aperture effects, filter plane resolution, lens diffraction limits, and thermal effects to name a few.[10] As such, we have a major difference between schlieren geometries and other imaging processes. Schlieren applies to imaging of fields alone, or to fields external or emmanating from an object of interest. Current ultrasound scan systems reconstruct an object of interest. The schlieren spatial image, however, is one which maps qualitatively only, as from Equations (2) and (5), we note it is both a function of the acoustic source width, the integrated optic effect, as well as acoustic pressure.

Digital reconstruction of the spatial pattern allows us to apply processing to the image and thereby invert, correct and enhance features of interest. Additionally, with proper configuration of a digital processing system, one can also apply digital processing concepts to simulate image reconstructions through the schlieren process utilizing simple acousto-optic models to learn how strongly various variables each affect the formation of a complex image. Further, with combined application of video acquisition and data processing, one can achieve a three dimensional reconstruction of a sound field volume using techniques of projection reconstruction. With this in mind, we have constructed a laboratory digital processing system, which, although small, appropriately attempts to solve as well as benchmark some of the data problems encountered when working with large frames of data from acousto-optic experiments.

3.0   THE DATA ACQUISITION PROBLEM

    Heart of the visualization system is the computer processor
as shown in the diagram of Figure 2.   The system is controlled by
the medium scale processor, the PDP-11/40, a 900 nsec cycle time
machine.   More important to the architecture of the acquisition
system, however, is the central asynchronous data path, the Unibus, as this is the limiting element in our data flow. Systems
such as this are configured to operate at high speeds, both in
terms of their real-time throughput and computation. Looking initially at the configuration as shown, we attempt to depict how
the multiprogramming capability of the system allows the
real-time processing of experimental data while yet maintaining
other tasks of data manipulation, simulation, and picture display.   All these functions revolve about maintenance of the large
data base, a 300 Mbyte SMD. Experimental input is through the
C1000[12] video camera, data display through the CVD/2[13] RBG color
monitor, archival output via hardcopy printer/plotter and mag-
tape.

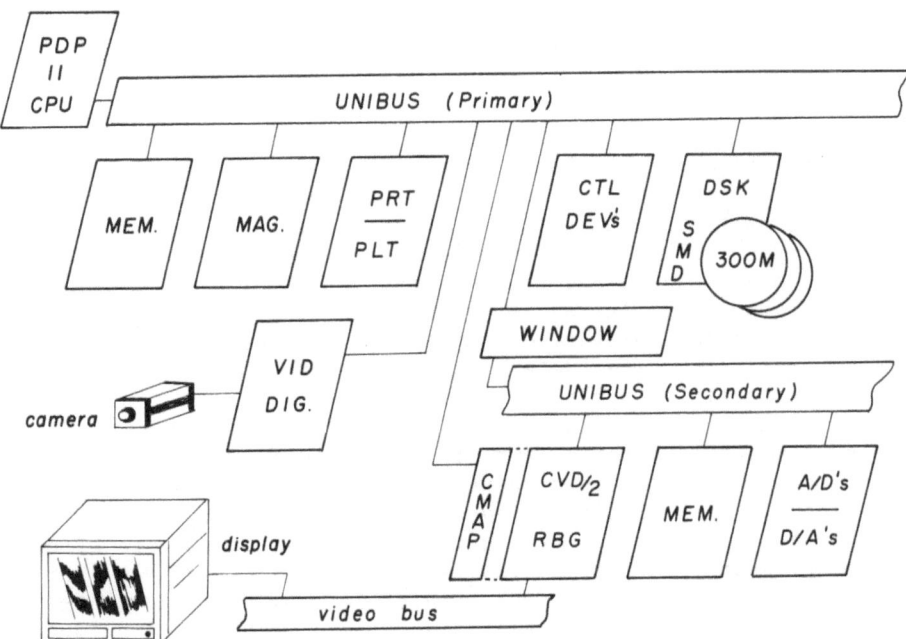

Fig. 2. Digital video processing system.

Under control of a modified RSX-11M[11] executive, we have been able to achieve both the experimental control and yet allow for an interactive simulation environment in which ideas and possible visualization mechanisms can be checked. This system relies heavily on the use of Fortran compilation, and offers considerable power in terms of editors, debuggers, library facilities, and fast access to files. Maintenance of this structure actually encourages users to attempt differing simulations without being mired in the systems structure; they need only become proficient in the workings of the Fortran compiler as all system hardware is handled in a device independent fashion.

While our computational ability is limited by the CPU, our data flow structure for real-time I/O is not. Video input requires data sampling and turnaround times on the order of frame speeds, in the 15 msec time window. Through proper design of the hardware and software, this can be easily achieved. Hardware can be configured with enough intelligence to assert control through DMA cycles and thus avert unnecessary CPU overhead. Drivers which control the devices through forked processes, can be made to operate at high speed. To assure that these real time processes operate at the highest speed possible, they can be additionally granted the highest scheduling priorities, even to the extent that they eliminate swapping and assure maximum transfer through memory residency. Other non-critical processes can be made to wait or assigned normal scheduling priorities.

The video camera operates off an internal crystal oscillator and allows digitization of 256,512,or 1024 square frames through 1,2,or 4 interleave operations. The high bloom characteristics and damage threshold of ordinary vidicon phosphors makes it imperative to switch to that of a chalnicon phosphor to allow for low burn possibility since coherent imaging systems are often plagued by high speckle and contaminants. Control of the camera is via the computer; in reading a complete single frame or reading columnar scans, however, one must read in synchronism to the frame raster or suffer data late possibility. Random position scan is not possible, a minor restriction. Each DMA transfer contains 512 bytes of data. Frames can be digitized by moving the scan line, column for column, in standard picture fashion, with up to eight active columns per raster scan. Single shot readout of individual pixels achieves a 5 bit accuracy. Further enhancement to achieve a true 8 bit video digitization requires the averaging of from 8 to 64 frames utilizing the signal processing capability of the CPU. One could, however, add minimal external hardware to achieve the same process. Since schlieren experiments primarily time sequence or freeze frame the video image, we rely on the CPU alone. Additional inputs to the system

are also provided and allow for synchronization of camera operations with external motions, mechanics, steppers, etc., thus presenting good laboratory versatility.

4.0   THE DATA DISPLAY PROBLEM

In addition to performance of the basic digital acquisition
and digital display functions, the system performs a number of
auxiliary functions, less time critical in nature, which allow
for the creation of modeled and simulated processes, along with
their resultant display. This involves usage of conventional
program development facilities, normal computational processes,
and usage of specifically developed software libraries and program displays. The centerpoint for this is the CVD/2[13] RBG color
display, a system capable of 640 x 480 pixel standard video display. Unlike conventional displays however, this device works in
a RLE, run-length-encoded fashion, which achieves greater picture
compaction. Computation of the complete RLE equivalent of a
frame is achieved in a few seconds. Once stored as a disk file,
playback and slow animation (approx. 3MHz BW) is possible. The
CVD display is under control of its own read-only microcoded processor capable of execution of the RLE instructions which are
from memory. Continual refresh of the display stored in computer memory, while under DMA control, removes both a significant
portion of the computer's address field from use by other programs and reduces overall bus bandwidth through its continual DMA
operation; both these operations preclude or slow concurrent operations under the multitasking environment. The limited address
field restriction can be eliminated with currently available
22-bit memory management CPU upgrades; nonetheless, operating in
an environment of continuous DMA activity reduces useable bus
bandwidth and seriously affects the real-time system operation.
It is therefore imperative that we simply attempt to off-load
this DMA activity.

To this extent we utilize a somewhat different approach,
creating what is akin to a bus 'T' to recreate the Unibus protocol on a secondary bus capable of supporting memory, peripherals,
and DMA activity; this bus, nonetheless, allows for the passage
of vector interrupts onto the main bus. Transactions to memory
on the secondary bus proceed through only four registers: These
act as program counter, direct, autoincrement, and autodecrement
data registers as depicted in Figure 3. Since all activity to or
from the secondary bus is serial and block structured, passage
through the window requires minimal overhead. DMA activity on
the secondary bus, however, proceeds unabated with no reduction
in primary bus bandwidth. Furthermore, due to the ability of the

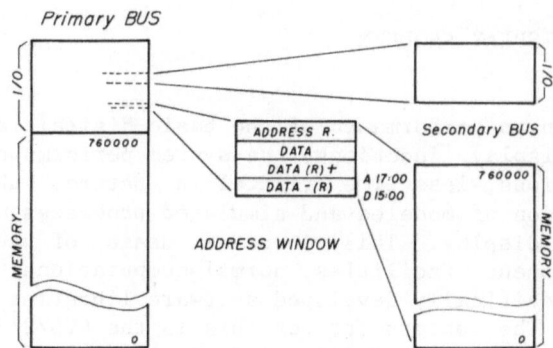

Fig. 3.   Address window mapping.

address window to execute autoincrement/autodecrement operations
through the address field of the secondary bus, block DMA
transfers to or from the disk can proceed without buffering
through main memory by inhibiting the disk DMA bus address increment circuitry.   This not only speeds transfer to or from the
disk of the RLE encoded images for the CVD/2 display, but allows
for the interesting capabilty of animated images without excessive hardware.   Pseudo-color display of both experimental or simulated data can be used to extend the researchers medium for
study of acoustic fields.   All references to the control of peripherals remains on the primary system, independent of whether the
peripheral is located on the primary or secondary bus.   This device allows for other interesting usage in addition to the display.   Other high speed DMA devices, such as A/D and D/A devices,
can be made to operate on the secondary bus.   It's memory can additionally serve as extremely high speed scratchpad for operations where main memory is restricted such as performance of DFFT
operations.

5.0   COLOR DISPLAY IMAGERY

    The pictures displayed by the CVD/2 contain up to 63 inde-
pendent colors in a color map; each color is specified by three
five-bit numbers representing the Red, Green and Blue color signals.   Selection of a specific color map of 63 distinct choices
from the 32K possible combinations required development of a ra-
tional selection method. The XYZ system from C.I.E. (depicted
in Figure 4) as well as the UVW system of McAdam were chosen as a
convenient color coordinate system because (1) they are a linear
combination of each other and of the RBG signals and (2) there is
a simple and satisfactory color metric associated with the McAdam
system. The color metric is a formula for determining the small
differences perceived between differening colors.   The literature
contains a number of formula developed from psychophysical
data.[14-16] Two metrics were used, the Wysecki U*V*W* system which
is related in a very simple way to the UVW coordinates, and the
Chickering-MacAdam-Friele color difference formula. The latter
corresponds more closely to experimental data, but the former is
much simpler to program. For our purposes the simpler metric
sufficed.[17]

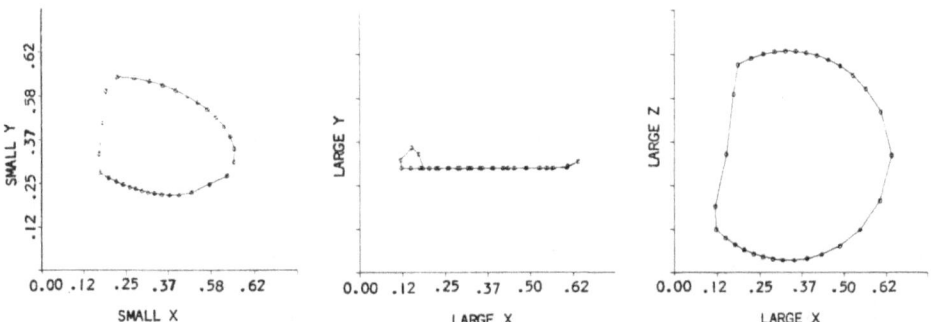

Fig. 4.   C.I.E. color coordinate system.

    After the metric was chosen, a trajectory in UVW color space
was constructed.   A trajectory corresponds to a set of colors
that change continuously in luminosity (brightness) and chromati-
city (hue and purity).   For example one trajectory chosen changed
from pure red at very low intensity to pure white at the highest
intensity. Another trajectory consists of colors all at the same
brightness and purity but changing hue from red to green to blue
and back to red again. Once the trajectory is chosen, equally

spaced colors can be selected along it using the selected metric. If 32 colors are required, then the trajectory is divided up 32 times; if 63 colors are needed, it can be divided 63 times, etc.

This method has considerable benefit over blindly choosing colors. A major factor is that the color map appears to go from some color to another color. The viewer is thus able to perceive the order of the colors. Another benefit is that a variable number of colors is available from the same color trajectory. For greater dynamic range, the trajectory is merely divided up more finely. There are two limits to this, however; quantization of the RBG as well as human perception limits the fineness in color differences.

This scheme was implemented as a set of Fortran-IV programs which generate color map files, start and stop the CVD/2 display, and load the color maps into the CVD/2 memory. Since these are performed on the main bus while the RLE picture is being manipulated by its own processor on the secondary bus, sequential viewing of one RLE picture with several color maps is quick and can be made interactive.

6.0   IMAGE SYNTHESIS AND MODELING

Most analytic solutions and numerical procedures attempt an exact formulation to depict the acoustic field from knowledge of the motion of the sound source.[18] These solutions are then evaluated to predict the behavior of the field at a point for known source motions. Optical experiments which monitor this behavior, however, probe the effect of all points along a sensor pencil beam and as such integrate to produce the cumulative effect. The capability to simulate the acousto-optic imaging process from various models and depict the resultant image under controlled conditions allows one to quantify and categorize the imaging processes under study. One can easily check beforehand concepts and test the limiting sensitivity of the imaging process as it applies to the problem under study, i.e.: simple fields, complex or flawed geomerties, etc. In this respect, the current system can be utilized in a number of ways with its interactive graphical output.

Data from simulated studies can be output on the same peripheral display devices as the experimental data to represent the computed spatial images quantitatively. Figure 5 is a representation of the computed integrated optic effect for a 5 cm. circular transducer assuming uniform sinusoidal motion of the face. This contour plot, as well as orthographic or line

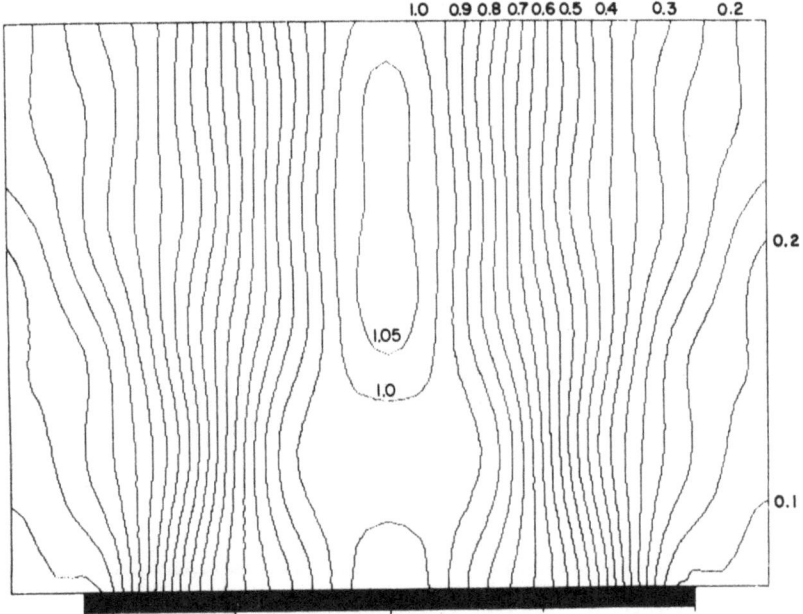

Fig. 5. Contour plot of computed integrated optical effect for circular transducer (ka = 10).

plots can aid in presentation of quantitative information. Further processing of the data to include the behavior of the schlieren imaging and filtering process allows one to simulate the expected image plane distribution. This can be displayed on the CVD/2 RBG color display quantatively as well as archived on the lineprinter plotter with a 30 gray scale dither plot. Figure 6 is a synthesized image of the image plane distribution for a strongly driven square transducer on one side operating as a plane piston source with uniform power output across its face. The simulation is for the schlieren geometry of Figure 1 with the zero-order removed.

7.0 EXPERIMENTAL DATA

Experiments to study both qualitative and quantitative data from the schlieren imaging process need to obtain video digitized data from the image plane to reconstruct the pressure in the object plane linearly. Figure 7 depicts a number of useful quali-

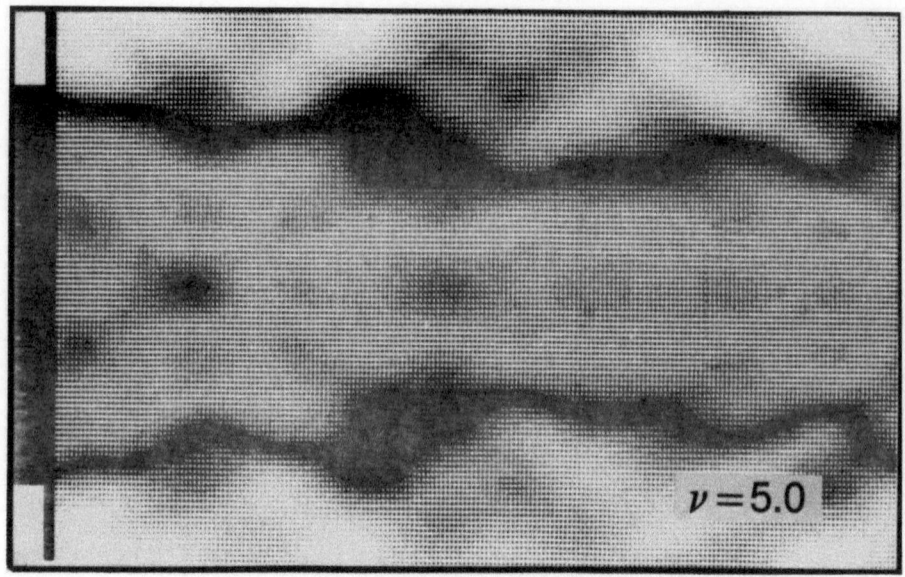

Fig. 6. Simulated schlieren image plane response for overdriven
         square transducer.

tative acousto-optic images recorded with the schlieren geometry
from the   video   camera;   the   system's   sensitivity and
diffraction-limited resolution are ideal in determining the behavior of fields or the mechanisms of scattering processes as shown
in this montage. One attempts to measure under controlled conditions   the   optical phase shift encountered by the probe beam and
establish as well an absolute   calibration   point.   Conventional
schlieren maps the acoustic pressure in a non-uniform fashion as
seen by Equation (5). Digital processing of the   obtained   image
allows one to apply simple corrections for the process up to mod-
erate powers and thereby achieve our goal of quantifying the   information.   Figure 8(a) is a dither plot of a linearized or inverted schlieren image of a 1 megahertz transducer; Figure 8(b)
is   a   line scan of this same image for a distinct column through
the image and shows the observed optical phase shift encountered.
This   particular scan coresponds directly to the columnar output
scan of the video camera digitizer circuit;   when swept across
the entire image plane, it reconstructs the complete frame.

     With simple digital processing of the acoustic field data we
can   thus achieve a linearization of the image plane data and relate it directly to the integrated acoustic pressure in   the   ob-

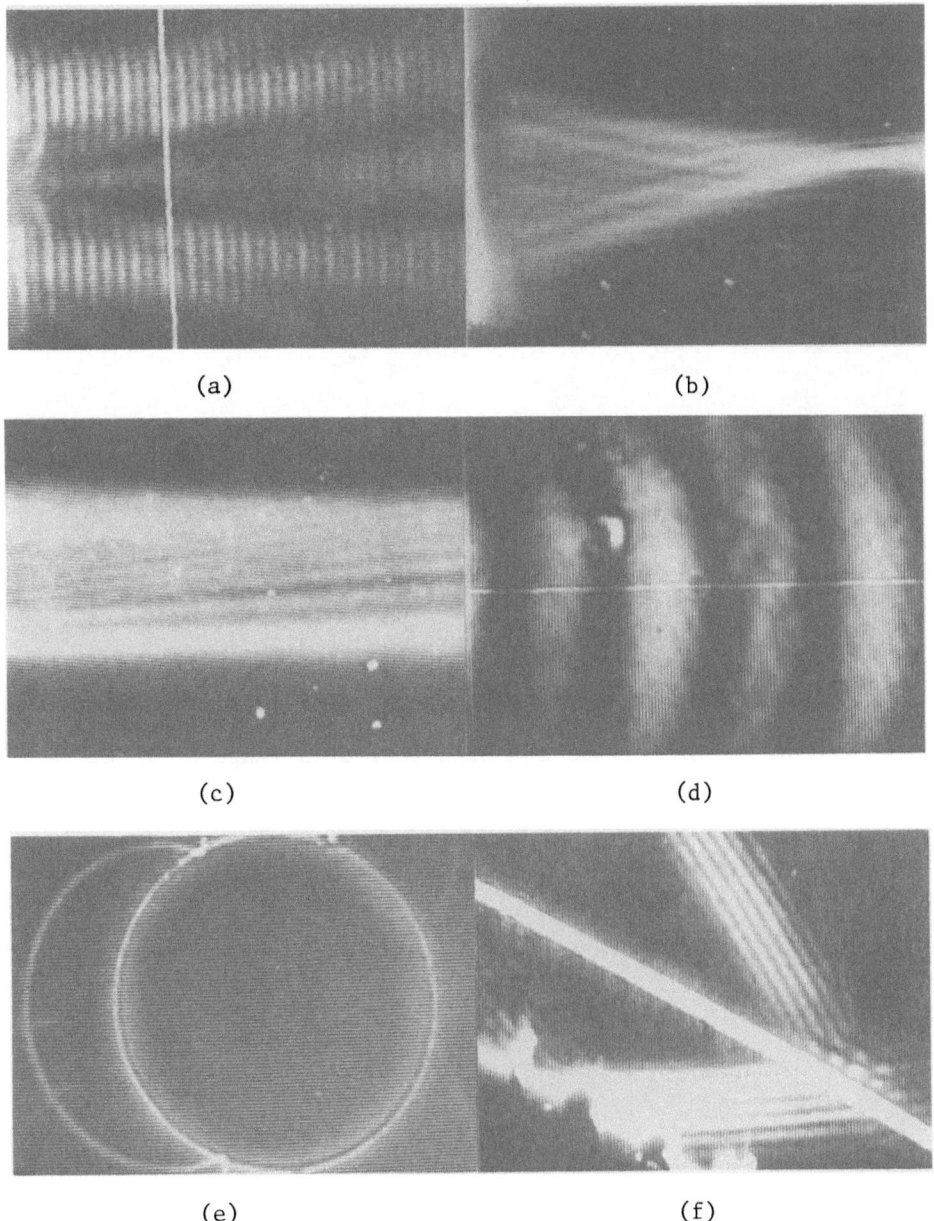

(a)   (b)

(c)   (d)

(e)   (f)

Fig. 7: Qualitative schlieren images: (a) phase resolved 1 MHz
transducer with absorbing strip, (b) 1 MHz focused Field from
plastic lens, (c) 2 MHz flawed field, (d) 40 KHz phase resolved
transducer image, (e) narrow guassian impulse scattered from
aluminum cylinder, and (f) scattering of phase resolved pulse
from complex structure.

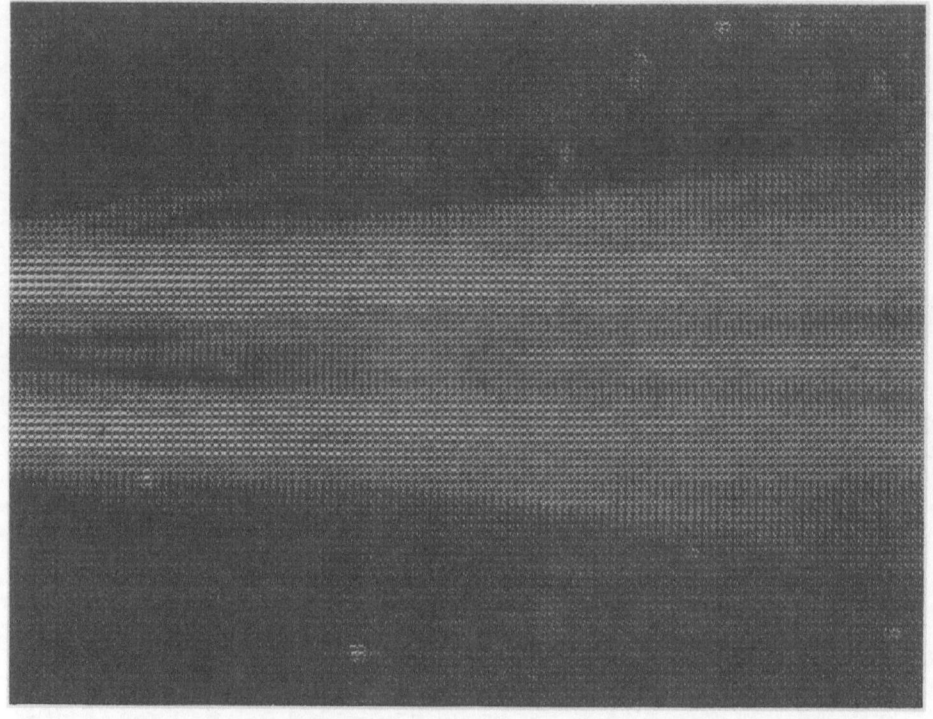

Fig. 8(a). Linearized inverted schlieren image of 1 MHz
            transducer;

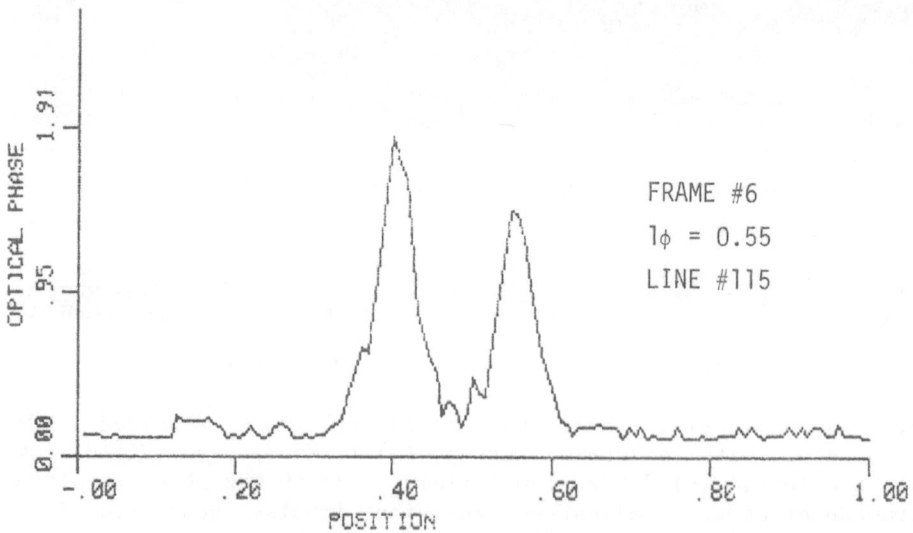

Fig. 8(b). line scan of frame through schlieren image.

ject plane. This process, however, is still one in which we retain the cumulative buildup of signal due to the finite source
width. As such, it is strongly dependent on assumptions of uniformity across its face and dependent on source geometry.
Certain oriented features, flaws or non-uniformities, for example, can image strongly or be completely hidden depending on
their orientation to the path of the probing optical beam. One
would optimumly like to reconstruct the acoustic field point for
point, through a specified plane of interest or through the entire volume. A single conventional schlieren geometry allows
only one single side view.

    Rotation of the acoustic sound source about its center al-
lows differing views of the sound field, each at a particular
side aspect. If one operates the schlieren imaging system in a
phase resolved mode, that is, freeze-frames the acoustic field so
as to resolve propagating wavefronts, one can utilize the digital
storage and processing power of the laboratory system to obtain
sufficient data so as to allow application of tomographic concepts to reconstruct the field.

    Application of backprojection algorithms to a select scan
line in a uniform phased plane of the acoustic field allows one
to then reconstruct tomographically an orthogonal or frontal view
of a slice of the field. Utilizing this process additionally removes the integrated optic effect of conventional schlieren to
achieve a direct point for point reconstruction. Figure 9 is a
reconstruction through simple backprojection of a 1 MHz transducer face, approximately 1.25 cm square, onto which we placed an
absorbing strip to simulate the effect of a non-uniform field in
one direction. Conventional schlieren would image this geometry,
with the strip oriented in the x-direction (refer to Figure 1) as
a uniform field, while in the y-direction, the field would contain a distinct hole. Tomographic reconstruction of the selected
plane clearly shows the frontal field composed of two hot regions
with the missing strip. Through selection and processing of additional scan planes, one could utilize the digital facility to
reconstruct the entire field volume in three dimensions.

8.0 CONCLUSIONS

    We have developed a laboratory digital video processing sys-
tem for the visualization of sound fields in a quantitative
fashion and are currently applying it to the schlieren imaging
process. A number of interesting concepts have been tested in
the design of the overall processing and interactive data display
enviromnent. The final system is both economic and productive

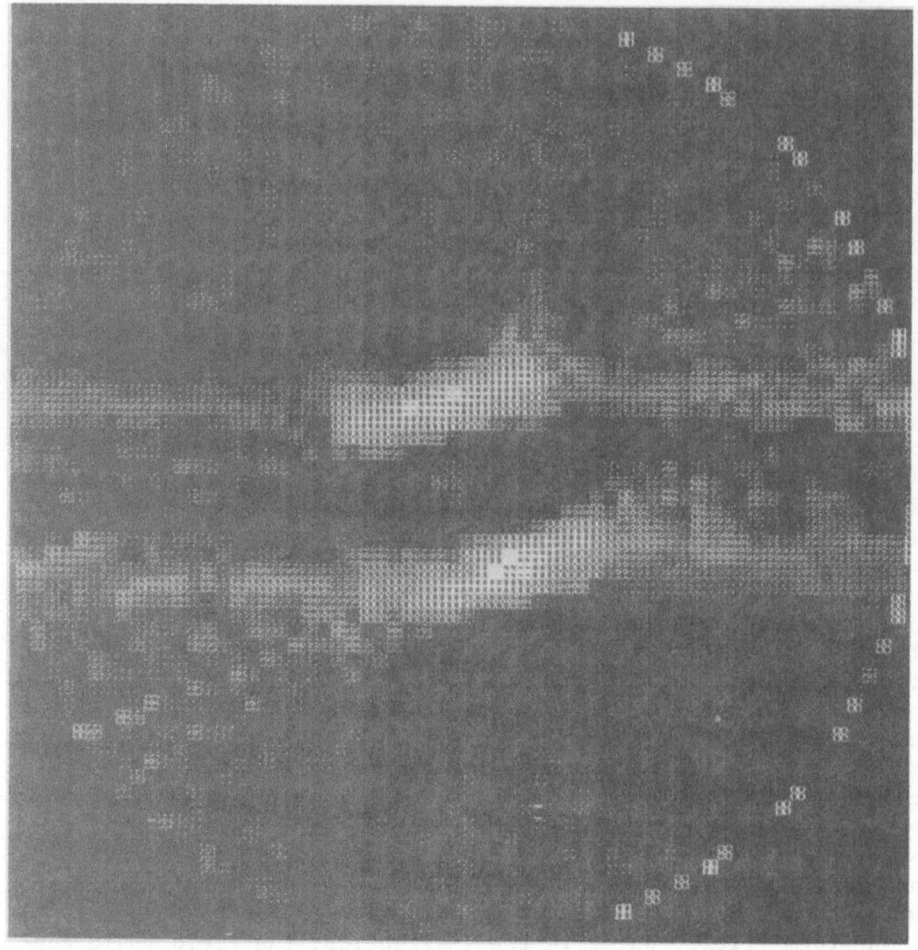

Fig. 9.   Tomographic reconstruction of 1 MHz square transducer
          with absorbing strip through center.

and contains interesting architecture concepts   to   off-load   the
overhead on the processor. More importantly, application of the
system to the linearization of schlieren images allows the inversion of optical images to their acoustic counterpart in a conventional fashion, and further application of processing to appropriately stored multiple images allows reconstruction of projections in tomographic fashion. The concepts learned here are applicable to other systems employing raster like output.

9.0 ACKNOWLEDGEMENTS

     The authors wish to thank Professor Bill D. Cook for his
encouragement and helpful discussions and to Stephen F. Shirron
for his excellent software design.

10.0 REFERENCES

1. R.B.Barnes and C.J. Burton, J. Appl. Phys. 20,
       286(1948).
2. K.L.Zankel and E.A.Hiedemann, IRE Trans. Ultrason. Engn.
       UE7, 71(1960).
3. D.R. Newman, IEEE Trans. Sonics Ultrason. SU-20,
       282(1973).
4. L.R. Dragonette, J. Acoust. Soc. Am. 51, 920(1972).
5. W.G. Neubauer in Physical Acoustics edited by W.P. Mason
       and R.N. Thurston (Academic Press, New York, N.Y. 1973),
       Vol. 10, Chap. 2.
6. O.I. Diachok and W.G. Mayer, Trans. Sonics Ultrason. 66,
       219(1969).
7. J.A. Bucaro, L. Flax, H.D. Dardy and W.E. Moore, J.
       Acoust. Soc. Am. 60, 1079(1976).
8. M. Born and E. Wolf, Principles of Optics (Pergamon Press,
       Oxford 1975) pgs. 422f.
9. F. Ingenito and B.D. Cook, J. Acoust. Soc. Am. 45,
       572(1969).
10. J.A. Bucaro and H.D. Dardy, J. Acoust. Soc. Am. 63,
       768(1978).
11. PDP-11 Processor Handbook, Digital Equipment Corp.,(Maynard,
       Mass. 1973). PDP, UNIBUS, RSX-11M are all registered
       tradenames of Digital Equipment Corporation.
12. Hamamatsu Systems, Inc., Waltham, Mass.
13. J. Teter, Mineapolis, Minn.; formerly produced by Three
       Rivers Computer Corp., Pittsburgh, Pa.
14. D. McAdam, Official Digest 37, 1487(1965).
15. K. Chickering, J. Opt. Soc. Am. 57, 537(1967).
16. R.W.G. Hunt, The Reproduction of Colour, J. Wiley Sons.,
       Inc., (New York, N.Y. 1968).
17. R. Ehrlich, C.F. Gaumond and H.D. Dardy, NRL Report (in
       process).
18. B.D. Cook, E. Cavanagh, and H.D. Dardy, IEEE Trans Sonics
       Ultrason. SU-27, 202(1980).

ACOUSTICAL IMAGING USING FOCALIZATION UNDER OBLIQUE INCIDENCE

THROUGH AN INTERFACE - APPLICATION TO NON DESTRUCTIVE TESTING[*]

J. Frohly, C. Bruneel, B. Bisiaux, J. Lefebvre

University of Valenciennes

59326 - Valenciennes Cedex - France

INTRODUCTION

The applications of acoustic imaging in the medical area have been the subject of a number of studies and have led to a great development.

The results obtained in this field may not however be applied as is to the non-destructive testing of materials for the following reasons.

First, the longitudinal acoustic wave velocity is generally three times higher in solid materials than in biological media : at a fixed frequency, this leads to a corresponding decrease in the spatial resolution for the images. The studies done about acoustic microscopy [1,4] have however shown that, at high frequencies, a very good resolution may be obtained, but over a small penetration depth.

Second, the acoustical impedances of the encountered media are also very different and may be thirty times higher for metallic materials than for biological samples. Such a parameter modifies drastically the transmission and reflection coefficients for the acoustical waves, and leads to a severe loss of useful ultrasonic energy. Moreover, the impedance discontinuities, coming from velocity mismatch and also from density variations, may induce phase rotations [5] which may sometimes disable the reconstruction of B sean mode images.

_____

[*] This work is supported by D.G.R.S.T.

Finally, owing to the geometry of the sample or to the physical nature of the defects to be detected, the examination may sometimes be only performed under oblique incidence, through plane, cylindrical or spherical interfaces. In such conditions, the problem becomes more complicated, since longitudinal and transverse waves must be considered and the non lineartity of the Snell's law of refraction leads to aberrations and defocusing. Such is particularly the case for the B mode non-destructive testing of a reactor tank .

Thus, for imaging the zone near the welding material the radial emission is not possible, owing to the curvature of the welding material itself, and a slanted emission is therefore needed.

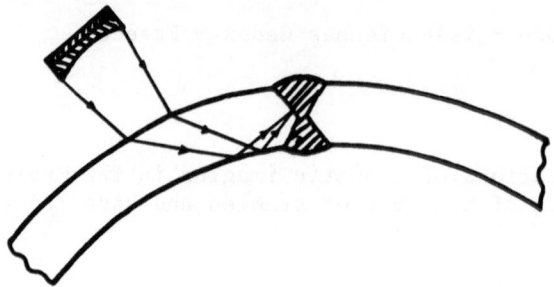

Fig. 1. Testing of welding cords using focalization under oblique incidence.

Our aim is to study the problem of focalization under oblique incidence in B scan imaging mode.

Using a geometrical approach, the characteristics of a lens giving a satisfactory tradeoff between resolution, aberrations and field of depth have been devised for focalizing under oblique incidence through a cylindrical interface. The images given by a manually swept prototype, built according to these principles, are shown.

However, a transducer array allowing for electronically synthetized scanning and focalization is to be prefered for its ease of use and flexibility. In this case, a geometrical model may no longer be applied, so the acoustic field distribution in the focal zone has been computed from the propagation equation and a widening of the focal spotwidth due to the aberrations has been evidenced.

The realization of a lens with the geometrical profile needed to compensate for the aberrations would prove very difficult, but a correction of the electronic delays needed for the focalization gives a very simple, efficient and elegant solution to the problem.

This efficiency is shown by comparing the focal spots theoretically obtained with and without correction.

GEOMETRICAL APPROACH FOR THE FOCALIZATION UNDER OBLIQUE INCIDENCE THROUGH A CYLINDRICAL INTERFACE

It may be seen on fig. 1, that, before accessing to the zone under study, the ultrasonic beam suffers from a refraction through the external cylindrical interface, followed by one or several reflection(s) over the internal cylindrical mirror. In order to restrict the problem to a single mode of vibration, the incidence angle is chosen greater than the critical angle for longitudinal waves in steel, that is to say 15 degrees if the external medium is water. In this case, a focused transverse wave is produced inside the metal.

Aberrations for a plane interface

The nonlinearity of Snell's law induces aberrations and a defocusing of the ultrasonic beam, these effects increasing in magnitude for increasing mean incidence direction angle.

Therefore, as seen in fig. 2, for a plane interface insonified under a mean incidence angle i_o which remains weak, the distance

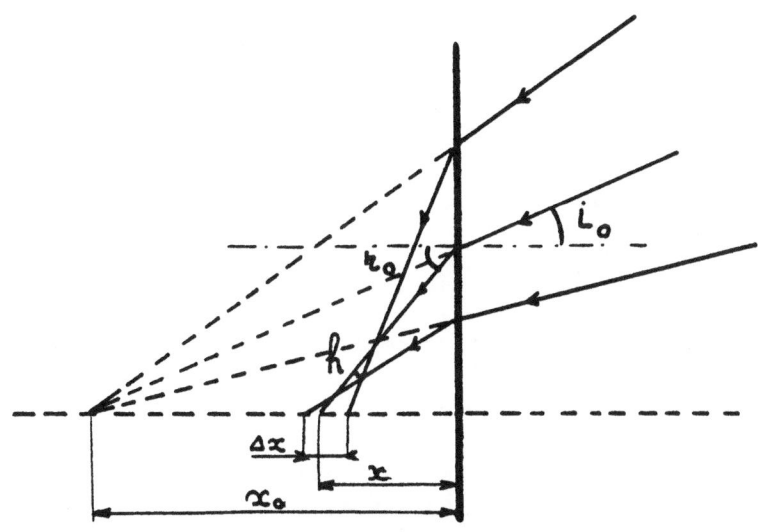

Fig. 2. Changes in Δx versus Δi.

$x = v_1 \cdot x_o / v_2$ doesn't depend on i_o and r_o the refraction angle.

However, when angles i_o and r_o increase over some ten degrees, one may show that for a variation Δi about i_o , the corresponding variation Δx is given by :

$$\frac{\Delta x}{\Delta i} = \frac{x_o}{tg\ r_o} \left[\frac{1}{\cos^2 i_o} - \frac{1}{\cos^2 r_o} \right]$$

For illustrating purposes, assume x_o = 5 cm, v_1 = 1500 m/s, v_2 = 3200 m/s and the curve of fig.3 is obtained.

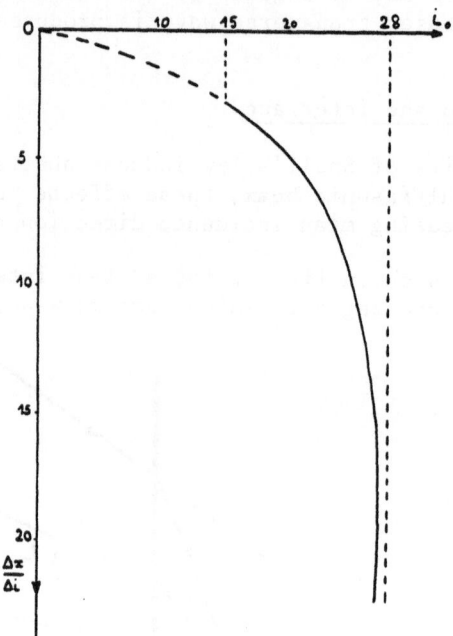

Fig. 3. Aberrations at a water-steel plane interface - transverse waves.

It is seen here that the focus goes towards the interface ($v_2 > v_1$) and a vertical asymptote is reached at the critical angle for transverse waves, 28 degrees in the present case. For small values of i_o the variation of x is quasi-linear and very small, but as i_o increases, it becomes highly nonlinear. As in our case, the angle i_o must lie somewhere between 15 and 18 degrees, severe aberrations result and it will prove interesting to choose an incidence angle as near as 15 degrees as possible. Moreover, as the ration $\Delta x/\Delta i$ increases

quasi-proportionally to x, the aberrations tend to increase with
the exploration depth for given i_o and r_o values. It is so necessary
to reduce the number of available reflections over the interfaces
to a minimum.

Cylindrical interface – Effect of the radius of curvature on the aberrations

In order to get a first geometrical evaluation of the aberrations, the distance h between the intersection of the extreme rays
with the median one has been computed for a cylindrical water-steel
interface. One transmission through the interface and one reflection over the internal surface have been assumed to occur before
the focalization (fig. 4) and the angular aperture of the beam in

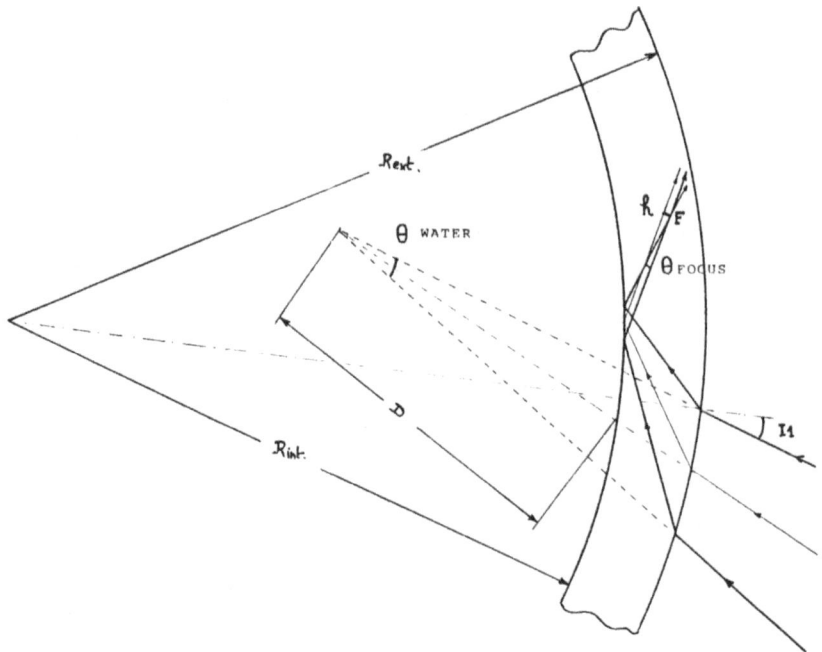

Fig. 4. Parameters used in the aberrations calculations.

the explored medium θf is taken as a parameter. This parameter defines the resolution of the system owing to the chosen depth of
field : for a 20 mm depth of field and transverse waves at a 4 MHz
frequency, this aperture θf is equal to 16 degrees. In the fig. 5,
the distance h has been drawn against the external radius of curvature R_{ext} for cylinders with a 20 mm thickness. It is seen that
the aberrations don't vary much with the external radius. In the

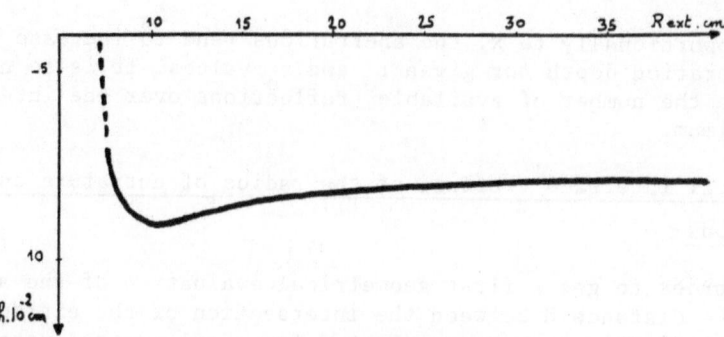

Fig. 5. Aberrations after transversal of a cylindrical interface
        and a reflection. Dependence from the radius of curvature.

fig. 6, the distance h has been drawn against the least slanted
angle i_1 for cylinders with external radius 170 mm and internal
radius   150 mm. The aberrations are so very sensitive to the least
inclinaison i_1. So, the sector scan technique (with angles varying
between 15 and 28 degrees) must not be applied in our case.

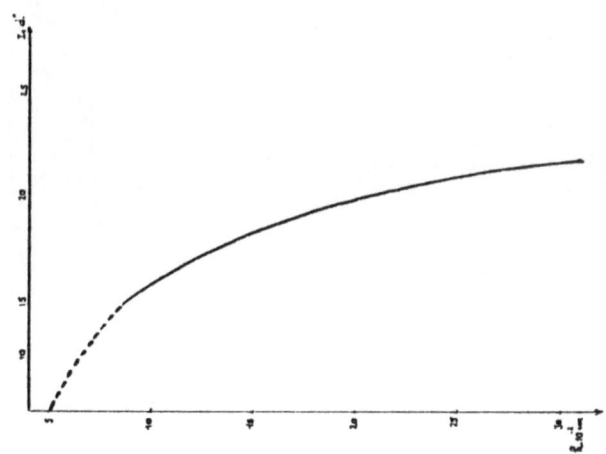

Fig. 6. Aberrations after transversal of a cylindrical interface
        and a reflection dependence from the incidence angle.

This very first study enables us to define the parameters for
building a prototype with an acoustic beam of 16 degrees angular
aperture and a least inclination of 15 degrees. For cylinders with
external and internal radius of, respectively, 170 and 150 mm, this

leads to an angular aperture of 12 degrees inside water and the distance D between the virtual focus in water and the external cylinder surface is equal to 80 mm, the median ray being slanted from
18 degrees (fig. 4). By using a 6 dB criterion for the resolution,
one gets, at a 4 MHz frequency, a 3 mm value. The calculated aberration h amounts then to 0.86 mm for on incidence angle of 15 degrees. It would increase to 2.05 mm if the incidence angle was
taken equal to 20 degrees.

Experimental set-up

    The experimental set up is shown at fig. 7. The ultrasonic ge-
nerator is an USL 32 from Krautkramer working at a 4 MHz frequency.

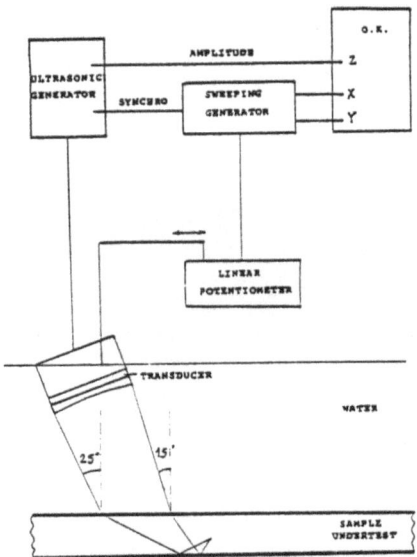

Fig. 7. Experimental set up.

    The plane transducer is built from a piezoelectric P1-60
ceramic (from Quartz et Silice, France), with an area of 25x10 mm^2
and an absorptive backing. On the front face is machined an acoustic lens with a 67.5 mm radius of curvature, leading to a theoretical focusing distance of 150 mm inside water.

    A linear potentiometer is coupled to the mechanical motion of
the probe and delivers an electrical signal for the position. This,
together with the synchronisation pulses from the ultrasonic emitter, allows generation of the horizontal and vertical sweep signals for the visualization monitor and display of the real zig-
zag path of the ultrasonic wave taking the material thickness,
the water path length, the transverse wave speed inside the material and the incidence angle into account.

The output signal from the receiver is used to modulate the
intensity of the light spot when ultrasonic echo jumps over a
predefined threshold.

The probe axis is inclined by 20 degrees using a mechanical
rotation stage, so that the minimum and maximum incidence angles
are, respectively, 15 and 25 degrees. This mean 20 degrees incidence inside water corresponds to 45 degrees inside steel.

Using the preceding apparatus, some results have been obtained
for plane leafs of steel with a 15 mm thickness on which artificial
defects have been produced. On the first view, fig. 8, the 1 mm

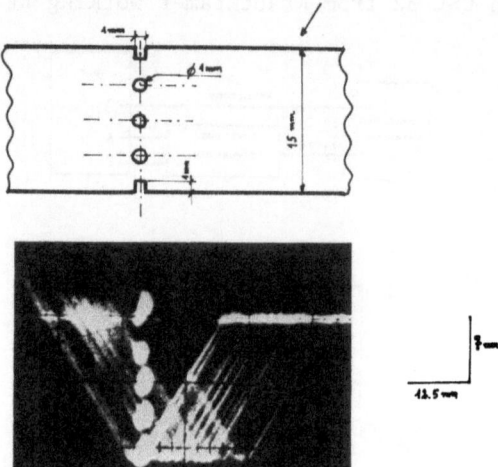

Fig. 8. View showing the scanning done for observing circular holes
        without gain control.

diameter holes are regularly spaced inside the thickness with two
1 mm^2 cuts at the extremities. The horizontal and vertical scales
are, respectively, 12.5 and 7 mm per division. The second view, fig. 9
shows now 1 mm diameter holes lying at helf thickness with center
to center distance of 2,3,4 and 5 mm respectively. The scales are
now 7 and 5 mm per division along horizontal and vertical axis, respectively. The third view, fig. 10, shows longitudinal cuts of
1 mm width lying on the back face with center to center distances
of 2,3,4,5,6,7,8 and 1 mm. The second cut is not observable, being
masked by the first one. The scales are the same as for the second
view.

One sees from these views that the resolution over the position of the holes and cuts is 2 mm along X and Y axis, which is in
good agreement with the theoretical calculations.

The fourth view shows, fig. 11, 3 holes with 1.5 mm diameter
and 3 mm center to center separation lying at half thickness of a
2 mm thick leaf.

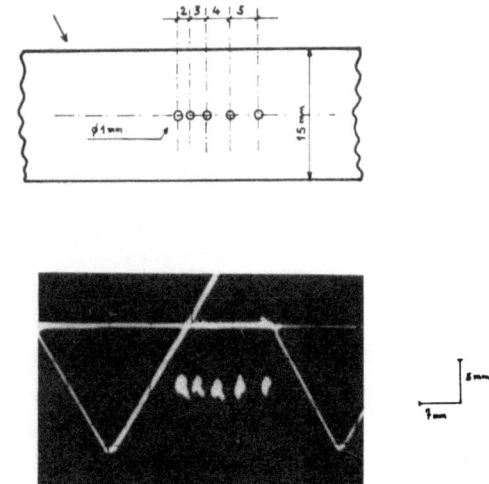

Fig. 9. View of the second test plate.

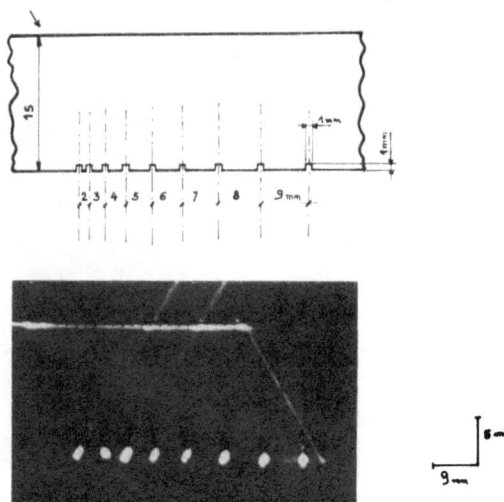

Fig. 10. View of the third test plate.

These artificial defects may be positionned relative to the extreme internal and external lips of the welding. The fifth view, fig. 12, shows similar holes, but lying in a plane slanted with respect to the surface.

The resolution and positioning of the defects is also satisfactory here.

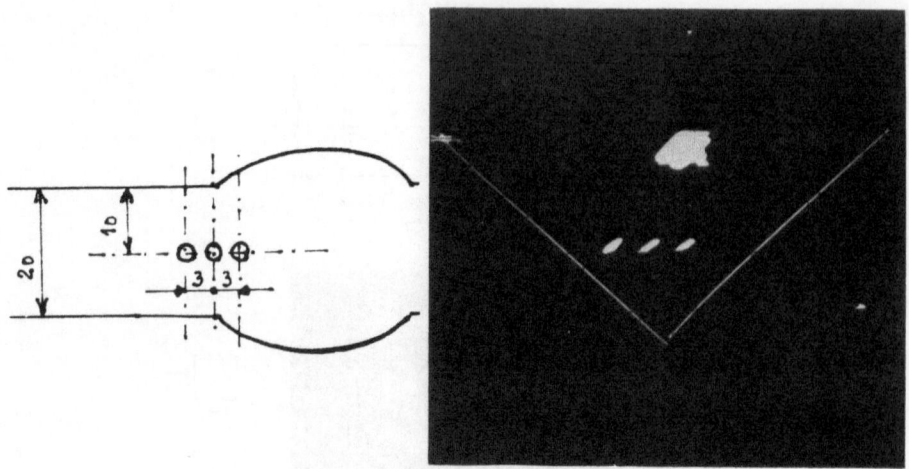

Fig. 11. Image of three holes lying in an horizontal plane.

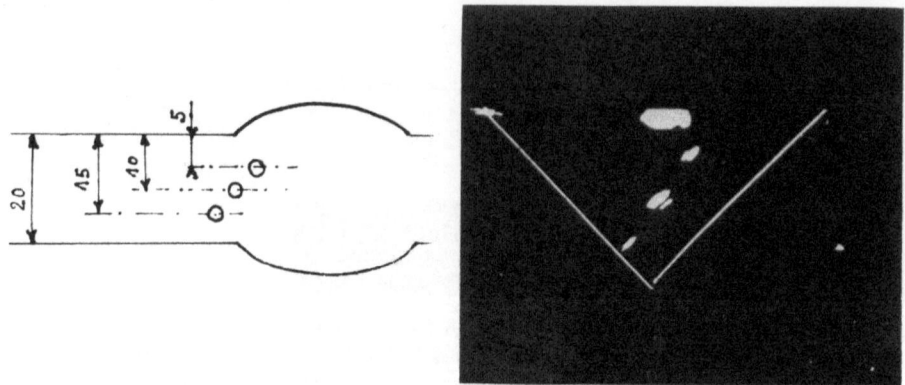

Fig. 12. Image of identical holes lying in a slanted plane.

The results obtained using this prototype show the usefulness of such an apparatus when applied to non-destructive testing. However, their use is somewhat difficult in an industrial environment, partly because of the manual displacement of the probe, needed to get the second axis of the image.

By using an array of transducers, the scanning and focusing functions may be realized electronically, so that the resulting apparatus will prove more interesting for industrial use. The

theoretical study of such an apparatus may however no longer be done
using a meometrical approach.

MULTITRANSDUCER ARRAY-ACOUSTIC FIELD COMPUTATION

Description of the model

In order to account for the spatial sampling induced by a
transducer array, it is necessary to start from the wave propagation equation. The detailed calculation of the acoustic field has
been performed in the case of a plane water-steel interface for a
19 transducer array, each 1.2 mm wide, launching an inclined beam.

Using the Sommerfeld's integral [6,7], the amplitude and phase
distributions over the plane interface are computed.

From these distributions, those obtained in the focal plane
may then be computed, a single component vector potential being
assumed for the transverse wave. The results so obtained take the
aberrations due to the traversal of the plane interface into
account.

Results without any correction

Figure 13 shows the theoretical amplitude distributions in
different planes surrounding the focal one. The reference axis is
that for the mean ray. The calculation shows that the propagation
happens along a weakly slanted direction. The fig. 14 shows the
3 and 10 dB widths obtained for different planes. One sees a signi-

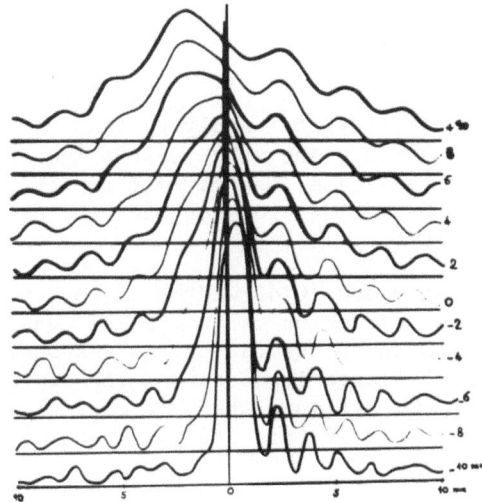

Fig. 13. Acoustical field distribution inside the focal zone
         without correction.

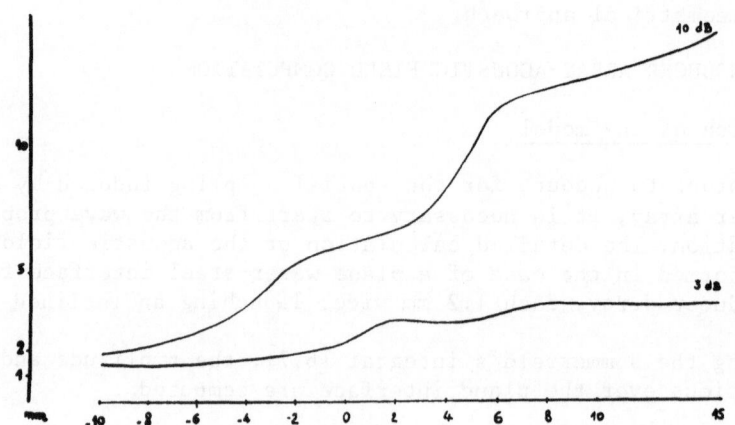

Fig. 14. The 3 and 10 dB widths for different planes without
         correction.

ficant broadening of the focal spot, particularly for the 10 dB case
where the rippling plays a role. The depth of field is therefore
decreased.

By varying the electronic delays at the emission stage, the
focusing may be easily corrected for the aberrations. For this
purpose, the calculation may be performed by starting from the
theoretical focus F and computing the new phase distribution law
needed at the emitter in order that all wavelets come in phase ac-
cordance at point F.

Results after correction

The results obtained after applying this correction process
are given in fig. 15. They show the suppression of the parasitic
ripples of the previous case. The maximal amplitude remains nearly
constant over all the exploration zone and the depth of field is
therefore much longer.

In fig. 16, the 3 and 10 dB widths of the spot in planes near
the focal one after correction is shown. The quasi-disparition
of the rippling induced by aberrations is responsible for the
attainment of a much flater 10 dB curve. The depth of field is
nearly equal to 20 mm, a figure which is more close to the criterion chosen at the start of this study.

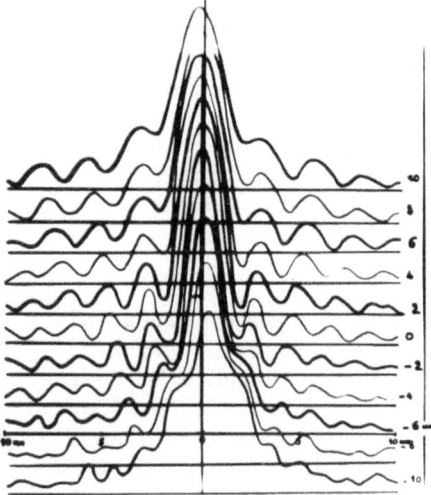

Fig. 15. Acoustical field distribution inside the focal zone
         after correction.

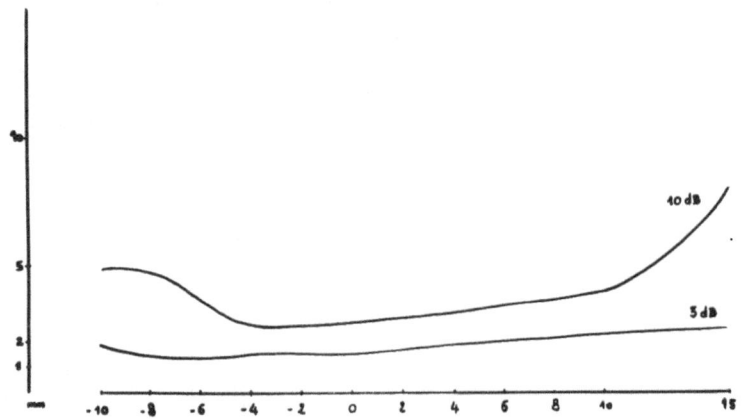

Fig. 16. The 3 and 10 dB widths for different planes after
         correction.

CONCLUSION

From the classical model of geometrical optics, a first
approach of the problem of focusing under oblique incidence through
an interface has been taken. The important role of aberrations has
then been evidenced and the parameters of a prototype giving satisfactory results have been devised.

In a second step, a much easier to use multitransducer array has been studied using a more detailed theoretical model. The broadening of the focal spot and consequent shortening of the depth of field have so been shown to occur. The same model has been applied to a system which implements an aberration correction scheme whose efficiency has been proved theoretically. This effectiveness is now under experimental verification.

REFERENCES

1. R.A. Lemons and C.F. Quate, Ultrasonic Symposium Proceedings (I.E.E.E, New York 1973), pp. 18-21 (1973).
2. Acoustic Microscopy, Lawrence W. Kessler, 9th international symposium on acoustical imaging. (Dec. 1979) Houston Texas.
3. Attal and Quate, J. Acoust. Soc. Am., 59,69 (1976).
4. B. Nongaillard, J. M. Rouvaen, E. Bridoux, R. Torguet and C. Bruneel. Visualization of thick specimens using a reflection acoustic microscope. J. Appl. Phys. 50 (3) (March 1979).
5. Effects of phase and amplitude quantization on the focusing aperture. Kenneth N. Bates, 9th international symposium on acoustical imaging (Dec. 1979) Houston Texas.
6. J.W. Goodman, Introduction to Fourier optics (Mc Graw - Hill,(New York, 1968)
7. B. Delannoy, H. Lasota, C. Bruneel, R. Torguet and E. Bridoux, J. Appl. Phys. 50 (8) (1979) pp. 5189-5195.

GENERATION OF ACOUSTICAL IMAGES FROM THE ABSORPTION OF PULSED MICROWAVE ENERGY

Richard G. Olsen*

Naval Aerospace Medical Research Laboratory

Pensacola, FL 32508

INTRODUCTION

One of the biological effects of pulsed microwave irradiation that has received much attention from physical scientists is the so-called microwave auditory effect in which subjects actually "hear" the pulses of microwave energy that are beamed to their heads. Several hypotheses have been advanced to explain the phenomenon; a direct neuronal interaction was originally suspected (Frey, 1961), but after it was demonstrated in animals that a distinct cochlear microphonic followed each microwave pulse (Chou et al, 1975), the hypothesis of thermo-elastic wave generation gained credibility (Lin, 1978). It is now accepted by many that the microwave auditory effect arises from a small but rapid rise in temperature and thermal expansion of the tissues of the head that absorb the microwave pulse.

Foster and Finch (1974) used a hydrophone suspended in a plastic container of saline solution to record the passage of waves produced by absorption of pulsed 2.450 GHz energy. Previous experiments in this laboratory used a Navy type E-27 standard hydrophone to show the acoustical waves generated in a relatively large mass of muscle-equivalent material by the absorption of pulsed 5.655 GHz energy (Olsen and Hammer, 1980). In that study, a radar transmitter was used to irradiate the mass at close range, and the hydrophone response showed a well defined acoustical pulse emanating from the region that absorbed the microwave

---

*Opinions or conclusions contained in this report are those of the author and do not necessarily reflect the views or the endorsement of the Navy Department.

energy. The observed acoustical pulse width was shown to correspond
to the measured penetration depth of the microwave energy analogous
to that previously shown for the absorption of optical energy in
organic liquids by Gournay (1966).

The purpose of this study was to investigate pulsed microwave
energy as used in the generation of acoustical images of biological
material. It was felt that the well defined, microwave-induced
acoustical pulses observed in previous experiments could be used
in the area of diagnostic ultrasound. The scope of this study
was limited to initial results obtained from a rather crude, low-
resolution system that lacked sophisticated image processing
capability. Experimental results to date have been encourging
and have indeed exhibited the potential usefulness of a concept
that could eventually become, along with microwave-induced hyperthermia, another beneficial use of microwave irradiation in man.

PROCEDURE

The procedures used in this study were kept simple and
straightforward in order to present only the most salient features
of the subject. Accordingly, experimental results are also
presented in a very unembellished form.

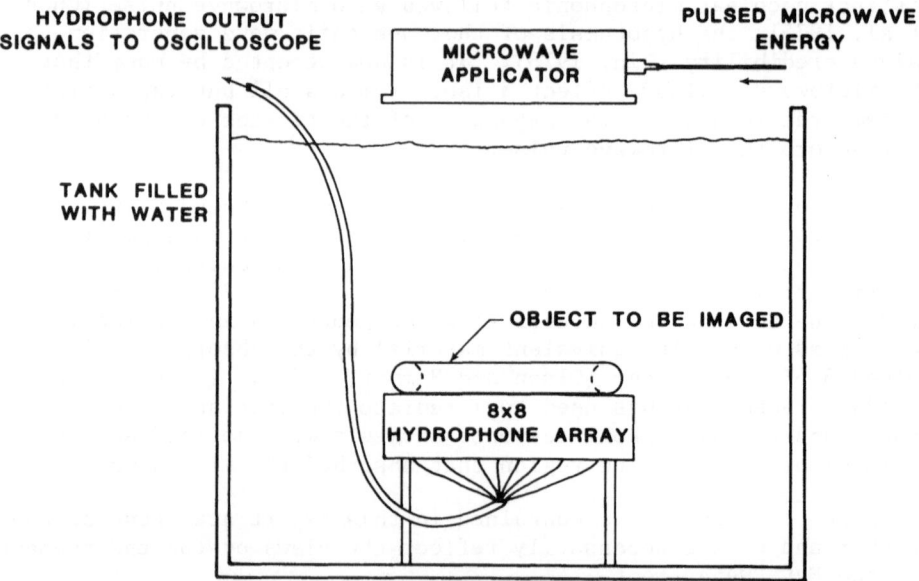

Fig. 1. Simplified diagram of the microwave-induced
         acoustical imaging apparatus.

Hydrophone Array Configuration

An array of hydrophone transducers was procured from International Transducer Corporation (ITC-5196) and was submerged in a 33 X 37 X 74.5-cm tank of water facing upward from a depth of 20 cm. The array was composed of a square pattern of 64 lead zirconate-titanate devices each measuring 1.9 X 1.9 cm on the top surface which was diced into four sections. All transducers had a free-field voltage sensitivity of -201 $\pm$ 0.1 dB (re:1V/μPa) at 100 kHz, and each device had a coaxial (RG-174) output cable that exited from the bottom. The array was sealed within a square metal frame that measured 22.9 cm on a side. A simplified diagram of the experimental configuration is shown in Figure 1.

Microwave Irradiation System

The irradiation system consisted of a microwave pulse generator (Epsco PG5KB) that fed a standard L-band waveguide-to-coax adaptor through a length of RG-214 coaxial cable. The adaptor aperture was spaced 1.9 cm above the free surface of the water and served as the microwave applicator shown in Figure 1. Single microwave pulses of 14.5 and 20 μs were used at a carrier frequency of 1.100 GHz with peak power of 4 kW.

Data Collection and Presentation

After each microwave pulse, output waveforms from the hydrophones were observed on an oscilloscope and were recorded on Polaroid film. Peak-to-peak amplitude of the first wave to reach the transducers was taken as the dependent variable. In the absence of the object to be imaged, a series of waveform photographs was first obtained to be used as a reference . Typical response of the transducers to passage of the initial wavefront was a 10-25 μV signal with a half-amplitude width of about 25 μs. An averaging factor was derived for each element in the array as a method of smoothing the overall unperturbed response and enhancing the resultant image. This factor was obtained by calculating the mean amplitude of response for the array and forming a ratio of the mean response to the element response, for all elements.

With the object to be imaged placed atop the array as shown in Figure 1, another set of amplitudes was obtained from the array. Those amplitudes were subtracted from the ones originally obtained on an element-by-element basis and each difference was multiplied by the averaging factor. The resulting set of values was made into a two-dimensional, graphical representation in the form of darkened circles in a matrix corresponding to the array where the net difference values were made proportional to the

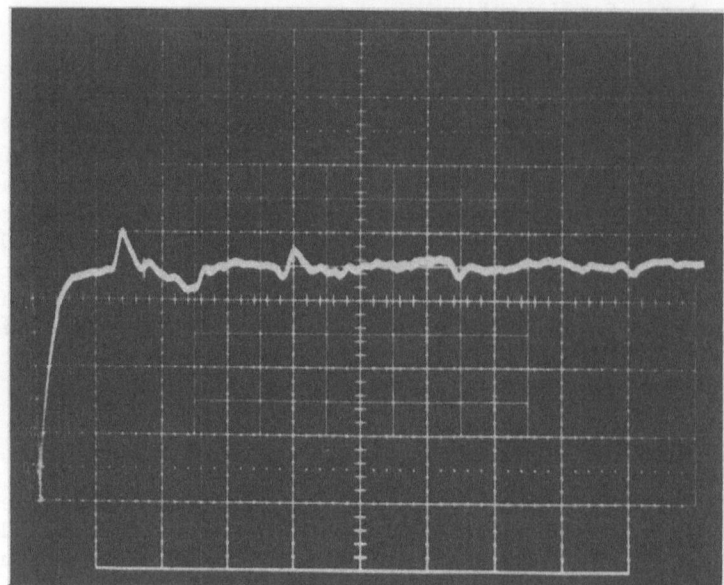

Fig. 2.   Typical hydrophone output signal obtained in the
          absence of the object to be imaged. Hydrophone
          location was #36; vertical scale was 20 µV/div;
          horizontal scale was 100 µs/div.

radii. In so doing, elements in the array that exhibited nearly
the same hydrophone response both with and without the object
yielded only small dots; whereas, elements in which there were
large differences produced large darkened circles.

RESULTS

     Figure 2 shows an example of the raw data collected in this
study. The microwave artifact is seen to disappear before arrival
of the first acoustical pulse. A typical free-field illumination
pattern is given in Figure 3 and was obtained by plotting circles
with radii proportional to the peak amplitudes recorded in the
absence of the object. Figure 4 shows a photograph and image
representation of a circular plastic tube filled with muscle-
equivalent material consisting of water, salt, powdered polyethylene,
and gelling agent (Guy, 1971). A second object that was imaged
consisted of a 35-cm length of sausage, 1.5 cm in diameter, that
was formed into an irregular shape. A photograph and hydrophone
image of the sausage are shown in Figure 5.

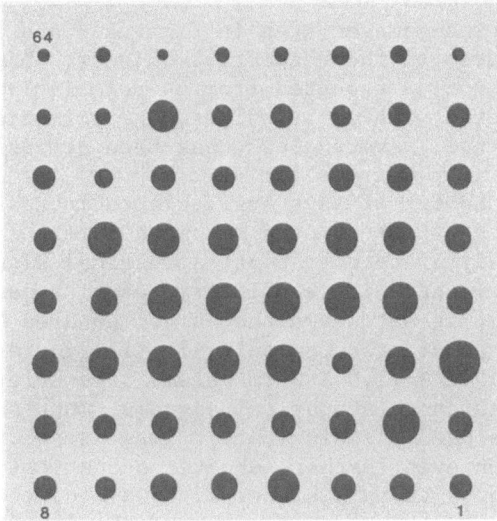

Fig. 3. Typical pattern of the free-field, microwave-induced
         acoustical illumination of the submerged hydrophone
         array.

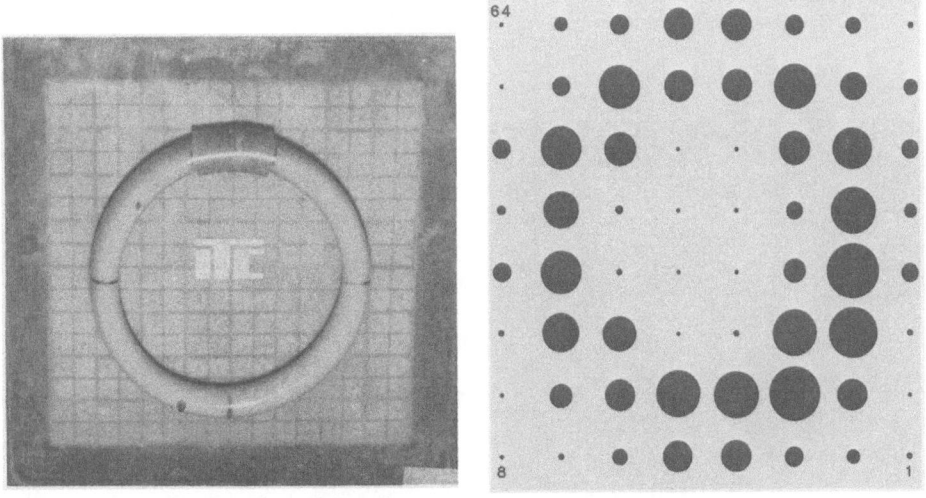

Fig. 4. Photograph and microwave-induced acoustical image of
         a circular mass of muscle-equivalent material.

DISCUSSION

   The rather crude images seen in Figures 4 and 5 clearly show
the overall features of the respective objects. Although high
resolution could not be expected from 64 relatively large array
elements illuminated with 25 μs pulses, the potential usefulness
of using this method, nevertheless, has been demonstrated.

   There are a number of ways available to increase the imaging
power of this kind of system. First, more powerful microwave
pulses could safely be used to boost acoustical signal levels.
In previous experiments with a military radar transmitter (Olsen
and Hammer, 1980), it was shown that a two hundred kilowatt,
half-microsecond pulse produced only 122 millijoules of energy
absorption in a 15.2 X 11.4 X 2-cm volume of muscle-equivalent
material. With stronger acoustical signals, much smaller transducer
elements could be used allowing better image resolution. To
improve resolution even further, shorter acoustical wavelengths
could be obtained by using a higher microwave carrier frequency
to deposit energy in thinner regions near the irradiated surface.
Alternatively, two short microwave pulses, applied arbitrarily
close in time, could be transmitted to generate ultrasound that
was arbitrarily high in frequency. Sophisticated image processing
techniques could then be adapted to transform both phase and
amplitude information of the signals into highly refined images.

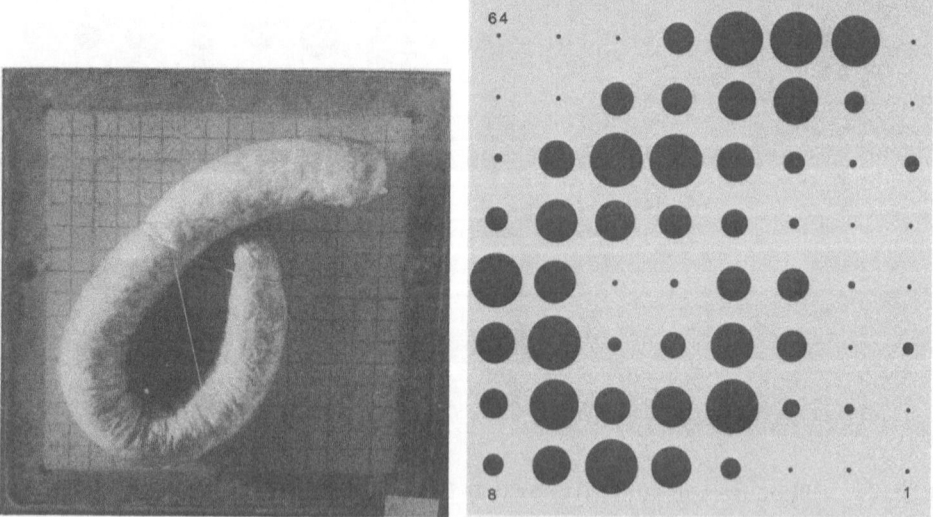

Fig. 5.   Photograph and microwave-induced acoustical image of
          a length of sausage formed into an irregular shape.

It is envisioned that a compact device might someday be available for clinical diagnostic applications wherein a single, brief burst of one or two microwave pulses would provide either deep, surface, or wide-area acoustical illumination according to a selected microwave absorption pattern. The microwave-induced ultrasound would then be instantly imaged by a sophisticated electronic display system connected to the transducer array positioned to receive the acoustical waves on the patient's skin.

REFERENCES

Chou, C.K., Galambos, R., Guy, A.W., and Lovely, R.H., 1975, Cochlea microphonics generated by microwave pulses, J. Microwave Power, 10:361.

Foster, K.R. and Finch, E.D., 1974, Microwave hearing: Evidence for thermoacoustical auditory stimulation by pulsed microwaves, Science, 185:256.

Frey, A.H., 1961, Auditory system response to radio frequency energy, Aerospace Med., 32:1140.

Gournay, L.S., 1966, Conversion of electromagnetic to acoustic energy by surface heating. J. Acoust. Soc. Am. 40:1322.

Guy, A.W., 1971, Analysis of electromagnetic fields in biological tissues by thermographic studies on equivalent phantom models, IEEE Trans. Microwave Theory Tech., MTT-19:205.

Lin, J.C., 1978, "Microwave Auditory Effects and Applications," Charles C. Thomas, Springfield.

Olsen, R.G. and Hammer, W.C., 1980, Microwave-induced pressure waves in a model of muscle tissue, Bioelectromagnetics, 1:45.

A THERMOPLASTIC ACOUSTICAL HOLOGRAPHY RECORDING DEVICE

D.B. Rivers and W.F. Ranson

College of Engineering
University of South Carolina
Columbia, South Carolina 29208

W.F. Swinson

Department of Mechanical Engineering
Auburn University
Auburn, Alabama 36849

H.K. Liu

Department of Electrical Engineering
University of Alabama
Tuscaloosa, Alabama 35486

INTRODUCTION

This paper describes the experimental evaluation of a device
to record real-time in situ acoustical holograms. The recording
concept investigated, as shown in Figure 1, consists of a layer of
thermoplastic (TP) spread over the surface of a rigid substrate.
This report describes attempts to optimize both substrate and TP.
In addition, an effective heating device for the TP is described.

The basic acoustical holography system is shown in Figure 2.
It was developed by Irelan, et al. [1] with improvements described
by Swinson [2,3]. Additional insight into the problems associated
with the TP came from Liu [4].

The use of deformable TP to record optical holograms is well
documented in the previously referenced reports. The idea of
applying this method to record acoustical holograms was introduced
by Young and Wolfe [5] in 1967. Such a technique could have

61

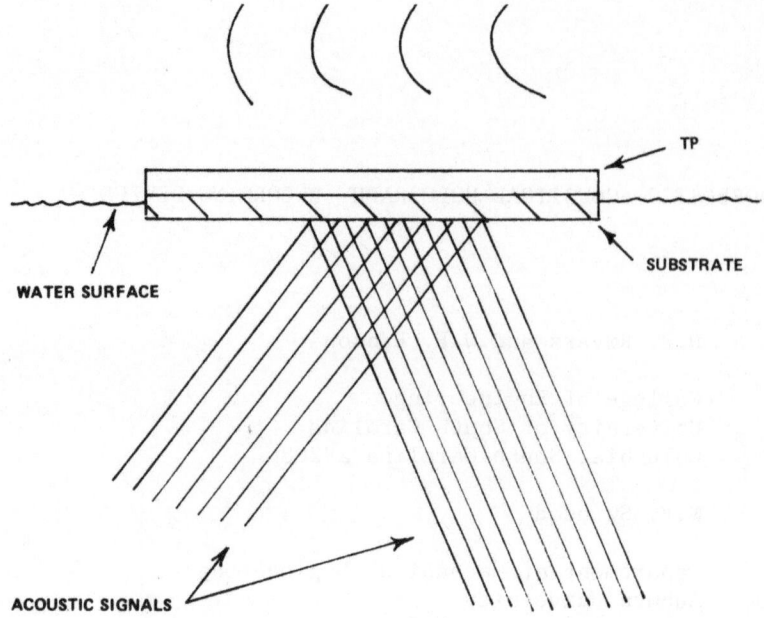

Figure 1.   Schematic Representation of a TP Recording Device

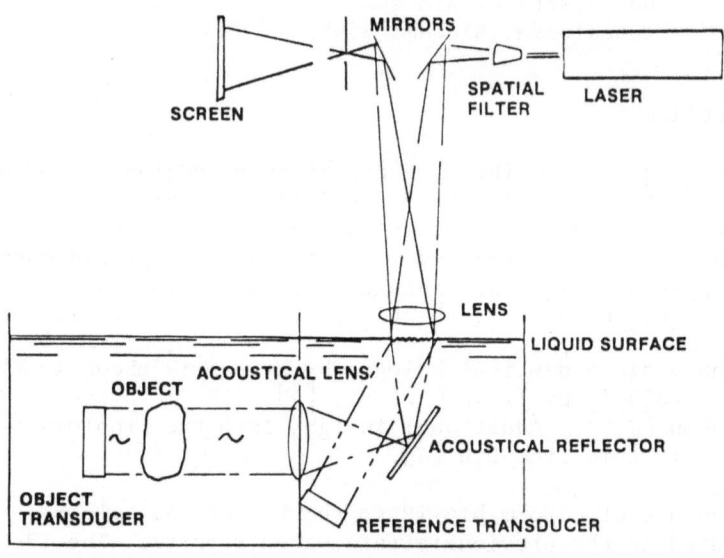

Figure 2.   Acoustical Holography Layout

widespread nondestructive testing applications where information concerning interior flaws and voids is needed. There are also potential medical imaging applications for such a device.

DISCUSSION OF IDEALIZED SYSTEM

Although there are many possible arrangements and materials which might generate acoustical holographic recordings, this report deals with just one, the arrangement shown in Figures 1 and 2. Ideally, the system would record in situ real-time acoustical holograms as follows.

Two identical underwater collimated ultrasonic beams are aimed at the same area of water surface in the imaging tank. One, the object beam, passes through the solid object to be viewed and, therefore, is changed spatially both in amplitude and phase. The second beam, the reference, is unaffected. When both beams combine at the water surface, an interference pattern, the acoustical hologram, is formed by distorting the water surface. Although these acoustical beams are quite powerful, they can only cause microscopic surface distortions due to the weight of the water and its surface tension. However, these distortions are greater than the wavelength of laser light reflected off the water surface, so that a diffraction pattern is formed at the focal plane of the light and, by the use of a slit, the zero, first, second, or higher orders of the diffraction orders can be selectively viewed on a screen. The mathematical formulation has been developed by Swinson [3].

To permanently record this interference pattern, some substance other than water must be used since it is obvious that the water levitation will collapse when the ultrasonic signals are removed. This replacement substance must be able to form a liquid surface levitation but be capable of retaining the distortion by some means. The use of TP seems ideal for this application since it is liquid when heated but quickly returns to its solid phase when the heat is removed. The need for a substrate is purely structural.

ANALYSIS OF AN IDEAL SUBSTRATE

Before an acoustical wave pattern can be recorded on the TP surface, the waves must first pass through two boundaries: from water to substrate; then from substrate to TP. For maximal energy transfer from water to substrate, assuming acoustic signals normal to the surface, the following condition must be met.

$$\rho_{water} \times C_{water} = \rho_{substrate} \times C_{substrate} \qquad (1)$$

Equation (1) implies there are an infinite number of combinations
of ρ and C for some substrate so that characteristic impedance
matching is accomplished. Practical limitations make this a much
more difficult problem, however. This same type of reasoning can
also be applied to the substrate – TP interfere. It is complicated
by the fact that, as explained later, the TP lower surface is
approximately equal to the water tank temperature (typically 22°C)
while the TP upper surface is heated to improve its recording
characteristics. Therefore, the velocity of longitudinal waves in
the TP changes as the phase of the TP changes from a solid to a
liquid state. Table 1 shows impedances of water and several
potential substrates.

EXPERIMENTATION

Initial Recordings

An initial attempt to record an acoustical wave pattern on
TP was performed. Figure 3 shows the test fixture, a plexiglass
base plate with a 114 mm square interior wall to contain the TP,
and a high outer wall to keep any water from the imaging tank out
of the test fixture. The three leveling screws and the two bubble
levels aid in the positioning of the test fixture on the water

Table 1.   Characteristic Impedances of Various Materials

Material	Density (ρ) g/cm^3	Velocity (C) cm/s	ρC
Water	1.0	1.45×10^5	1.45×10^5
Plexiglass	1.18	2.68×10^5	3.2×10^5
Plate Glass	2.51	5.77×10^5	15.4×10^5
Pyrex	2.23	5.57×10^5	12.4×10^5
Teflon	2.2	1.35×10^5	3.0×10^5
Nylon	1.20	2.0×10^5	2.40×10^5
Aluminum	2.7	6.4×10^5	17.3×10^5
Steel	7.8	6×10^5	46.8×10^5

Figure 3. Plexiglass Recording Fixture

surface. Figure 4 shows the initial TP recording made with this
device. Only the 3 MHz reference transducer was activated. The
3 MHz object transducer was not activated. The TP was S-25, (a
thermoplastic supplied by Hercules, Inc., Wilmington, Delaware).
This particular TP was sufficiently fluid at room temperature so
that additional heating was not necessary for image formation.
The concentric pattern of the fresnel zone rings in clearly
visible. This image began to fade after two minutes and was completely gone at the end of five minutes. This image fading was
caused by the S-25 TP being too fluid at room temperature and
indicated that TP with a higher melting temperature would be needed
for permanent recordings.

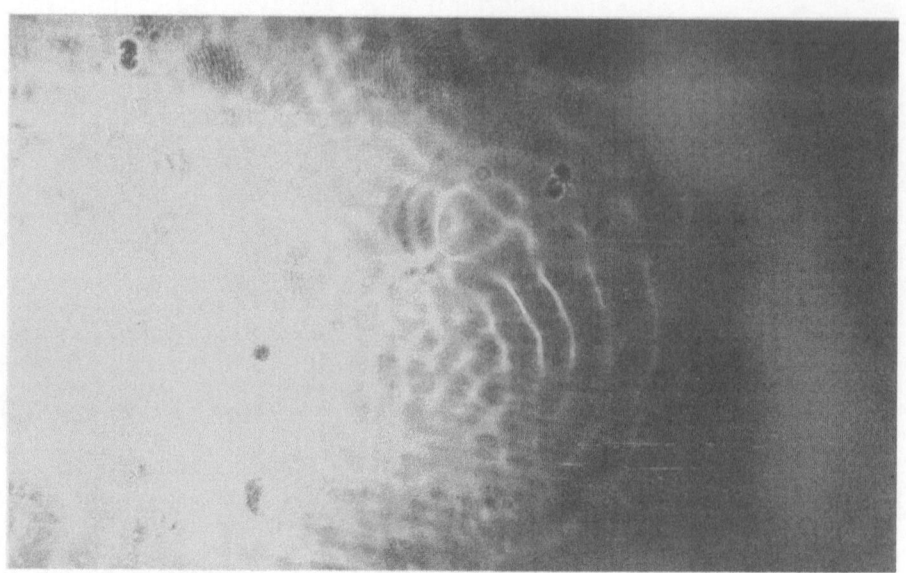

Figure 4.   TP Recording of Fresnel Rings

        Potential problems with higher temperature TP are shown as
follows. Figure 5 is the imaging object used for acoustical
imaging in the remainder of the experiments in this paper. A flat
washer was taped to a 12.7 mm thick sheet of plexiglass. The
washer was surrounded by a taped triangle. Since the tape was not
expected to image well, a thin strip of metal foil lined the underside of the tape. The object was located in the imaging tank
according to Figure 2. The square window recorder was covered
with a 2.5 mm layer of S-25 TP previously heated to 120°C. This
high temperature was needed to lower the TP viscosity sufficiently
so the TP could easily be poured to form an even surface across
the bottom of the recording device. Previous experimentation had
shown that S-25 TP needed to be heated to at least 90°C so that air
bubbles caused by pouring the TP would not be trapped in the liquid.
The heat from the melted TP caused the 1.5 mm thick plexiglass
base of the recording device to warp. Consequently, it was difficult to align the recording device so that a flat recording area
could be found. Figure 6 shows the unevenness of the upper surface
of the TP prior to activation of the object transducer. This unevenness of the recording TP distorted the acoustical image when
the object beam was activated as shown in Figure 7. However, this
recorded image was still superior to any previous attempts and the
object target was clearly visible.

Figure 5.   Acoustical Imaging Object

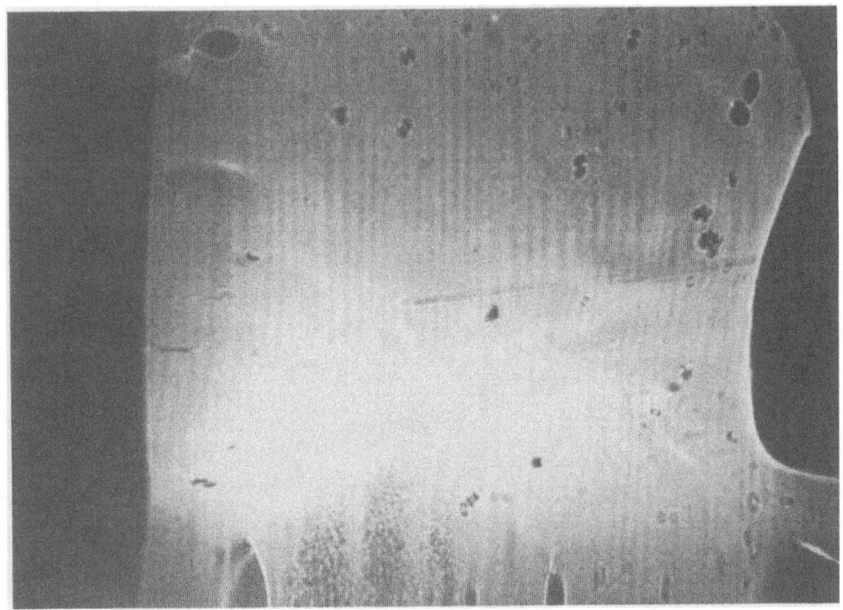

Figure 6.   TP Surface Before Acoustical Signal

An attempt was made to further enhance the image shown in
Figure 7 by heating the TP surface. The heating device consisted
of a 127 x 152 mm glass plate laid over the 114 mm square borders
of the recording device. The bottom side of the glass was coated
with InO, a transparent but resistive material. It was found that
the coated glass could not generate sufficient heat to change the
temperature of the TP before the glass broke from excessive thermal
stresses. This might have been corrected by using coated pyrex,
but such was not available. A different type of heater is discussed later in this paper.

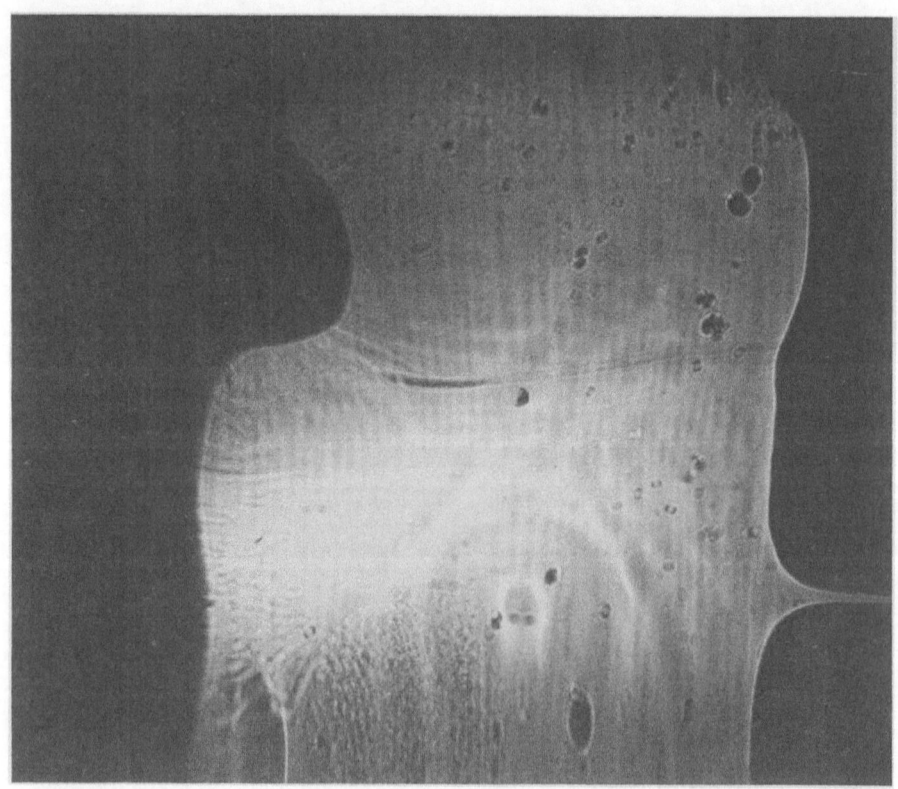

Figure 7. TP Surface After Acoustical Signal

Effect of Substrate Thickness

Since the 1.5 mm thick plexiglass base of the recording device was not capable of maintaining the required flatness after the liquified TP was poured due to the thermal warping, a study was undertaken to determine how much image quality would be lost by using a thicker plexiglass base. The recording device was modified so that the base area inside the 11.4 cm square was removed. Plexiglass squares of various thickness up to 6.4 mm were then temporarily attached to the underside of the device and a 6.4 mm water layer was used as an imaging medium instead of TP. The images became increasingly poorer as the substrate bases were progressively thickened up to 6.4 mm. In addition, this last and thickest substrate still thermally warped when heated with a heat lamp in an attempt to liquify a TP layer.

To find a more heat resistant substrate, the water imaging experiment was repeated using both mylar and glass. The 0.09 mm mylar deformed significantly, not from the heat but from the water pressure on its lower side when the recording fixture was placed in the imaging tank, even though the mylar was positioned less than 6 mm below the tank water surface. This indicated that rigidity of the substrate was also a desirable quality. A 1.3 mm thick glass plate was sufficiently rigid but, as shown earlier, the glass reflected most of the ultrasound so that insufficient energy was available to form a water surface image.

Effect of TP Surface Smoothness

Up to this point, all images were recorded with only one acoustical transducer activated. By activating both transducers, an acoustical hologram can be generated at the liquid surface, although there is some argument that the images already shown were Gabor in-line holograms. Nevertheless, by activating both trans- ducers, a true interference pattern is formed. This interference image, when viewed at the optical focal plane, produces a bright dot which represents the zero order diffraction and a series of diffraction lines above and below the central dot which represent the first, second and higher diffraction views as mathematically shown as Swinson [3]. The contrast of the images was significantly improved by viewing the object at higher diffraction views. These water surface images were recorded at increasing laser power to compensate for the decreasing brightness of each successively higher diffraction order.

To create similar diffraction images on a TP layer, the sur- face of the TP was required to be optically flat. All attempts to record diffraction patterns on the TP were unsuccessful due to a lack of flatness which prevented the reflected light from forming

a spot on the optical focal plane. Instead, an unfocused folded
image was created at the focal plane, making it impossible to
separate diffraction orders. This uneven upper TP surface was
caused by a combination of several separate problems. Besides the
rigidity question which has already been raised, the surface of
the TP formed a convex meniscus after pouring. This tended to
expand the reflected light beam and destroy the potential for sepa-
rating diffraction orders at the focal plane. Of course, as the
distance from the boundaries increased, the surface more closely
approximated a flat surface. In fact, a previous experiment had
shown that water in the 114 mm square recording device formed a
less than flat surface due to meniscus curvature and this made
separation of diffractions more difficult. By filling the recording
device with enough water to cover over the 114 mm square retaining
walls, thereby causing the outer walls to become retaining walls,
the boundary effects were sufficiently remote from the central recording area and a normal diffraction pattern was observed at the
focal plane.

     A new recording device was designed in a further attempt to
create a flat TP recording surface. As shown in Figure 8, the new
device differed from the old in several ways. The plexiglass base
was replaced with 6.4 mm thick aluminum for rigidity. The 114 mm
square central acoustical opening was replaced by a 76 mm diameter
round hole to maximize the rigidity of the substrate insert. A
76 mm diameter window was the smallest acoustical window thought
practical and the round shape of an acoustical substrate insert
minimized deflections due to underside water pressure. The interior
retaining walls were removed so that the outer walls acted as a
retaining wall for the TP and a water barrier as well, thereby
causing the TP to approximate a flat surface near the central acoustical window since this window was far removed from the boundary
effects.

Effect of Substrate Surface Roughness

     Plugs (76 mm in dia.) of various substrate materials were
placed in these new recording devices. Water surface holograms
were taken to compare these materials. The zero and first order
diffraction images through a 3 mm thick nylon substrate plug
covered with 12.7 mm of water were observed. The quality of the
images was significantly below that of the water surface holograms
with no substrate due to the relaxation absorption and attenuation
of the ultrasound through the substrate material. Observation of
the zero and first order images through a similar nylon plug which
had been sanded to a thickness of 1.8 mm showed that no real improvement of the previous images resulted, even though the
substrate thickness had been more than halved. In fact, there was
some deterioration of image quality in the first order images.

Figure 8.   Improved Recording Device

It was theorized that substrate surface roughness might account
for these results, since the thinner nylon plug had a rougher
surface caused by the sanding. The wavelength of 3 MHz ultrasound
in water is 0.51 mm. Sanding the nylon plug with 100-count sandpaper might have caused surface variations in the substrate of the
same order of magnitude as the wavelength of the ultrasound, thereby
causing the image to distort. To test this theory, the thicker
nylon plug was sanded with 100-count paper until rough and another
first order water surface holograph was attempted. The image
quality was greatly reduced. In a further test the surfaces of the
thinner nylon plug were sanded with increasingly fine sandpaper up
to 600 count. The surfaces were then polished with a polishing
compound and a first order hologram was imaged. Some improvement
in the image was seen, indicating that an ideal substrate should
have acoustically smooth surfaces.

A first order hologram of a 1.6 mm thick teflon substrate plug
was observed. This image was clearly superior to any previously
generated, although the teflon surface was not particularly flat.
This indicated that teflon had a lower degree of attenuation and
relaxation absorption than other materials investigated.

An Improved Heating Device

A device was designed and constructed to heat high melting
point TP to a liquid for image formation. As seen in Figure 9, the
device consisted of a plywood top with a central opening for the
spatially filtered laser beam. Thin (0.25 mm dia.) stainless steel
wire was wound between two sets of hooks, which was attached to
spring loaded movable supports. Heat was generated by passing an
AC current from a variable transformer through the resistive device
(57 ohms, cold). As the wire heated, significant thermal expansion
occurred, but this was compensated for by the expansion of the
springs on each movable support, thus keeping the wires taut.
Experimentation showed that 160V could be tolerated by this device
before the wires broke (500W). Figure 10 shows the heating device
installed in the recording fixture. The construction of the heater
is such that the wires are positioned approximately 6.4 mm above
the TP layer.

Additional TP Recordings

The round window recording device was prepared for TP recording.
The substrate plug used was Teflon, since it had previously produced
the best water images. The plug was attached to the aluminum base
of the recorder with silicon rubber so that the Teflon upper surface
was level with the aluminum upper surface. Small gaps between the
edges of the two materials were filled with automotive body putty
so that the smoothest possible continuous surface could be achieved.
The recorder base was then coated with 2.5 mm thick layer of S-55
TP, a material which is rigid at room temperature. The TP had to
be heated to 125°C before pouring in order for it to flow to an
even layer on the recorder. The heating device was then installed.
The assembly was first tested with no acoustic signals to verify
that the TP would adequately liquify with the heater activated.
This test was necessary since the effect of having the bottom of
the TP coupled through a heat conducting medium (the aluminum
base or the thin substrate) to a heat sink (the tank water)
was not not known. One could assume that the TP would be in a
multiple phase during heating since the top would liquify from the
radiant heat but the bottom might always remain close to the temperature of the tank water and not liquify. The test showed that
the TP liquified completely. However, the square viewing hole had
to be covered with another wood block so that enough heat could be
retained in the vicinity of the TP for proper liquification. This
does not, as one might assume, destroy the possibility of in situ
real-time recording. However, previous experimentation had shown
that the laser light image was temporarily distorted when the heater
was activated due to air convection currents. These distortions
disappeared immediately when the heater was deactivated. By
modifying the heater so that the central window was permanently

Figure 9. Bottom View of Improved Heating Device

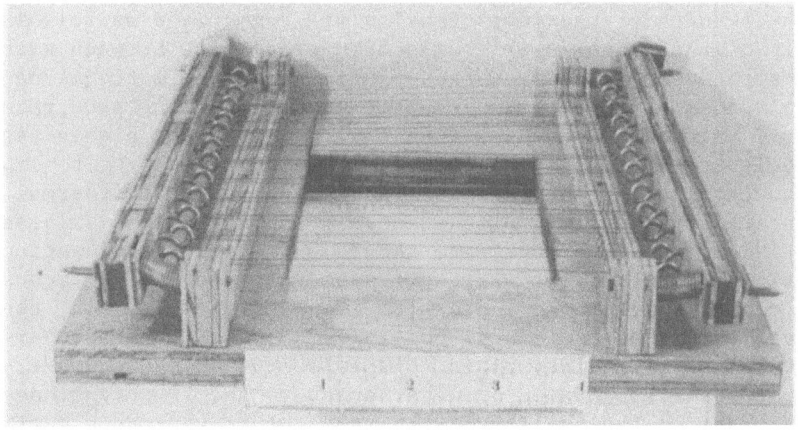

Figure 10. Heating Device Installed in Recording Fixture

covered with optical grade pyrex, one could use the heater without
adding and removing a heat retaining cover.

    When the heat was removed, it was observed that the TP surface
was slightly distorted around the perimeter of the substrate disk.
Although great care had to be taken to insure a smooth surface
continuum between aluminum base and substrate disk, there were still
slight depth variations. When the TP was liquified or when water
was used as an imaging medium, no upper surface imperfections
existed since the shear forces caused by these depth variations
could not be transmitted from the bottom of a liquid imaging media
to the top. Consequently, the top of these liquids always was
plane. However, as the TP cooled when the heater was deactivated,
the TP changed from a pure liquid phase toward a solid phase.
During this nonliquid period, the TP could, and in fact did, transmit these imperfections to the top TP surface so that a slight
boundary effect was evident all around the substrate plug on the
TP surface. This meant that higher order diffraction views could
not be generated due to a nonplane imaging surface on the TP.
However, zero order views of the target were generated as seen in
Figures 11 and 12. This sequence demonstrated an interesting
property of TP which bears additional study. As can be seen, the
images were slightly distorted, especially the heating wires, since
the reflecting TP surface was not adequately plane. However, as
the TP cooled the images grew in size. This appears to be caused
by a slight curvature change of the TP surface, so that the curvature increased as the TP cooled, thereby causing a lens-like
expansion of the light image. Whether or not this effect would be
present had the TP surface originally been plane is not obvious.

CONCLUSIONS

    The results of the investigation are summarized as follows.
Analysis of the transmission-reflection properties between water
and substrate show that a characteristic impedance match is necessary to transmit the maximum acoustic energy into the substrate TP
complex. Materials such as Teflon, nylon, mylar and plexiglass are
relatively transparent to ultrasound. The heat required to liquify
the TP necessitates that the substrate be resistant to thermal
deformation up to 125°C. Of the above group, all but plexiglass
are satisfactory. The substrate should have minimum relaxation
absorption and attenuation properties so that the maximum acoustical
energy can be transmitted to the TP. For a similar reason, the
substrate should be as thin as possible. However, the thinner a
given substrate, the less rigid. Experimentation showed that slight
surface deformation caused by water pressure and TP weight has a
severe impact on the ability of the recorder to reproduce diffraction order views. The recorded image is also subject to distortion
if the substrate and TP surfaces are not acoustically pure; i.e.,

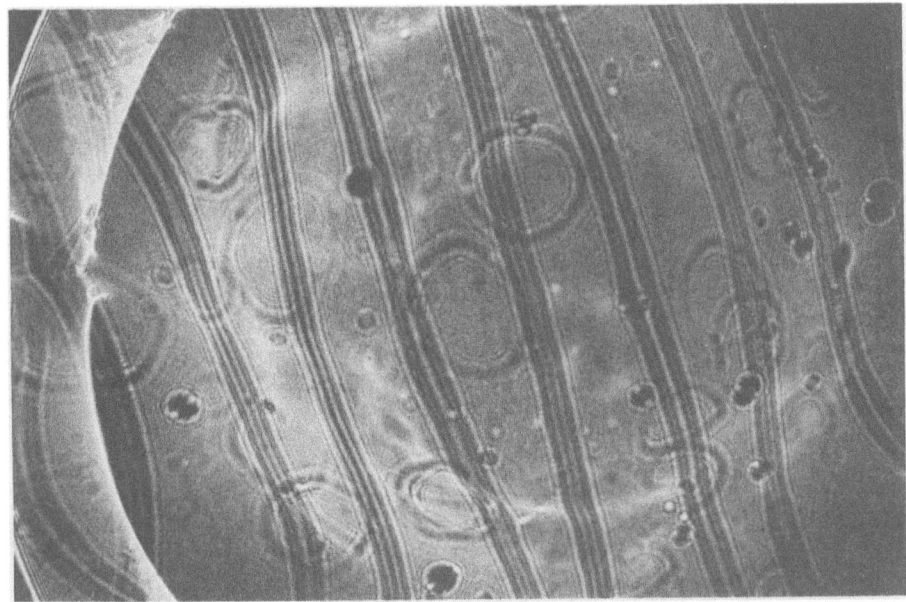

Figure 11. Zero Order Image After One Minute

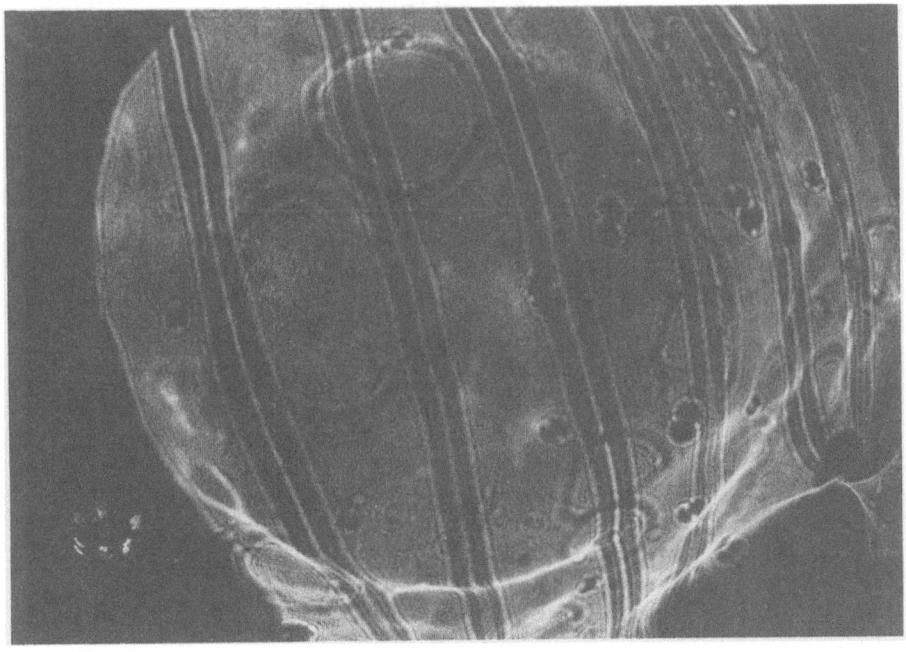

Figure 12. Zero Order Image After Thirty Minutes

if the surface roughness is not at least an order of magnitude
below the wavelength of the ultrasound.

The TP layer needs to be no thicker than that necessary to
free the central recording area from boundary effect distortions.
Thickness of TP above this causes a lessening of the visual image
since the TP itself exhibits significant relaxation absorption and
attenuation of ultrasonic signals. The type of TP used was shown
to be a trade-off between a sufficiently high melting point (so
that the TP was rigid at room temperature) and some maximum temperature (dependent on the substrate) so that the substrate would not
thermally deform when the TP was heated to a liquid during holographic recording.

This work was supported by Scientific Program Delivery Order
No. 1506, United States Army Missile Command, Redstone Arsenal,
Alabama.

REFERENCES

1. V. G. Irelan, B. R. Mullinix, J. G. Castle, "Real-Time
    Acoustical Holography Systems," U.S. Army Missile Research
    and Development Command, Technical Report T-78-10,
    Redstone Arsenal, Alabama, October 1977.
2. W. F. Swinson, "Optimizing a Real-Time Acoustical Holography
    System," U.S. Army Missile Command, Technical Report
    RL-80-2, Redstone Arsenal, Alabama, 1 October, 1979.
3. W. F. Swinson, "Improving a Real-Time Acoustical Holographic
    System for Flaw Detection," U.S. Army Missile Command,
    Technical Report RL-80-3, Redstone Arsenal, Alabama,
    1 October, 1979.
4. H. K. Liu, "A Thermoplastic Device for Real-Time In-Situ
    Recording of Acoustical Holograms, "U.S. Army Missile
    Command, Technical Report RL-80-4, Redstone Arsenal,
    Alabama, 1 October, 1979.
5. J. D. Young, J. E. Wolfe, "A New Recording Technique for
    Acoustical Holography," Applied Physics Letters, Vol. 11
    (1967), p. 294.
6. J. A. Schaeffel, Jr., "Acoustical Speckle Interferometry,"
    Technical Report T-79-39, U.S. Army Missile Research and
    Development Command, Redstone Arsenal, Alabama, March 1979.
7. L. M. Brekhovskikh, Waves in Layered Media, Academic Press,
    New York, 1960.
8. B. P. Hildebrand, B. B. Brenden, An Introduction to Acoustical
    Holography, Plenum Press, 1972.

TWO-DIMENSIONAL IMAGING WITH A HIGH RESOLUTION PVF_2/Si OPTICALLY-SCANNED RECEIVING TRANSDUCER

S. O. Ishrak and C. W. Turner

Department of Electronic and Electrical Engineering
King's College London
Strand, London, WC2R 2LS

INTRODUCTION

The use of optically-scanned composite piezoelectric/semiconductor receiving transducers as monolithic faceplates for two-dimensional acoustic imaging at low MHz frequencies has already been demonstrated[1]. Furthermore, recent experimental and theoretical work[2] has shown that the overall performance of the piezoelectric polymer PVF_2 in an imaging transducer of this type is superior to that of other available materials.

In this paper, results of detailed experiments aimed at quantifying the ultimate two-dimensional performance of a PVF_2/Si optically-scanned transducer will be presented. In addition, amplitude and phase images of test objects will be used to evaluate the image fidelity available with this type of device. This will be followed by a brief discussion of different signal processing methods to overcome the limitations of commonly used image displays. An example of an image obtained after non-linear processing will also be included.

The basic operating principle of the optically-scanned transducer has already been reported. However, a brief review of the working mechanism serves as a useful starting point for this paper.

BASIC BACKGROUND THEORY

The optically-scanned transducer shown in Fig. 1, consists of a monolithic receiving transducer with its front face electrode removed and replaced by a semiconductor layer which has a semi-transparent metal film evaporated on it. Current generators at the

77

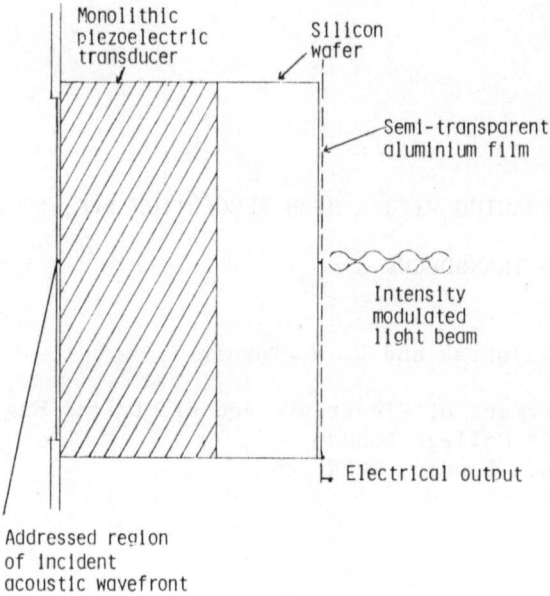

Fig. 1 The composite piezoelectric/semiconductor optically-scanned
       transducer

acoustic frequency, corresponding to the local amplitude and phase of
the incident acoustic wavefront, are created in the piezoelectric
layer. A scanning light beam samples these current generators
sequentially by photoconductively switching the adjacent semiconductor
layer. The problem of the large background reference signal at the
acoustic frequency, present in practical devices, has been largely
overcome by brightness modulation of the addressing beam; this process
generates sidebands in the semiconductor layer originating only from
the switched or illuminated part of the transducer. Typically, a
modulation frequency of 50kHz is used with an acoustic wave frequency
of 2.9MHz.

     In general, however, the set of current generators at the side-
band frequency does not faithfully represent the acoustic wavefront.
The quality of the final image is influenced by the spatial response
of the piezoelectric layer, the spatial response of the photoconductive layer and the size of the addressing light spot. It has been
shown that a low Q piezoelectric transducer with a good acoustic
impedance and velocity match to the load medium (water) is essential
for faithful and reliable imaging. The mechanical loss of PVF_2 and
its relatively low acoustic impedance makes it well suited for use
in the optically-scanned transducer[2]. Moreover, the use of comparatively thick, half-wave resonant samples has ensured that an adequate
acoustic sensitivity can be achieved.

The area over which the current generators at the acoustic frequency in the piezoelectric layer are optically sampled is determined by the size of the light spot and the photocarrier diffusion length in the semiconductor layer. Optical spreading effects are relatively unimportant if the effective instantaneous sampling diameter
is less than the acoustic wavelength in PVF$_2$. In silicon the carrier
diffusion length is approximately 300µm which is about $\lambda/2$ in PVF$_2$
at 2.9MHz; this means that the spot size must be no greater than
300µm to allow diffraction-limited imaging.

EXPERIMENTAL ARRANGEMENTS

The device consists of a 13kΩ-cm silicon wafer, 2.5 cm in
diameter and 150µm thick, coated with a 150Å semitransparent aluminium
film, mechanically clamped to a 350µm thick and 2.5 cm square PVF$_2$
transducer, half-wave resonant at 2.7MHz. Useful imaging was performed
over a 1.8 cm diameter circular aperture on the composite transducer
at 2.95MHz.

The transducer assembly was mounted on the wall of a water tank
and the scanning optical source was provided by a short-persistence
oscilloscope CRT, modulated at 50kHz, as shown in Fig. 2. A lens
was used to focus the light spot produced by the CRT onto the silicon
wafer. The electrical output from the transducer contains several
frequencies in addition to the sidebands; this signal was filtered,
mixed and detected as indicated in the system block diagram of
Fig. 3.

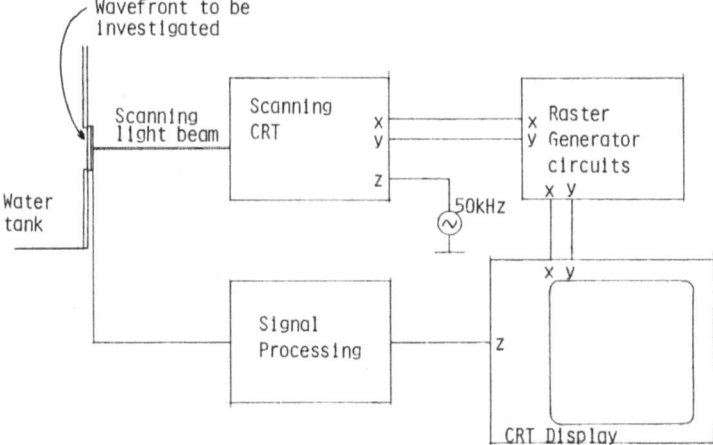

Fig. 2 Experimental arrangement for optically-scanned acoustic imaging

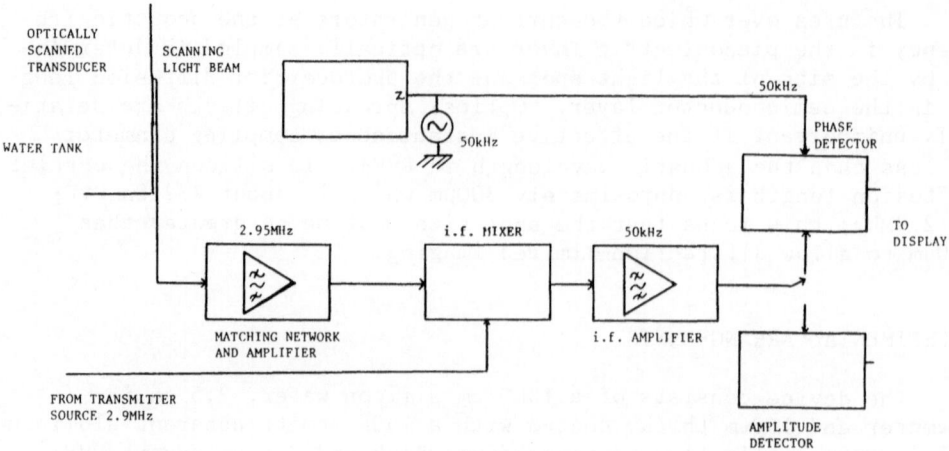

Fig. 3 Signal processing employed for phase and amplitude imaging

Phase and amplitude acoustic images were then obtained either as intensity modulation of a CRT display or in two-dimensional line scan format. A useful feature of the present system is the facility for examining particular lines of the amplitude or phase image quantitatively under slow-scan, narrowband conditions. The frame rate of the 64 line image could be varied from 0.1Hz to several Hz depending on the required image bandwidth and sensitivity.

In the following sections, various imaging results will be presented confirming the performance of the optically-scanned transducer and demonstrating its usefulness as a versatile two-dimensional imaging faceplate.

EXPERIMENTAL RESULTS

Spatial Frequency Response Tests

It was found that acoustic images of different spatial frequencies could be conveniently presented to the optically-scanned receiving transducer by varying the angles of incidence of acoustic beams from transmitting transducers. Before proceeding with the experimental results, however, it is perhaps appropriate to present some simple background theory as an aid to interpreting the amplitude and phase distributions in the experiments.

Consider a plane acoustic wave of amplitude A_1, at frequency f with a wavelength λ_w in water, incident at some angle θ on the

receiving transducer. The acoustic field distribution at the faceplate is then given by

$$F_1(x,t) = A_1\cos(2\pi ft - \frac{2\pi}{\lambda_w}\sin\theta\ x) \qquad\qquad \ldots\ldots(1)$$

F_1 is analogous to a travelling wave at the surface of the imaging
transducer, propagating in the x direction with a wavelength $\lambda_w/\sin\theta$
(see Fig. 4a). The corresponding steady-state acoustic amplitude and
phase distributions are therefore as shown in Fig. 4b.

A plane wave of amplitude A_2 incident at an angle $-\theta$ may likewise
be modelled by a travelling wave in the $-x$ direction expressed by

$$F_2(x,t) = A_2\cos(2\pi ft + \frac{2\pi}{\lambda_w}\sin\theta\ x) \qquad\qquad \ldots\ldots(2)$$

The steady-state amplitude and phase behaviour of F_2 is thus similar
to that for F_1.

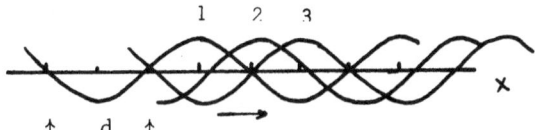

Fig. 4a Acoustic Field at instants 1, 2 and 3 (travelling wave)

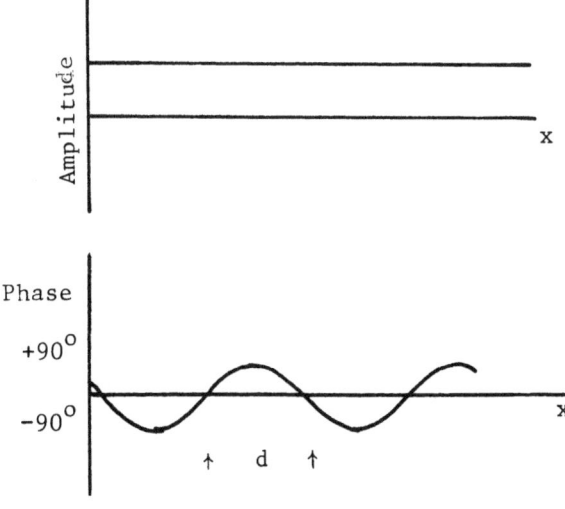

Fig. 4b Steady state amplitude and phase distribution for
$\theta = \sin^{-1}(\frac{\lambda_w}{2d})$ (travelling wave)

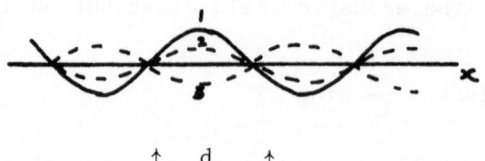

↑   d   ↑

Fig. 5a   Acoustic field at instants 1, 2 and 3 (standing wave)

Amplitude

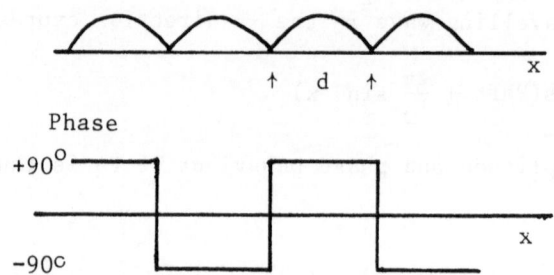

Phase

Fig. 5b   Steady state amplitude and phase distributions for

$$\theta = \sin^{-1}(\frac{\lambda_w}{2d}) \quad \text{(standing wave)}$$

When the travelling waves F_1 and F_2 are incident on the faceplate together a standing wave is created given by

$$F_s(x,t) = (A_1 + A_2)\cos(2\pi ft)\cos(\frac{2\pi}{\lambda_w} \sin\theta\ x)$$

$$+ (A_1 - A_2)\sin(2\pi ft)\sin(\frac{2\pi}{\lambda_w} \sin\theta\ x) \qquad \ldots\ldots(3)$$

If $A_1 = A_2$, equation 3 reduces to

$$F_s(x,t) = 2A_1\cos(2\pi ft)\cos(\frac{2\pi}{\lambda_w} \sin\theta\ x) \qquad \ldots\ldots(4)$$

The resulting instantaneous and steady-state acoustic fields are sketched in Fig. 5. Note that the effective fundamental spatial frequency of the amplitude distribution is twice that of the phase distribution. Note also that the distance, d, between zero crossings in both the distributions is given by

$$d = \frac{\lambda_w}{2\sin\theta} \qquad \ldots\ldots(5)$$

(this is also true for the phase distribution of the single travelling wave).

It is important to realise that if $A_1 \neq A_2$ (as is the case in general) the second term in equation 3 is non-zero; the amplitude minima do not then represent zero crossings. In the extreme case, if $A_1 \ll A_2$, the distributions revert to those for the single travelling wave F_2.

Two circular PZT-5A transmitting transducers, each 2.5 mm in diameter, were used in experiments based on the simple theory discussed above. Initial tests were concerned with a single transducer transmitting towards the optically-scanned receiving faceplate through a range of angles of incidence. Results showing good correlation with equation 5 are illustrated in Fig. 6, where the distance between zero crossings, d, as plotted against $1/\sin\theta$ calculated from the geometry of the arrangement.

Fig. 7a and b are examples of typical two-dimensional amplitude and phase images from a single transducer at oblique incidence. Practical considerations necessitated working in the near field of the transmitting transducers; as a result the amplitude distributions were non-uniform.

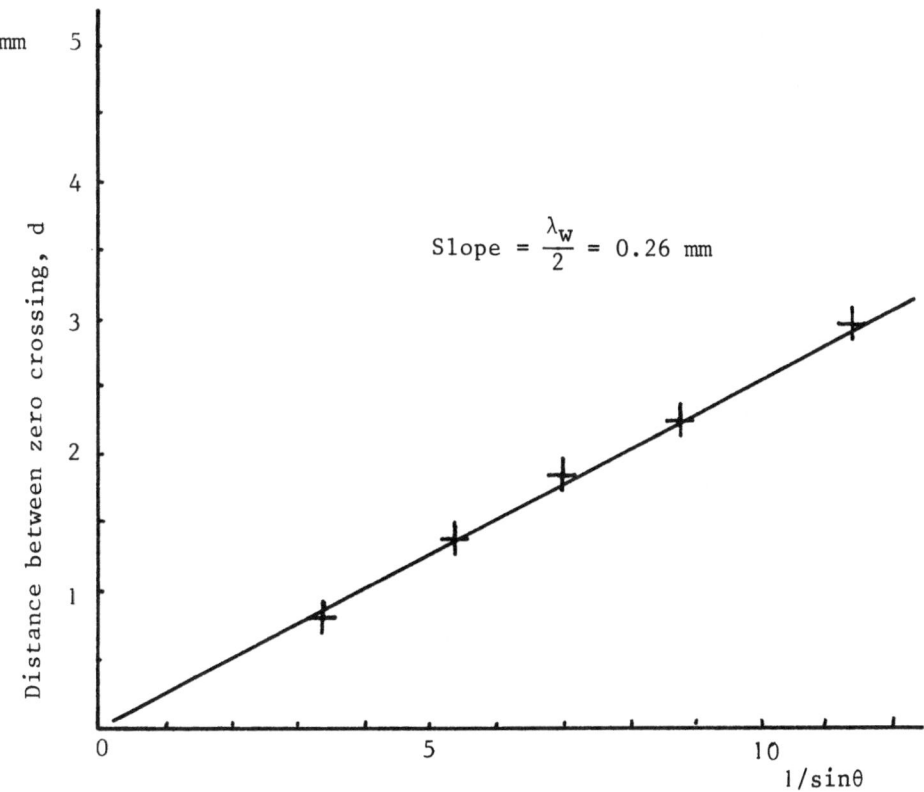

Fig. 6 Spatial wavelength plotted against angle of incidence

AMPLITUDE

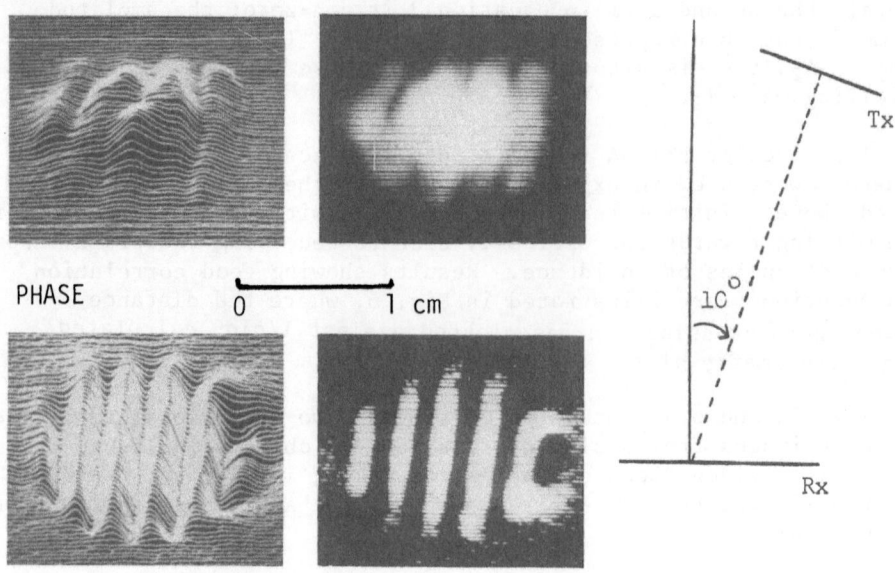

Fig. 7a   Two-dimensional amplitude and phase images from a single
          transducer at θ = 10°

AMPLITUDE

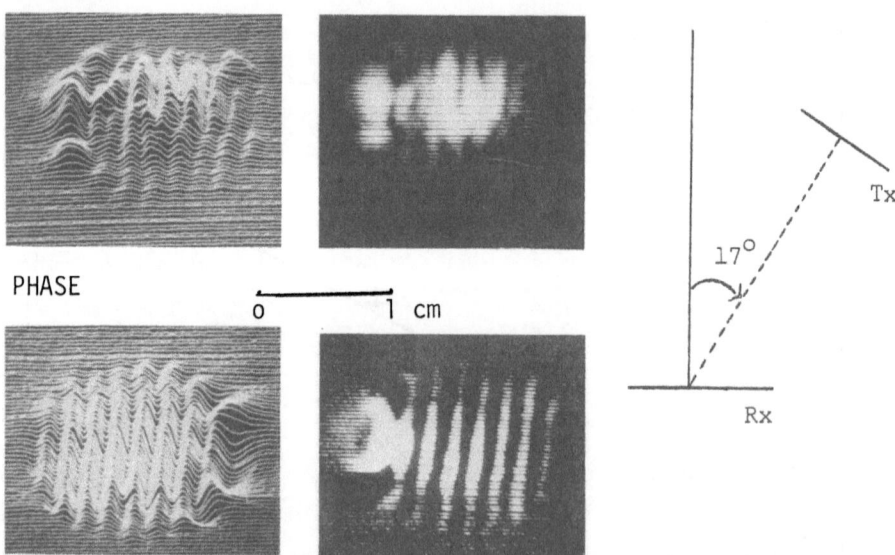

Fig. 7b   Two-dimensional amplitude and phase images from a single
          transducer at θ = 17°

Amplitude fringes were then produced as described by equation 3 by using two transmitting transducers. The phase and amplitude images of Fig. 8a to Fig. 8f illustrate the qualitative agreement of these results with theory. (Note, in particular, that the fundamental spatial frequency of the amplitude is twice that of the phase image.)

AMPLITUDE

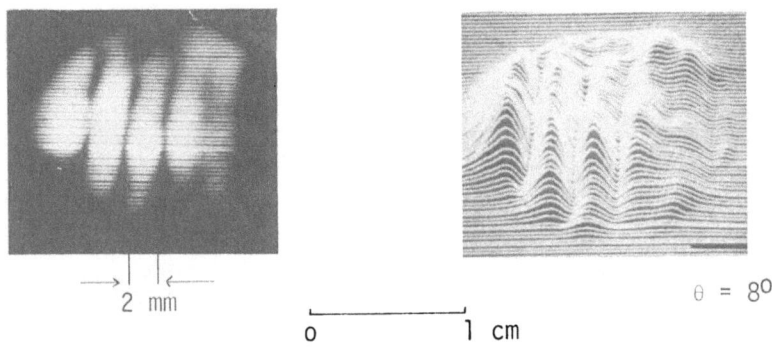

2 mm

0   1 cm

θ = 8^0

PHASE

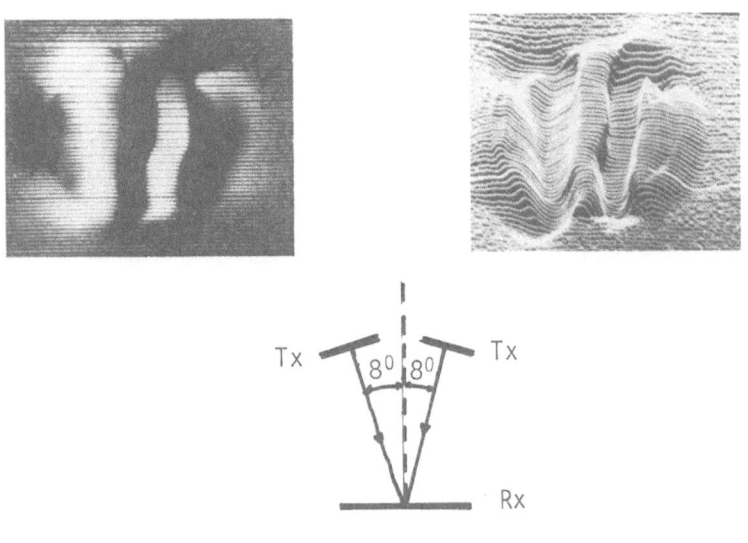

a

Fig. 8 Two-dimensional amplitude and phase images from two transducers transmitting at oblique incidence

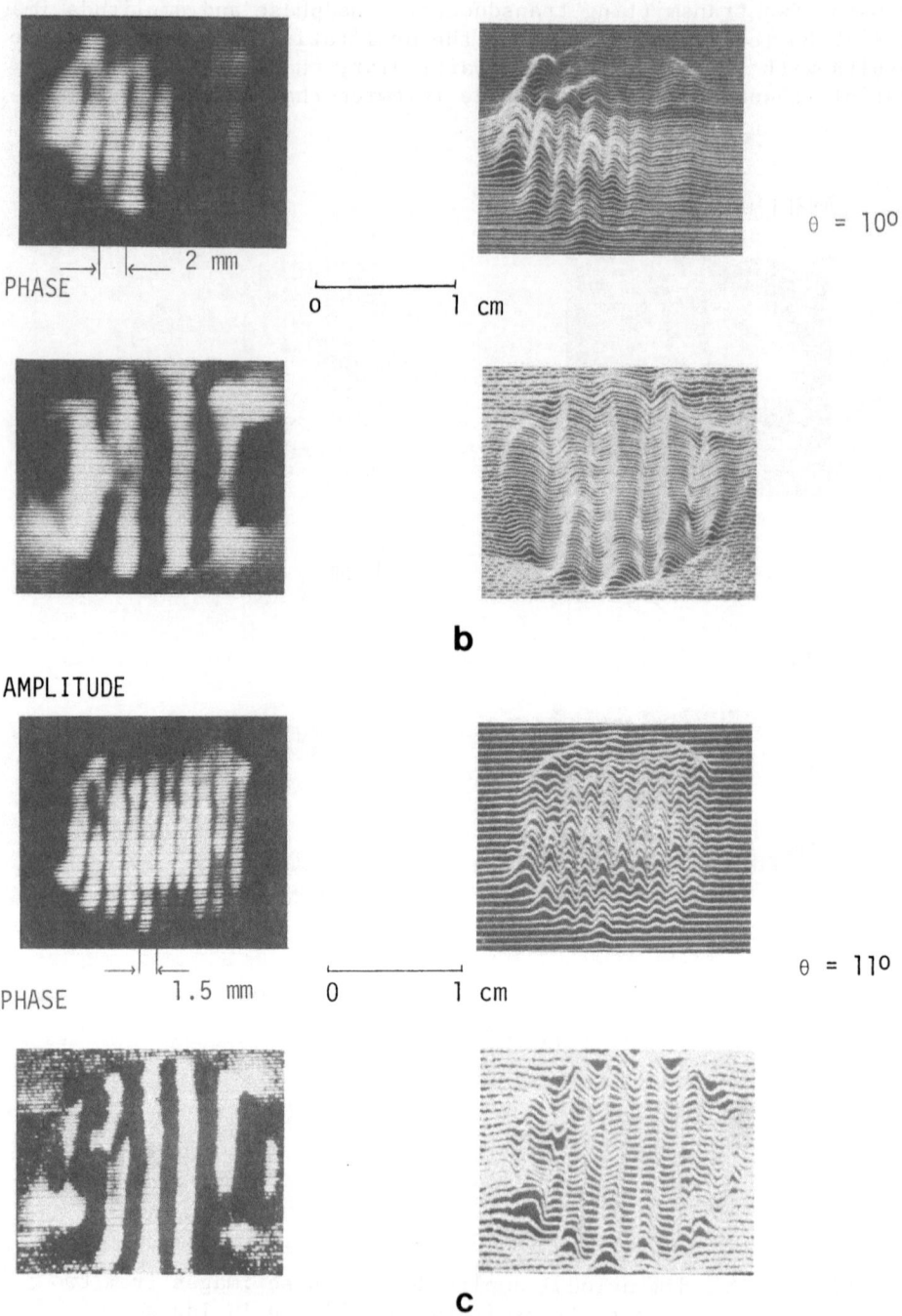

Figure 8   (Continued)

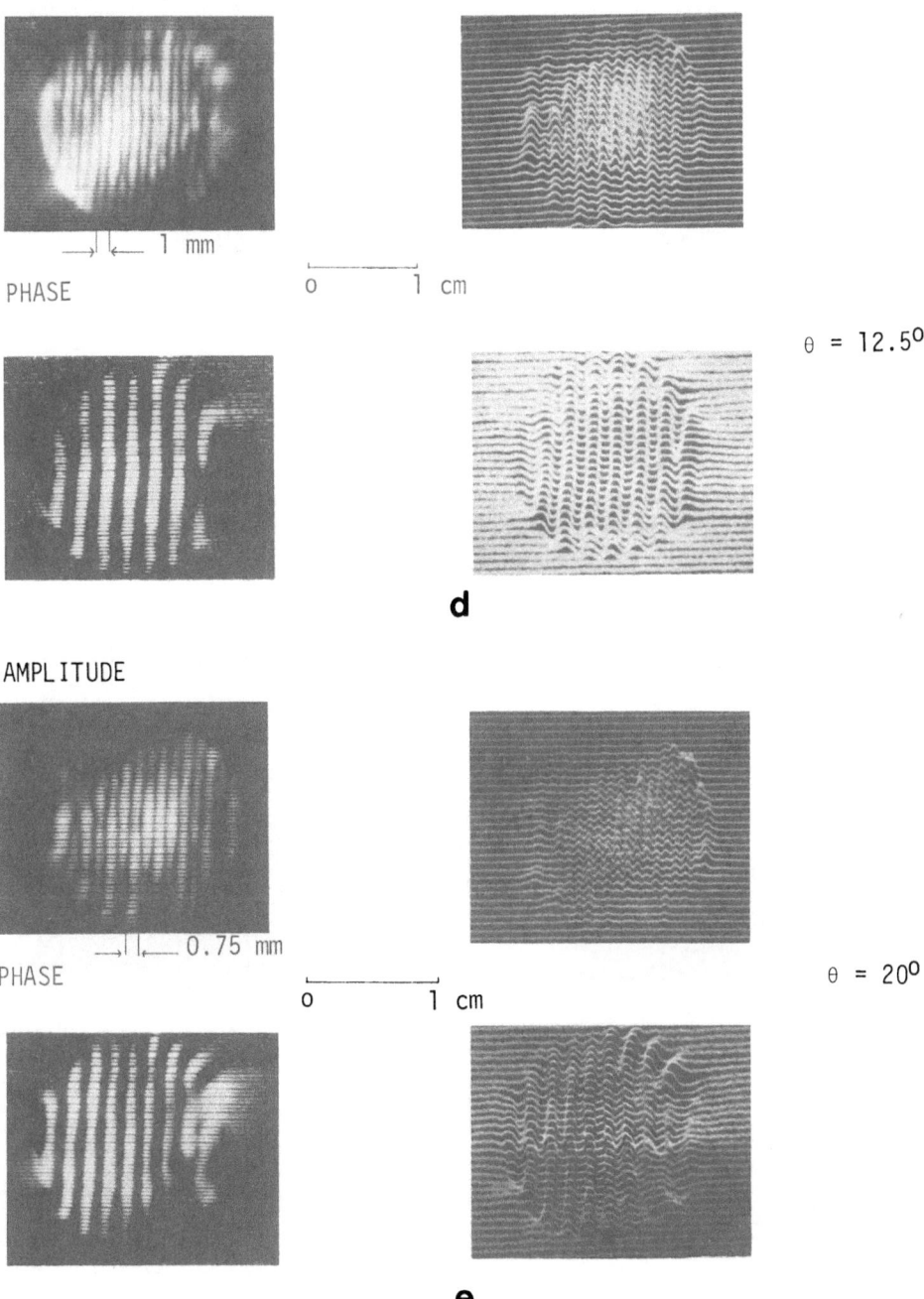

Figure 8   (Continued)

AMPLITUDE

PHASE

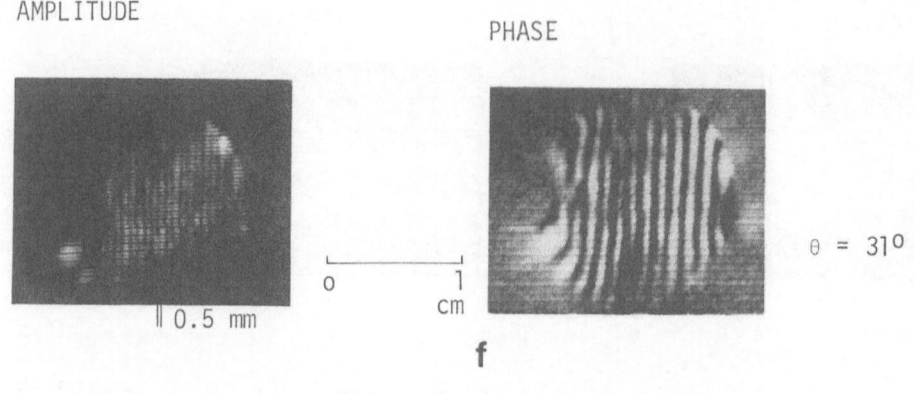

$\theta = 31^o$

‖ 0.5 mm

0   1
cm

f

Figure 8   (Continued)

Some distortions are, however, present in all the images, but these are mainly because of non-ideal experimental conditions. It is, therefore, useful to examine particular line scans in detail to obtain an improved understanding of the results.

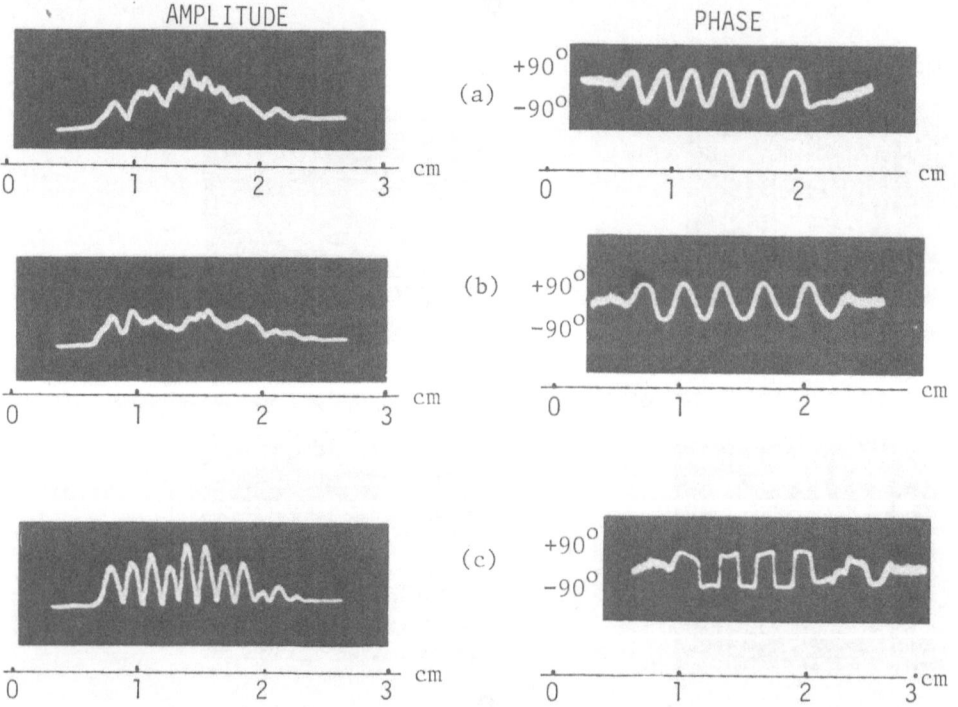

Fig. 9   Phase and amplitude line scans for each individual transducer
(a) and (b), and for both transmitters at oblique incidence
(c) ($\theta = 11^o$)

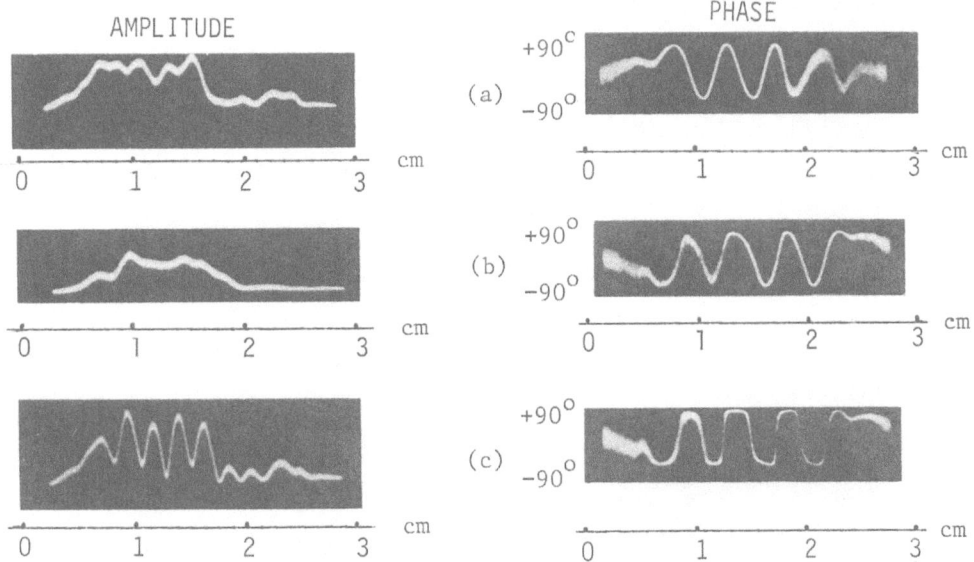

Fig. 10   Phase and amplitude line scans for each individual transducer
          (a) and (b), and for both transmitters at oblique incidence
          (c) ($\theta = 8^{0}$)

     Fig. 9 shows examples of line scans obtained from each
individual transducer separately together with scans derived from a
line of Fig. 8c. Note that the peaks of the standing wave amplitude
distributions are modulated by the local non-uniformity of the constituent travelling wave amplitudes. Note also that the amplitude
fringes deteriorate on one side of the distribution where the
individual transducer amplitudes are unequal. This is confirmed in
the phase line scan by the departure from the sharp 180^{0} oscillations
characteristic of standing waves.

     Fig. 10 is a further example of line scans obtained from
different angles of incidence. It is especially clear from Fig. 10
that the amplitude minima are non-zero for unequal travelling wave
amplitudes, accompanied by corresponding non-ideal behaviour of the
standing wave phase distribution

     The approximate resolution limit of the composite device is
illustrated in Fig. 8f. The 700μm separation of the amplitude peaks
roughly corresponds to 1.5 λ_{w}.

Images with Test Objects

     Acoustic field distributions were obtained with a perspex block
with 3 mm holes drilled in it, 3 mm apart. The phase and amplitude
images of Fig. 11 show that perspex is largely transparent to acoustic

AMPLITUDE

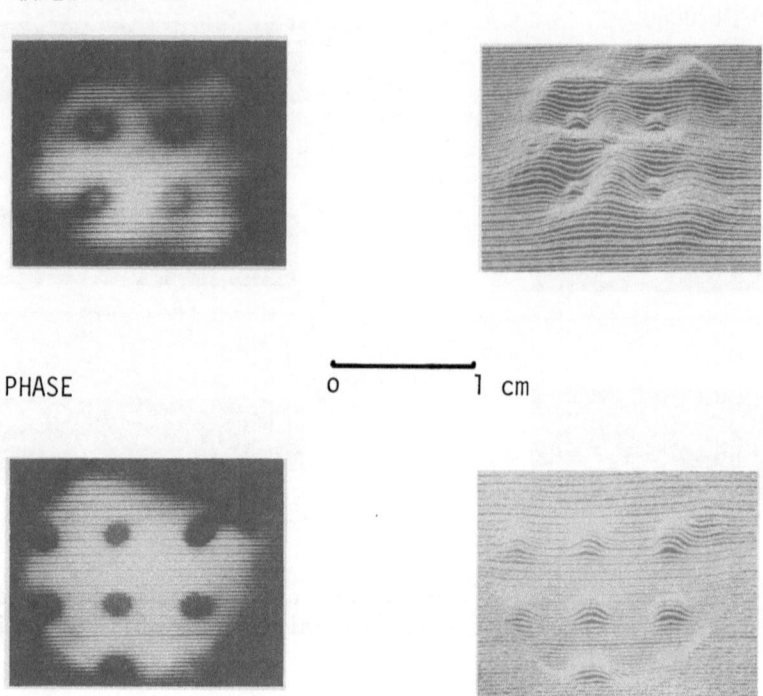

PHASE

Fig. 11   Acoustic images of a 3 mm thick perspex block with 3 mm
          diameter holes drilled in it, 3 mm apart

waves. Scattering at the circular edges of the holes, however, gives
rise to coherent 'spatial ringing' in the amplitude images. The
phase images demonstrate clearly the finite phase shift introduced
by the perspex block. A similar image (Fig. 12), but of higher
spatial frequency, can be obtained with a piece of circuit board as
a test object.   Fig. 12 is a further example of the high resolution
of the device.

AMPLITUDE

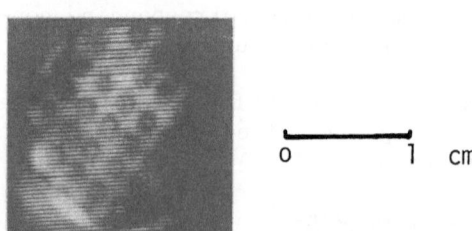

Fig. 12   Transmission amplitude image of a piece of circuit board
          with 1 mm diameter holes, 1 mm apart

Fig. 13 shows the focusing action of a polythene lens. Ideally, a uniform phase response is expected in the focal regions; phase variations in the image are probably a result of imperfections in the lens.

AMPLITUDE

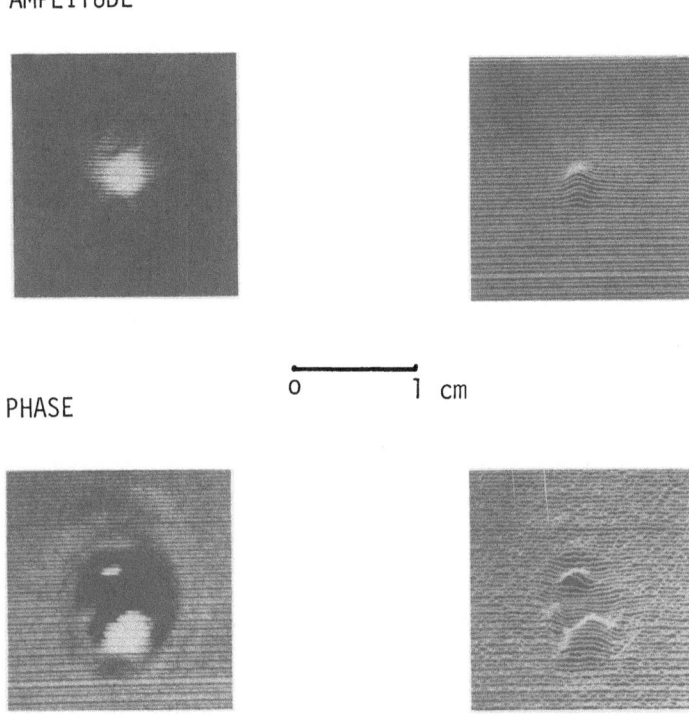

PHASE

Fig. 13   Acoustic image of the field distributions in the focal plane of a polythene lens

SIGNAL PROCESSING METHODS

It is apparent from a comparison of the two-dimensional images, already presented in this paper, with their corresponding one-dimensional line scans, that the final display limits the ultimate quality of the acoustic image to a large extent. This is especially true in amplitude images where a large dynamic range is required to use all the available information from the device. For low spatial frequencies, the two-dimensional line scans provide improved presentation (see Fig. 8a), but their effectiveness degrades at high spatial frequencies (see Fig. 8e).

AMPLITUDE

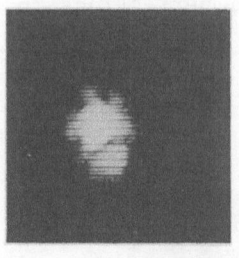
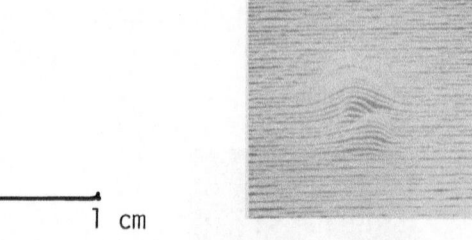

0   1 cm

Fig. 14 Acoustic image of the polythene lens after logarithmic
         processing

A common method for achieving dynamic range compression is
through logarithmic processing of the image. However, this tends to
'wash-out' low contrast images because of decreased gain at higher
signal levels. Furthermore, a loss in S/N ratio can result from the
selective amplification of low level noise. The use of log processing is thus limited to high contrast images with no low-contrast
image information.

The images of Fig. 13 obtained from a lens are well-suited for
log amplification. The clearer visibility of the low-level sidelobes
of the lens in the processed image of Fig. 14 illustrates the usefulness of log processing for certain types of images. It is evident
that while non-linear processing can improve the qualitative visual
presentation of the image, care has to be taken in the interpretation
of quantitative information, so that non-linearities are fully and
correctly accounted for.

A useful method for obtaining quantitative, reliable and easily
interpretable information is to examine sections of the image over a
limited dynamic range; the primary disadvantage of this viewing
technique is that an overall view of the acoustic field is not
readily available.

Two-dimensional line scans similar to the ones presented here,
but coupled with hidden line elimination perhaps provide the most
versatile two-dimensional visual display technique. Digital processing of the image can be used to take full advantage of the amplitude

and phase information available from the optically-scanned transducer, but only at the expense of the relative speed and the simplicity inherent in this class of imaging transducer.

CONCLUSIONS

In principle the aim of this paper has been to provide evidence in the form of two-dimensional images of known objects, that the PVF_2/Si optically-scanned transducer is capable of high resolution and reliable phase and amplitude imaging with good fidelity. Furthermore, it has also been shown that quantitative data can be simultaneously obtained from the device which can assist enormously in the interpretation of qualitative images.

Another significant advantage of the device lies in its capability of quasi real-time operation, which allows continuous adjustment of the object (e.g. transmitters). While the image frame-rate may be further increased (up to 12Hz for 50kHz optical modulation frequency), a consequent degradation in S/N ratio results; a typical sensitivity of 10^{-9}W/cm^2 is obtained at a frame rate of 0.2Hz.

In summary, the work reported has demonstrated that high resolution amplitude and phase images can be obtained with an optically-scanned transducer requiring relatively simple instrumentation.

ACKNOWLEDGEMENTS

The contribution of Thorn-EMI (Plastics Division), who provided the plate of thick film PVF_2 used, is gratefully acknowledged.

REFERENCES

1.   Turner, C. W., Ishrak, S. O. and Fox, D. R., "Optically-scanned Amplitude and Phase probing of acoustic fields", paper presented at the 1979 Ultrasonics Symposium.

2.   Turner, C. W. and Ishrak, S. O., "Comparison of different piezoelectric materials for optically-scanned acoustic imaging", Acoustical Imaging X, 1981.

AN ULTRASONIC DETERMINATION OF CARDIAC MUSCLE STRUCTURES

D. Nicholas, A. W. Nicholas and R. Greenbaum

Physics Department, Institute of Cancer Research/Royal
Marsden Hospital, Sutton, Surrey, U.K., and
Cardiology Department, London Chest Hospital, London E2
U.K.

INTRODUCTION

Left ventricular disease is a major cause of mortality and
morbidity in the Western world. Principal causes include ischaemic
heart disease, hypertension and aortic valve disease, leading to
angina, myocardial infarction or sudden death. Hypertension and
aortic valve disease are important causes of left ventricular
hypertrophy. In this condition it is not clear whether there is a
generalised increase in all elements of left ventricular structure
of whether there are selective regional increases. The few studies
which have been undertaken suggest that the latter is the more
likely possibility.

In myocardial infarction, part of the myocardial muscular
tissue dies. There has been much interest by clinicians in the
estimation of the extent of the affected myocardium. To date, in
the acute stage this determination has been indirect, for example,
relating size of infarct to elevation of enzymes in the blood
following their release from the necrotic tissue or by relating to
the extent of ECG change. A non-invasive technique to accurately
determine the location and size of the affected region should prove
to be clinically useful.

The basic components of heart muscle tissue are the sarcomeres
- units composed of the contractile proteins actin and myosin -
which are filamentous. It is now widely accepted that it is the
interdigitation of actin and myosin that ultimately mediates the
contractile process in the cardiac muscle. Just as the properties
of an inanimate body are not solely determined by the properties
of its component parts, but also depend on their geometric

95

arrangement, so the properties of the ventricle are not merely an extension of those of the individual sarcomere, but are significantly determined by the arrangement of the macroscopic fibres.

Since many diseases of the heart are related to structural changes in the cardiac muscle fibres it is of interest to establish a non-invasive technique which is capable of assessing structural changes in the myocardium at an early stage. The purpose of this paper is to examine the feasibility of using specific ultrasound techniques to characterize myocardial muscular tissue. The primary objective is to discover a technique which is capable of monitoring orientation and separation changes in the fibres as a function of distance between the endocardium and epicardium surfaces and to relate these findings to different regions of the heart wall. It is hoped that by mapping the structural configuration of normal myocardial tissue the structural changes associated with diseased tissue may be detected at a relatively early stage.

DIFFRACTION TECHNIQUE

It has previously been reported (Nicholas and Hill, 1975; Nicholas, 1977) that ultrasonic diffraction techniques are capable of characterizing tissue types whose differences are determined by structural variations at a submillimetre level. Latterly, the success of these techniques in in vivo clinical trials (Nicholas, 1979; Merton et al., 1981) has indicated that they would be worth applying to an investigation of cardiac muscle structures.

Diffraction techniques, to date, can be implemented in two ways: the interference effects are either monitored by considering the full frequency spectrum of the scattered acoustic echoes or by selecting a specific frequency and changing the orientation of the specimen with reference to the incident acoustic beam. The relative merits and advantages of the two techniques are dependent upon the tissue to be examined and the information required (Nicholas, 1981). It is helpful, therefore, to consider the structural variations to be monitored.

The fibrous anatomy of the ventricle is complex in that fibre orientation varies not only across the thickness of the wall in an individual region, but also between regions of the myocardium (Streeter et al., 1969; Greenbaum and Gibson, 1980). This variation is illustrated in Figure 1, where an histological section through an excised human heart is displayed. In this study the muscle fibres are modelled as monofilament parallel scatterers of regular separation with a preferred orientation that varies radially through the ventricular wall. The plane of the fibres will be considered relative to the surface of the heart.

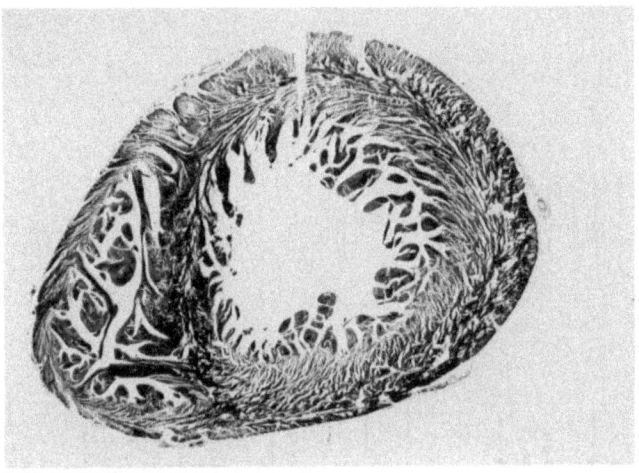

Fig. 1.   Histological section through the left ventricle illustrating
          the layer of circumferentially directed muscle in the mid-
          wall, with longitudinally directed fibres both inside and
          outside.

     Since our primary interest is in the orientation of the fibres
rather than their separation it is necessary that the fibres should
be investigated from a variety of directions.   Our novel diffraction apparatus, discussed elsewhere in this volume (Nicholas et al.,
1981) is especially suited to this task.

Muscle Model

     If we assume an array of parallel equally-spaced monofilaments
as our model for muscle tissue and orientate it such that it is
orthogonal to the ultrasound beam, then all of the scatterers will
return echoes in phase with each other.   If the scanning direction
is changed such that the model is effectively rotated about its long
axis, then the returned echoes will arrive at differing time intervals.   The resulting summation of the individual waves ψ will,
therefore, experience changes in intensity dependent upon the interference of the individual waves.   Thus, as illustrated in Figure 2,
scanning our model from different directions will produce an interference pattern showing intensity changes in the received wave ψ
as a function of scanning angle.   Since the scanning angle is
directly related to the phase difference between adjacent scatterers
' δ ', it is also related to the scatterer separation.   Changes in

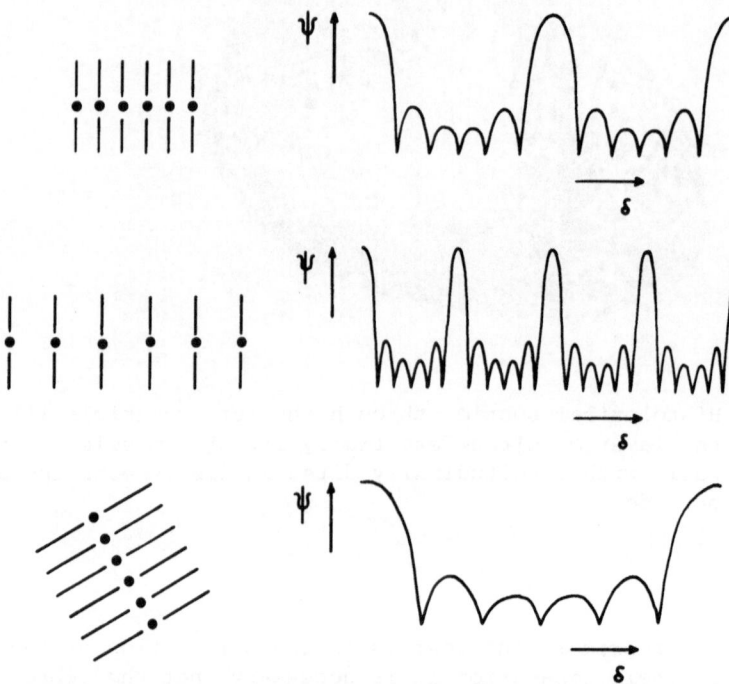

Fig. 2. Schematic relationship between monofilament models and
         ultrasonic diffraction patterns obtained by scanning in
         a plane orthogonal to the plane of the model.

the scatterer separation will, therefore, produce corresponding
changes in the interference pattern (also illustrated in Figure 2).
However, by altering the original orientation of the array of mono-
filaments we are effectively changing their separation as measured
along the transducer beam axis.   Thus changes in fibre orientation
should produce changes in the recorded interference pattern in a
manner similar to that portrayed in Figure 2.

     This simplistic argument is not meant to be definitive but
merely suggests a technique for measurements which may monitor
changes in fibre orientation.

Apparatus

The principle of the mechanical scanning facilities has been reported elsewhere (Huggins and Phelps, 1977; Nicholas, 1979) and will only be briefly mentioned here.   The transducer is attached to a 'pantograph' arm incorporating a linear B-scan movement, either of which can be clamped whilst the other is in operation. Once a tissue region of interest has been located on the linear B-scan the pantograph is adjusted so that the transducer moves in such a way that the region of interest is at the isocentre of the beam axis.   The imaging facility is now restricted to an "isocentric B-scan" but, by means of appropriate electronics, to be described, a diffraction pattern corresponding to the tissue region of interest is recorded simultaneously.   This diffraction information is accumulated on-line in a PDP/8e mini-computer for subsequent analysis.

By electronically time-gating the backscattered echoes from an organ, a specific volume of tissue, determined by the gate length and transducer beam-width, can be investigated independently of the rest of the organ.   This gated signal contains a broad spectrum of frequencies because of the short pulse used in the investigation. To relate to the continuous waves associated with interference phenomena the gated signal is frequency filtered to limit the information to a specific (narrow band) frequency; usually the "working frequency" of the transducer.   A diffraction pattern is then obtained by examining the selected tissue region from various angles and recording the backscattered intensity (frequency filtered) as a function of orientation.

TISSUE PREPARATION

Four normal human hearts were studied post-mortem.   They were removed from patients who had died of non-cardiovascular disease and who had no past history of cardiovascular disease or hypertension.   Immediately following excision the hearts were lightly inflated and fixed in 10% formalin for a period of one week.   This maintained a smooth curvature to the myocardial walls and prevented the ventricles from collapsing.   Following fixation the superficial layers of fatty tissue were removed from the epicardium to reveal the musculature structures comprising the heart wall.   Blocks of tissue were selected from sites that were known to exhibit regional variation of fibre orientation and marked by the insertion of small sutures, through the thickness of the myocardium, at the edges of the blocks.   Corresponding sites were selected from each heart and a total of eight different regions chosen from each tissue (Table I specifies the sites chosen).   The hearts were then used for in vitro estimation by ultrasound diffraction techniques of variation in fibre angle across the myocardial wall.

Table 1.   Anatomical Sites for Diffraction Measurement
            of Myocardial Muscle Structure.

Block Number	Anatomical Situation
1	Obtuse margin of left ventricle - near the base.
2	Obtuse margin of left ventricle - at the level of the papillary muscle.
3	Obtuse margin of left ventricle - towards the apex.
4	Diaphragmatic wall of left ventricle - at the level of block 2.
5	Acute margin of right ventricle - at the level of block 2.
6	Middle of interventricular septum.
7	Diaphragmatic wall of right ventricle - at the level of block 2.
8	Anterior surface of left ventricle - at the level of block 2.

Fibre Angle Assessment

     At the conclusion of our diffraction measurements the blocks
were removed from the specimen and routinely mounted in paraffin
wax.   Serial sections were then cut from epicardium to endocardium
using a microtome.   The sectional thickness was 10 μm and every
25th section was routinely stained and mounted.   A series of histological slides representing quarter millimeter intervals through
the heart wall were then available for fibre angle assessment.

     Microscopic measurement of fibre angle, using a graticule cali-
brated in degrees, showed a continuous variation across the wall.
This was found to vary from approximately -60 degrees at epicardium
to +60 degrees at endocardium, zero degrees corresponding to a
circumferential orientation.   For each section ten measurements of
angle were made and the mean and standard deviations calculated.
Figure 3 graphs the fibre angle against depth into the heart wall
for a block taken from the left ventricle close to the mitral annulus.   Each block in each of the four hearts was measured in a
similar manner and the results collated for comparison with the
diffraction findings.

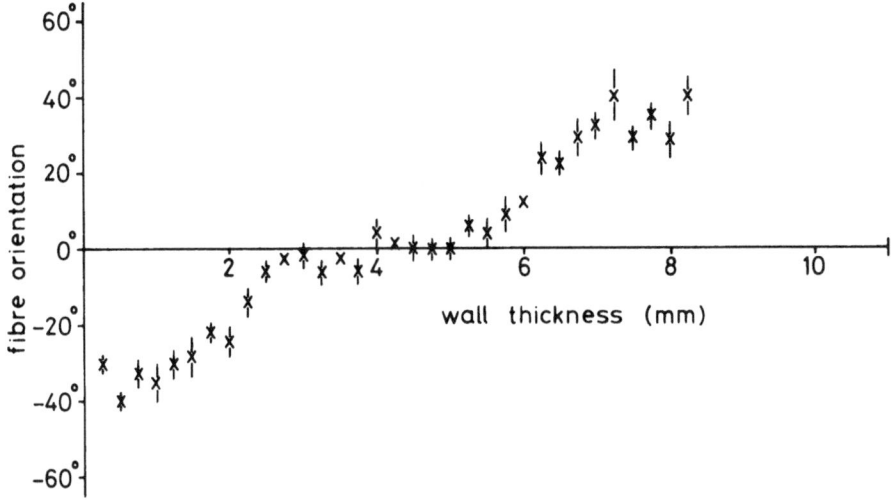

Fig. 3.   Histological measurement of fibre orientation versus depth
          into the heart wall.

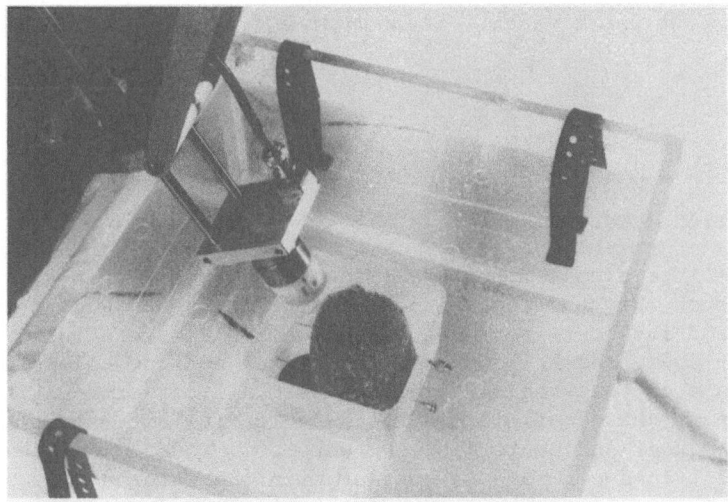

Fig. 4.   Arrangement of tissue and scanning apparatus.

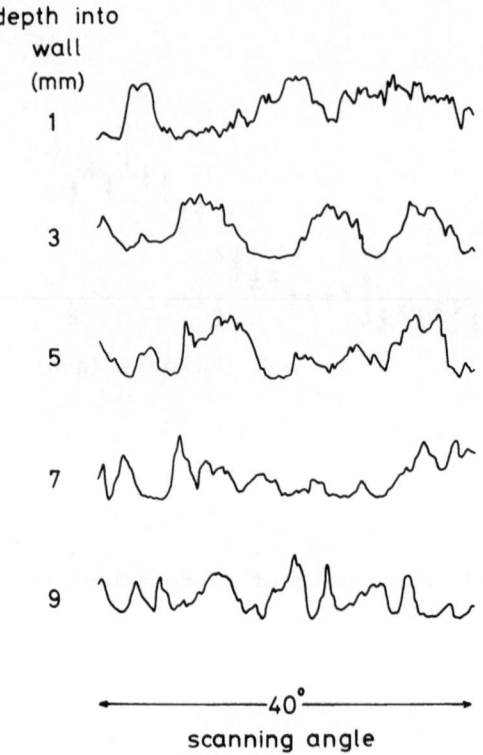

Fig. 5.   Diffraction patterns from one site at various depths into
          the heart wall.   Scanned at a frequency of 2.5 MHz.

DIFFRACTION RESULTS

     The intact post-mortem hearts were secured within a small con-
tainer such that the region to be investigated was uppermost. The
tissue and its container were placed within a water-filled sound
tank (at 20°C) and the depth of the tissue surface below the water
adjusted such that the measurements were all performed at the same
position in the far-field of the investigating ultrasound beam.
The transducer and pantograph arm were positioned above the tissue
with the transducer face in contact with the water.   Figure 4
depicts the actual arrangement of the tissue and scanning apparatus.
The point of isocentre for the pantograph motion was then adjusted
to occur on the surface of the heart wall and a 2 μs time gate used

to define a tissue region of 1.5 mm depth.   The backscattered
echoes were frequency filtered at 2.5 MHz and collected as a function of incident beam direction.   This direction was varied by 20°
either side of normal orientation to the heart wall surface.   The
position of isocentre was then readjusted to centre on a region
deeper into the heart wall.   Figure 5 illustrates a series of
diffraction patterns collected from a block of tissue at different
depths into the myocardial wall.

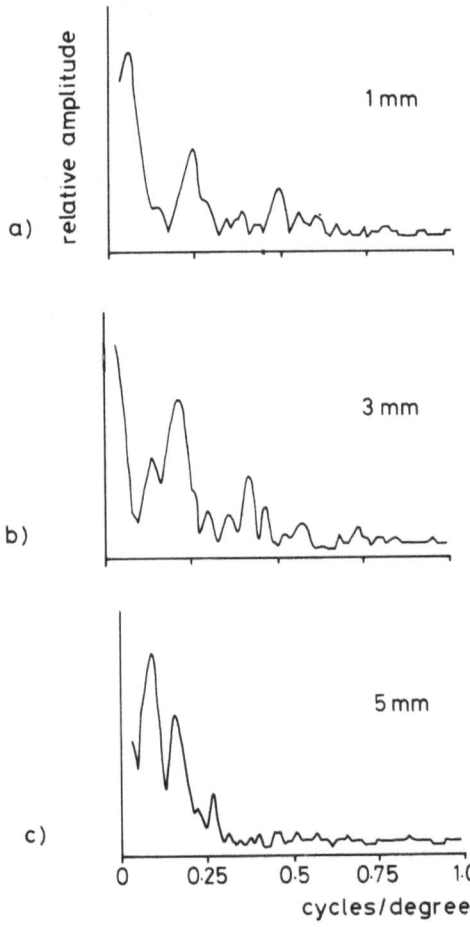

Fig. 6.   Amplitude angular frequency spectra derived from three
          of the diffraction patterns displayed in Figure 5.

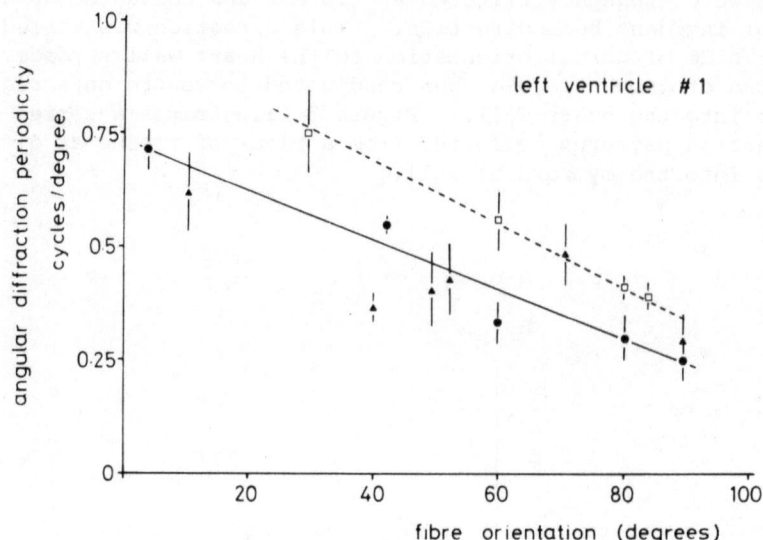

Fig. 7.   Angular frequency separation as a function of fibre orientation for three different hearts examined at position 1.

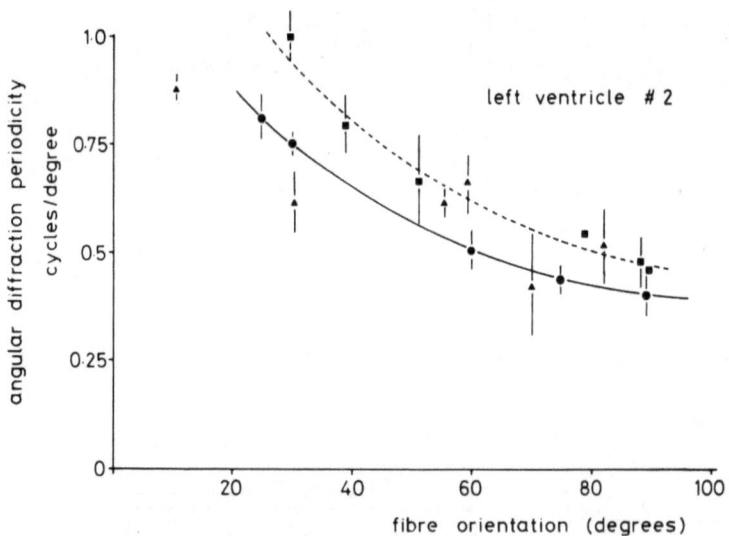

Fig. 8.   Angular frequency separation as a function of fibre orientation for three different hearts examined at position 2.

Data Analysis

Although the diffraction patterns collected by this technique can be assessed visually the main objective is a quantitative characterisation for relating to the microscopic evaluation of fibre angle. Our early applications of this technique indicated that the diffraction patterns originating from different histological structures differ particularly in the rate of angular variation of echo amplitude (as seen in Figure 5) and this suggests the application of spectrum analysis techniques for objective characterisation (Nicholas and Hill, 1975). Our present approach involves the Fourier transformation of the angular diffraction pattern into the angular frequency domain. Figure 6 illustrates the amplitude angular frequency spectra associated with three of the diffraction patterns displayed in Figure 5.

The objective of this paper is to find a means for monitoring fibre orientation non-invasively and as such any quantitative description of the diffraction patterns which meets this requirement is adequate. We shall describe, therefore, one measure extracted from the Fourier domain which has proved to be useful in other applications and apply it here. This is the spectral position of significant peaks in the angular frequency domain; other descriptors may be equally applicable but will not be discussed here.

Since the fibre orientation is known to exhibit regional variations (Greenbaum and Gibson, 1981) the different selection sites have been analysed independently. By plotting our spectral parameter (an angular frequency periodicity) against fibre orientation we can determine whether any correlation exists between the ultrasonic findings and the geometry of the myocardial fibres. Figures 7, 8 and 9 show the results for three specific sites in normal human heart muscle. For normalisation purposes a fibre orientation of zero degrees is specified when the fibres lie at right-angles to the scanning motion of the transducer. Most of the points on these curves are the mean of three measurements from adjacent sites, within a block, at the same depth and the error bars depict the standard deviations.

Although eight blocks were selected for each of the four hearts not all were suitable for scanning. Some of the selected sites had to be omitted due to surface irregularities such as creasing, which occurred during fixation, or damage when the fatty epicardial layer was removed. These problems reduced the total number of blocks investigated to: 10 from the left ventricle, 5 from the right ventricle, and 2 from the septum.

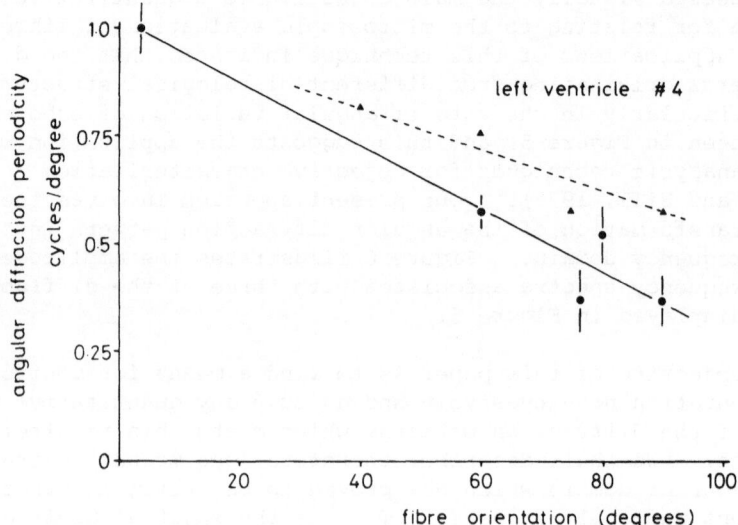

Fig. 9. Angular frequency separation as a function of fibre
orientation for three different hearts examined at
position 4.

DISCUSSION

Although the results presented are based upon a small data set
they do suggest that a correlation does exist between our diffraction measurements and the fibre orientation of the muscle structures
within the ventricular wall. Of the curves presented only the
measurements from one heart (portrayed by the triangle symbol) show
that positions 1 and 2 yield random results. All the other curves
show a marked degree of correlation (greater than 0.91). However,
it is interesting to note that for these two regions the epicardial
fat layer had not been completely removed, unlike region 4.

A further point to note is that the different sites do not
show the same dependence on fibre orientation. A possible explanation for this is that we have yet to account for differences in
fibre separation. Examination of the histological sections confirms that macroscopic fibre separation not only differs for
different regions of the heart but can also vary with depth into
the myocardial wall. It is therefore necessary that fibre separation be measured together with the fibre orientation.

The general trend for the angular frequency separation to be
inversely related to the fibre orientation suggests that our

simplistic model, discussed previously, may be feasible. Large
values for an angular frequency separation suggest a low value for
separation in the time domain, i.e. the echo-producing structures
are close together (see Figure 2).   Conversely lower angular
frequency separations can correspond to larger time domain separations where the structures are spaced further apart.   Other models
could also be postulated but the limited data collected to date
precludes any detailed analysis and merely suggests possibilities.

SUMMARY

The results presented here indicate that this ultrasound diffraction technique is capable of assessing empirically the fibre
orientation of ventricular structures as a function of depth into
the heart wall.   The possibility also exists that the technique
will provide quantitative measurement of both macroscopic fibre
spacing and orientation angle.   Since the technique, as described,
only evaluates the scattering from small tissue volumes (typically
10 mm^3) it should be capable of detecting local changes in structure
such as might occur with infarction.   However, until further
studies have been made on both normal and diseased excised hearts
the limitations of the technique can not be evaluated.

In anticipation of eventual in vivo clinical trials these
studies have been duplicated on a 'real time' phased array scanner
where the diffraction is recorded across the separate receiving
elements of the transducer array (Nicholas et al., 1980). This
machine can collect the diffraction information within 40 μs thus
enabling the heart wall to be investigated at various stages of
the heart cycle.

Although only a preliminary study has been made the indications
are that ultrasound diffraction may play a significant role in the
early diagnosis of cardiac disorders, especially when one considers
that it is a structural change rather than any abnormality of
muscle function that is primarily expressed in left ventricular
disease (Greenbaum and Gibson, 1981).

REFERENCES

Greenbaum, R. A., and Gibson, D. G., 1980, Myocardial structure of
    the left ventricle in health and disease, in: "Recent Advances
    in Cardiology", J. Hamer and D. J. Rowlands, eds, Churchill
    Livingstone, London.
Greenbaum, R. A., and Gibson, D. G., 1981, Regional non-uniformity
    of left ventricular wall movement in man, British Heart
    Journal, 45: 264
Huggins, R. W., and Phelps, J. V., 1976, Bragg diffraction scanner
    for ultrasonic tissue characterisation in vivo, Ultrasound
    Med. Biol., 2:271.

Merton, J., Nicholas, D., Hill, C. R., Grover, S., Queenan, M.,
    and Cosgrove, D. O., 1981, Ultrasound diffraction scanning of
    the thyroid, Ultrasound Med. Biol. (in press).
Nicholas, D., 1977, Orientation and frequency dependence of back-
    scattered energy and its clinical application, in: "Recent
    Advances in Ultrasound in Biomedicine", D. N. White, ed.,
    Research Studies Press, Oregon.
Nicholas, D., 1979, Ultrasonic diffraction analysis in the investi-
    gation of liver disease, Brit. J. Radiol., 52:949.
Nicholas, D., 1981, Time-frequency-domain analysis and its applica-
    tion to ultrasonic tissue characterization (submitted to
    Phys. Med. Biol.).
Nicholas, D., and Hill, C. R., 1975, Acoustic Bragg diffraction
    from human tissues, Nature, 257:305.
Nicholas, D., Cosgrove, D. O., and Pussell, S., 1980, Real-time
    diffraction scanning, Brit. J. Radiol., 53:1025.
Nicholas, D., Merton, J., and Hill, C. R., 1981, The application of
    diffraction analysis to liver and thyroid disease, in:
    "Acoustical Imaging, Vol. 11", Plenum Press.

THE APPLICATION OF DIFFRACTION ANALYSIS TO LIVER AND THYROID DISEASE

D. Nicholas, J. Merton, and C. R. Hill

Physics Department, Institute of Cancer Research/Royal
Marsden Hospital,
Sutton, Surrey, U.K.

INTRODUCTION

Ultrasonic imaging is now widely accepted as a useful technique for non-invasive diagnosis of the pathological condition of
various regions of human anatomy. Despite its undoubted success,
however, there are limitations to its ability which warrant careful
investigation. In this paper we shall limit our initial discussion
to hepatic disorders and report on a new diagnostic technique, to
complement existing B-scanning procedures, which provides extra
information for improved diagnosis. Furthermore, as a test of the
usefulness of this novel technique we will report on its success
in differentiating between thyroid disorders which are, at present,
poorly diagnosed by existing non-invasive procedures.

Although the ultrasonic detection of hepatic lesions is
usually reliable, provided they exhibit a diameter greater than
about 2 cm, no single appearance is characteristic of all metastases.
Varying descriptions in the literature indicate that the 'echo-
genicity' of metastases compared with general liver parenchyma is
inconclusive as an indication of primary neoplasm type (McArdle,
1976). For this reason it is hoped that a quantitative characterization of the lesion, based upon its structural composition,
will eventually lead to a more definitive diagnosis.

Another major problem in ultrasonic diagnosis occurs when one
attempts to diagnose diffuse infiltrations of the liver. With
the possible exception of cirrhosis, which is usually recognised
by the generally increased level of echoes within the liver parenchyma (Dewbury and Clark, 1979), the majority of diffuse disorders
are only registered (if at all) as a slight change in the general

pattern of parenchymal echoes, as for example, in the case of fatty
infiltration.

It seems that clinical diagnosis using ultrasound is still
very much dependent upon the diagnostician and his particular
machine, and is thus very subjective in its approach. Although
much useful and accurate diagnosis is being achieved, a more desirable method would be one in which a quantitative relationship
between the scattered echoes and the structural composition of the
interrogated tissues was obtained. This, of course, demands that
any pathological abnormality in a tissue must be related to a
structural variance from normal: a fairly realistic premise.

PULSE ECHO DIFFRACTION

Although conventional ultrasound primarily provides a location
of suspect areas of abnormality, its interactive behaviour with
tissue structures suggests that the returning echoes contain information relating to the nature of the scattering sites. Our objective
has been to usilize such information (using a non-imaging technique)
in an attempt to provide additional information on the scattering
structures.

It has previously been demonstrated (Nicholas and Hill, 1975)
that human soft tissues can be characterised by utilising a diffraction phenomenon analogous to the Bragg diffraction of X-rays by
crystalline structures. These results showed that, in a pulse
echo system, if a sample of tissue is rotated relative to the beam
axis, the amplitude of the frequency-filtered backscattered signal,
from a small tissue volume located on the axis of rotation, varies
with angle. Moreover, the diffraction patterns (relationships
of echo amplitude to orientation angle) thus formed were found to
characterise differing tissue types and pathologies (Nicholas,
1976, 1977a).   It has been postulated by Hill et al. (1978) that
the tissues examined comprise a semi-regular array of scatterers
over the small volume being interrogated, and that the diffraction
patterns result from constructive and destructive interference of
the backscattered signal from each scatterer. Since many diseases
only result in a subtle change in the structural composition of a
tissue it seems likely that any technique which is capable of
monitoring such changes could have immense diagnostic implications.

Apparatus

Incorporating such a technique into a clinical machine requires
that it should be capable of providing a conventional B-scan image
in addition to the diffraction information. The prototype equipment described by Huggins and Phelps (1977) provided a specialised
scanning motion for the transducer to permit the investigation of a
small tissue region from a variety of directions. This has

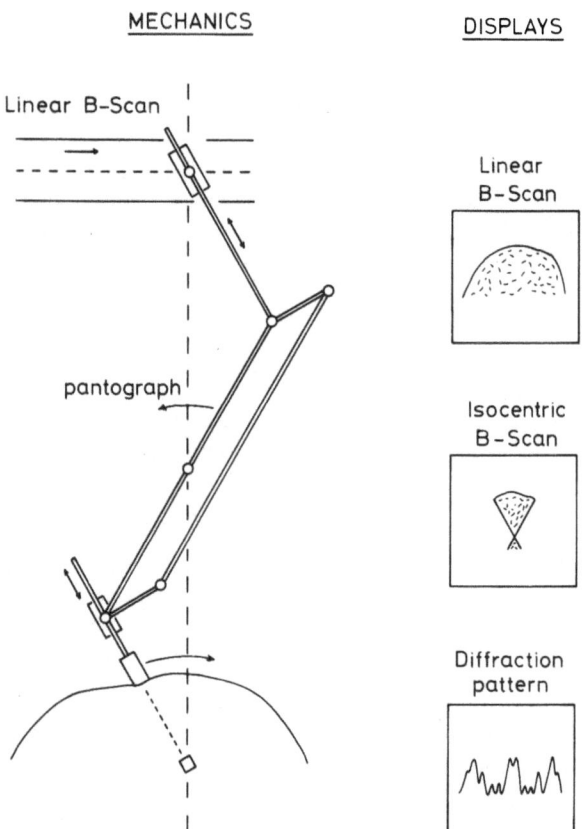

Fig. 1.   Mechanical arrangement of scanning arm and related
          imaging and analytical modes (Nicholas, 1979).

subsequently been modified (Nicholas, 1979) to facilitate the
addition of a more conventional B-scan image.

The principle of this arrangement is indicated in Figure 1,
further details of the geometry and electronics being given in
earlier papers.   The mechanical scanning facilities consist of a
pantograph and a linear B-scan arrangement, either of which can be
clamped whilst the other is in operation.   With the pantograph
movement clamped to leave the transducer in a vertical orientation
an exploratory linear B-scan can be performed.   Having located a
tissue region of interest the linear movement is clamped and the
pantograph adjusted so that the transducer moves in such a way that
the region of interest is at the isocentre of the beam axis.   The
imaging facility is now restricted to that of an "isocentric B-scan"
but, by means of appropriate electronics, previously described, a
diffraction pattern corresponding to the tissue region of interest
is simultaneously recorded.   In our present arrangement the diffraction information is accumulated on-line in a PDP/8e mini-
computer for subsequent analysis.

Method

The _in vitro_ diffraction experiments exploited the fact that,
by electronically time gating the backscattered echoes from a tissue
organ, a specific volume of tissue, determined by the gate length
and transducer beam-width, could be investigated independently of
the rest of the organ.   Although the _in vivo_ requirements are
identical a complicating factor is the duration of the investigating
pulse.   Fourier analysis shows that the shorter the pulse duration
the broader is its corresponding frequency spectrum.   This leads
to an immediate conflict in requirements.   For scanning purposes
the axial resolution necessary for adequate imaging demands as short
a pulse as is possible so as to separate, in time, the echoes
returning from closely spaced scatterers.   The relationship of
diffractive phenomena to structural arrangement, however, is normally considered to be associated with continuous waves.   This
apparent disparity can be overcome by frequency filtering the time-
gated portion of the backscattered echo train.   Derivation of the
frequency spectrum of a gated pulse results, analytically, in a
convolution of the spectrum of the pulse with that of the time gate.
Hence, for example, any single component of the frequency spectrum
of a signal selected by a 10 μs time gate will have an associated
bandwidth of 100 kHz.   By recording only the amplitude of a specific
frequency component the investigation is limited to a single
frequency (a continuous wave) convoluted with a 100 kHz bandwidth
(the time gate).   The overall effect is that the interrogating
pulse is essentially a time-gated continuous wave.   This technique
allows a great flexibility in the control of the investigating parameters and in particular makes possible the simultaneous use of a
short pulse for conventional B-scan imaging and the continuous wave

analysis of backscattering for quantitative assessment.

The three parameters determining the outcome of the diffraction procedure are the ultrasonic frequency, transducer beam-width and time gate duration.   For all the results presented here the transducer used was weakly focussed, with a 6 dB beam width of 2.5 mm between 4 and 9 cm from the transducer face (as measured in water), and had a nominal working frequency of 2.5 MHz.   The time gate duration was kept constant at 10 μs, allowing approximately 7.5 mm depth of tissue to be investigated.   Furthermore, the signals scattered back to the transducer, originating from the selected volume, are filtered at a frequency of 2.6 MHz (the most sensitive frequency of our transducer) and then recorded for 128 positions evenly spaced throughout the 28° movement of the transducer.   The resulting time-gated, frequency-filtered signal is displayed on the oscilloscope as a function of the angle of transducer movement. A repetition of the movement demonstrates the reproducibility of the diffraction patterns and confirms that they are not influenced significantly by random noise or tissue movement.   The data are digitised and stored on a PDP/8e mini-computer.

RESULTS FROM HEPATIC TISSUE

Some preliminary diffraction patterns associated with human calf muscle, normal liver and a metastatic deposit in the liver were first reported by Huggins and Phelps (1977).   Subsequently observations have been made on patterns from various tissues and pathologies (Nicholas, 1977a).   In this paper the results of a clinical trial involving 70 patients are summarised.   Investigations have been performed on the two categories of disease best described as diffuse disorders and focal neoplasms.   In the former case, organs which exhibit a uniform parenchymal echo pattern (though abnormal) on the conventional B-scan images are examined over as wide a region as possible with diffraction patterns being collected from randomly selected regions of the tissue.   For the second case, regions of interest (suspected lesions) are first located using standard scanning techniques, and then diffraction patterns obtained from within the suspected lesions are compared with those from adjacent, supposedly normal, regions of tissue.   As this latter case involves a differential analysis it permits an immediate appraisal of tissue normality.

Figure 2 illustrates, in detail, the procedure and results for one particular patient.   Here the longitudinal B-scan of the patient shows an interesting area of about 4 cm diameter in the posterior region of the right lobe of the liver.   The enlarged picture of this region indicated that it consisted of a ring of diminished echoes within which the echo pattern was visually indistinguishable from that of the supposedly normal adjacent parenchymal echoes. The clinical diagnosis based on this conventional scan and others

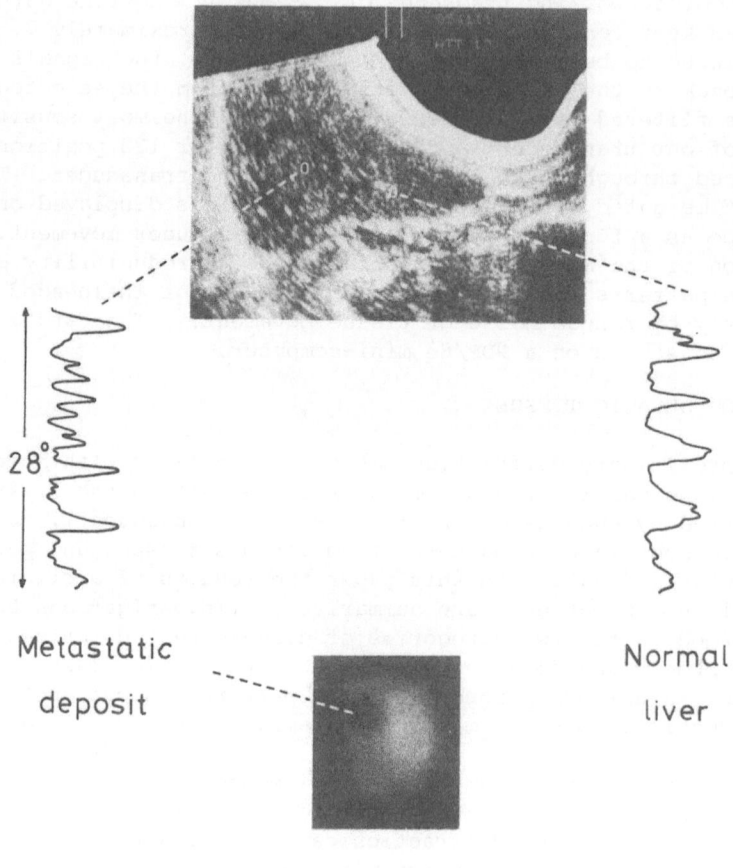

Metastatic
deposit

Normal
liver

Liver   scintigram

Fig. 2.   Longitudinal B-scan of the right lobe of the liver depict-
ing an anechoic ring of approximately 3 cm diameter.
Accompanying diffraction patterns and liver scintigram
(anterior view) confirm this as a solid lesion (Nicholas,
1979).

was inconclusive as to whether this region was metastatic or normal.
The ring of diminished echoes was suggestive of a 'target' lesion
but these usually display an abnormal pattern of internal echoes.
When investigated with the diffraction scanner at a frequency of
2.5 MHz and with a 10 µs gate length the region of interest produced
a diffraction pattern which differed significantly from the scans
associated with the supposedly normal regions (see Figure 2). This
difference, an increase in the number of diffraction peaks per unit
angle of scanning movement, is typical of both the in vitro findings
of Nicholas (1976) and previous in vivo results where similar
patterns have been associated with neoplastic involvement.   Confirmation of these findings was obtained on the following day when
a liver scintigram, also illustrated in Figure 2, clearly indicated
the presence of a large single lesion in the posterior border of
the right lobe of the liver.

Although these diffraction patterns can be visually assessed
the main objective is for a quantitative characterization of tissue
pathology.   The present approach (Nicholas, 1979) involves the
Fourier transformation of the angular diffraction patterns into the
angular frequency domain.

Figures 3 and 4 illustrate the results for a patient presenting
with a primary hepatoma in the right lobe.   Visually, the diffraction patterns originating from within the neoplasm are significantly
different from that associated with normal liver parenchyma illustrated in Figure 2.   The amplitude angular frequency spectra
derived from these patterns (see Figure 4) indicate the presence of
high frequency components, suggestive of neoplastic tissue, and
show a similarity of spectra (two predominant spectral peaks).

The examination of diffuse disorders of the liver presents a
similar problem to that experienced using conventional ultrasonic
techniques in that no internal tissue comparisons can be made.
Thus one is forced to attempt a differentiation based upon the
general appearance of the diffraction patterns associated with
normal liver tissue in other individuals.   Figure 5 shows examples
of diffraction patterns obtained from the right lobe of a patient
suffering from alcoholic cirrhosis.   Visually the patterns seem to
exhibit the same high periodicity as those for neoplastic disease;
however, the associated angular frequency spectra displayed in
Figure 6 also indicate the presence of a low frequency modulation
of the traces.   It is this feature which has occurred in 23 of the
28 diffraction patterns associated with 5 different cases of
cirrhosis.

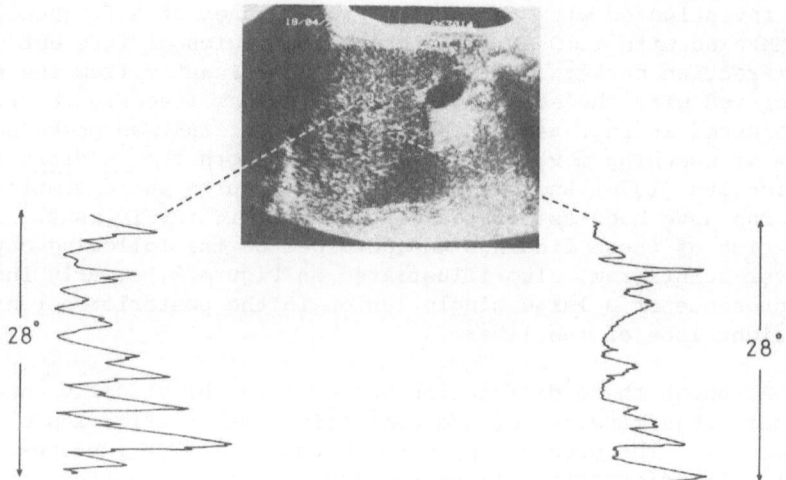

Fig. 3.   Liver scans from a patient with a primary hepatoma in the
          right lobe.

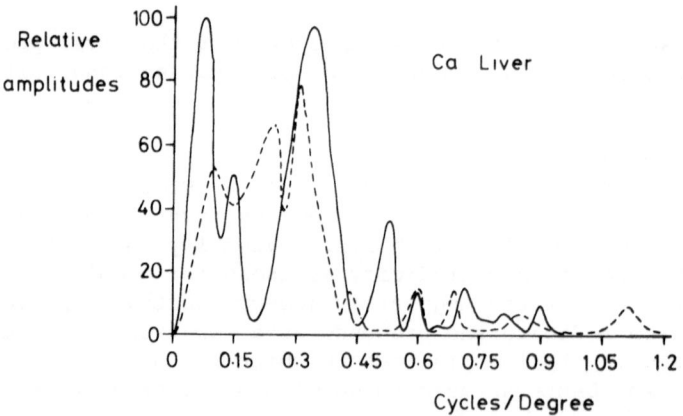

Fig. 4.   Amplitude angular frequency spectra derived from the
          diffraction patterns illustrated in Figure 3.

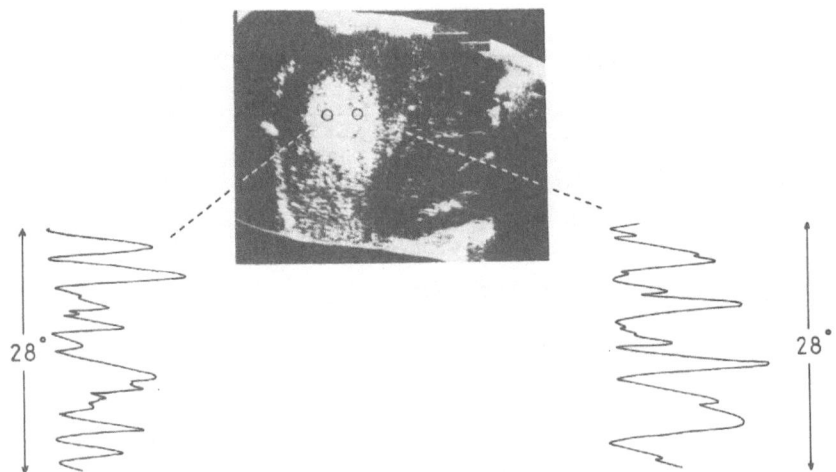

Fig. 5. Liver scans from a patient exhibiting alcoholic cirrhosis.

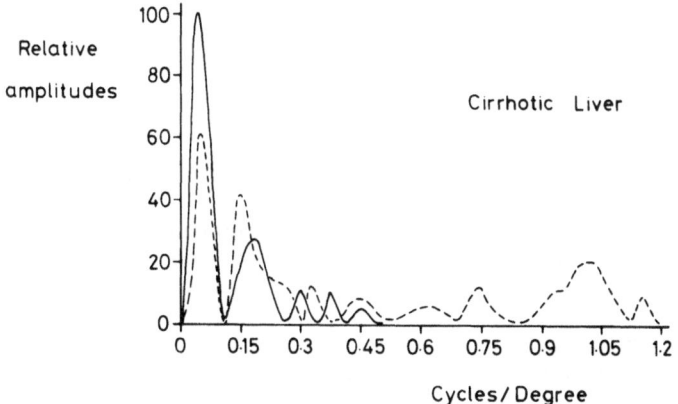

Fig. 6. Amplitude angular frequency spectra derived from the
diffraction patterns displayed in Figure 5.

Table 1.   Summary of Liver Conditions Examined in vivo.

Pathology of primary liver	Liver findings	No. of cases	No. of diffraction patterns
Normal	Normal	22	135
Teratoma	Focal metastases	3	19
Melanoma	Focal metastases	1	6
	diffuse infiltration	3	17
Lymphoma	Focal metastases	1	6
	diffuse infiltration	3	18
Sarcoma	Focal metastases	1	7
Carcinoma	Focal metastases	12	70
	Widespread infiltration		
	"patchy"	5	26
	Hepatoma	3	18
Hodgkin's	Focal metastases	3	18
	Cirrhosis	5	28
	Fatty liver	3	19
	Hepatomegaly	5	29
		70	416

ANALYSIS OF HEPATIC INVESTIGATIONS

Several analytical approaches are obviously available for the
quantitation of these scans and many have been discussed in detail
elsewhere (Nicholas, 1977b).   Our present approach utilises the
most predominant angular spectral frequency and the width of the
spectral envelope, and has been applied to all the 416 diffraction
patterns reported here.   Table 1 lists the categories of liver
disease examined by the diffraction technique to date.   Confirmation of disease has been achieved by biopsy, surgery or post mortem
findings in 16 of the 48 abnormal livers.   Of the remaining 32,
30 were examined by other techniques such as liver scintigraphy and
X-ray computer tomography.   These examinations and subsequent

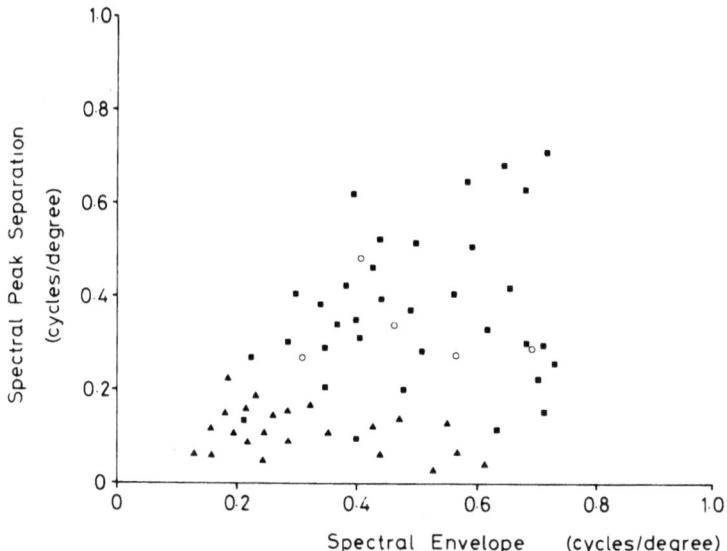

Fig. 7. Scatter diagram of the predominant spectral angular
         frequency component versus spectral envelope width for
         normals ▲ , cirrhosis O , and malignant secondary
         involvement ■ .

follow-ups certainly confirmed all the cases where a malignant condition was suspected, although the categorisations of specific
secondary deposits may be open to question.   This uncertainty has
prompted the broad classifications of liver conditions into four
classes: normals, neoplastic livers (including primary hepatomas,
diffuse and focal secondary involvement), cirrhosis and secondary
liver involvement associated with Hodgkin's disease. Unfortunately
our technique in its present form has proved to be incapable (using
these parameters) of differentiating fatty or non-specific enlarged
livers from normal.

     Figure 7 summarises our findings by plotting on a scatter map
the parameters of predominant angular frequency separation and
spectral width (a fuller description of this analysis is reported
elsewhere (Nicholas, 1979)).   The data represent the mean value
of scans taken in a specific abnormality within the liver and
clearly show the separation of the normal and abnormal pathologies.
Discriminant analysis was applied to these results to establish an
accuracy fot tissue differentiation.   For the 70 subjects an overall accuracy of 91% was achieved in differentiating normal from
abnormal. If the neoplastic diseases were considered separately
then our differentiation from normal pathology increased to a 96%

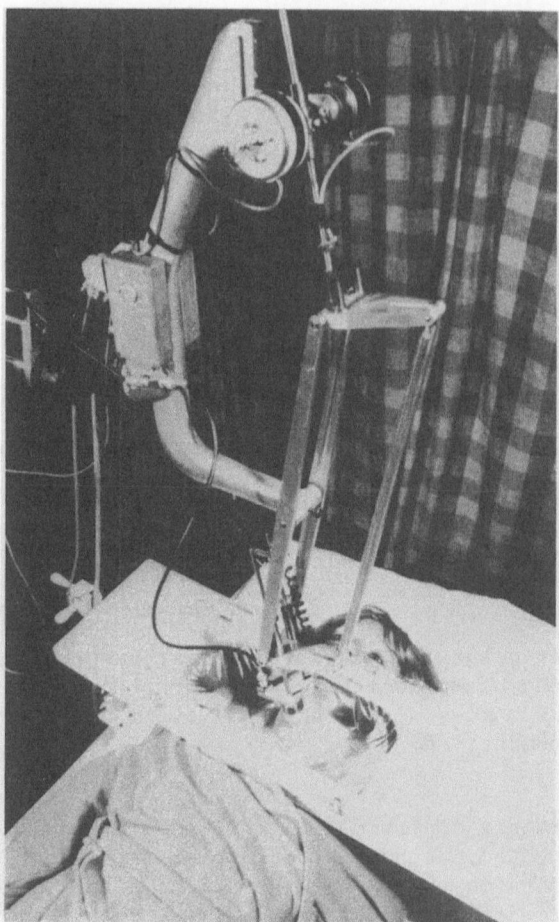

Fig. 8. Arrangement of diffraction scanning arm and water bag for
        scanning thyroids.

accuracy. Similarly the cirrhotic cases also provided a good discrimination from normal pathology although they overlapped considerably with the malignant conditions.   Here an improved separation
has been achieved (Nicholas, 1979) by using a third discriminant
variable based upon the envelope of the angular diffraction pattern.

THYROID APPLICATIONS

     Techniques currently used for diagnosing suspect abnormalities
of the thyroid may be classified as invasive or non-invasive.   The
former includes biopsy and surgery, which permit a histological
definition of the abnormality based on changes in the normal structural configuration. Although this constitutes the most conclusive

diagnostic procedure, information derived from non-invasive means
is preferable, and for the thyroid, radionuclide imaging and ultrasonic B-scanning are two often-used techniques.   Unfortunately,
although ultrasonic scanning of solid, partially cystic or degenerating masses easily locates these abnormalities, it cannot provide
accurate discrimination or specific information on histology
(Rasmussen, 1978).

Our primary goal in applying ultrasonic diffraction techniques
to thyroid disorders was to assess the ability of the technique in
its application to an alternate region of human anatomy.   It is
recognised that the thyroid is capable of producing a wide variety
of tissue patterns when it undergoes neoplastic or other pathological alterations (Hazard, 1964).   However, the existence of structural differences in pathological tissue does suggest that ultrasound
diffraction may complement existing diagnostic procedures.

For the in vivo thyroid studies the same diffraction scanner
and electronics, as described previously, were used.   The only
addition to the apparatus was a water bag, which solved the problem
of maintaining transducer/skin contact for a scanning arc of
approximately 30 degrees.   The arrangement of the scanning arm and
water bag in relation to the patient is depicted in Figure 8.

THYROID RESULTS

A total of 70 diffraction scans were performed on nine volun-
teers that had no known history of thyroid disease.   These scans
were used to provide information on the normal thyroid diffraction
patterns.

Fifty-two patients, with various suspect thyroid diseases,
underwent conventional ultrasound (linear B-scans) to locate regions
of abnormality and were then scanned using our diffraction technique.
Four hundred and seventeen scans were taken in areas that appeared
abnormal on radionuclide imaging and/or on grey-scale ultrasound
scans. Whenever possible, scans were also taken in any normal
appearing areas of the thyroid for intrapatient comparison. This
was especially useful where the patient had an abnormality in one
lobe only.

Figures 9 and 10 illustrate, respectively, the diffraction
patterns obtained from a patient with multinodular goitre and from
a patient where the right lobe contained a large follicular adenoma.

To date, only 23 cases have resulted in histological confirmation following surgery and only these will be included in our survey.
The analysis is again based on the use of the Fourier transform
and two parameters will be discussed here. The width of the spectral envelope is again used although the second parameter is

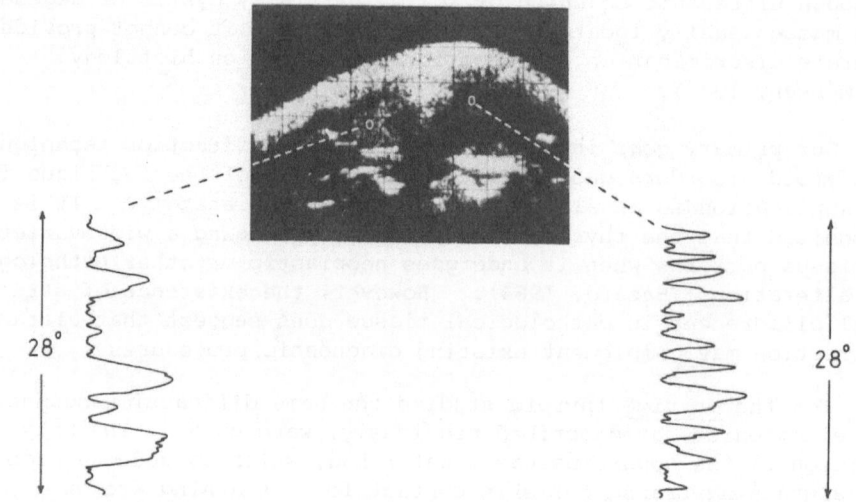

Fig. 9.   Thyroid scans from a patient exhibiting multi-nodular
          goitre.

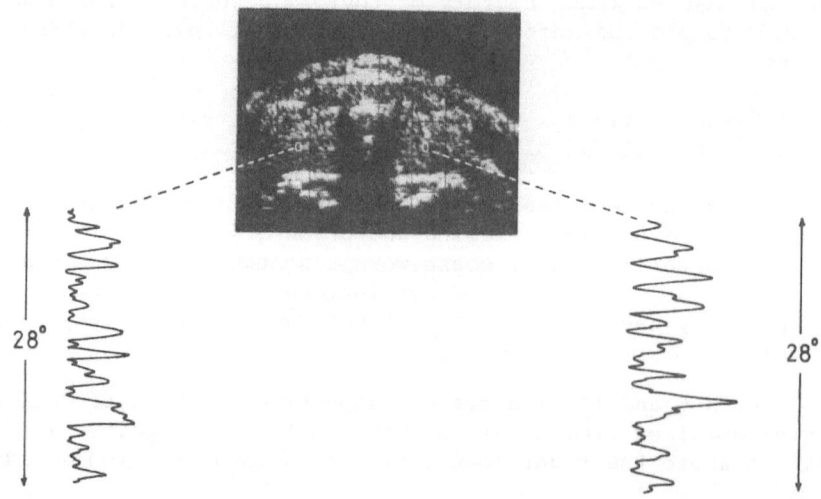

Fig. 10. Thyroid scans from a patient with a normal left lobe
          and an adenoma present in the right lobe.

Table II. Histological Descriptions of Thyroid Pathologies.

Histological Disease Description – Some Structural Features	Number of Patients
Adenoma - ho histological type	1
Adenoma - follicular - encapsulated, small, medium or large follicular cells	5
Adenoma - follicular degenerating	3
Adenoma - foetal - encapsulated, a combination of small and closed follicular cells or closed follicular cells only	4
Adenoma - papillary - not completely encapsulated distinct well formed papillae, fibrous stroma	1
Adenoma - colloid	1
Adenoma - colloid degenerating	1
Single nodule with fibrosis	2
Multi-nodular with fibrosis	1
Multi-nodular goitre - colloid nodules, local haemorrhage, fibrous structure, calcification	3
Hashimoto's/papillary - thickened interlobular septa, interlobular and interfollicular infiltration by lymphocytes/ well-formed papillae and fibrous stroma	1
Normal volunteers	9
Patients with clinically normal thyroids	2
	34

simplified to a "normalised high frequency index" (Merton et al., 1981) which is the computed area in the angular frequency domain for frequency components above the arbitrarily chosen cut-off of 1.15 cycles/degree of scanning motion. This value is then normalised to the total energy within the Fourier domain.

Table II classifies the 23 abnormal cases in terms of an histological description. Because of the small number of patients

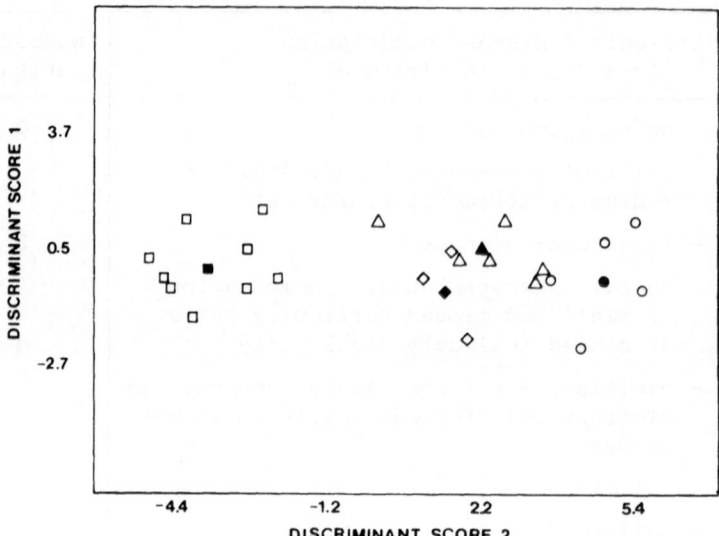

Fig. 11.   Plot of computer calculated discriminant scores for
           normals □ , follicular adenomas O , degenerating
           follicular adenomas ◇ , and colloid masses △
           The solid symbols ■ ● ◆ ▲ indicate the mean discriminant
           value for each group.   Score 1 is calculated from the
           high frequency index and score 2 from the percentage of
           scans with a spectral envelope width ≯ 0.5 cycles per
           degree (Merton et al., 1981).

in the study it is not realistic to attempt to identify the individual diseases.   Instead three disease groups were considered;
the follicular adenomas, degenerating follicular adenomas, and
colloid masses (colloid nodules and multi-nodular goitres).   On
these 3 groups and the normals a multi-group discriminant analysis
was applied.   A plot of the data as a scatter diagram, with the
two variates plotted on the axes, is displayed in Figure 11.
The solid symbols represent the mean discriminant values on the two
variates for each group.   The analysis showed 88% of the cases from
the three chosen groups and the normals to be correctly classified.
When the data from a fourth abnormal group, the foetal adenoma, was
added, correct diagnosis dropped to 72.4%.   This suggests that a
third discriminating variable would be necessary to improve the
identification of abnormal groups.

CONCLUSION

     The results presented here can be summarised according to the
method of scanning and disease state.   Where a differential method

can be applied, as in focal disorders of the liver and thyroid,the
technique involves making a diffraction record within the focal
abnormality and comparing this with the corresponding record for,
supposedly, normal tissue.   In this situation an overall 95% success
rate was achieved in determining the existence of an abnormal condition.   Only 2 hepatic cases were incorrectly classified as normal
and in each case only a single deposit of diameter between 1.5 and
2.5 cm was involved.   The error in diagnosis in these cases may
well have been associated with incorrect positioning of the scanning
arm, and consequent failure to locate the suspected region.   Likewise, only 2 thyroid conditions yielded results compatible with
those from normal tissue, these were both histologically identified
as foetal adenomas.   As discussed elsewhere (Merton et al., 1981)
the structure of these lesions differs from the other abnormalities
examined and is of a scale which is too small for detection by
this technique.

The technique also seems particularly useful in differentiating
between fluid filled lesions (containing debris) and degenerating
tumours, both of which can produce similar appearances on conventional sector B-scans.   The diffraction patterns associated with
the former are non-repeatable due to the motion of the debris within
the lesion, whilst the latter present some degree of rigidity in
structure and thus give patterns which exhibit some repeatable
features.   Simple cysts have not been discussed as they present
little problem using conventional B-scanning.

The use of diffraction scanning to characterise specific
diffuse disorders has met with varied degrees of success.   Diffuse
neoplastic liver involvement is easily characterised, with diffraction patterns similar to those associated with focal neoplasms.   So
far a 100% success rate in detecting malignant infiltration of the
liver has been achieved in our investigation of 11 cases (including
5 cases of widespread 'patchy' metastatic involvement (see Table 1))
involving 61 diffraction patterns.   Of the other types of diffuse
disorder only cirrhotic tissue has yet produced a specific disease
signature, where a single cycle modulation of the diffraction pattern
occurs.

Although many patterns have been collected, the variations in
disease types have resulted in too few examinations of specific
tissue pathologies to do other than group them into broad categories.
It is to be hoped that when a larger set of data has been accumulated a more specific differentiation of tissue abnormalities may
prove possible.

The clinical value of our present apparatus can only be
postulated at this stage, but the results indicate that it is
capable of providing useful quantitative data to complement the high
quality grey-scale scans that are at present attainable.

REFERENCES

Dewbury, K. C., and Clarke, B., 1979, The accuracy of ultrasound
    in the detection of cirrhosis of the liver, Brit. J. Radiol.
    52:945.
Hazard, J. B., 1964, in: "Neoplasia in the Thyroid", J. B. Hazard
    and D. E. Smith eds, The Williams and Wilkins Company,
    Baltimore.
Hill, C. R., Chivers, R. C., Huggins, R. W., and Nicholas, D.,
    1978, in: "Ultrasound: its Applications in Medicine and
    Biology", F. J. Fry, ed., Elsevier.
Huggins, R. W., and Phelps, J. V., 1977, An ultrasonic scanner for
    recording the angular dependence of echoes from a fixed tissue
    volume in vivo, Ultrasound Med. Biol., 2:2.
McArdle, C. R., 1976, Ultrasonic diagnosis of liver metastases,
    J. Clin. Ultrasound, 4:265.
Merton, J., Nicholas, D., Hill, C. R., Grover, S., Queenan, M.,
    and Cosgrove, D. O., 1981, Ultrasonic diffraction scanning of
    the thyroid, Ultrasound Med. Biol. (in press).
Nicholas, D., 1976, The application of acoustic scattering para-
    meters to the characterisation of human soft tissues, in:
    "Proc. 1976 IEEE Ultrasonics Symposium", Cat. 76 CH1120-5SU,
    64.
Nicholas, D., 1977a, An ultrasonic diffraction scanner for in vivo
    tissue characterisation, in: "Ultrasonics International 1977"
    IPC Science and Technology Press.
Nicholas, D., 1977b, in: "Recent Advances in Ultrasound in Biomedi-
    cine", D. N. White, ed., Research Studies Press.
Nicholas, D., 1979, Ultrasonic diffraction analysis in the investi-
    gation of liver disease, Br. J. Radiol., 52:949.
Rasmussen, S. N., 1978, Thyroid Gland, in: "Handbook of Clinical
    Ultrasound", M. de Vlieger, ed., John Wiley and Sons, New
    York.

TISSUE PARAMETER MEASUREMENT AND IMAGING

Lynda Hutchins and S. Leeman

Department of Medical Physics, Royal Postgraduate
Medical School, Hammersmith Hospital, London, W12 OHS
U.K.

INTRODUCTION

The attenuation and velocity of ultrasound in tissues are often measured in attempts to uncover an experimental basis for some tissue characterisation methods. A knowledge of these parameters is essential also to describe fully the propagation of ultrasound waves in tissues, and to correct for certain image artefacts. However, even a cursory glance at the literature will reveal that published values for the observed attenuation, in particular, show a wide discrepancy. Much of this is known to result from, for example, differences in sample preparation and handling, but it is equally true that little attention has been paid to the influence that the experimental techniques themselves have on the measured parameter values. Methods for assessing "in vivo" attenuation and velocity are still in their infancy, and experiments with tissues are generally conducted with in vitro specimens. Such water-tank experiments usually rely on the interpretation of pulse transmission measurements in order to arrive at the desired propagation parameters. It is to this type of experiment that we will confine ourselves here. The analysis and interpretation of the measurements are made difficult by the twin circumstances of diffraction and dispersion. The former arises from the inevitably finite aperture size of the pulse generating transducer, and the latter arises from the relatively poor current understanding of the behaviour of propagating pulses within tissue-like dispersive media. A further complication is that the receiving transducers in most experiments are of the phase-sensitive type, but this is only briefly alluded to here.

We consider the situation depicted in Figure I. A bounded pulse

127

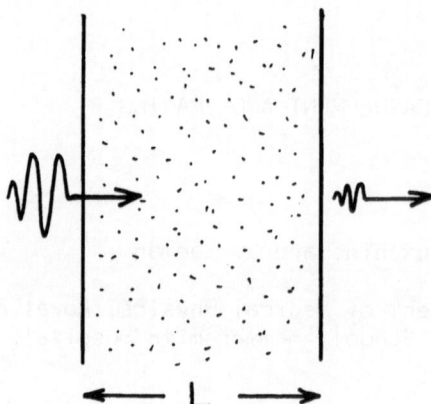

Figure I: Schematic diagram of experimental arrangement.

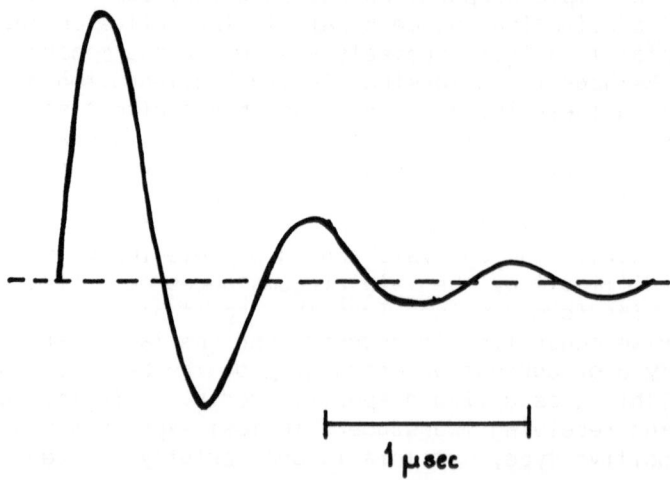

Figure 2:   Assumed form of incident pulse.

is assumed perpendicularly incident upon a plane-parallel sided block
of uniformly, but dispersively, attenuating material, of thickness L.
The lateral extent of the material is considered larger than that of
the pulse. We address ourselves to the problem of establishing what
information about the propagation properties of the material can be
extracted from measurements on the emerging (on-axis) pulse, assuming
that the material exhibits dispersive attenuation not too unlike that
of many soft tissues.

PULSES IN DISPERSIVE MEDIA

We adopt the following equation as modelling the propagative
properties of tissues.

$$\nabla^2 p - \frac{1}{c^2}\ddot{p} - 2A\dot{p} = 0 \tag{1}$$

Here $p(\underline{r},t)$ denotes the pressure at location $\underline{r}$ at time t, with A and
C characteristic constants of the medium. The validity of this model
has been discussed at some length (Leeman, Hutchins and Jones, 1981;
see also the accompanying paper in this volume: "Pulse scattering in
dispersive media"). The form of the dispersive attenuation predicted
by this model, in the frequency range 1 - 10 MHz is shown in Figure 3,
for the choice α_o = AC = 213 m^{-1}. To fix ideas, we adopt the following simple axially symmetric form for the incident pulse (using
cylindrical polar co-ordinates (h, Θ, Z))

$$p(h,\Theta,0;t) = H(R - h)\, f(t) \qquad\qquad t \geqslant 0$$

$$= 0 \qquad\qquad t < 0$$

where $f(t)$ is the incident pulse shape, and H denotes the unit step
function. A uniform incident beam profile has been assumed, equal
to unity over a disk of radius R, and zero outside that. The pulse
emerging from the sample of thickness L may be calculated (Leeman,
Hutchins and Jones, 1981) to be given by:

$$p(h,L;t) = f(t)*P(h,L;t)$$

where * denotes convolution, and where

$$P(h,L;t) = cH(R - h)\, e^{-\alpha_o L}\, \delta(\tau - L)$$

$$+ cLe^{-\alpha_o\tau}H(\tau-L)\, \frac{R}{\xi}\frac{\delta}{\delta\xi}\int_0^\infty d\lambda\, J_o(h\lambda)J_1(R\lambda)J_o(\xi\sqrt{\lambda^2-\alpha_o^2}) \tag{2}$$

with α_o = AC, τ = ct, $\xi = \sqrt{\tau^2 - L^2}$, and J_n is the Bessel function
of order n. $P(h,L;t)$ may be regarded as the impulse response of the
medium, which is defined as the effect of a slab of thickness L on an
incident pulse of the form $H(R - h)\, \delta(t)$.

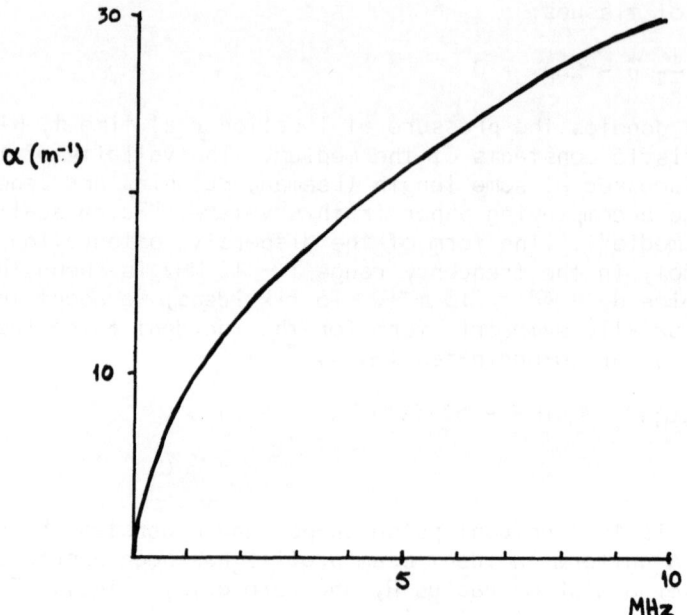

Figure 3:   Attenuation assumed in calculations.

If measurements are conducted at the face L with a circular phase-sensitive transducer of radius D and sensitivity function S(h), the appropriate impulse response is given by

$$\overline{P}(L;t) = c \int_0^D dh\, h\, H(R - h)\, S(h)\, e^{-\alpha_0 L}\, \delta(\tau - L)$$

$$+ cLe^{-\alpha_0 \tau} H(\tau-L) \frac{R}{\xi} \frac{\delta}{\delta\xi} \int_0^\infty d\lambda\, \sigma(\lambda) J_1(R\lambda) J_0(\xi\sqrt{\lambda^2-\alpha_0^2}) \tag{3}$$

where $\sigma(\lambda)$ is the Hankel transform of $S(h)$. We note the expression (3) merely for reference, and consider further only the information to be gained from the axial pulse alone. The axial impulse response is

$$P(L;t) = ce^{-\alpha_0 L}\, \delta(\tau - L)$$

$$+ cLe^{-\alpha_0 \tau} H(\tau-L) \frac{R}{\xi} \frac{\delta}{\delta\xi} \int_0^\infty d\lambda\, J_1(R\lambda) J_0(\xi\sqrt{\lambda^2-\alpha_0^2}) \tag{4}$$

This may be rewritten in a rather more informative form, by noting that by Sonine's second integral (Sneddon, 1972)

$$\int_0^\infty d\lambda J_1(R\lambda) J_0(\xi\sqrt{\lambda^2-\alpha_0^2}) H(\lambda-\alpha_0) =$$

$$= \frac{1}{R} \int_0^\infty dt\, t(t^2+\xi^2)^{-\frac{1}{2}} J_0(\alpha_0 t) \int_0^\infty d\lambda\, \lambda\, J_1(\lambda R) J_1(\lambda\sqrt{t^2+\xi^2})$$

$$= \frac{1}{R} \int_0^\infty dt\, t(t^2+\xi^2)^{-\frac{1}{2}} J_0(\alpha_0 t)\, \delta(\sqrt{t^2+\xi^2} - R)$$

$$= \frac{1}{R} J_0(\alpha_0 \sqrt{R^2-\xi^2})\, H(R - \xi)$$

We have used a well-known representation of the δ-function. Equation (4) now reduces to

$$\frac{1}{c}P(L;t) = e^{-\alpha_0 L}\, \delta(\tau-L) - \frac{L}{\sqrt{L^2+R^2}} e^{-\alpha_0\sqrt{L^2+R^2}}\, \delta(\tau-\sqrt{L^2+R^2})$$

$$+ \alpha_0 Le^{-\alpha_0 \tau} \frac{J_1(\alpha_0\sqrt{R^2+L^2-\tau^2})}{\sqrt{R^2+L^2-\tau^2}} H(\tau-L)\, H(R-\xi)$$

$$+ RLe^{-\alpha_0 \tau} \frac{1}{\xi} \int_0^{\alpha_0} d\lambda\, \sqrt{\alpha_0^2-\lambda^2}\, J_1(R\lambda)\, I_1(\xi\sqrt{\alpha_0^2-\lambda^2})\, H(\tau-L) \tag{5}$$

I_1 is the modified Bessel function of order 1.

STRUCTURE OF THE TRANSMITTED PULSE

The axial impulse response has an interesting multi-component structure, some elements of which are familiar from the theory of transient fields in loss-less media (Stephanishen, 1971). The first term on the right-hand side of (5) can be identified as the "direct wave", giving rise to a non-dispersively, exponentially, attenuated replica of the incident pulse. The second term is the "edge-wave" - a diffractive contribution from the edge of the aperture. This component gives rise to a negative replica pulse, whose strength depends on distance and does not exhibit simple exponential attenuation. The arrival of the edge-wave is delayed somewhat behind that of the direct wave, but the delay decreases with distance, leading to eventual destructive interference with the direct wave. These first two terms are the only components of $P(L;t)$ that survive in the limit that the attenuation vanishes. The third component of $P(L;t)$ contributes only during the time interval between the arrivals of the direct and edge-waves. Geometrical constructions indicating the structure of the field (see, for example, Leeman, Hutchins and Jones, 1981) lead us to name this the "aperture wave". The last term in (5) is the "rumble" associated with the dispersive character of the attenuation. In the limit of infinite R (one-dimensional case) this rumble reduces to that associated with the passage of a plane-wave pulse (Leeman, 1980). The front of the rumble travels with the direct wave, but its tail extends out beyond the edge-wave. With the values of α_o, R and L adopted for the computations below, the aperture wave and rumble make rather small contributions to the shape of the propagating pulse.

IMPLICATIONS FOR VELOCITY MEASUREMENT

In the model discussed here both the phase and group velocities are dispersive. However, it is clear from (5) that the signal velocity - viz. the speed at which the first detectable signal travels through the medium - is a frequency-independent constant, C. Although this is proved above only for the axial pulse (h = 0), it may be shown to be valid for any value of $h \leqslant R$. Note that the front of the (axial) edge-wave actually travels faster than the signal velocity, and it approaches closer to the direct wave, with distance. But it is clearly the signal velocity that is measured when ultra-sound velocities are determined from the time-of-flight of the axial pulse front. Although the axial pulse has a complicated structure, with many interfering components, it does not seem unreasonable to infer that there are circumstances in which the above considerations are valid also in the case that the major peak in the pulse is monitored.

The signal velocity predicted for soft tissue by this model is not in agreement with the value $\sim$ 1500 m/sec usually associated with

ultrasound transmission in such media, and some of the above conclusions may be less relevant for tissue parameter measurements.
However, they are in accord with the non-dispersive behaviour
observed for ultrasound tissue velocities, and certainly indicate
that there are circumstances in which apparently non-dispersive
velocities may be measured in a dispersive medium.

ATTENUATION MEASUREMENTS

     We have performed explicit calculations of the propagating
pulse and its spectrum for the specific case: α_0 = 213 m^{-1}; R =
8 mm; L = 8 mm, 32 mm and 56 mm, for the incident pulse shape shown
in Figure 2. The central frequency of the pulse is at 1.5 MHz, and
the signal velocity has been rescaled, in the final result, to 1540
m/sec. It may be shown that this last, seemingly arbitrary, procedure is mathematically justified, and it obviates the need to
introduce more complicated terms into the model such as those discussed by Leeman, Hutchins and Jones ("Pulse scattering in dispersive
media", in this volume). As a yardstick against which to measure the
importance of diffractive effects, we show in Figure 4 the magnitude
spectrum of the incident pulse, and the changes with L in this
spectrum for the one-dimensional case (R = ∞). As expected, the
preferential damping of the higher frequencies leads to a gradual
reduction in the central frequency of the pulse spectrum. A dramatic
change occurs with finite aperture incident pulses. In Figure 5 we
show the spectral changes of the axial pulse in a loss-less medium
(water), and in Figure 6 the corresponding spectra in a lossy medium
characterised by the dispersive attenuation shown in Figure 2. We
have been unsuccessful in recreating the attenuation behaviour from
the axial pulse spectra, even when applying the usual diffraction
correction (expressing the attenuation relative to the apparent
attenuation in water).

DISCUSSION

     The axial pulse appears to be a singularly poor indicator of
attenuation in dispersive media. Attenuation recovered from measurements of the axial pulse, even in an idealised situation (computer
experiment) bears little resemblance to the known properties of the
medium, when diffractive effects are prominent. We have enhanced the
severity of the effects by the use of a rather narrow incident beam,
with a strong discontinuity at its edge. In practice, for attenuation measurements, a rather wider incident beam might be employed,
with a beam profile that is more smoothly varying. With increasing
R, the interference between the rumble and direct waves becomes more
marked, and the results more in accord with plane-wave predictions
are expected. Thus, it would be expected that recovered attenuation
values will be beam-profile dependent, even when diffractive effects
are corrected for in the conventional manner. Moreover, measured

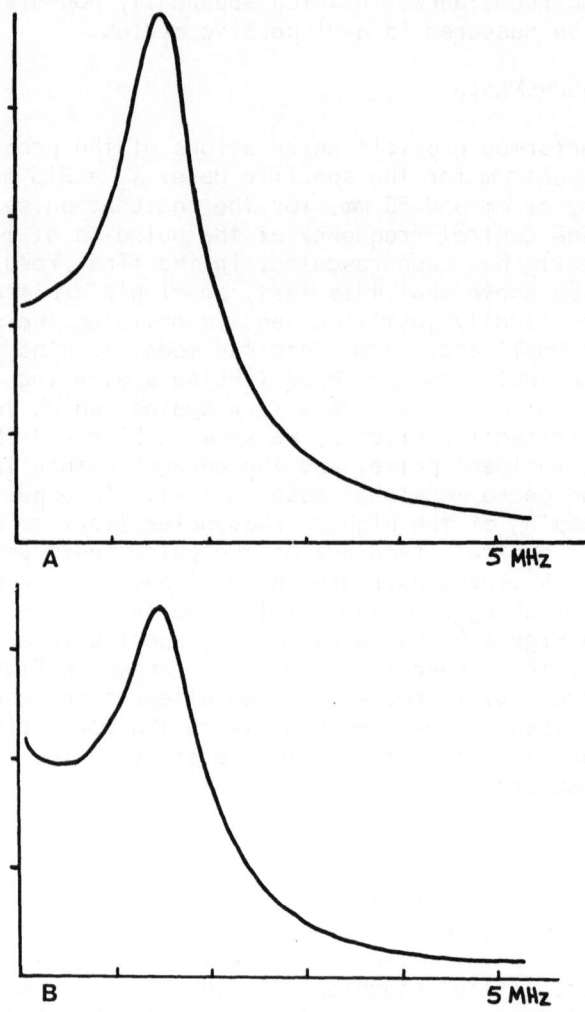

Figure 4:   Amplitude spectra of one dimensional pulses, of shape as
            in Figure 2, incident (a) on attenuating medium, and at
            depth 56 mm in medium (b).   Medium chosen to have α_o =
            213 m^{-1}.

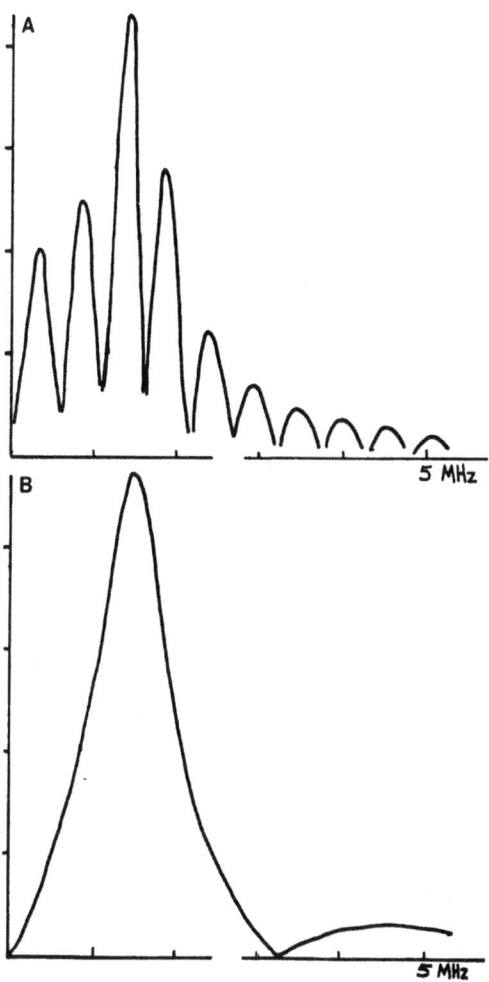

Figure 5: Axial amplitude spectra of bounded pulses with incident
axial shape as in Figure 2, at depths 8 mm (a) and 56 mm
(b) in a non-attenuating medium. (α_o = 0). Incident
aperture radius is 8 mm.

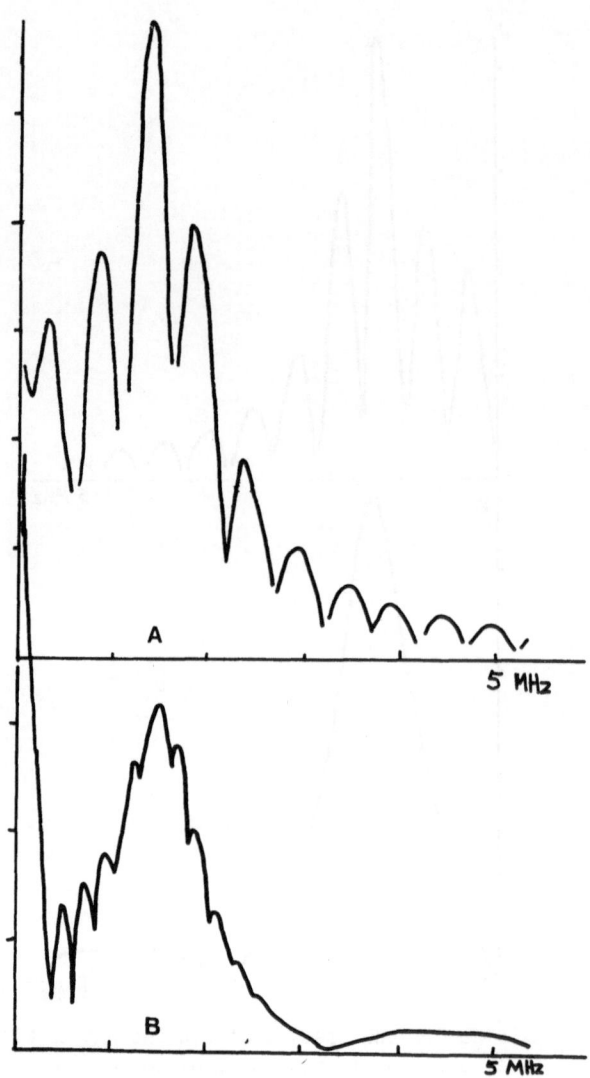

Figure 6: Axial amplitude spectra of bounded pulses with incident
           axial shape as in Figure 2, at depths 8 mm (a) and 56 mm
           (b) in an attenuating medium with α_o = 213 m^{-1}. Incident
           aperture radius is 8 mm.

attenuation values show a dependence on sample thickness. It may be anticipated that some of the difficulties are obviated by utilising information from the entire transmitted pulse - e.g. by utilising (3) instead of (4) - but we have not investigated this yet. It would appear that rather more powerful diffraction correction procedures need to be developed for experiments of the type described here.

While it is certainly possible to devise an experimental set-up to give meaningful results, there are certain situations in which the use of narrow incident beams may be mandatory, e.g. in transmission ultrasound tomography for attenuation reconstruction. It does appear that diffractive effects should be carefully investigated, and fully corrected for, in such investigations.

ACKNOWLEDGEMENTS

S.L. wishes to thank The Wellcome Trust for generously making a research travel grant available. Miss Ann Freemantle made the illegible script readable.

REFERENCES

Leeman, S., 1980, Ultrasound pulse propagation in dispersive media, Phys. Med. Biol., 25:481.
Leeman, S., Hutchins, L. and Jones, J. P., 1981, Bounded pulse propagation, in: "Acoustical Imaging 10", P. Alais, Ed., in press.
Sneddon, I. N., 1972, "The Use of Integral Transforms", McGraw-Hill, N.Y.
Stephanishen, P. R., 1971, Transient radiation from pistons in an infinite planar baffle, J. Acoust. Soc. Amer., 49:1629.

PULSE SCATTERING IN DISPERSIVE MEDIA

S. Leeman, L. Hutchins and J. P. Jones*

Department of Medical Physics, Royal Postgraduate
Medical School, Hammersmith Hospital, London, W12 0HS
U.K., and *Department of Radiological Sciences,
University of California, Irvine, California 92717

INTRODUCTION

Human tissues are manifestly dispersively attenuating, and a
knowledge of the propagation and scattering of ultrasound pulses in
such media is fundamental to progress in both tissue characterization
techniques and novel imaging methods, such as impediography (Jones,
1977), wave extrapolation (Berkhout, Ridder and van der Wal, 1981),
diffraction- (Mueller, Kaveh and Iverson, 1980) and transmission-
(Greenleaf, Johnson, Samayoa and Duck, 1975) tomography. Treatments
of ultrasound pulse scattering in tissues have varied from those
involving gross simplifications, such as disregarding attenuation or
the three-dimensional character of the problem, to advanced applications, such as in wave extrapolation methods, where recourse to
extensive computer calculations is mandatory.

Here, the problem is tackled in three distinct stages. First,
the propagation of a three-dimensional pulse in a uniformly attenuating medium is considered. Next, the form of a suitable tissue model
which includes scattering from inhomogeneities is specified.
Finally, the first two stages are combined to solve the problem of
interest via the Born approximation - i.e. weak scattering is assumed.
This last assumption is indicated by the observed weakness of back-
scattered echoes from tissues, and it does not seem unreasonable to
extrapolate this observation to the assumption of the validity of the
Born approximation itself (weakness of the total scattering). This
is particularly so if the relatively strong, near-specular, reflections from major tissue interfaces are not evident, or are treated
separately. The Born approximation provides a drastic simplification

in scattering problems, and is implicit in many of the computerised
imaging methods referred to above; it is, fortunately, probably
quite adequate for many situations of practical interest. Other
simplifying assumptions to be made are that the initial, incident
pulse has a suitably uncomplicated form on entering the medium, and
that the backscattering is measured far from the scattering region.
Only backscattering is considered here, but the treatment is easily
extended to scattering into any direction.

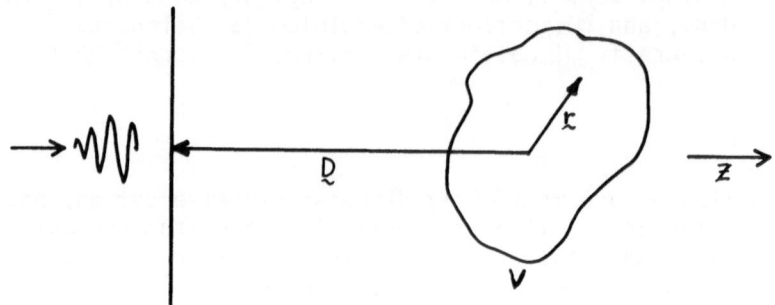

Figure I: Definition of co-ordinates used in text.

PULSE PROPAGATION IN A UNIFORMLY ATTENUATING MEDIUM

Consider the situation depicted in Figure I: a three-dimensional pulse is incident, at time $t = 0$ on a uniformly attenuating
medium which fills the half-space $Z \geqslant 0$. The pulse propagates in
the Z-direction, and at some depth in the medium encounters a region
of scattering, V. It is convenient to fix co-ordinates as shown in
the figure, viz. the origin situated inside V, r denoting the position vector, and D indicating the location of the incident pulse at
$t = 0$. It is assumed that the mean acoustical properties of the
scattering region are equal to those of the uniform imbedding medium,
so that there is no interface reflection or attenuation change as the
pulse enters V. The scattered waves generated within V are described
as propagating through a uniformly attenuating medium (no multiple
scattering), and are measured at D (back-scattering). This disregard

of multiple scattering events, and the uncoupling of the propagation
and scattering components of the problem, are the direct benefits
accruing from the (first) Born approximation.

It is clear that a wave equation describing the propagation of
ultrasound waves in a uniformly attenuating medium is the first
requisite. The attenuating properties of this medium must obviously
closely mimic those of real tissues. A possible linear wave equation
for the propagation of longitudinal waves in such a uniformly attenuating medium is

$$\nabla^2 p - \frac{1}{c^2}\ddot{p} - 2A\dot{p} - 2B|\nabla|\dot{p} = 0 \tag{1}$$

where $p(\underline{r},t)$ denotes the pressure, and A, B and C are constants
characterising the medium. The operator, $|\nabla|$, is defined by

$$|\nabla|p \equiv \{\nabla^2\}^{\frac{1}{2}}p$$

and is comprehended by transforming to the Fourier domain, where

$$\nabla^2 p(\underline{r},t) = \nabla^2 \int d^3\underline{k}\phi(\underline{k},t)e^{-i\underline{k}\cdot\underline{r}}$$

$$= \int d^3\underline{k}(-k^2)\phi(\underline{k},t)e^{-i\underline{k}\cdot\underline{r}}$$

Hence,
$$|\nabla|p(\underline{r},t) = \int d^3\underline{k}(ik)\phi(\underline{k},t)e^{-i\underline{k}\cdot\underline{r}}$$

The positive root of $\{-k^2\}^{\frac{1}{2}}$ has been chosen, without any loss
of generality.

The wave equation (1) then implies a dispersive attenuation, ,
of the form

$$\alpha = \frac{\omega}{\sqrt{2b}}\{(1+\frac{4A^2b^4}{\omega^2})^{\frac{1}{2}} - 1\}^{\frac{1}{2}} \tag{2}$$

where ω is the circular frequency, and $b=(B^2+1/c^2)^{-\frac{1}{2}}$. It has already
been demonstrated that such a frequency-dependent attenuation
adequately describes known tissue values in the $\sim$ .5 - 10 MHz frequency range (Leeman, 1980a; Leeman, Hutchins and Jones, 1981).

A fundamental quantity in the evaluation of both propagating
pulses and scattered fields, is the 'outgoing' Green function,
$g(\underline{r},\underline{r}';t,t')$, associated with (1). This satisfies

$$\nabla^2 g - \frac{1}{c^2}\ddot{g} - 2A\dot{g} - 2B|\nabla|\dot{g} = \delta(\underline{r}-\underline{r}')\delta(t-t')$$

By transforming to the Fourier-space representation of this equation,
it is easily shown that

$$g(\underline{r},\underline{r}';t,t') = \frac{c^2}{(2\pi)^4} \int d^3\underline{k}\, d\omega\; \frac{e^{i\underline{k}\cdot(\underline{r}-\underline{r}')}e^{-i\omega(t-t')}}{\omega^2 - 2(Bk-iA)\omega c^2 - k^2 c^2}$$

The evaluation of this integral is not trivial. A fairly lengthy computation finally results in

$$g(\underline{R},\tau) = \frac{c}{4\pi\sqrt{1+\beta^2}}\; \frac{e^{-\alpha_o ct}}{R}\; \frac{\partial}{\partial R}\left\{ J_0\left(\frac{\alpha_o c}{1+\beta^2}\sqrt{(\frac{R}{c}+\beta\tau)^2 - (1+\beta^2)\tau^2}\right)\cdot\right.$$

$$\left.\cdot e^{\frac{-\alpha_o\beta c}{1+\beta^2}(\frac{R}{c}+\beta\tau)} H(\tau - R/c_0)\right\} \tag{3}$$

with $R=\underline{r}-\underline{r}'$, $\tau=t-t'$, $\alpha_o = Ac$, $\beta=Bc$, $c_0=c(\sqrt{1+\beta^2}-\beta)$.

J_o denotes the Bessel function of zero order, and H is the unit step function. A suitably modified form of this Green function may be substituted in the Kirchoff integral appropriate to (1) in order to calculate the pulse emanating from an aperture set in a planar baffle. Also, the expression (3) is used directly to describe the wavelets emanating from the scattering inhomogeneities in V. This rather exhausting calculation may be short-circuited by noticing that the attenuation in (1) is determined essentially by the parameter A; another parameter, B, is basically a velocity-modifying term. Much of the complexity of (1) stems from the presence of the B-term. If this is dropped, the attenuation (2) is still obtained, but with b replaced by c. Hence, the model

$$\nabla^2 p - \frac{1}{c^2}\ddot{p} - 2A\dot{p} = 0 \tag{4}$$

provides the identical fit to attenuation data, as does (1). In this case, the outgoing Green function has the simpler form

$$g_o(\underline{R},\tau) = \frac{c}{4\pi}\;\frac{e^{-\alpha_o c\tau}}{R}\;\frac{\partial}{\partial R}\left\{ J_0(\alpha_o\sqrt{R^2-c^2\tau^2})\; H(\tau-R/c)\right\} \tag{5}$$

The model (4) is well-known as describing the propagation of damped, scalar electromagnetic waves. Indeed, the accuracy of the calculations leading to (3) may be verified by ascertaining that (5), to which (3) reduces in the special case B = 0, corresponds correctly to that given by a direct calculation from (4), by somewhat different methods (Morse and Feshbach, 1953). The model (4), while providing a good fit to the observed attenuation data, does not accurately predict the signal velocities measured in tissues. However, all the main features of propagation in dispersive media are retained, and the velocity error may be corrected by an appropriate rescaling procedure. In this spirit, therefore, we accept (4) as the basic propagative wave equation, even though (I) is the more accurate model.

In order to reduce the problem in hand to a reasonably tractable one, some simplifying assumption about the form of the incident pulse will be made. Assume an axially symmetric pulse incident at $t = 0$, travelling along the Z-direction, such that, in cylindrical polar co-ordinates (h,e,z),

$$p\,(\underline{r};t) \;=\; p(h,z;t)$$

with
$$p(h,-D;t) \;=\; W(h)f(t) \qquad\qquad t \geq 0$$

$$= \;0 \qquad\qquad t < 0$$

and
$$p\,(h,z;o) \;=\; \overset{\bullet}{p}(h,z;o) \;=\; 0 \qquad z > -D$$

Here, $W(h)$ is the incident beam profile, normalised to $W(o)=1$, and $f(t)$ is the incident pulse shape. The incident pulse is assumed to be of finite extent, i.e. $W(h)=0$ for $h > a$, $f(t) \neq 0$ only for $0 \leq t \leq t_m$. The pulse propagating in the attenuating medium may be shown (Leeman, Hutchins and Jones, 1981) to be of the form:

$$p\,(n,z;t) \;=\; \frac{z+D}{c} \int_{\frac{z+D}{c}}^{t} dt_o \int_{o}^{\infty} d\lambda\lambda\omega(\lambda)J_o(h\lambda)\,\frac{e^{-\alpha_o c t_o}}{t_o}\,f(t-t_o) \;\cdot$$

$$\cdot \;\frac{\partial}{\partial t_o}\,\{J_o(\sqrt{\lambda^2-\alpha_o^2}\,\sqrt{c^2 t_o^2-(z+D)^2}\,)H(t_o-\frac{z+D}{c})\} \qquad (6)$$

with $\omega(\lambda) = \int_o^{\infty} dh.hW(h)\,J_o(h\lambda)$, the Hankel transform of the incident beam profile.

For future reference, we note that the Laplace transform (denoted $\mathcal{L}$) of the propagating pulse is given by:

$$\Pi(h,z;s) \;=\; \mathcal{L}\,\{p\,(h,z;t)\}$$

$$= \int_{o}^{\infty} d\lambda\lambda\omega(\lambda)J_o(h\lambda)F(s)\exp\{-(s^2+2\alpha_o cs+\lambda^2 c^2)^{\frac{1}{2}}(z+D)/c\}$$

with $F(s) \equiv \mathcal{L}\{f(t)\}$

Also, it may be shown that

$$G(\underline{R};s) \;\equiv\; \mathcal{L}\{g_o(\underline{R};t)\} \;=\; -\frac{1}{4\pi R}\;\exp\,\iota - (s^2+2\alpha_o cs)^{\frac{1}{2}}R/c\}$$

The relative simplicity of the propagating pulse and Green function in the Laplace domain will be used to good effect in the following.

SPECIFICATION OF THE SCATTERING SOURCE FUNCTION

Although the treatment above regards the pulse as propagating in an attenuating medium, it is an entirely phenomenological description incapable of generating scattered waves as such. The latter may be formally introduced by specifying the nature of the medium inhomogeneities. One natural extension of (4) is:

$$\nabla^2 p - \frac{1}{c^2}\ddot{p} - 2A\dot{p} = -\frac{1}{c^2}\gamma(\underline{r})\ddot{p} + \nabla\cdot(\mu(\underline{r})\nabla p) \qquad (7)$$

with $\gamma(\underline{r})=(k-k_\theta)/k$ and $\mu(\underline{r})=(\rho-\rho_0)/\rho$

where $k(\underline{r})$ and $\rho(\underline{r})$ are the elasticity and density variations within V. k'_0 and ρ_0 refer to the imbedding medium surrounding the scattering volume, and take on the same magnitudes as the mean values of the appropriate parameters within V. The left-hand side of (7) may be interpreted as the "propagator", and the right-hand side as the "reflector", in the propagator-reflector formalism, first introduced to derive the impediography equations (Leeman, 1980b). The form of the scattering source function is suggested by the wave equation for acoustic propagation in loss-less, inhomogeneous media (Morse and Ingard, 1968). In lossy media, absorption fluctuations also give rise to scattering (Leeman, 1980c), but the effect is weak for the absorption variations in soft tissues, and is neglected here. Strictly, (4) should pertain to the absorption properties of the medium, with the form of the attenuating pulse being ascertained from the forward propagating solution of (7) (with V now assumed to fill the entire volume of interest). However, as the propagator-reflector method makes clear, the approach followed here is quite acceptable if multiple scattering is neglected, and provided that the pressure waveform appearing in the scattering source function is taken to have the form (6). In this case, must be chosen to be appropriate to the attenuation (not absorption) of the medium.

BACK-SCATTERING IN THE BORN APPROXIMATION

In the Born approximation, the field back-scattered to the point $\underline{D}$ is given by:

$$p_s(\underline{D},t)=\int_v d^3\underline{r}\int_{(D+z)/c}^{t-R/c} dt'\{-\frac{1}{c^2}\gamma(\underline{r})\ddot{p}(\underline{r},t')+\nabla\cdot(\mu(\underline{r})\nabla p(\underline{r},t')\,\}g_0(\underline{R},t) \qquad (8)$$

Here, $\underline{R} = \underline{D} - \underline{r}$, and the time differentiation is assumed to operate with respect to the t'-variable. A considerable simplification is effected by assuming that

$$\frac{r}{D} \ll 1 \text{ for all } \underline{r} \text{ in } V$$

Then $R \approx D + Z$ and $\frac{1}{R} \approx \frac{1}{D}$

Equation (8) then takes on a form reminiscent of a convolution integral, and it is readily shown that, on taking Laplace transforms

$$\mathcal{L}\{p_s(\underline{D},t)\} = \int_V d^3\underline{r}\{-\frac{1}{c^2}\gamma(\underline{r})s^2\pi(\underline{r},s)+\nabla\cdot(\mu(\underline{r})\nabla\pi(\underline{r},s)\}G(D+z,s)$$

It is extremely tedious to write out the full expression for p_s, so, for reasons of economy, we consider further only the contributions from the γ-fluctuations (monopole scattering). It should be emphasized, though, that there is no essential difficulty in handling the contributions from the μ-fluctuations (dipole scattering).

Thus,
$$\mathcal{L}\{p_s(\underline{D},t)\} = \frac{1}{4\pi c^2 D}\int_0^\infty dh.h\int_0^{2\pi}d\theta\int_{-D}^\infty dz\int_0^\infty d\lambda.\lambda\omega(\lambda)J_0(h\lambda)\gamma(h,\theta,z)s^2F(s)\cdot$$

$$\cdot\exp\{-(\sqrt{s^2+2\alpha_0 cs+\lambda^2 c^2}+\sqrt{s^2+2\alpha_0 cs})(z+D)/c\}$$

$$= \frac{1}{2Dc}\int_0^\infty d\xi\int_0^\infty d\lambda.\lambda\omega(\lambda)\ \Gamma(\lambda,\xi)s^2F(s)\exp\{-\sqrt{s^2+\alpha_0 cs+\lambda^2 c^2}+\sqrt{s^2+2\alpha_0 cs})\xi\}$$

with $\xi = (z+D)/c$ ('transit time' variable)

$$\Gamma(\lambda,\xi) = \int_0^\infty dh.hJ_0(h\lambda)\overline{\gamma}(h,z)$$ (Hankel transform)

$$\overline{\gamma}(h,z) = \frac{1}{2\pi}\int_0^{2\pi}d\theta\gamma(h,\theta,z)$$ (angle-average)

In the above, the extension of the h-, z- and ξ- integrations to the indicated limits is made permissible by observing that γ vanishes smoothly beyond V.

The back-scattered pressure then reduces to the particularly simple form:

$$\mathcal{L}\{p_s(\underline{D},t)\}_\gamma = \frac{1}{2DC}s^2T(\sqrt{s^2+2\alpha_0 cs})\ F(s) \tag{9}$$

with $$T(u) = \int_0^\infty d\xi E(u^2,\xi)\ e^{-u\xi}$$

and $$E(u^2,\xi) = \int_0^\infty d\lambda.\lambda\omega(\lambda)\Gamma(\lambda,\xi)e^{-\sqrt{u^2+\lambda^2 c^2}\xi}$$ (Beam profile "smoothing")

The implication of (9) is that the back-scattered signal may be expressed as the convolution of the <u>incident</u> pulse shape with (the second derivative of) a one-dimensional equivalent "tissue", T. The form of the latter is determined by the incident beam profile and the actual distribution of the scattering inhomogeneities.

DISCUSSION AND SUMMARY

It has been shown possible to construct a realistic tissue model in the form of a wave equation exhibiting dispersive attenuation which can be fitted closely to measured tissue values. All the main features of this model are retained in a simpler equation, which is solved to provide an exact expression for a propagating pulse. This solution also provides useful clues as to the nature of the "propagator" to be used in wave extrapolation methods, and may also be useful in interpreting the measurements made in transmission tomography studies.

Scattering by density and elasticity fluctuations may be incorporated, and the back-scattered echoes in the far-field may be calculated. It is comforting to note that analytic treatments, such as the one presented here, can go so far in handling the apparently complex problems involved in ultrasound pulse scattering from human tissues. If the far-field simplification is dropped, the above treatment (for scattering into the general direction) is relevant to the diffraction tomography technique, which is basically an inverse scattering problem.

In general, signal processing and resolution enhancement methods, as well as certain impediographic techniques, are applied to the one-dimensional voltage-time signal generated by the back-scattered echoes impinging on the transducer. Real tissues and pulses are three-dimensional, and the relationship of these one-dimensional analyses to the full three-dimensional problem has not always been clearly mapped out. It is apparent from the above that one-dimensional "equivalent" models may, indeed, be constructed, and conventional processing options be applied. However, the one-dimensional equivalent tissue impulse response function bears a complicated relationship to the distribution of scattering inhomogeneities and, unfortunately, depends on the interrogating pulse as well (through the incident beam profile). The whole question of equivalent one-dimensional models certainly demands more attention than it has received in the past.

ACKNOWLEDGEMENTS

S.L. wishes to thank The Wellcome Trust for generously making a research travel grant available. Miss Ann Freemantle generated the machined sections of the text!

REFERENCES

Berkhout, A. J., Ridder, J. and van der Wal, L. F., 1981, The
        potential of wave field extrapolation methods in medical imag-
        ing, in: "Acoustical Imaging 10", P. Alais, Ed., in press.
Greenleaf, J. F., Johnson, S. A., Samayoa, W. F. and Duck, F. A.,
        1975, Algebraic reconstruction of spatial distributions of
        acoustic velocities in tissue from their time-of-flight pro-
        files, in: "Acoustical Holography 6", N. Booth, Ed., Plenum
        Press, N.Y., pp 71-90.
Jones, J. P., 1977, Ultrasonic impediography and its applications to
        tissue characterisation, in: "Recent Advances in Ultrasound in
        Biomedicine I", D. N. White, Ed., Research Studies Press,
        Forest Grove, pp 131-156.
Leeman, S., 1980a, Ultrasound pulse propagation in dispersive media,
        Phys. Med. Biol., 25:481.
Leeman, S., 1980b, The impediography equations, in: "Acoustical Imag-
        ing 8", A. F. Metherell, Ed., Plenum Press, N.Y., pp 517-525.
Leeman, S., 1980c, Impediography revisited, in: "Acoustical Imaging
        9", K. Wang, Ed., Plenum Press, N.Y., pp 513-520.
Leeman, S., Hutchins, L. and Jones, J. P., 1981, Bounded pulse propa-
        gation, in: "Acoustical Imaging 10", P. Alais, Ed., in press.
Morse, P. M. and Feshbach, H., 1953, "Methods of Theoretical Physics",
        McGraw-Hill, N.Y.
Morse, P. M. and Ingard, K. V., 1968, "Theoretical Acoustics", McGraw-
        Hill, N.Y.
Mueller, R. K., Kaveh, M. and Iverson, R. D., 1980, A new approach to
        acoustic tomography using diffraction techniques, in:
        "Acoustical Imaging 8", A. Metherell, Ed., Plenum Press, N.Y.,
        pp 615-628.

AN ABSOLUTE MEASUREMENT OF BACKSCATTERING COEFFICIENTS FOR

EXCISED HUMAN TISSUES AND ITS RELEVANCE TO ABDOMINAL IMAGING

D. Nicholas

Physics Department
Institute of Cancer Research/Royal Marsden Hospital
Sutton, Surrey, U.K.

INTRODUCTION

Although the major proportion of diagnostic ultrasound techniques are based upon the exploitation of scattering processes a
rigorous quantitative evaluation of the scattering of ultrasound
by human tissues is a topic which has received very little attention to date.   Not only are the data sparse but the existing references tend to be implicit rather than explicit.   Despite a recent
interest in the frequency dependence of scattering measurements
only a few workers have evaluated an absolute measurement for the
differential scattering coefficient as a function of frequency for
soft biological tissues (Nicholas, 1976b, 1981; Shung and Reid,
1977; Freese and Lyons, 1977, and Kadaba et al., 1980).   Other
researchers have investigated scattering both as a function of
frequency (Chivers et al., 1973; Chivers and Hill, 1975b, and
Lizzi and Elbaum, 1979) and angle of scattering (Waag et al., 1976;
Lele et al., 1976, and Nassiri et al., 1978) but have only produced
relative measurements.

The intention of this paper is to provide a rigorous description of the techniques necessary for calculating the differential
scattering coefficient, as a function of the frequency of the
incident ultrasonic waves,and to report on measurements made on
freshly excised soft human tissues.   These measurements will be
compared with complementary frequency dependent attenuation measurements (conducted on the same tissues) to assess the contribution of
scattering to attenuation.   Furthermore, it will be shown that
careful inspection of the frequency dependence of the backscattering
suggests that a single power law fit to the data is insufficient
and that a combination of such functions both fit the data more

149

accurately and can be related to a theoretical description of
scattering.   This latter aspect has been suggested previously
(Nicholas, 1977b) and indicates that such measurements are representative of the scale of structure responsible for the scattering
processes.

MEASUREMENT TECHNIQUE

     It has previously been shown that the scattering of ultrasound
by small tissue volumes is dependent upon the orientation of the
scattering elements in relation to the incident wave front
(Nicholas and Hill, 1975a,b).   This effect of orientation is due to
the superposition of the scattered waves from each individual homo-
geneity which can add up constructively or destructively depending
upon the relative positions of the scatterers.   In order to establish
an absolute value for the frequency dependence of the bulk scattering coefficient (cross-section per unit volume) it is necessary to
average out these diffraction effects.   Other workers, notably
Shung et al. (1976,1977) and Lizzi and Elbaum (1979) have accounted
for these effects by averaging the scattering results from adjacent
or overlapping volumes of tissue.   Such an approach relies upon
the different volumes being spatially uncorrelated.   The approach
reported here involves averaging the scattered energy from the same
region of tissue for different orientations of the volume.

     A cylindrical tissue specimen is positioned vertically within
a water-filled sound tank such that its long axis is centred perpendicular to the incident sound beam (see Figure 1).   A time
gated portion of the backscattered echo train is selected such that
the tissue volume under investigation is centred within the cylindrical sample.   A set of (echo intensity) values is then recorded,
covering the frequency band of the transducer for each of 200
orientations obtained by rotating the sample about its long axis
in increments of 1.8°.   The rotation interval chosen is sufficiently
small to average the diffraction variations associated with a 360°
rotation of the tissue (Nicholas, 1976a) and sufficiently large
that adjacent samples are substantially uncorrelated.   Finally
the 200 readings at each specific frequency are averaged to yield
an average backscattering frequency spectrum.

     The basic requirement for obtaining accurate scattering
measurements is that the properties of the tissue must be separated
from the properties of the equipment.   This may be accomplished in
two ways; by the use of a substitution method of measurement
(reported here), or by the use of exact equations expressing the
scattered signals as a function of both the apparatus parameters
and tissue properties.   The latter method, however, is unlikely to
be accepted until a more detailed knowledge of the interaction of
ultrasound with biological tissues is obtained.

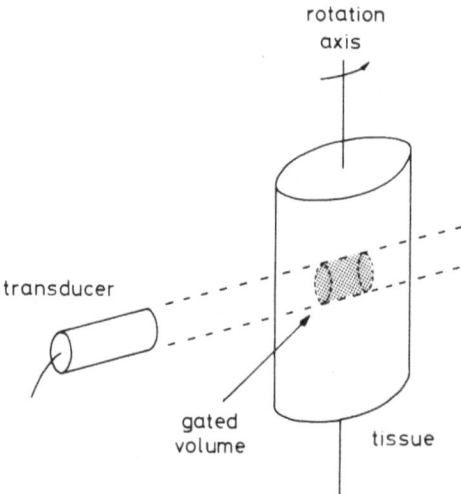

Fig. 1. Positioning and orientation of tissue specimen.

        The substitution method for scattering adopted here requires
measurement of the average volume backscattering from the tissue
sample and its comparison with the power reflected from a plane
reflecting interface.   The scattering will be defined in terms of
a differential scattering coefficient $\mu_d(\emptyset)$ where $\emptyset$ is the
scattering angle (180° for backscattering).   This is necessary
because practical measurements involve an assessment of the
scattering coefficient over a limited solid angle.   The total
scattering is now the integral of this differential scattering
coefficient over all space :

$$\mu_s = \int_{sphere} \mu_d(\emptyset) \; d\Omega \qquad (1)$$

where the differential scattering coefficient has units of $cm^{-1}sr^{-1}$.

Calculation of the mathematical expression for the back-scattering coefficient ($\emptyset = 180^{\circ}$) using the substitution technique has been rigorously described elsewhere (Nicholas et al., 1981) and is reproduced here as :

$$\mu_d(180^{\circ}) = \frac{<P_r>}{P_i} \cdot \frac{\eta}{4\beta^2} \cdot \frac{x^2}{A} \cdot \frac{2\mu}{\exp(-2\mu x_1)(1-\exp(-\mu c\tau))}$$

$$\cdot \frac{\int_o^{r_t} rD(2x,r,\lambda)\ dr}{\int_o^{r_s} rD^2(x,r,\lambda)dr} \tag{2}$$

where   A = effective receiving area of the transducer face.
$D(x,r,\lambda)$ = the directivity distribution for the whole transducer face, assuming axial symmetry.
λ = wavelength.
μ = attenuation coefficient of scattering medium.
η = intensity reflection coefficient of reference target.
β = intensity transmission coefficient for specimen/water interface.
c = velocity of sound in the scattering medium.
τ = the duration of the time-gate selecting the scattering volume.
r_s = radius of specimen.
r_t = radius of effective transducer face.

(the terminology is described in detail in the aforementioned publication, whilst Figure 2 illustrates the system of coordinates for the backscattering formulation).   Here $<P_r>$, the averaged back-scattered power from the tissue, and P_i, the reflected power from the plane target, are measured as a continuous function of frequency over the bandwidth of the transducers used. The remaining terms are either constants relating to the transducer and target characteristics or variables associated with the acoustic properties of the tissue under investigation.

The electronics is basically that of a conventional pulse-echo imaging system where the broadband frequency contributions associated with a short pulse are utilised.   Thus, by feeding the time gated signal into an analogue spectrum analyser the relative contributions of scattering at different frequencies can be measured.   Although this technique may be less sensitive than the tone burst method originally employed by Sigelmann and Reid (1973), its advantage is its ability to provide a continuous measure of frequency for the

Transducer   Tissue

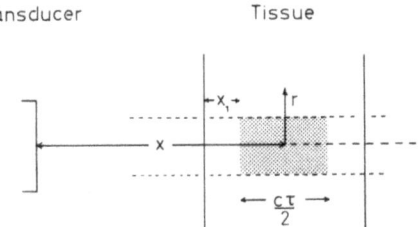

Fig. 2. Coordinate system adopted for backscattering formulation.

bandwidth of the incident pulse (Chivers and Hill, 1975a).

Associated Techniques

     The complexity of Equation (2) illustrates that various other
measurement procedures need to be performed before an accurate
assessment of the backscattering coefficient can be made. All of
the quantities expressed are either known or easily measurable,with
the exception of the directivity function.   Ideally the effective
directivity (Bamber and Phelps, 1977) of the transducer can be
measured by using a hydrophone as a 'point receiver' at various
positions within the beam. Absolute measurement, however, requires
that both the sensitivity and frequency response of the hydrophone
should be known. An alternative method (adopted here) is to
position a small steel ball (less than 1 cm diameter) within the
sound beam and, by reciprocity, measure the reflected power as a
function of position and frequency to obtain D^2. Since the function is normalised to its on-axis position the reflection capabilities of the target are unimportant provided they remain constant
over the range of frequencies measured. This method of measurement
also incorporates the electronics used in the experiments and thus
accounts for any amplifier or gating effects.

     The method employed for the attenuation measurements has been
described previously (Papadakis et al., 1973; Chivers and Hill,
1975a). A wideband transducer is operated in pulse-echo mode and

used as both transmitter and receiver.   The pulse of ultrasound is
reflected by a plane liquid-solid interface oriented normal to the
beam axis and positioned in the far-field (Fraunhofer zone) of the
transducer.   The received signal is isolated by an electronic time-
gate, amplified, and fed into an analogue spectrum analyser (Hewlett
Packard model 8552/3) where the frequency spectrum of the gated
signal is displayed on a logarithmic scale.   The spectra of this
signal, with and without a parallel sided slab of tissue of known
thickness interposed between the transducer and target, are recorded
and the attenuation in dB, resulting from a double traverse of the
specimen, is calculated as the difference between these two spectra.

A further measurement extracted from the technique employed in
making attenuation measurements was that of an average velocity of
sound in tissue 'c'*.   This is achieved by noting the time of
arrival of the pulse reflected from the plane target with and without
the tissue present.   The difference in the times allows calculation
of the average velocity over the thickness of tissue traversed relative to the speed of sound in the surrounding medium (degassed
water at 20°C).   The accuracy of both this measurement and that of
attenuation is limited primarily by the uncertainty as to the
thickness of tissue traversed.   This is due to the lack of precision
in sectioning the tissue to a plane parallel sided slab.

The concluding measurement on the parallel sided specimens was
that of an average value for the intensity transmission coefficient
β.   This was achieved by noting the intensity of the echo reflected
from the reference interface, when the target was positioned in the
far-field of the transducer and orientated normal to the incident
pulse, and comparing this with an average value for the intensity
reflected from a water/polythene/tissue interface positioned and
oriented in a like manner.   Knowledge of the reflection properties
of the reference target then allows calculation of the intensity
reflection coefficient for the water/polythene/tissue interface and
hence the intensity transmission coefficient β.   Although this
method requires a flat tissue surface, simple angular reflection
measurements from the water/polythene/tissue interface suggest that
this simple calculation is accurate to within 0.02%.

RESULTS

Attenuation

Three types of pathologically normal biological tissue have
been examined: freshly excised human liver, spleen and brain (white
matter).   Eight tissue specimens of each type are reported on where

---

* The correction term incorporating the velocity of the ultrasonic
waves in tissue 'c' is small enough to warrant the adoption of a
constant value for velocity over the frequency range reported here.

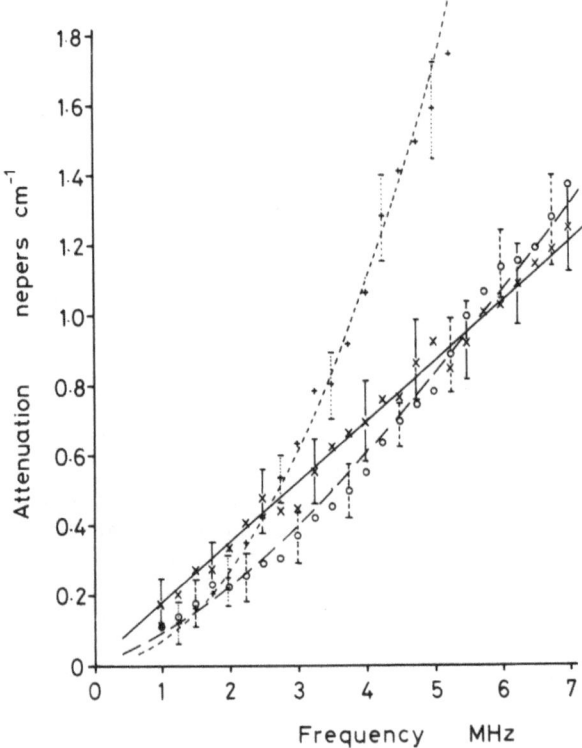

Fig. 3. Average intensity attenuation coefficient for three types
of freshly excised normal human tissues; x liver,
o spleen, + brain (white matter).   Average of eight
tissues with three samples from each (Nicholas, 1981).

several samples were sectioned from each tissue.   Ideally, attenuation and velocity measurements should be performed on the same tissue
sample as the backscattering measurements.   However, the use of a
cylindrical sample for backscattering complicates any attenuation
measurements due to the geometry of the sample.   The specimens
for attenuation assessment were sectioned from the same tissue as
those for backscattering evaluation and were slabs having plane,
parallel surfaces of approximately 40 cm^2 and a thickness in the
range 1.5 to 2.5 cm.   An uncertainty in sample thickness, due to
cutting errors and lack of rigidity in the tissue, was estimated at
7%.   The specimens were mounted in a holder with thin polythene

walls (∼70 μm); the effects of the polythene on both the attenuation and backscattering results were accounted for in the measurement
of the transmission coefficient β for a water/polythene/tissue interface.

Five different measurements were made on each sample at different sites, for each of four transducers having nominal working
frequencies of 1.0, 2.0, 4.0 and 6.0 MHz.   This enabled continuous
coverage of the frequency range 0.5 to 7 MHz.   Owing to the overlap
of the transducer spectral windows some of the attenuation values
are the average of 240 independent spectral readings, the others
being an average of 120.   Corresponding measurements of tissue
thickness were used to calculate the intensity attenuation coefficient
'μ' in nepers cm^{-1}.

Figure 3 displays the mean attenuation values as a function of
frequency where the curves are a best fit to the data values and
the error bars represent the standard errors on the means.   For the
sake of clarity only a few of the 68 independent frequency measurements are portrayed.   The indicated standard errors are those
corresponding to the variations observed between the average values
found for the different specimens.   In all cases however these are
similar in magnitude to the variations found between individual
measurements made at different sites in a given specimen.   This
strongly indicates that the overall results are not seriously
influenced by errors from other sources such as phase cancellation
(Miller et al., 1976), trapped air bubbles and specimen thickness
measurement (Nicholas, 1976a).

Backscattering

For the backscattering measurements the same tissues were
investigated although a separate cylindrical specimen was sectioned
with an approximate radius of 1.5 cm and length of 6 cm.   The
cylinder of tissue was tightly enclosed and supported by a polythene
cylinder with rigid brass end plates.   The brass end pieces acted,
respectively, as a weight to ensure vertical alignment of the
cylinder and as a connection to a shaft attached to a stepping
motor capable of rotating the specimen about its long axis.

The polythene containers used in all the measurements reported
here provided rigidity to the specimens and ensured that the water
of the sound tank, by which both transducer and specimens were
coupled, remained free of any blood seepage from the tissues.   As
mentioned previously these containers also allowed the specimens to
be degassed (by subjecting them to a gentle vacuum) prior to
placement in the sound tank.

Three different measurements were made on each sample, at
different sites, for each of three transducers having nominal working

frequencies of 1.0, 2.0 and 4.0 MHz. This enabled continuous coverage of the frequency range 0.5 to 4.3 HMz. The higher frequency
transducer only permitted a couple of measurements at around 5 MHz
due to its poor signal to noise ratio at these higher frequencies.
This was purely a limitation of the transducers available at the
time. As before, some of the average backscattering values are not
only the average of 200 individual specimen readings but also the
average of 48 independent volume measurements due to overlap of
transducer bandwidths.   Thus the average backscattering coefficient
measured for a specific tissue type at any one frequency is the
result of averaging either 9600 or 4800 readings.

     Utilising equation (2) and incorporating the correction factors
permits the calculation of the differential scattering coefficient
(for 180° scattering) as a function of incident ultrasound frequency.
The averaged values for three tissue types are portrayed in Figure 4,
where the error bars indicate the standard errors on the means. For
the sake of clarity only a few of the 51 independent frequency
measurements are displayed.

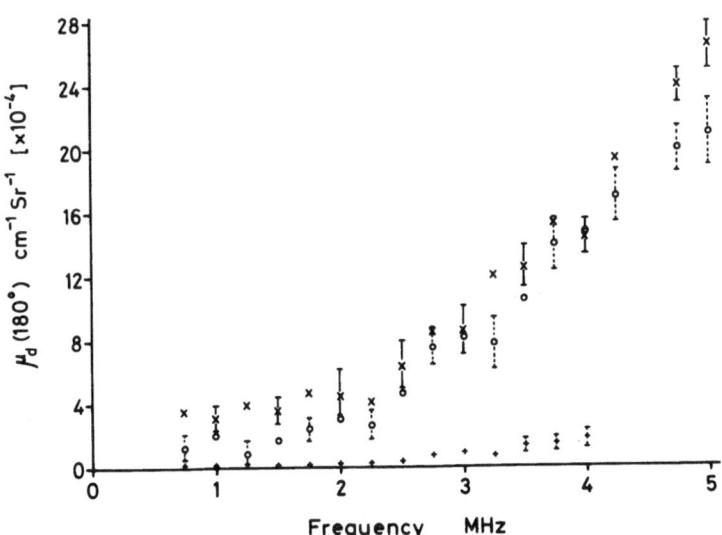

Fig. 4.   Average backscattering coefficient as a function of
          frequency for three types of freshly excised normal human
          tissues;   x liver, o spleen, + brain (white matter).
          Average of eight tissues with three samples from each
          (Nicholas, 1981).

Table 1.   Summary of Tissue Measurements

Tissue type	Intensity Attenuation Coefficient nepers cm^{-1}	Intensity Backscattering Coefficient cm^{-1}Sr^{-1} (x10^{-3})	Average Velocity at 20°C ms^{-1}	Intensity transmission Coefficient
Liver	0.17 f$^{1.0}$	3.3 + 0.2 f^3	1570 + 12	0.987 + 0.001
Spleen	0.09 f$^{1.5}$	1.2 + 0.2 f^3	1573 + 12	0.985 + 0.001
Brain (white matter)	0.07 f$^{2.0}$	0.23 + 0.006 f^4	1530 + 10	0.980 + 0.001

In order to summarize these tissue measurements it is helpful to fit a single power law dependence to the results.   Table I lists the attenuation and scattering results in this manner although (Nicholas, 1981) a multi-term polynomial provides a more accurate mathematical description of the latter.   Included in Table I are the average values for the velocity of ultrasound in tissue over the frequency range 0.5 - 7 MHz, and the average intensity transmission coefficient for a water/polythene/tissue interface oriented normal to the incident pulse.

INTERPRETATION

The interpretation of ultrasound scattering data is complicated by the present lack of knowledge concerning the tissue structures responsible for the scattering.   This is noticeable in conventional B-scan imaging where the 'parenchymal echoes', pertaining to the internal regions of organs, cannot be related to any specific tissue structure.   Indeed, it seems probable that these echoes are due to a complicated interference of the scattered waves originating from structures separated by less than the incident pulse dimensions (Burckhardt, 1978; Nicholas, 1978; Bamber and Dickinson, 1980).

As an aid to interpreting and improving clinical sector B-scans any scattering measurements which can elucidate the small scale tissue structures interacting with the acoustic pulse are to be encouraged.   Existing theoretical descriptions of the acoustic scattering by biological material (Chivers, 1977; Nicholas, 1977a; Gore and Leeman, 1977) indicate that the scattering coefficient is dependent upon the size and separation of scatterers, their scatter-ing strength, and the frequency of the incident wave.   Thus a measure of the frequency dependence of the scattering coefficient should indicate the scale of the structures responsible for the scattering.

In a previous publication (Nicholas, 1977a) the author derived an expression for acoustic scattering from an 'inhomogeneous continuum' utilising the work of Chernow (1960) and Morse and Ingard (1968). Although this theory has its limitations (as discussed elsewhere) it will suffice to provide an interpretation of the scattering data published here.

The basic expression for backscattering was given by :

$$I(180^{\circ}) = \frac{I_o \, V \, K_o^3 \, Y_{\varkappa}^2}{\pi x^2} \int_o^d N(R) \, R \, \sin(2K_o R) \, dR \qquad (3)$$

where I_o = incident intensity.

K_o = incident wave vector.

$Y_{\varkappa}$ = the root mean square derivation of the compressibility of the inhomogeneities from its mean value.

$N(R)$ = the correlation coefficient for the separation of the inhomogeneities.

V = the scattering volume.

and the integration is performed over a distance 'd' corresponding to the radius of the scattering volume.

In this expression density variations have been neglected since for soft mammalian tissues they appear to be an order of magnitude smaller than compressibility variations (Fields and Dunn, 1973). Though this is a limitation to the adopted theory it does not detract from a general appraisal of the variation of backscattering with wave vector 'K' and average scatterer separation '$\bar{a}$'. However, should one attempt to extend the discussion to angular scattering in general then the form of Equation (3) must be adapted to account for the dipole term associated with the density fluctuations.

In order to solve Equation (3) it is necessary to specify the form of the autocorrelation function $N(\underline{R})$ which defines the characteristic separation of the scattering structures within the tissue. By utilising a spherical coordinate system in the expression for backscattering it is assumed that the statistical properties of the scattering medium are isotropic such that the correlation coefficient depends solely on the modulus of $\underline{R}$.

The fundamental problem in theoretical modelling of ultrasonic scattering by tissue is to determine the form of the autocorrelation function.   For the purpose of this publication it will suffice to adopt the functions which have been described in the literature to date, namely the Gaussian $(\exp(-R^2/\bar{a}^2))$ and exponential $(\exp(-R/\bar{a}))$ functions.   The former, when substituted into Equation (3), necessitates the evaluation of the standard integral

$$\int_{0}^{\infty} \exp(-R^2/\bar{a}^2) \, R \sin(2K_o R) \, dR$$

$$= \frac{K_o \bar{a}^3 \pi^{\frac{1}{2}}}{4} \exp(-K_o^2 \bar{a}^2) \tag{4}$$

where by assuming periodicity about the dimension d, or zero scattering outside the volume of dimension d, the limits of integration have been extended to infinity.

The second case where $N(R) = \exp(-R/\bar{a})$ again results in a tabloid integral.

$$\int_{0}^{\infty} \exp(-R/\bar{a}) \, R \sin(2K_o R) \, dR$$

$$= \frac{2K_o \bar{a}^3}{\left[1+4K_o^2 \bar{a}^2\right]^2} \tag{5}$$

a form which is to be found in Pekeris (1947).

From these solutions we can plot the variation of backscattering as a function of the wave vector K and mean structural separation $\bar{a}$.   Figure 5 portrays these scattering functions on a logarithmic scale thus emphasising any power law dependence.

For small scale fluctuations, i.e. $K_o \bar{a} \ll 1$, these formulae exhibit a $K_o^4 \bar{a}^3$ Rayleigh dependence: the backscattering coefficient is proportional to the fourth power of the frequency.   However, for large scale inhomogeneities, i.e. $K_o \bar{a} \gtrsim 1$, the dependencies differ: the Gaussian function adopts an exponentially decaying dependence on frequency whereas the simple exponential function remains constant.   Comparison of these functions with the simple power law descriptions of the scattering data (see Table I) indicates

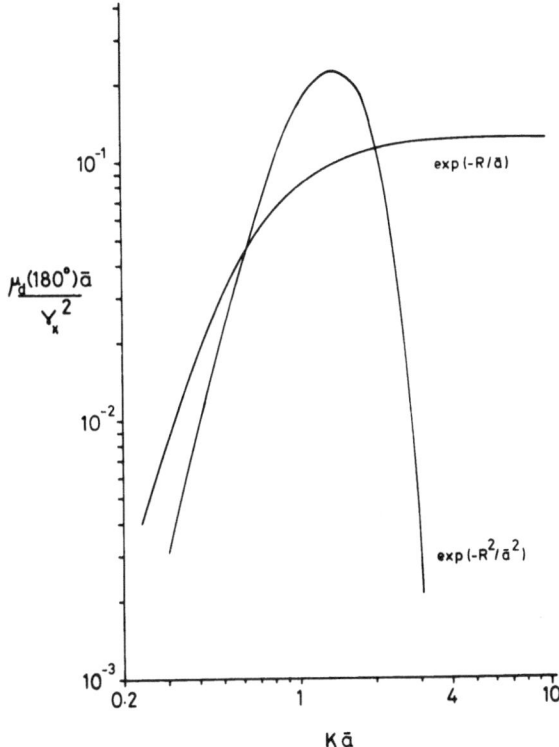

Fig. 5.   Logarithmic plot of the backscattering coefficient
($\mu_d \equiv I(180^\circ) x^2/I_o V$) as a function of $K_o\bar{a}$ adopting two
forms for the correlation coefficient $N(R)$.

that a single description for the correlation coefficient $N(R)$ is
not a realistic assessment of the acoustic scattering structures
present in human soft tissues.

Histological examination of the internal regions of organs
indicates that they are likely to possess a wide range of scattering
structures most of which exhibit a degree of regularity over short
ranges.   For the volumes of tissue examined in the experiments
reported here (typically 0.1 cm^3) human liver tissue will possess
structures varying from small capillaries at a few millimetres
separation down to liver cells with a typical separation of 10-20
μm.   A further complication will also result due to the variation

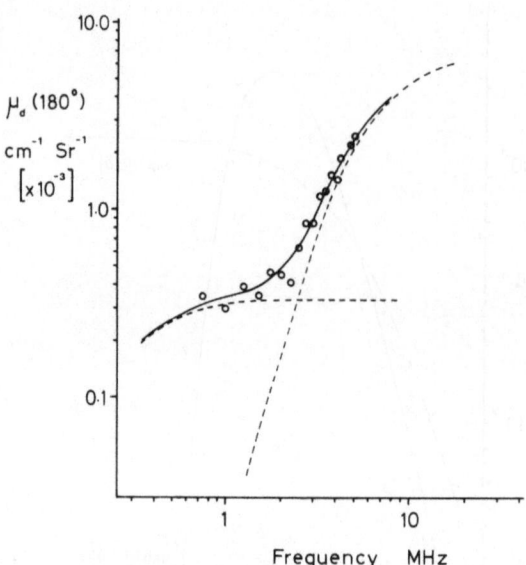

Fig. 6.   Logarithmic plot of the average backscattering coefficient
          for liver tissue as a function of frequency with combina-
          tions of the theoretical curves fitted to the data.

in scattering strength of structures of different sizes and/or
biological composition.

    In an attempt to interpret the scattering data more fully the
backscattering coefficient as a function of frequency has been
plotted on a logarithmic scale and combinations of the theoretical
scattering curves fitted to the data.   Figure 6 portrays the
backscattering data for freshly excised liver tissue and the two
theoretical curves whose summation best fits this data.   Attempts
at fitting larger combinations of curves has not resulted in any
improvement to the accuracy of data fitting for the frequency ranges
examined here.   Although the exponential form for the correlation
coefficient has been adopted here it must be stressed that other
descriptions of the correlation coefficient can also be made to
fit the data in a like manner.

    The backscattering data for spleen and brain (white matter)
tissue has been interpreted in the same manner and the predicted

Table II. Scatterer separations and strengths suggested by a theoretical fit of the correlation coefficient $N(R) = \exp(-R/\bar{a})$ to the backscattering coefficient $\mu_d(180^{\circ})$. The scattering strength is normalised to that necessary to describe the backscattering coefficient for liver tissue at 1 MHz.

Tissue	Scatterer separation $\bar{a}$ mm	Relative scattering strength $\gamma_{\varkappa}^{2}$
Liver	$\geqslant 1$ 0.025	1 0.75
Spleen	$\geqslant 0.7$ 0.025	0.4 0.6
Brain (white matter)	$\geqslant 1$ 0.012	0.08 0.45

values for scatterer separation ($\bar{a}$) and scattering strength ($\gamma_{\varkappa}^{2}$) tabulated in Table II.

CONCLUSIONS

This paper has reported on measurements of the differential scattering coefficient (for 180° scattering) for three types of freshly excised, normal human tissues using techniques and calculations which attempt to provide an assessment of soft tissue scattering (Nicholas et al., 1981). By using a spectral analysis approach both the attenuation and backscattering coefficients have been evaluated as a function of incident acoustic frequency for the range 0.75 - 5 MHz. Since the measurements are over a continuous frequency range it has proved possible to fit accurate polynomial descriptions to the backscattering data and relate them to theoretical descriptions of the scattering structures. Tentative explanations for the form of the data suggest that two basic tissue structures are responsible for causing ultrasonic scattering over the reported frequency range. Large scale structures of the order of a millimeter separation and small scale tissue components at the level of 20-40 μm. Tests of the accuracy of these predictions can only be achieved by supplementing the existing data with measurements at both higher and lower frequencies and by direct observation of the tissue structures to accurately evaluate the relevant correlation coefficients $N(R)$ (ideally by acoustic microscopy).

   Discrepancies within the limited literature suggested that the
calculation of an absolute value for backscattering had yet to be
fully achieved.   In this publication criteria for such calculations
have been specified which are believed to represent the most
definitive evaluation to date.   Until a universal acceptance of a
technique and theoretical description for scattering measurements
is achieved little sense can be made in relating the different
scattering measurements reported in the literature.

   The interpretation of these results in the diagnostic context
raises some interesting speculation.   In abdominal scanning most
commercial B-scanners operate at frequencies around 3 MHz which
suggests that the parenchymal echoes or 'speckle' present in
diagnostic B-scans are the result of scattering from small-scale
acoustic structures with sizes of the order of 20-40 μm.   Since
these sizes are far smaller than the resolution limitations imposed
by the wavelength of the incident energy (0.5 mm at 3 MHz) the
diagnostic advantages of using higher acoustic frequencies may only
improve the detection and display of the larger vascular structures
and tissue boundaries.   It will be interesting to perform similar
analyses on pathological tissues where histology indicates a
structural variation from normal tissues at this microscopic level.

   To date little accurate data on the scattering of ultrasound
by human soft tissues has been reported, yet that which is available
suggests that such measurements have a great significance in
improving our understanding of the interaction of ultrasound with
soft tissue.   It is hoped that future work will remedy this situation and enable us to improve our understanding of clinical B-scans
and utilise such measurements (Nicholas, 1979) as a means for
tissue characterisation in their own right.

REFERENCES

Bamber, J. C., and Phelps, J. V., 1977, The effective directivity
      characteristics of a pulsed ultrasound transducer and its
      measurement by semi-automatic means, Ultrasonics 15:169.
Bamber, J. C., and Dickinson, R. J., 1980, Ultrasonic B-scanning:
      a computer simulation, Phys. Med. Biol. 25:463.
Burckhardt, C. B., 1978, Speckle in ultrasound B-mode scans,
      IEEE Trans Sonics and Ultrasonics 25:1.
Chernow, L. A., 1960,"Wave propagation in a random medium", Dover,
      New York.
Chivers, R. C., 1977, The scattering of ultrasound by human tissues
      - some theoretical models, Ultrasound in Med. and Biol.
      3:1.
Chivers, R. C., and Hill, C. R., 1975a, Ultrasonic attenuation in
      human tissue, Ultrasound in Med. and Biol. 2:25.
Chivers, R. C. and Hill, C. R., 1975b, A spectral approach to ultra-
      sonic scattering from human tissue : methods, objectives and

backscattering measurements, Phys. Med. Biol. 20:799.

Chivers, R. C., Hill, C. R., and Nicholas, D., 1973, Frequency dependence of ultrasonic backscattering cross-sections: an indicator of tissue structure characteristics, in: "Ultrasonics in Medicine", Excerpta Medica:300.

Fields, S., and Dunn, F., 1973, Correlation of echographic visualizability of tissue with biological composition and physiological state, J. Acoust. Soc. Am. 54:809.

Freese, M., and Lyons, E. A., 1977, Ultrasonic backscatter from human liver tissue: its dependence on frequency and protein/ lipid composition, J. Clin. Ultrasound 5:307.

Gore, J. C., and Leeman, S., 1977, Ultrasonic backscattering from human tissue: a realistic model, Phys. Med.Biol. 22:317.

Kadaba, M. P., Bhagat, P. K. and Wu, V. C., 1980, Attenuation and backscattering of ultrasound in freshly excised animal tissues, IEEE Trans. Biomed. Eng. 27:76.

Lele, P. P., Mansfield, A. M., Murphy, A. I., Namery, J., and Senepati, N., 1976, in: "Ultrasonic Tissue Characterisation", NBS Special Publication 453, US Dept. of Commerce : 167.

Lizzi, F. L., and Elbaum, M. R., 1979, Clinical spectrum analysis techniques for tissue characterisation, in: "Ultrasonic Tissue Characterisation II", NBS Special Publication 525, US Dept. of Commerce : 111.

Miller, J. G., Yukes, D. E., Mimbs, J. W., Dierker, S. B., Busse, L. J., Laterra, J. J., Weiss, A. N., and Sobel, B. E., 1976, Ultrasonic tissue characterisation: correlation between biochemical and ultrasonic indices of myocardial injury, Ultrasonics Symp. Proc., IEEE Cat. No. 76 CH1120-5SU : 33.

Morse, P. M., and Ingard, K. V., 1968, "Theoretical Acoustics", McGraw-Hill, New York.

Nassiri, D. K., Nicholas, D., and Hill, C. R., 1978, Scattering and attenuation in anisotropic human tissue (Abstract), Proc. 3rd European Congress Ultrasound in Medicine : 381.

Nicholas, D., 1976a,"Ultrasonic Scattering and the Structure of Human Tissues", Ph.D. Thesis, University of London.

Nicholas, D., 1976b, The application of acoustic scattering parameters to the characterisation of human soft tissue, Ultrasonics Symp. Proc. IEEE Cat. No. 76 CH1120-5SU : 64.

Nicholas, D., 1977a, in: "Recent Advances in Ultrasound in Biomedicine", D. N. White, ed., Chapter 1, Research Studies Press.

Nicholas, D., 1977b, in: "Recent Advances in Ultrasound in Biomedicine", D. N. White, ed., Chapter 2, Research Studies Press.

Nicholas, D., 1978, Clinical application of the diffractive scattering of ultrasound, in: "Proc 3rd International Symp. on Ultrasonic Tissue Characterisation", US Dept. of Commerce: 112.

Nicholas, D., 1979, Ultrasonic diffraction analysis in the investigation of liver disease, Brit. J. Radiol. 52:949.

Nicholas, D., 1981, Differential scattering coefficients for excised
    human tissues: results, interpretation and associated measure-
    ments, Ultrasound in Med. and Biol. (in press).
Nicholas, D., and Hill, C. R., 1975a, Acoustic Bragg diffraction from
    human tissues, Nature 257:305   and   261:330.
Nicholas, D., and Hill, C. R., 1975b, Tissue characterisation by an
    acoustic Bragg scattering process, Proc. Ultrasonics
    International 1975, IPC Science and Technology Press,
    Guildford, England : 269.
Nicholas, D., Hill, C. R. and Nassiri, D. K., 1981, Differential
    scattering coefficients for excised human tissues: principles
    and techniques, Ultrasound in Med. and Biol. (in press).
Papadakis, E. P., Fowler, K. A., and Lynnworth, L. C., 1973, Ultra-
    sonic attenuation by spectrum analysis of pulses in buffer
    rods: methods and diffraction correction, J. Acoust. Soc. Am.
    53:1336.
Pekeris, C. L., 1947, Note on the scattering of radiation in an
    inhomogeneous medium, Phys. Rev. 71:268.
Shung, K. K., Sigelmann, R. A., and Reid, J. M., 1976, Scattering
    of ultrasound by blood, IEEE Trans. Biomed. Eng. BME-23:460.
Shung, K. K., Sigelmann, R. A., and Reid, J. M., 1977, Angular
    dependence of scattering of ultrasound from blood, IEEE
    Trans. Biomed. Eng. BME-24:352.
Sigelmann, R. A., and Reid, J. M., 1973, Analysis of ultrasound
    backscattering from an ensemble of scatterers excited by
    sinewave bursts, J. Acoust. Soc. Am. 53:1351.
Waag, R. C., Lerner, R. M., and Gramiak, R., 1976, Swept-frequency
    ultrasonic determination of tissue macrostructure, in:
    "Ultrasonic Tissue Characterisation", NBS Special Publication
    453, US. Dept. of Commerce : 213.

THE DESIGN AND APPLICATION OF SOFTWARE FILTERS TO IMPROVE LATERAL

RESOLUTION IN A B-MODE ULTRASONOGRAM

Carolyn Kimme-Smith and Joie Pierce Jones

Department of Radiological Sciences
University of California, Irvine
Irvine, CA 92717

This study is a small part of a larger study which is evaluating the effects of B-mode processing on ultrasonograms of the abdomen. Since we examine innovative as well as conventional processing algorithms, we have recently studied the effects of lateral beam profile filters on ultrasonograms of objects which simulate the texture found in abdominal scans.

Finite beam widths, associated with all conventional ultrasonic transducers, significantly degrade the lateral resolution obtained in a B-mode ultrasonogram. Software filters, constructed from transducer beam profiles, have been proposed as a means for improving lateral resolution. Unfortunately the filters designed in this manner perform poorly when applied to textured medical ultrasound images. In 1974 McSherry and Keller (1) briefly described a minimum mean-square error filter designed for application to bistable cardiac images. If this filter were applied to conventional gray scale ultrasonograms, the large side lobes characteristic of the filter would have produced ringing in the resulting image. Noise and windowing problems in beam profile filter design were handled heuristically in a recent paper by Hundt and Trantenberg (2). They collected beam profile data from a B-mode display and then derived a frequency domain filter from the data. B-mode display data is characterized by poor spatial resolution and non-linear A to D gray level assignment. These were therefore incorporated in the beam profile measurements, and increased the noise content of the data. This clearly hampers successful filter design.

In this study we wish to examine the effects of lateral beam profile filters applied to A-line data from textured objects. This is accomplished by collecting r-f waveforms from test objects and then using this data to simulate on a computer the functions

167

performed in hardware by a B-mode processor (5). The data was collected
from a large "elephant ears" natural sponge and a commercially
available gray scale test object. Unlike previous workers who
applied a single filter to a wide depth range, we have found that it
is more appropriate to design range specific filters.

    Beam profiles are produced by recording the echoes from a small
reflector moving at a known depth from the transducer. To produce
the lateral beam profiles in Figures 1 and 2, we measured peak echo
amplitudes with the reflector at 4 and at 7 cm distances from the
transducer. The reflector is a 1.5 mm diameter wire and the transducer is a 3.5 MHz, 13 mm unfocused piston source. To eliminate
noise from our filter design, we model these profiles with sections
of Gaussian curves. The result of this modeling is superimposed on
the noisy beam profiles in the two figures. In addition to beam
profiles, our ultrasound data acquisition system (see Figure 3) can
acquire rf waveforms in various test objects at prescribed intervals.
Using such waveforms, our present study simulates in software the
following B-mode processing: 1) rf signal rectification and envelope
detection, 2) peak detection of those samples in an A-line which fall
within a single pixel of a B-mode display, 3) convolution of the peak
detected samples with the filter, 4) gray scale mapping, described
below, and 5) peak weighting algorithm for storage in the digital
scan converter. The resulting simulated B-mode ultrasonogram is
viewed and evaluated on a Genesco 512 x 512 display (5,6).

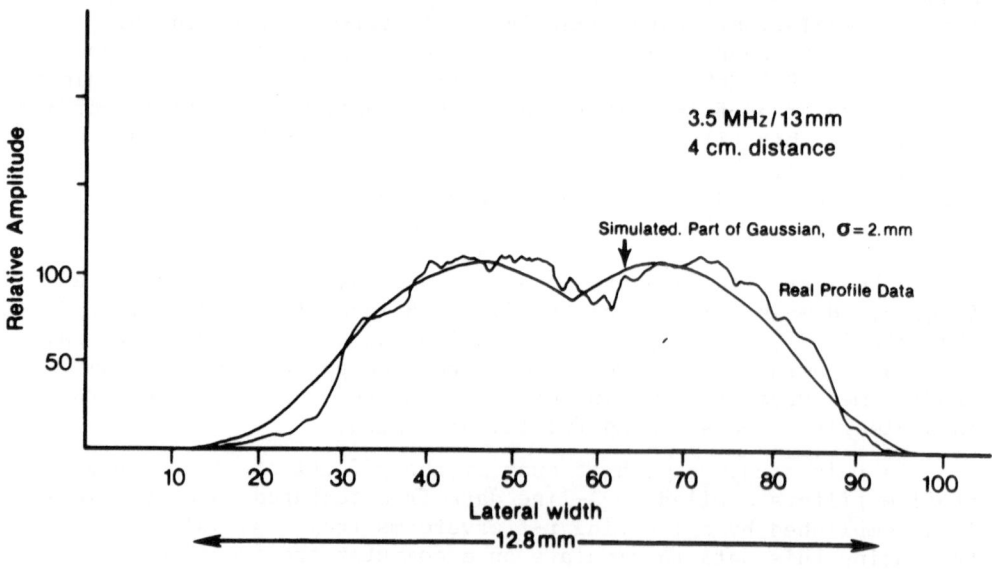

Figure 1

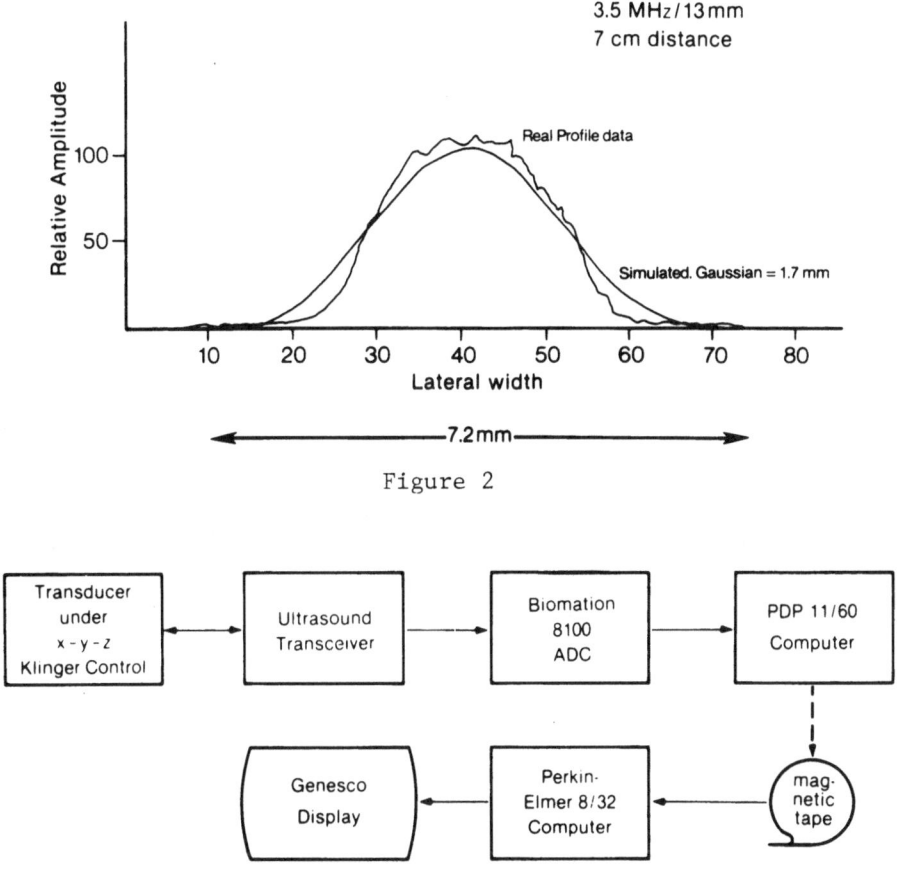

Figure 2

Figure 3.   UCI Ultrasound Data Acquisition System

We have collected two data sets of rf signals for this study. In each set, the waveforms are sampled at 20 MHz to yield 2048 samples over 8 cm of depth. The 8 cm region begins 4 cm from the transducer face. The Biomation 8100 yields only 6 to 7 bits of reliable gray scale information, although 8 bits are stored. The TGC and system gain are set within those ranges used in conventional clinical studies. To avoid non-linear receiver effects, we have removed some hardware filters in the ultrasound transceiver. The natural sponge data set was collected so that the spacing between

individual A-lines is .16 mm. This gives a 4.4 cm wide image. The
second data set was collected with the A-lines spaced at .25 mm to
give a 6.8 cm wide image. The second data set only gives two A-lines
per B-mode pixel (at 2 cm/division), which tends to limit resolution
improvement compared with the more finely spaced natural sponge data
set.

If we identify either of the two data sets at a given depth as
$I(x)$, and if we define the lateral beam profile for a given depth as
$P(x)$, then the problem is to design a filter, $F(x)$, which will yield
the deconvolved object, $O(x)$. In the spatial domain we have

$$I(x) = P(x) * O(x)$$

We wish to find $F(x)$ so that

$$O(x) = F(x) * I(x)$$

A simplistic frequency domain filter design would suggest that, if
$\hat{I}(f)$, $\hat{P}(f)$ and $\hat{O}(f)$ are the Fourier transforms of I, P and O then

$$\hat{I}(f) = \hat{P}(f)\ \hat{O}(f)$$

$$\text{and}\quad \hat{O}(f) = \hat{I}(f)/\hat{P}(f)$$

so we have $\hat{F}(f) = 1/\hat{P}(f)$ as our filter. Unfortunately, even without
ultrasound noise in the beam profile, we have singularities for $f > f_0$
in $1/\hat{P}(f)$. Figure 4 illustrates the frequency domain representation
of the simulated beam profile at a 4 cm distance, as well as a cosine
window which will control $1/\hat{P}(f)$ beyond f_0. The cosine window illustrated is a Blackman window (3) and is used throughout this study.
With the window, our filter now becomes

$$\hat{F}(f) = \hat{W}(f)/\hat{P}(f)$$

$$\text{and}\quad \hat{W}(f) = \hat{F}(f)\ \hat{P}(f)$$

so that in the spatial domain we have

$$W(x) = F(x) * P(x)$$

If we wish to test the accuracy of a deconvolution process, we
can convolve the filter with the point spread function. The difference between this result and a delta function gives us a measure of
the error in the deconvolution process. We have obtained $W(x)$,
rather than a delta function, so our process cannot be considered a
deconvolution. The $W(x)$ required by our filter design does give a
measure of the resolution improvement expected from the filter's
convolution with the image. Rather than computing the inverse transform of $\hat{W}(f)$ for this resolution measure, we compute $F(x) * P(x)$ so
that computational round off and truncation errors are included.
As example, the profile and window in Figure 4 produce the filter and
resolution measure of Figure 5. If we compare the resolution measure
with the 4 cm profile in Figure 1, we can see little improvement in
resolution for this filter design at the -3dB point.

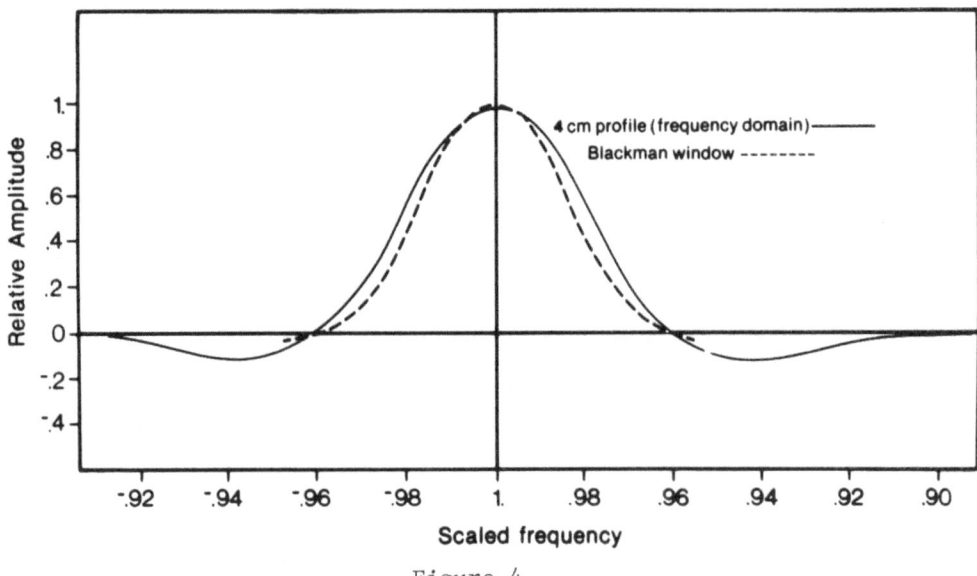

Figure 4

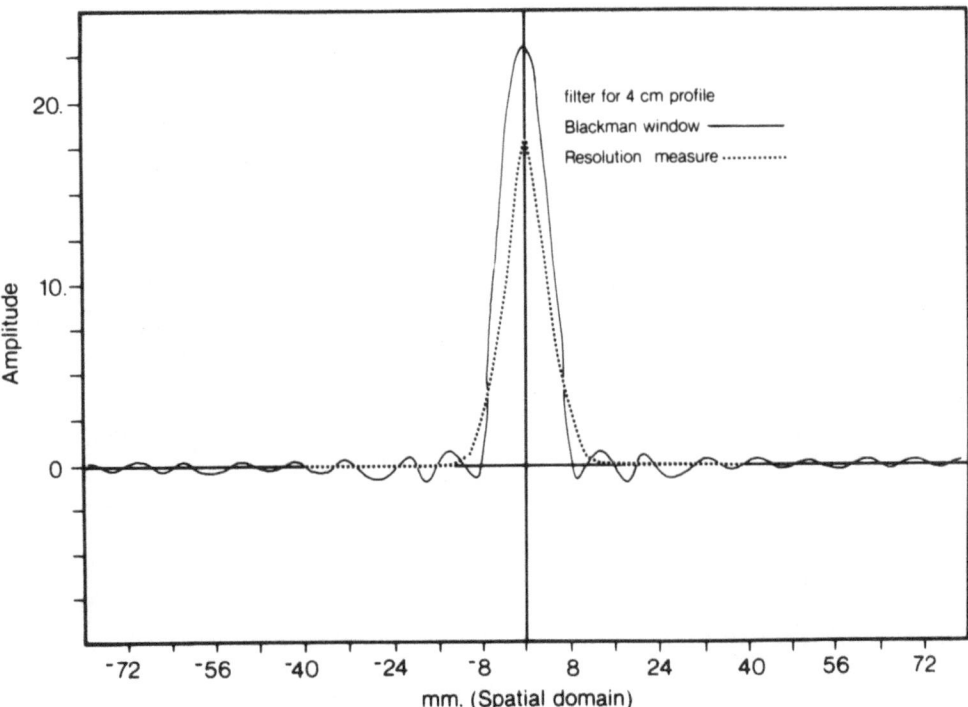

Figure 5.   1024 point filter designed with a Blackman window.

We repeat the filter design process for a simulated profile of
$\sigma=1.7$ and for a third far field profile, with straight sides, filling
out a Gaussian curve of $\sigma=2.25$ mm. We continue to use Blackman
windows of different sizes to control the filters. In addition to
these three filters, we computed filters from the noisy 4 and 7 cm
beam profiles in Figures 1 and 2. We found that these beam profiles
had a phase change which decreased the frequency cut off and window
size for the filter. This gave a wider resolution measure in the
spatial domain. We then convolved the two noisy beam profile filters
with the natural sponge data in the near (4 cm to 6 cm) and the far
field (6 cm to 12 cm). We also simulated the unfiltered B-mode
image of the natural sponge data. The unfiltered image is illustrated
in Figure 6, with the top of the image representing the near field.
The noisy beam profile filters convolved with the natural sponge
data taken in the two different regions is illustrated in Figure 7.

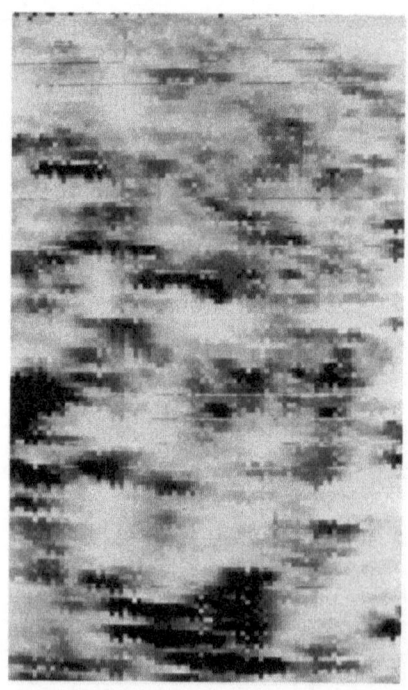
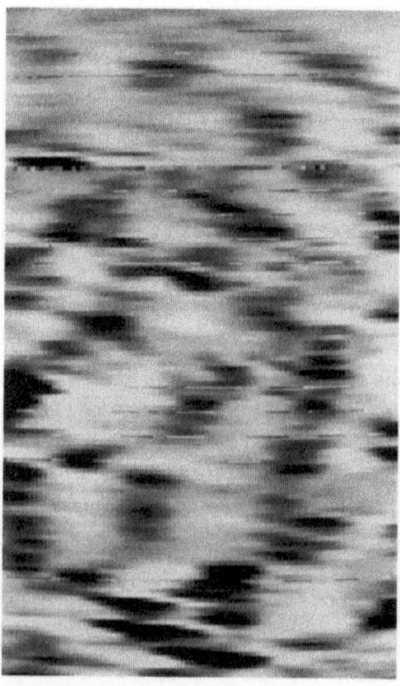

Figure 6. Unfiltered simulated   Figure 7. Same as Figure 6 with
B-mode ultrasonogram of a   filters designed from experimen-
natural sponge.   tally obtained beam profiles.

The blurred streaky upper third of this image substantiates the predictions of the window resolution measure. The remainder of the filtered image has lost texture information because of the high frequency cut-off caused by the window's cut-off in the frequency domain. In order to compare the real and simulated beam profile filters, we also applied the three filters derived from Gaussian curves to the natural sponge data. The near field, double topped curve (σ=2) produced the filter illustrated by Figure 5. This was convolved with the near field natural sponge data. The Gaussian curve of σ=1.7, producing a smaller filter than the real data, was convolved with the focal zone natural sponge data. The far field filter was convolved with the remaining 3 cm of sponge data. The resulting image, in Figure 8, scarcely differs from Figure 7.

These experiments seem to verify that this filter design method is unsatisfactory. If we can reduce the main lobe width of the inverse window function, we should increase the difference between the beam profile width and the resolution measure. Hamming (4) shows that the main lobe width can be reduced by half if a rectangular rather than a cosine window is used in filter design. However, the side lobe will increase in amplitude for such a window. The side lobe amplitudes can be reduced somewhat if we use a modified window

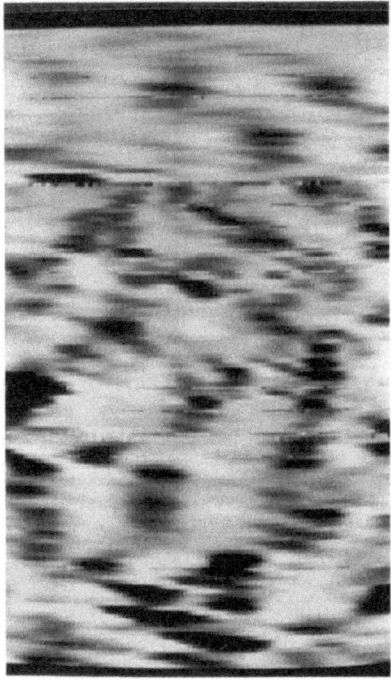

Figure 8.   Same as Fig. 6 with three filters designed
            from the simulated beam profiles.

such as:

$$W_R(f) = \begin{cases} 0 & f>f_o \\ \tfrac{1}{2} & f=f_o \\ 1 & f<f_o \end{cases}$$

If we design filters using the same f_o as we used for the Blackman
window cut-offs, we should reduce our resolution measure by one-half;
we hope the side lobes will not contribute significant distortion
to the filtered image. When we implemented filters with W_R and a
cut-off of f_o, we found that the filters did not coverge; that is,
they continued to oscilate for the full range of transform values.
By decreasing the frequency cut-off, we obtained convergent filters
with smaller resolution measures than we had obtained using Blackman
windows. The three filters that we designed correspond to the beam
profile simulations applied in Figure 8. The 4 cm beam profile
filter produced the filter and resolution measure illustrated in
Figure 9. If this is compared to Figure 5, we can see that the main
lobe in Figure 9 is not quite half that of the one in Figure 5.
Rather than apply these filters to the natural sponge data, we
prefer to apply it to the GSTO data, since this phantom contains a
similated circular cyst in the far field. Figure 10 illustrates a

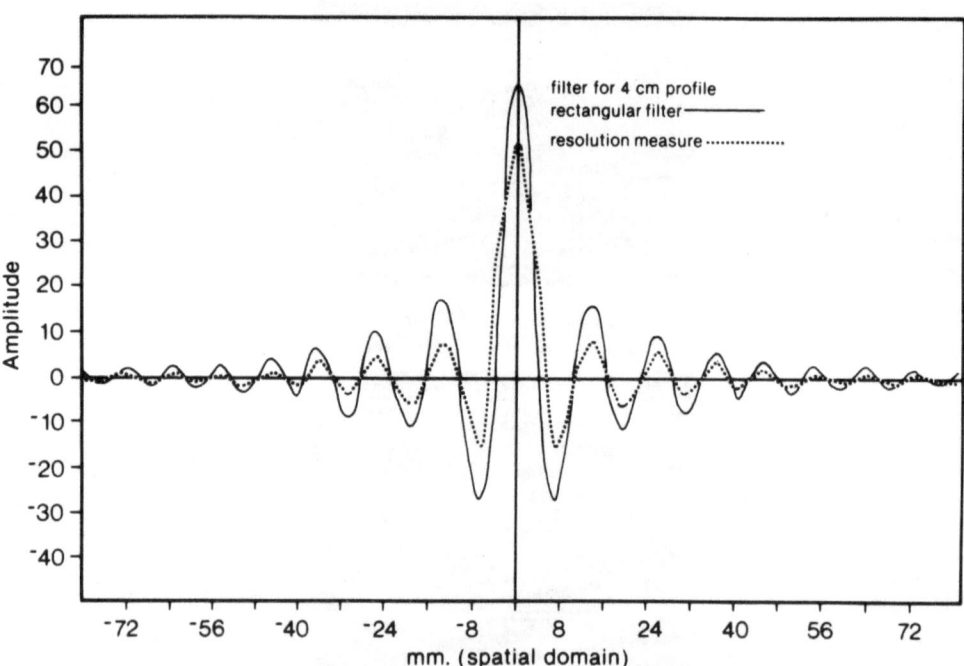

Figure 9.   1024 point filter designed with a modified
             rectangular window.

Figure 10. Unfiltered simulated B-mode ultra-sonogram of a gray scale phantom (GSTO).

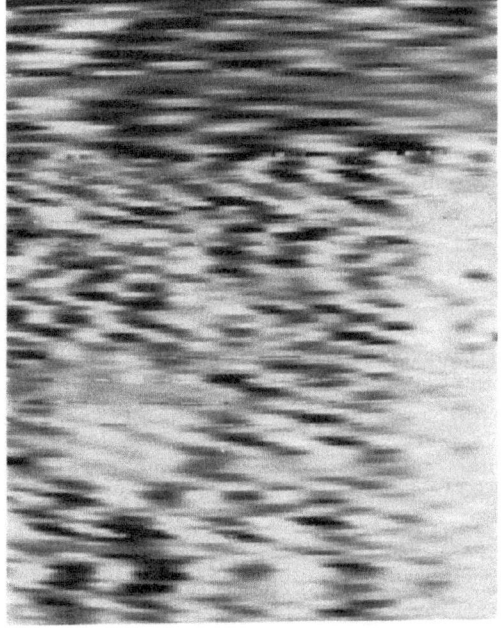

Figure 11. Same as Fig. 10 with three filters designed from simulated beam pro-files and modified rec-tangular windows.

B-mode simulation of the unfiltered GSTO. Figure 11 illustrates
the filtered GSTO. The side lobes contribute so much ringing that
the cyst in the phantom is almost obscured. For comparison, we
applied Blackman window filters, similar to those shown in Figure 8
to the GSTO (see Figure 12). Because the spacing between the
A-lines of the natural sponge data and those of the GSTO data are
different, we recomputed the three filters used in Figure 12 rather
than interpolate the filters in Figure 8. Round off error and
decreased frequency domain resolution contribute to the numerical
error for this type of filter design. The A-line spacing in the
GSTO also contributes to increased errors in the Fast Fourier Transform algorithm.

We used a Fast Fourier Transform of 1024 points, even though
the largest beam profile was only 90 points wide. Since our data
acquisition scheme had a dynamic range of less than 20 dB, we took
the -20 dB point in the frequency domain as the zero point of the
inverse profile. This caused the frequency domain filters to vary
between 40 and 120 points. The spatial domain filters usually con-
verged in 200 points, but were all truncated to 128 points before
being convolved with the image data. In addition to this truncation, the filters were biased so the total power was zero. After
the filter was convolved with the image data, we remapped the gray
levels so that the gray levels in the near field of the filtered
image matched the gray level distribution in the far field. Since
we also compare filtered and unfiltered images, the mean gray levels
and contrast level of the two images must be similar. The following
mapping fulfills these criterea:

Let $\mu_j^{(o)}$ be the jth section's mean gray level before filtering.

$\mu_j^{(f)}$ be the jth section's mean gray level after filtering.

$\sigma_j^{(o)}$ be the standard deviation of the unfiltered section's
gray levels.

$\sigma_j^{(f)}$ be the standard deviation of the filtered section's
gray levels.

Then if $g_{k,j}$ is the filtered gray-level value of the (k,j) pixel,
the new gray level value will be

$$g'_{k,j} = \mu_j^{(o)} + \frac{\sigma_j(o)}{\sigma_j(f)} [g_{k,j} - \mu_j^{(f)}]$$

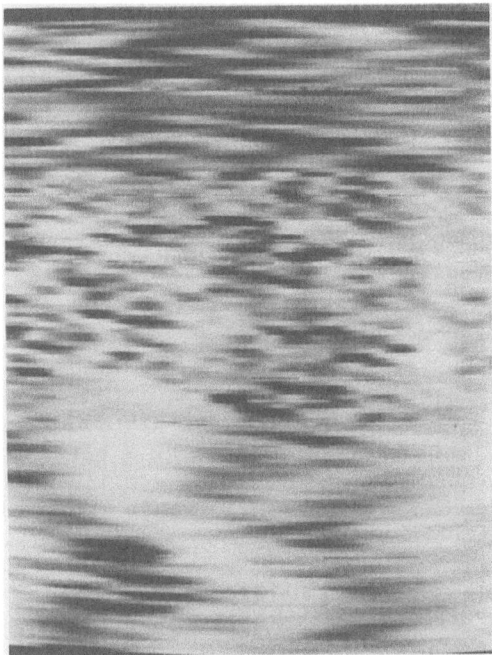

Figure 12. Same GSTO and
filters as Figure 11, but
with Blackman windows.

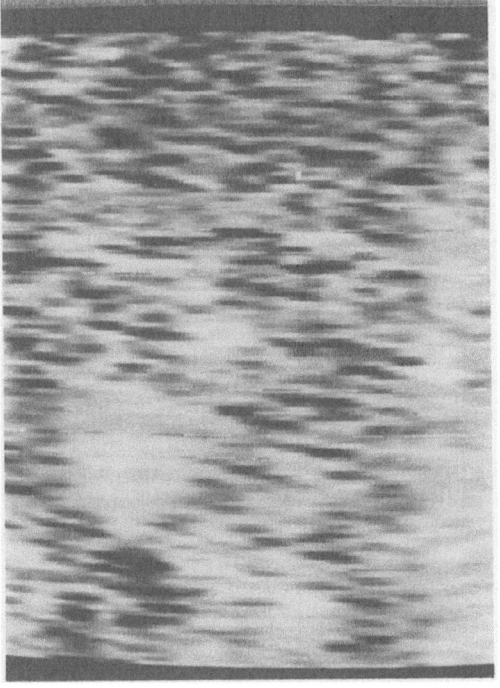

Figure 13. Same as Fig. 10
with 3 edge enhancement
filters.

This mapping was applied to all the filtered images. Because the
filtered images have an order of magnitude greater contrast (the
standard deviation ratio is typically below .09), we might be able
to improve the filtered images, and restore some of the texture information, by nonlinear gray scale mappings. However, this mapping
will not significantly improve the resolution.

Since filters based on the beam profile do not appreciably
improve resolution, we experimented with filters which are regarded
as edge enhancement filters in image restoration. They are used to
decrease the width of edges in the images. We observed that as the
beam profile size decreases, the frequency domain windows can be
larger; and the resulting filters converge rapidly. We therefore
constructed three filters for the three distance ranges of the image
data; the near field range had a μ of 1.7 mm; the focal distance
range had a $\mu=1.$ mm; and the far field was represented by a $\mu=1.3$ mm.
We designed filters for these values as though the beam profiles
were 10. mm, 6. mm, and 8. mm wide, respectively. The main lobe and
negative portion of the secondary lobe for each filter were 13. mm.,
9. mm, and 12. mm wide, respectively which was the size of the actual
beam profiles in these regions. The negative portion of each filter,
when convolved with the data, decreased the main lobe contribution,
giving some improvement in clarity over the previous filters. Figure
13 illustrates the application of these filters on the GSTO data.

These experiments lead to the following conclusions. Beam
profiles which are nearly Gaussian perform better with inverse filtering than profiles which cause phase reversals in the frequency
domain. Such phase reversals, characterized by negative amplitude,
cause abrupt discontinuities in the reciprocal of the frequency
domain beam profiles. The window cut off frequency which controls
these discontinuities is less than the cut off frequency required
by a Gaussian profile.

Any frequency cut off in the filter causes us to lose texture
information in the filtered image. Since this texture contains
diagnostic information, an unfiltered image should be available to
the physician.

Texture acts like noise in images subjected to inverse filters;
therefore, statistical filters, which filter the noise as well as
deconvolve the image, will cause a loss of texture or ringing in the
filtered image. They will also be even more expensive than inverse
filters to implement, since they require statistical approximations
of each view in the ultrasound survey of the abdomen.

Even inverse filters are expensive to implement, since they
require convolution in a direction perpendicular to the direction of
A-line storage. The amplified dynamic range of the filtered image
will also require more gray level resolution than currently available

on digital scan converters. A better solution to resolution improvement may lie with one-dimensional edge enhancement filters. When
these are applied to images which have been smoothed in the lateral
direction by a noise filter, they help delineate small interfaces in
abdominal scans, such as the common bile duct and pancreatic duct.
However, they must also be accompanied by the matching unfiltered
image.

Textured abdominal ultrasound scans are poor candidates for
lateral beam profile filtering. Small lesions, hematomas, pancrea-
titis, cirhosis and some abscesses present as texture changes, sometimes embedded in large textured areas and only identified by subtle
gray scale changes. Diagnosis of these effects would be impaired
by using filtered images.

REFERENCES

1. D.H. McSherry and J.R. Keller, "Ultrasonic Cardiac Imaging and
    Image Enhancement Techniques", 1974 Ultrasonics Symposium Pro-
    ceedings, IEEE Catalog #74 CHO 896-ISU, pages 5-11.
2. E.E. Hundt and E.A. Trautenberg, "Digital Processing of Ultrasonic
    Data by Deconvolution", IEEE Transactions on Sonics and Ultrasonics,
    Vol. SU-27, No. 5, September 1980, pages 249-252.
3. A.V. Oppenheim and R.W. Schafer, "Digital Signal Processing",
    Prentice Hall, Inc., Englewood Cliffs, NJ, 1975, pages 241-243.
4. R.W. Hamming, "Digital Filters", Prentice Hall, Inc., Englewood
    Cliffs, NJ, 1977, Chapter 9.
5. J.P. Jones and C. Kimme-Smith, "An Analysis of the Parameters
    Affecting Texture in a B-mode Ultrasonogram", in Acoustical
    Imaging, Vol. 10, P. Alais (Ed.), Plenum Press, in press.
6. J.P. Jones and R. Kovack, "A Computerized Data Analysis System
    for Ultrasonic Tissue Characterization", in Acoustical Imaging,
    Vol. 9, K. Wang (Ed.), Plenum Press, 1980.

DIGITAL FILTERING OF ACOUSTIC IMAGES

B. Burgoyne and J. Pavkovich
Varian Associates 611 Hansen Way,
Palo Alto, Ca. 94303

G. S. Kino
Ginzton Laboratory, Stanford University
Stanford, Ca. 94305

INTRODUCTION

We describe in this paper the use of various types of filtering techniques to improve the resolution of a Varian V3000 real-time ultrasonic imaging system. This system, which operates at 2.25 MHz center frequency, employs a 32-element imaging array 1.28 cm wide. Thirty-two elements are used on transmit and the central sixteen elements on receive. After envelope detection, the image intensity obtained is stored in a solid state RAM and is available for digital processing, as shown in Fig. 1.

The system employs a cylindrical lens focused at approximately 8 cm range in the center of the field. In this region the transverse definition is approximately 3.2 mm and the range definition due to the use of a short pulse is 1.5 mm. The employment of time delays on both transmit and receive generates a radial sector scan, as illustrated in Fig. 2. It will be noted that in this radial sector scan scheme, the spacing between the radial sector lines is in equal increments of $x = z \tan \theta$ where z is the depth rather than the angle θ. A typical liver scan from this system is shown in Fig. 3.

It has been our aim in this work to make use of various types of filtering techniques to improve the quality of the image; in particular its transverse definition at all ranges. Each line of the image is sampled in 128 points. Hence the sampling interval in the transverse direction is approximately 1 mm at a range of 8 cm and resolution may therefore be considerably improved. We did not feel that it would be necessary to improve

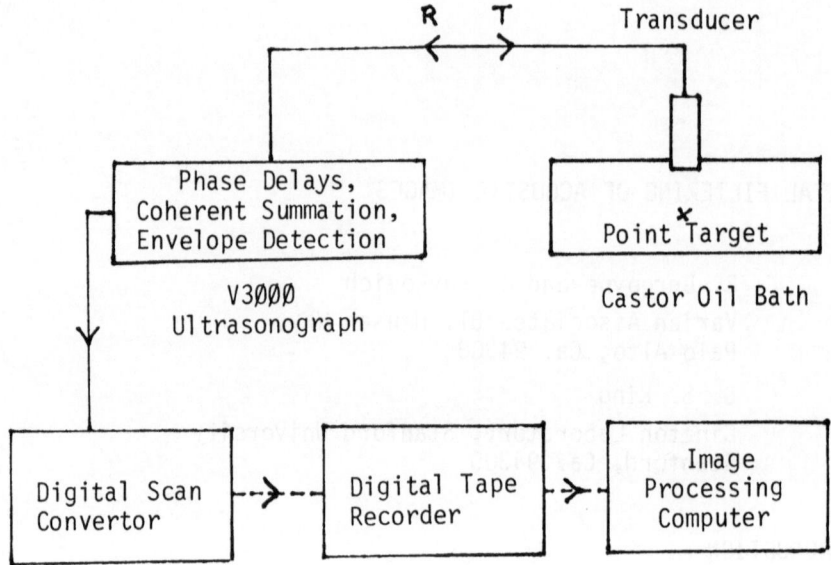

Fig. 1.   Image Reconstruction Flow Chart.

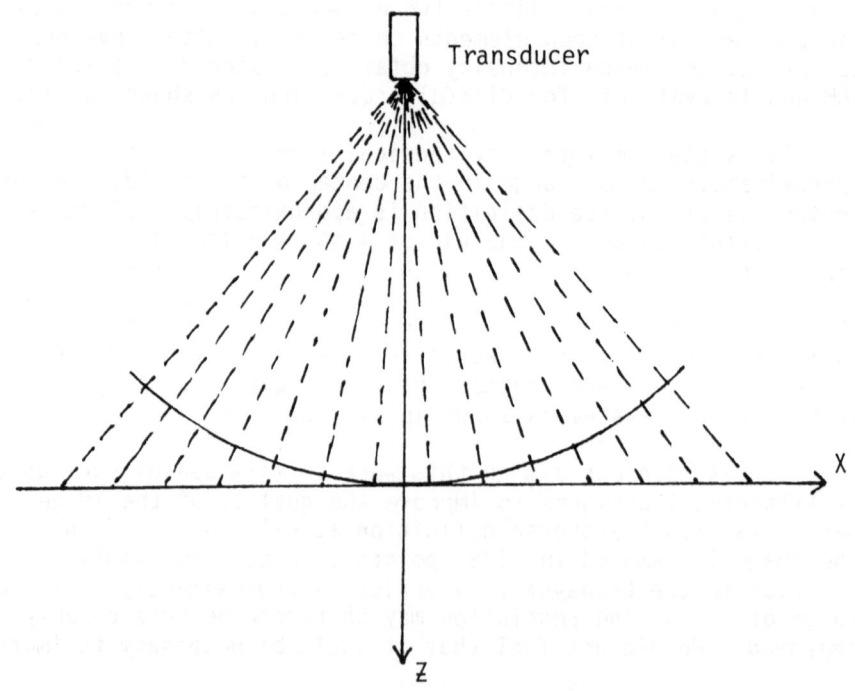

Fig. 2.   Sector scan display.

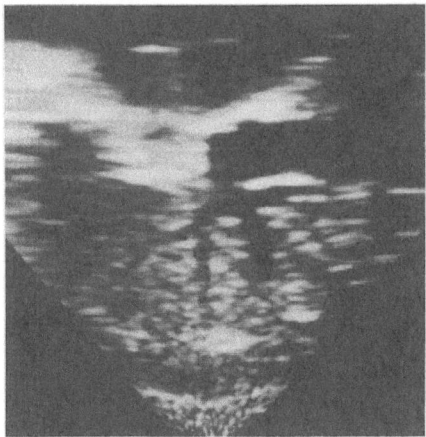

Fig. 3.   Detected Liver image

the range definition, which is adequate at the present time.   The
principal techniques which we have employed are pseudo-inverse
filtering and a modified form of the Jansson van Cittert iterative
algorithm.

The basic idea was then to make use of the theoretical
transverse line response of the detected image at the center of
the field, or alternatively, an experimental line response
obtained from a wire target in constructing a filter to
individually process each line of the detected image. Our initial
aim has been to use just one theoretical transverse impulse
response in processing all lines of the image. The resolution
enhancement has been carried out by convolution in the real domain
with a view to real-time hardware implementation.

FILTERING TECHNIQUES EMPLOYED

As we are concerned only with improving resolution in the
transverse direction, we have used a rectangular format and
changed the software in the processing system for this purpose.
The horizontal lines of our images now no longer correspond to
constant depth but represent data at a constant range. Fig. 4
shows a plot of the theoretical impulse response at a range of
8 cm and, as noted previously, the abscissa is divided into equal
increments of $\tan \theta$.

The detected image of the Varian V3000 ultrasonic phased
array can be modeled as a linear convolution of the transducer's
lateral impulse response $t(x)$ with an object $o(x)$.[1] As may be

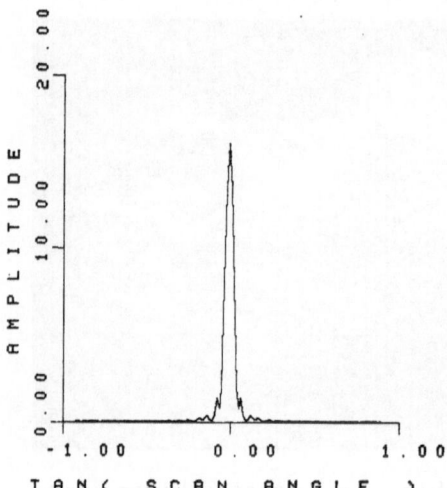

Fig. 4.   Theoretical impulse response.

expected, our experimental data shows that the lateral impulse
response does not exactly conform with theoretical predictions due
in part to the presence of a focusing lens.   There are also range
and angular variations of the lateral impulse response which
degrade the images.   In addition, there is a strong component of
"multiplicative" noise which is not just simple sampling noise.
This is the dominant noise source in the system.   The source of
this noise has not yet been fully determined.

      The detected image spectrum $I(s)$ is related to the object
spectrum $O(s)$, the imaging transfer function $T(s)$, and the
"multiplicative" noise spectrum $N(s)$ by

$$I(s) = [O(s)T(s)] * [\delta(s) + N(s)] \tag{1}$$

where s is the spatial frequency.

LINEAR FILTERS

      A Wiener filter is the optimum linear filter for restoration
of noisy blurred images.   This filter has the characteristic

$$H_w(s) = \frac{T^*(s)\phi_0(s)}{|T(s)|^2\phi_0(s) + [|T(s)|^2\phi_0(s)] * \phi_N(s)} \tag{2}$$

where $\phi_N(s) = \langle|N(s)|^2\rangle$ and $\phi_0(s) = \langle|0(s)|^2\rangle$.

In practice, knowledge of the noise statistics, which are nonstationary is incomplete. For this reason a pseudo-inverse filter has been substituted. It is found in this application that the result is a mean squared error considerably greater than the minimum mean squared error, but with less sensitivity to error in the assumed image transfer function $T(s)$.

The pseudo-inverse filter we choose has a characteristic spectrum given by the relation

$$H_p(s) = \frac{T^*(s)W(s)}{|T(s)|^2 + A} \qquad (3)$$

where A is a constant which has typically been chosen to be between 0.004$T(0)$ and 0.001$T(0)$ and it has been assumed that $T(0)$ is the maximum value of $T(s)$. The parameter $W(s)$ is an apodization function which was chosen to be a Hanning weighting with the sampling frequency as the upper frequency.

The filter is first generated in the spatial frequency domain, preferably using the analytical theoretical impulse response, and then transformed to give a deconvolution kernel. To allow for a simple $1/\cos\theta$ dependence of the beam width, the kernel is interpolated and resampled at intervals of 10^0 and used in filtering the corresponding angular sector of the image.

It will be observed that the spectrum of the reconstructed object $0_r(s)$, with the use of the pseudo-inverse filter, is

$$0_r(s) = \frac{T^*(s)W(s)}{|T(s)|^2 + A} \{0(s)T(s) + [0(s)T(s)] * N(s)\} \qquad (4)$$

When $|T(s)|^2 \gg A$ and $\delta(s) \gg |N(s)|$, it follows that $0_r(s)$ tends to $0(s)$ with good resolution improvement. Where $A \gg |T(s)|^2$, the filter approximates a matched filter giving the object estimate the optimal signal-to-noise ratio.

Figure 5 demonstrates the result of the pseudo-inverse filter upon the theoretical impulse response with $A = 0.002T(0)$.

One of the problems with linear filters is the introduction of "ringing" artifacts caused by reconstructing within a finite band limit. This implies poor image reconstruction and may also give rise to negative outputs when restoring a detected image, a condition which is obviously nonphysical.

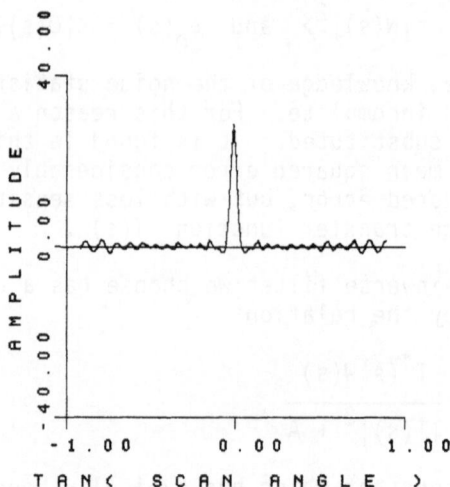

Fig. 5. Theoretical impulse response filtered by pseudo-inverse
         filter. A = 0.002T(0).

We have investigated alternative techniques based on the
iterative approach of van Cittert.[2] Van Cittert's algorithm uses
a least squared criterion for convergence to the principal
solution, a result equivalent to using a linear inverse filter.
However, incorporating a nonlinear point constraint, similar to
that of Jansson,[2] which operates on the error term, gives
considerable improvement in the quality of the image.

CONSTRAINED ITERATIVE ALGORITHM

The modified algorithm, which we shall call the constrained
iterative algorithm, is described by the following relations.

The response of the simulated imaging system at the k'th
iteration is

$$i_k(x) = o_k(x) * t(x) \tag{5}$$

A modified error function $e_k(x)$ is generated from the
difference between the detected image and the kth simulation of
the detected image

$$e_k(x) = [i(x) - bi_k(x)] \tag{6}$$

where b is a "weighting" factor.

In the simple van Cittert algorithm, the error and object estimate terms are summed giving an updated object estimate. We have employed the additional step of backprojecting the error term by convolution with t(-x) , followed by constrained summation with the object estimate. This procedure is closely related to the algebraic reconstruction technique (ART) in the field of tomography.[3] With a constant constraining function, this would also be equivalent to the steepest descent technique of the LMS method used in quadratic optimization.[4] As discussed below, the inclusion of the backprojection operation together with proper normalization ensures convergence of the algorithm. Thus we write

$$o_{k+1}(x) \; = \; d\{o_k(x) + r_k(x)[e_k(x) * t(-x)]\} \tag{7}$$

where d is a scaling factor and $r_k(x)$ is a constraining function.

The scaling factor, d , multiplies the updated object estimate so that the average value of the subject estimate stays constant for all iterations, i.e., $\overline{o_{k+1}(x)} = \overline{o_k(x)}$, where

$$\overline{o_k(x)} \; = \; \frac{1}{N} \sum_{n=1}^{N} o_k(x_n) \tag{8}$$

where N is the number of sampling points.

The constraint is chosen so that the output is in agreement with a priori information. It is reasonable to constrain the restored object of a detected image to be positive. Also, since we expect resolution improvements on the order of a factor of two, the restored object should be constrained to have less than twice the amplitude of the detected image at each sample point. This process tends to limit noise build-up, as well as eliminate "ringing" artifacts.

The constraining function, $r_k(x)$, used is triangular in shape, having a maximum value wherever $o_k(x)$ has a value equal to i(x) and decreasing linearly to zero at values of $o_k(x)$ equal to 0 and 2i(x) . Thus, the object estimate may be constrained to lie between twice the detected value and zero.

We can write the constraining function explicitly in terms of the detected image and restored object in the form

$$r_k(x) = c_k \frac{\{1.0 - |o_k(x) - i(x)|\}}{i(x)} \qquad (9)$$

Should $r_k(x)$ become less than zero, it is set equal to zero. c_k is maximized in Eq. (7), independently at each iteration and for each line of image data subject to the conditions that the updated object estimate must be positive and have a maximum value less than or equal to twice that of the detected image at any point.

Next, k is incremented and the entire procedure is repeated with a new simulation of the detected image.

We note that initially an estimate of the object must be chosen. In our case, the detected image is used for the first try.

It is not possible to solve this nonlinear iterative algorithm analytically. However, a great deal of insight into its behavior may be obtained by solving a closely related linear algorithm. To do this, we assume that c_k is a constant c for each iteration, that $r_k(x)$ may be replaced by c for all x, and that $d = 1$. The iterative algorithm can be written in the spatial frequency domain in the form

$$O_{k+1}(s) = O_k(s) + c[I(s) - bT(s)O_k(s)]T^*(s) \qquad (10)$$

This has the complementary solution

$$O_k(s) = A[1 - bc|T(s)|^2]^k \qquad (11)$$

and particular solution $O_p(s)$ given by the relation

$$O_p(s) = \frac{I(s)}{bT(s)} \qquad (12)$$

Thus, the complete solution is

$$O_k(s) = \frac{I(s)}{bT(s)} + A[1 - bc|T(s)|^2]^k \qquad (13)$$

The initial condition $O(s) = I(s)$ when $k = 0$ allows us to solve for A giving the result

$$A = I(s)\left[1 - \frac{1}{bT(s)}\right] \tag{14}$$

Hence, we have

$$O_k(s) = \frac{I(s)}{bT(s)} + I(s)\frac{[bT(s) - 1][1 - bc|T(s)|^2]^k}{bT(s)} \tag{15}$$

By appropriate scaling of $T(s)$, e.g., $T(0) = 1.0$, it follows that if

$$[1 - bc|T(s)|^2] < 1 \tag{16}$$

the algorithm converges to the solution

$$O_\infty(s) = \frac{I(s)}{bT(s)} \tag{17}$$

We note that the convergence condition, Eq. (16), may be satisfied by the use of a positive factor $|T(s)|^2$ arising through the use of the backprojection operation and is valid even for $T(s)$ complex.

Figure 6 shows the result after 40 iterations of the constrained iterative algorithm in filtering the theoretical impulse response. A "weighting" factor of unity has been chosen. The cost in decreasing the sidelobe levels and suppressing "ringing" artifacts has been poorer resolution improvement than shown with a pseudo-inverse filter.

We have found that introduction of a "weighting" factor b greater than unity, causes an increasing derivative component of the image to appear in successive object estimates. This is useful in speeding convergence while the problem of noise amplification, which one usually associates with differentiation, is controlled by the constraint.

The effect of the "weighting" factor may be seen by comparisons of the residual error after k iterations. The residual error at the k'th iteration is given by the relation

$$E(s) = \frac{I(s)[bT(s) - 1][1 - bc|T(s)|^2]^k}{bT(s)} \tag{18}$$

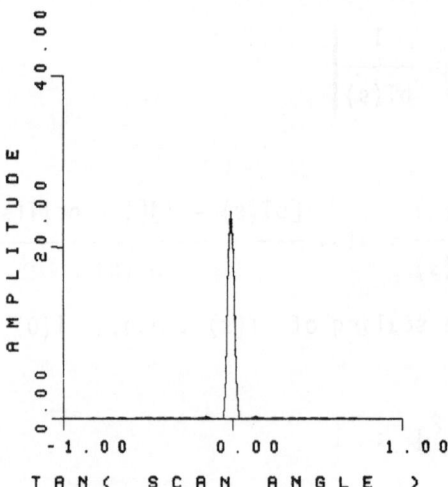

Fig. 6. Theoretical impulse response filtered by constrained
         iterative algorithm. 40 iterations, "weighting factor"
         is 1.

     The factor bc is determined solely by the deconvolving
impulse response and the detected data. In other words, an
increase in b causes a proportional decrease in c . Hence, the
difference in the error term when choosing a "weighting" factor
of b and a "weighting" factor of unity is determined by the
factor

$$M(s) \; = \; \frac{I(s) \; [bT(s) \; - \; 1]}{bT(s)} \tag{19}$$

     For a "weighting" factor equal to unity, this error term is
greatly dependent on the spatial frequency. In moving to higher
spatial frequencies, the transfer function has a general tendency
to decrease in value and so the error increases. On the other
hand, with very large "weighting" factors, the error is frequency
independent over the range $|bT(s)| \; \gg 1$. Thus, the error in
the object estimate at high spatial frequencies is less when using
large "weighting" factors.

     Therefore, convergence is rapid. However, as we increase the
high-frequency response of the system, there will be a general
tendency toward differéntiation.

     Figure 7 demonstrates the effectiveness of increasing b .
The resolution is greatly improved over the previous result, in
40 iterations, using a "weighting" factor of 100 . Table 1
compares the performances of the filters upon a theoretical
impulse.

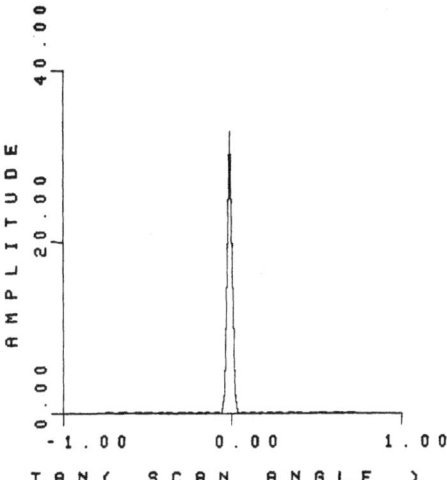

Fig. 7. Theoretical impulse response filtered by constrained
iterative algorithm. 40 iterations, "weighting factor"
is 100

TABLE 1

Comparison of Filtering Results from a Theoretical Impulse

	3 dB Width Filtered Original (%)	Principal Lobe Level Maximum Sidelobe Level	Normalized Mean Squared Error (%)
Pseudo-inverse A = 0.002T(0)	71.6	20.28	54.0
Constrained Iterative Algorithm iter = 40, b = 1	77.8	110	56.6
Constrained Iterative Algorithm iter = 40, b = 100	43.4	635	30.14

Not only has the "ringing" artifact in the object estimate
been eliminated using the constrained iterative algorithm, but
also the sidelobes of the original estimate have been decreased
remarkably.

The squared error has been found by integrating, over the real domain, the squared error between the restored object and the actual object which is an impulse. The normalized mean squared error is then obtained by dividing the squared error by the integral of the square of the restored object response.

Next we investigate the relative performances of the filters upon experimental data with noise present. An experimental "noisy" impulse response was measured, Fig. 8, and many such impulses, taken exactly from the same position, averaged to give a "noiseless" impulse, Fig. 9. The width of these impulses are in close agreement with theory although the sidelobes are notably absent.

A Wiener filter can be determined from the "noisy" and "noiseless" impulse response and is useful in giving the optimum solution for purposes of comparison. Figure 10 shows the results of Wiener filtering the "noisy" impulse response.

The constrained iterative algorithm, gives resolution comparable with the Wiener filter in eight iterations with a "weighting" factor of 100, Fig. 11. Table 2 compares the results of the two filters.

For comparable resolution, the SNR is 15% higher and the MSE is 25% lower using the constrained iterative algorithm.

The definition of the "noisy" impulse response has been increased in only eight iterations, much more than has the definition of the theoretical impulse response in forty iterations. The sidelobes of the theoretical impulse give rise to regions with amplitudes very close to zero and this limits the size of the parameter c_k to comparatively small values. Thus, the absolute rate of convergence of the algorithm is dependent on the characteristics of the detected data being filtered.

TABLE 2
Comparison of Filtering Results From a "Noisy" Impulse

	3 dB Width Filtered Original (%)	SNR	Normalized Mean Squared Error (%)
Wiener Filter	22.8	2.9	43.1
Constrained Iterative Algorithm iter = 8, b = 100	24.0	3.4	31.0

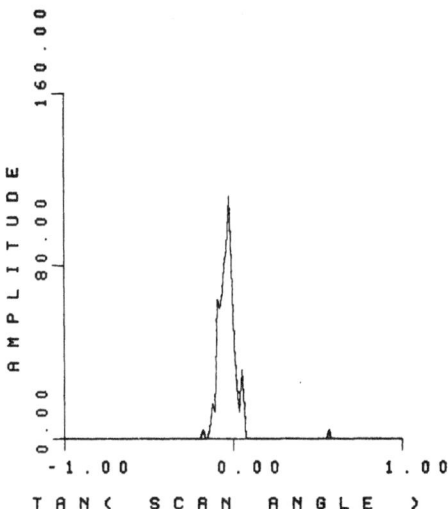

Fig. 8. "Noisy" experimental impulse response.

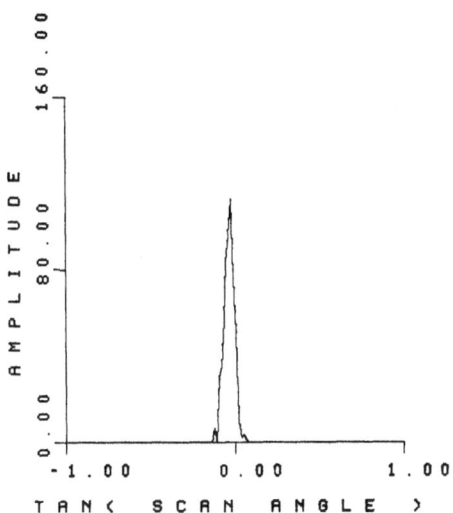

Fig. 9. "Noiseless" experimental impulse response.

The result of pseudo-inverse filtering is shown in Fig. 12. Again deconvolution is with the "noiseless" experimental impulse response using a value of A = 0.002T(0). It is to be noted that the pseudo-inverse filter gives poorer resolution than the Wiener filter but also less "ringing" artifact.

Resolution comparable with the pseudo-inverse filter is achieved in three iterations by the constrained iterative algorithm (Fig. 13). Table 3 compares the results of these two filters.

The constrained iterative algorithm gives a SNR which is 25% greater than obtained by the pseudo-inverse filter.

Table 3

Comparison of Filtering Results From a "Noisy" Impulse

	3 dB Width Filtered / Original (%)	SNR	Normalized Mean Squared Error (%)
Pseudo-Inverse A = 0.002	59.3	6.24	64.1
Constrained Iterative Algorithm iter = 3, b = 100	54.5	7.81	64.7

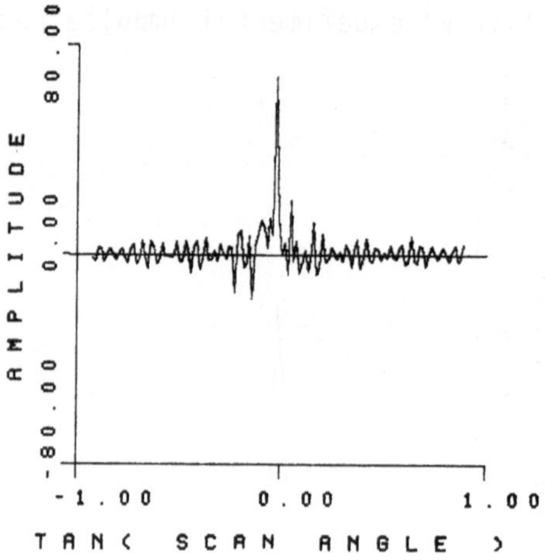

Fig. 10. "Noisy" experimental impulse response filtered by Wiener filter.

A linear filter will multiply both the noise and the signal spectra in the image by the same factor. As a consequence, the SNR of the restored object from a linear filter other than the matched filter will necessarily be lower than the original image SNR.

The constrained iterative algorithm is able to some degree to differentiate between the noise and signal spectra. This is apparent since the restored object spectrum from the constrained iterative algorithm appears smooth, a property associated with the spectrum of a physical object. The noise spectrum, on the other hand, is likely to be uncorrelated and its amplification results

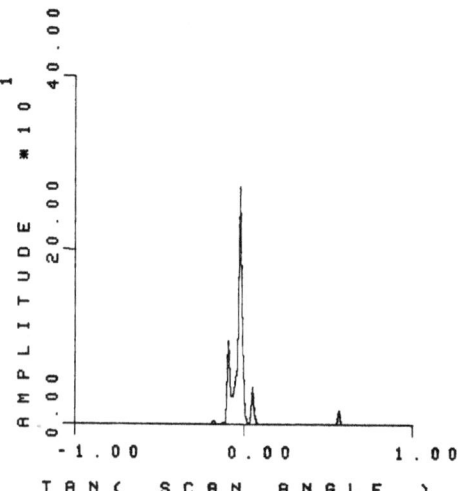

Fig. 11.   "Noisy" experimental impulse response filtered by
           constrained iterative algorithm. 8 iterations,
           "weighting" factor is 100, using "noiseless" impulse for
           deconvolution.

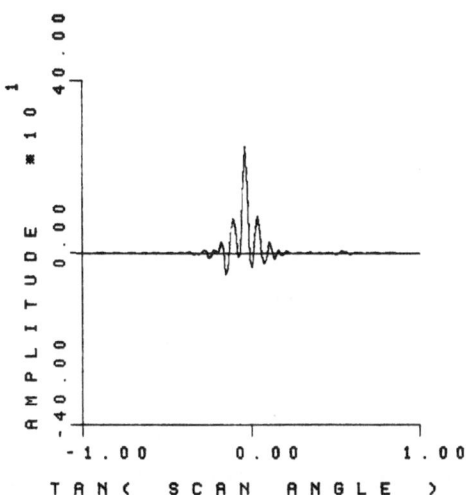

Fig. 12.   "Noisy" experimental impulse response filtered by
           pseudo-inverse filter. A = 0.002T(0), using "noiseless"
           impulse for deconvolution.

in a noisy appearing restored object spectrum. Such is the case
with Wiener and pseudo-inverse filtering. Figure 14 shows the
spectrum of the restored "noisy" impulse using the Wiener filter
and Fig. 15 shows the corresponding spectrum using the constrained
iterative algorithm.

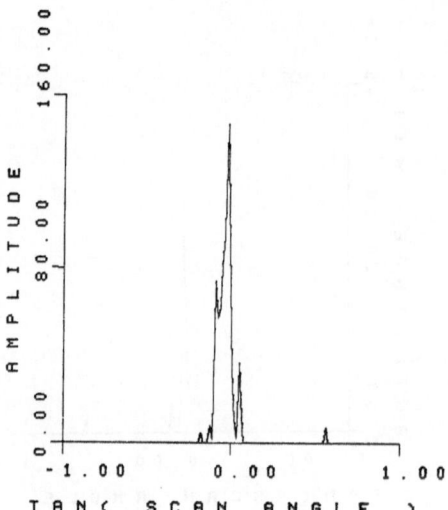

Fig. 13. "Noisy" experimental impulse response filtered by
constrained iterative algorithm. 3 iterations,
"weighting" factor is 100.

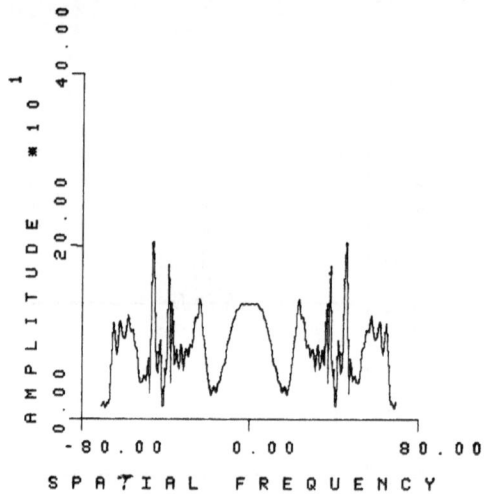

Fig. 14. Spectrum of Wiener filtered "noisy" experimental impulse
response.

Also of importance is the sensitivity of the filters to
errors in the impulse response used in deconvolution. Using the
theoretical impulse response for deconvolution gave the result
shown in Fig. 16 for pseudo-inverse filtering. The change from
using the "noiseless" impulse response is considerable. There has
been a relative change of 100% in the distribution of restored
object energy. The constrained iterative algorithm on the other
hand has a relative change of 3.5% in energy distribution after

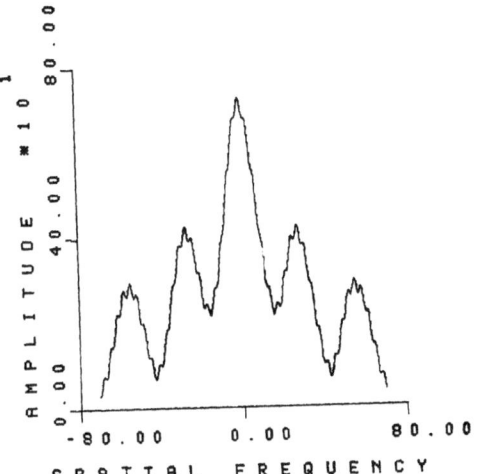

Fig. 15. Spectrum of restored "Noisy" experimental impulse
         response after 8 iterations of the constrained iterative
         algorithm. "Weighting" factor is 100.

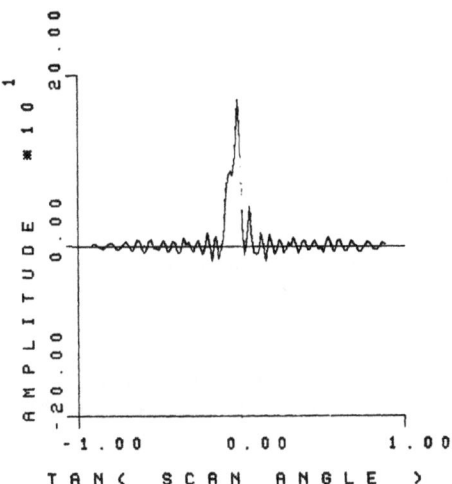

Fig. 16. "Noisy" experimental impulse response filtered by
         pseudo-inverse filter. A = 0.002T(0), using theoretical
         impulse for deconvolution.

eight iterations (see Fig. 17), and is extremely robust against
errors in the assumed lateral impulse response.

    When using nonlinear filters, we should also consider the
effects on cross-product terms which arise from multiple

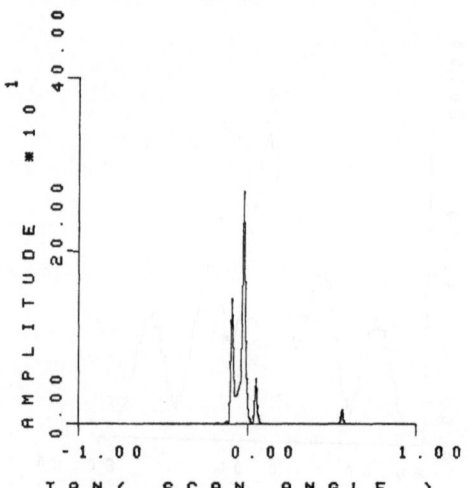

Fig. 17. "Noisy" experimental impulse response filtered by
        constrained iterative algorithm. 8 iterations,
        "weighting" factor is 100, using theoretical impulse for
        deconvolution.

targets. Figure 18 shows an original image taken of three pins at
a range of 8 cm , each with a separation of 6 mm , slightly
greater than the lateral 3 dB resolution. Figure 19 shows the
result of iterative filtering which has resolved the pins while
approximately retaining their relative amplitudes. The amplitude
of the central pin has decreased slightly as a consequence of
differentiation. To investigate the effect of the filter on the
integrity of edges, a block of metal has been imaged and is shown
in Fig. 20. The filtered result shown in Fig. 21 has remained
unaltered except for edge-enhancement again due to the derivative
action of the filter.

    It therefore appears that the constrained iterative algorithm
behaves in an acceptably linear fashion, as we shall see in its
application to the more complex liver images.

    The center of the grey level range of the CRT display for
each liver image has been chosen to coincide with the median grey
level of that image. The range of grey levels displayed by the
CRT is then adjusted to include the range of pixel values in the
image. Thus, each liver image, having a different range and
distribution of pixel grey levels, is optimally scaled over the
dynamic range of the CRT.

    The results of applying the constrained iterative algorithm
for 4 and 6 iterations to the liver image of Fig. 3 are shown
in Figs. 22 and 23, respectively.

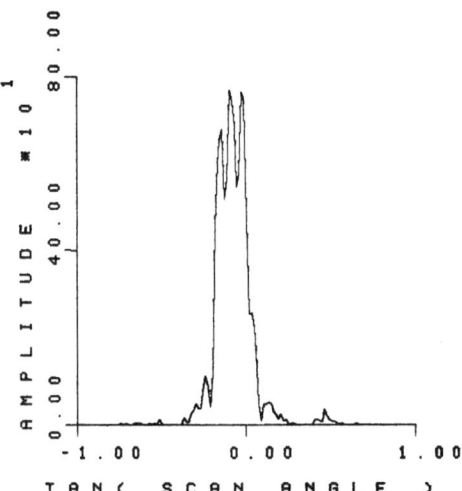

Fig. 18. Detected image of three pins with 6 mm separations.

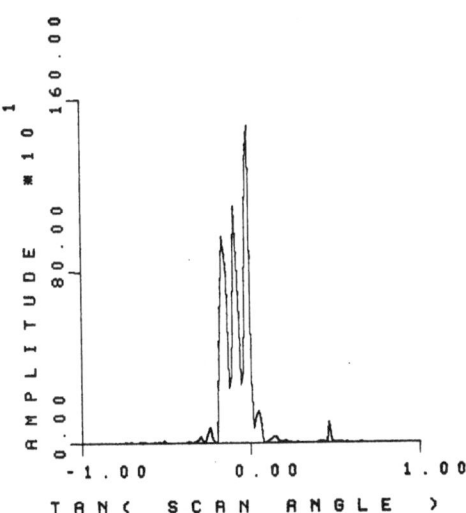

Fig. 19. Three pins filtered by constrained iterative algorithm.
16 iterations, "weighting" factor is 100.

There is increased definition with a high percentage of the
original structural detail appearing to be retained.
Low-frequency information, however, is de-emphasized with
increasing iterations due to the action of differentiation.

One-dimensional lateral plots of the liver data at a range
of 4 cm show this trend. The original liver is shown in Fig. 24
and the filtered liver after 4 and 6 iterations of the

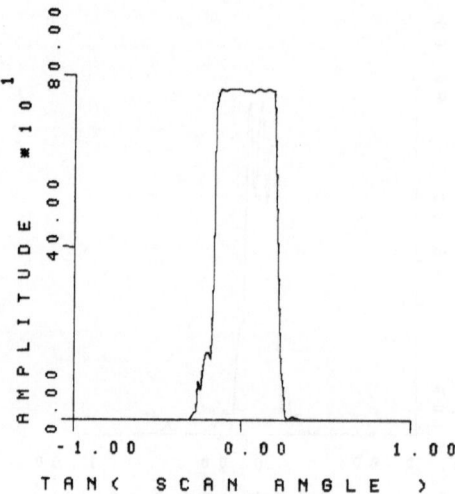

Fig. 20. Detected image of a block of metal.

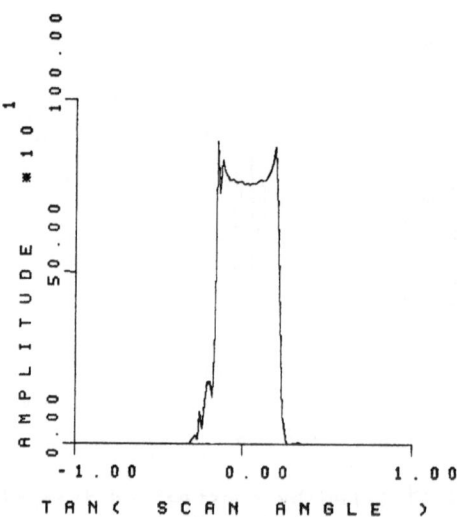

Fig. 21. Block of metal filtered by constrained iterative
         algorithm. 8 iterations, "weighting" factor is 100.

constrained iterative algorithm is shown in Fig. 25 and Fig. 26.
By six iterations, the definition appears to have increased by
approximately a factor of 3 . The low-frequency pedestal
supporting the structural detail of the original is now
considerably de-emphasized due to differentiation.

Fig. 22. Detected liver image filtered by constrained iterative
        algorithm. 4 iterations, "weighting" factor is 100.

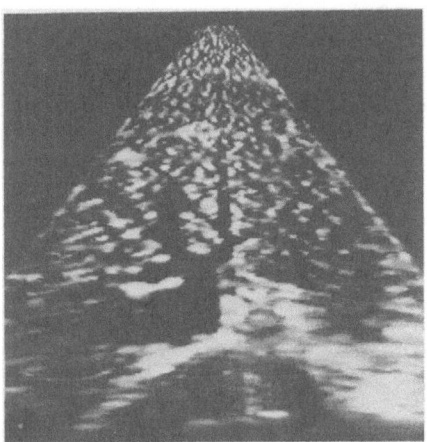

Fig. 23. Detected liver image filtered by constrained iterative
        algorithm. 6 iterations, "weighting" factor is 100.

      For comparison, Figs. 27 and 28 show the results of
pseudo-inverse filtering, with A = 0.002T(0) on the whole liver
and the one-dimensional lateral plot, respectively. There appears
to be little improvement in image quality.

      The original and filtered liver images have been magnified by
a factor of 4 to allow closer inspection of the liver image
detail. A region around the portal vessels has been chosen for
magnification and the histogram of the pixels within the
superimposed square is calculated. The detail of the liver images

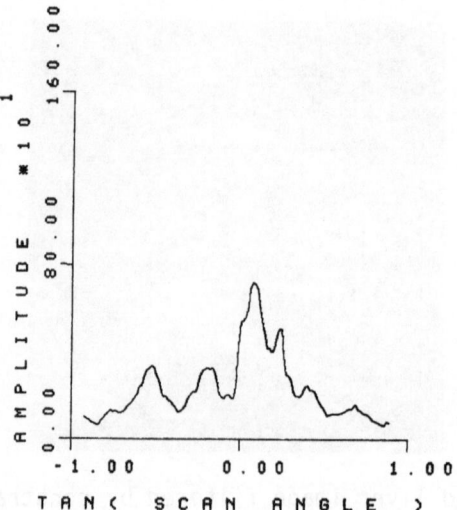

Fig. 24. Lateral plot of detected liver image, range 4 cm.

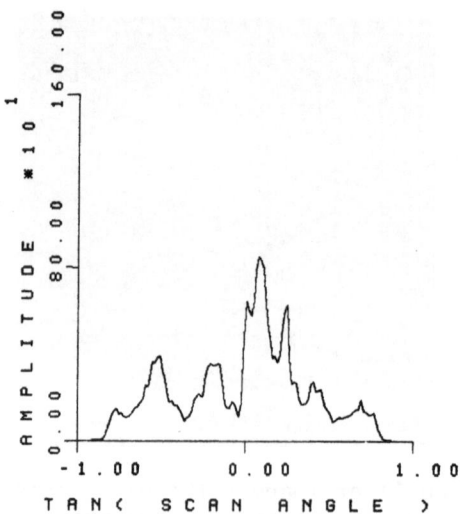

Fig. 25. Lateral plot of detected liver image filtered by
constrained iterative algorithm. 4 iterations,
"weighting" factor is 100.

before and after filtering have a close one-to-one
correspondence. Figure 29 is the original image while Figs. 30
and 31 are the filtered images after 4 and 6 iterations,
respectively.

The contrast of the image within the superimposed square is
equal to the ratio of the pixel variance to the pixel mean. There

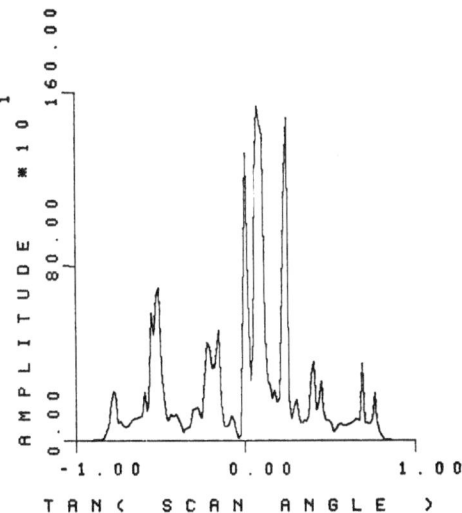

Fig. 26. Lateral plot of detected liver image filtered byconstrained iterative algorithm. 6 iterations, "weighting" factor is 100.

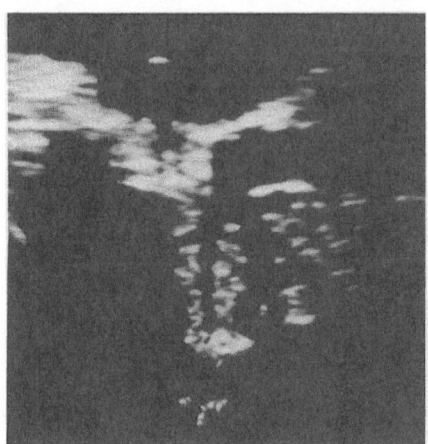

Fig. 27. Detected liver image filtered by pseudo-inverse filter. A = 0.002T(0).

is good correlation between the expansion of the grey scale evident from the histograms, and the increase in the image contrast.

The increase in image contrast is a direct result of increasing the definition of the liver detail. After 4 iterations the contrast has been increased by 50% and after 6

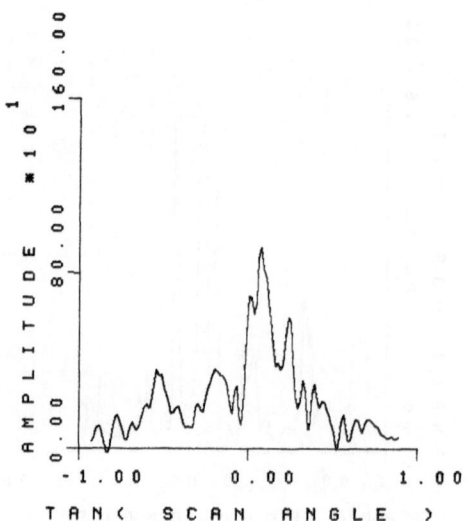

Fig. 28. Lateral plot of detected liver image filtered by
         pseudo-inverse filter. A = 0.002T(0).

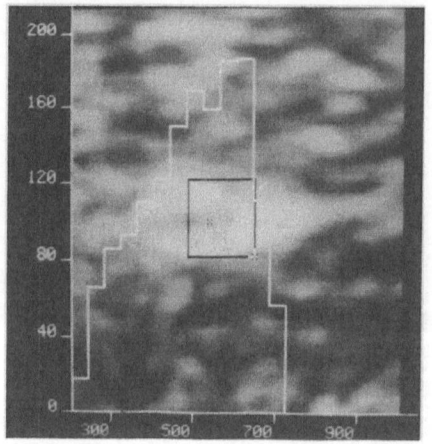

Fig. 29. Detected liver image, magnification 4.

iterations, it is increased by 200% . Also, with increasing
iterations, the histograms approach an exponential, a result also
associated with increased definition.

    For comparison, Fig. 32 shows the result of pseudo-inverse
filtering with A = 0.002T(0) . There is a small increase in
contrast of about 20% but with degradation of the image
quality.

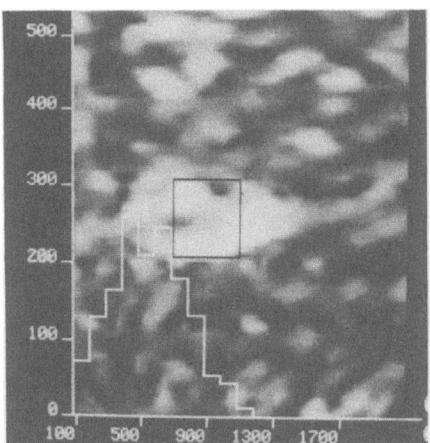

Fig. 30. Detected liver image filtered by constrained iterative
         algorithm, 4 iterations, "weighting" factor is 100,
         magnification 4.

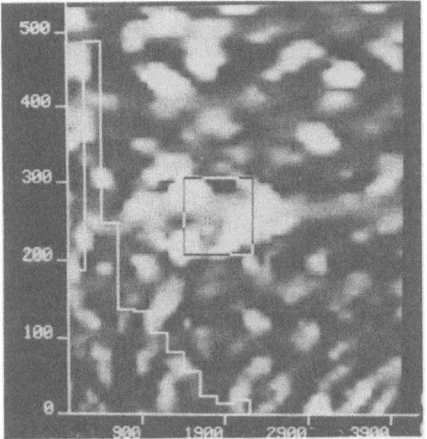

Fig. 31. Detected liver image filtered by constrained iterative
         algorithm, 6 iterations, "weighting" factor is 100,
         magnification 4.

DISCUSSION

     There is at least one undesirable feature of the constrained
iterative algorithm as presentd here. By allowing c to vary
independently for each new line of data, it is to be expected that
some lines will now be blurred relative to others. Streaks
therefore begin to appear (see Fig. 24). However, the majority of

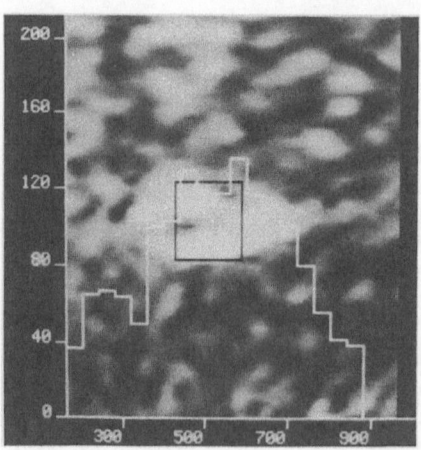

Fig. 32. Detected liver image filtered by pseudo-inverse filter.
         A = 0.002T(0), magnification 4.

the data lines appear to have equal resolution improvement. It
should be a simple matter to improve the uniformity of the
resolution enhancement from line to line by varying the number of
iterations until a desired contrast is reached.

    The design of filters with a lower mean squared error than
the Wiener mean squared error is routinely done in the field of
communications. In such cases, a priori knowledge not only of the
second order statistics of the object and noise but also of the
probability distribution functions is available. Bayesian
analysis then leads to the application of maximum a posteriori and
maximum likelihood estimation techniques.

    It is instructive to equate the constrained iterative
algorithm with maximum a posteriori algorithms developed for image
processing using the Bayesian approach.[5] The constraining
function used here is then equivalent to the object covariance
divided by the noise covariance of the parameter estimation
approach. This gives a useful insight into the physical
significance of the constraining function if we think of each
point of the image as possessing a local mean and variance.
Assuming that the noise variance is a constant, we have modeled
the object variance as decreasing linearly to zero as the value of
the object estimate, which is an approximate to a local object
mean, approaches the lower or upper bound.

    Alternative models for the object may lead to other
appropriate constraining functions. For instance, the modulus of
the amplitude information from a random walk follows a Rayleigh

distribution,[6] in which the object standard deviation is proportional to the object mean. In such a model, therefore, the appropriate constraining function would increase linearly from zero in proportion with the object estimate. However, the filtered result could very likely be too susceptible to noise.

Other iterative schemes for maximum a posteriori estimation have been found[7] to be quite insensitive to errors in the object and noise covariances. Inaccuracies in modeling the object and noise variances are therefore likely to be uncritical and a priori knowledge consisting simply of an upper and lower bound may be sufficient for an accurate solution.

One characteristic of object restorations using the constrained iterative algorithm supports the hypothesis that the constrained iterative algorithm estimate is closely associated with the maximum a posteriori estimate. The spectrum of an object estimate is much smoother using the constrained iterative algorithm than using either the Wiener or pseudo-inverse filter. Thus, the entropy of the restored object using the constrained iterative algorithm is higher. It is also noted, in contrast to the case of linear filtering, the constrained algorithm removes zeros introduced by the imaging system transfer function from the restored object spectrum.

Iterative approaches have characteristic advantages over the conventional linear Wiener filtering technique. In particular, less a priori information is required for a solution which is less sensitive both to noise and to errors in the a priori information. In addition, techniques which incorporate a consistent positivity constraint into their solutions give object restoration devoid of spurious oscillations.

REFERENCES

1. B. Burgoyne, Ph.D. Thesis, in preparation.
2. B. R. Frieden, "Image Enhancement and Restoration," Topics in Applied Physics 6, Ch. 5, Springer Verlag, New York (1978).
3. G. T. Herman and A. Lent, "Iterative Reconstruction Algorithms," Comput. Biol. Med. 6, Pergamon Press, 273-294 (1976).
4. B. Widrow and J. M. McCool, "A Comparison of Adaptive Algorithms Based on the Methods of Steepest Descent and Random Search," IEEE Trans. Antennas Propagat. AP-24, 616, 637 (1976).
5. G. T. Herman and A. Lent, "A Computer Implementation of a Bayesian Analysis of Image Reconstruction," State Univ. of New York, Buffalo, New York, Tech. Rep. 91, 1974.

6.   J. W. Goodman, "Statistical Optics," Class Notes, 1980.
7.   H. J. Trussell, "Notes on Linear Image Restoration by
      Maximizing the A Posteriori Probability," IEEE Trans.
      Acoustics, Speech and Signal Processing, <u>ASSP-26</u>, April,
      1978.

TWO-DIMENSIONAL NULL PROCESSING

Takeo Sawatari and R. Paul Gorman

Bendix Advanced Technology Center
Bendix Center
Southfield, Michigan 48037

ABSTRACT

A beamforming technique based on the simultaneous operation of the null processing method in both the spatial and temporal domains successfully enhances a weak target signal by rejecting a number of strong localized interferences. This two-dimensional null processing (2-DNP) technique is compared to the previously developed adaptive null processing (ANP) method and conventional beamforming methods, in part using computer simulation results. The application of the 2-DNP method to tracking a moving target is also discussed.

INTRODUCTION

The wide application of adaptive beamforming techniques in the areas of sonar and radar has been recognized for years and has sparked a good deal of research.[1] The work of Widrow and Griffiths on optimal array processing represents one of the earliest and more general treatments of adaptive filtering.[2,3] The technique, based on the Wiener filter, employs a covariance matrix formed from the input to different array elements and has proved both versatile and effective. However, the complexities involved in optimization have caused many "suboptimal" techniques to be investigated. Their speed of adaptation makes these methods attractive despite their usual loss in generality or performance.

One approach in particular limits its applicability to a
specific yet frequently occurring scenario. This technique,
described by Keating et al.[4] and known as adaptive null processing
(ANP), assumes a field in which strong localized interferences mask
a weak target. The ANP technique temporally processes the input,
then reduces the problem of inverting an MxM cross-correlation
matrix,[5] where M is the number of array elements, to one of
inverting a JxJ matrix, where J is the number of interfering
sources. Processing complexity is reduced without sacrificing
performance. In fact, the general optimal processing technique
reduces to the simpler ANP approach for this application.[6]

This paper presents an extension of the ANP technique into the
time domain. The resulting two-dimensional null processing (2-DNP)
technique initially estimates the orientation and frequency of the
localized interfering sources using conventional beamforming
techniques, then employs null processing to enhance the target in
both the spatial and temporal domains. The ANP approach enhances
the target only in the spatial domain, but the 2-DNP technique
provides the additional advantage of discriminating and rejecting
interfering sources which are spatially close to the desired
signal. Although two JxJ matrices must be inverted and one
additional parameter optimized per interfering source, a two-
dimensional optimal processing technique would require the
inversion of an additional NxN matrix, where N is the number of
samples taken. The customary gradient-search techniques, which can
become rather unwieldy as the number of parameters increase, have
been abandoned in favor of an optimization algorithm based on the
simplex method.[7] Fairly well-behaved functions involving less than
twenty parameters converge with greater processing efficiency.

The 2-DNP technique also has potential as a tracking filter.
It can compensate the independent motions of target and
interferences by re-optimizing the projection operators without
repeating the costly Fourier analysis.

The following section reviews the adaptive null processing
technique before developing its two-dimensional extension. In the
next section three techniques--conventional beamforming, ANP, and
2-DNP--are compared theoretically using a simple scenario, and for
performance using computer simulation results. Finally, the
function of the 2-DNP technique as a tracking filter is examined.

NULL PROCESSING

This paper uses the following notation: boldface lowercase
letters underscored by a tilde represent a complex column vector;

boldface uppercase letters represent a matrix; **I** represents the
identity matrix; † represents a conjugate transpose; and *
represents a complex conjugate.

Adaptive Null Processing

The adaptive null processing technique initially filters the
input temporally, selecting a particular frequency component ω for
analysis. Let the vector $\underset{\sim}{x}(\omega)$ represent the output from a
receiving array after spectral processing. Assuming far-field
sources and therefore plane waves at the linear equi-spaced
receiving array, we may express this vector as

$$\underset{\sim}{x}(\omega) = \rho_o \underset{\sim}{d}_o + \mathbf{D}\underset{\sim}{\rho} + \underset{\sim}{n} \tag{1}$$

where **D** is an MxJ direction matrix, M being the number of array
elements and J the number of interfering sources. Its elements are
given by

$$D_{mj} = \exp(i \frac{\omega_j}{c} a_m \sin \Theta_j) \tag{2}$$

where a_m is the position of the mth array element, Θ_j and ω_j are
the bearing angle and the central frequency of the jth interferences, respectively, and c is the propagation velocity of the
signals. The expression $\underset{\sim}{\rho}$ is a vector whose jth element $\rho_j(\omega)$ is
the complex amplitude of a particular frequency component ω emitted
by the jth source. The first term on the right of Eq. (1)
represents a similar contribution by the target signal where

$$\rho_o d_{mo} = A \exp(i \frac{\omega_o}{c} a_m \sin \Theta_o + \phi) \tag{3}$$

and the last term represents a Gaussian white noise vector.

To remove the interfering highlight terms **D**$\underset{\sim}{\rho}$ from the input
vector--the key to the ANP method--the bearing angles Θ_j of the
highlights are initially assumed to correspond approximately to the
location of the peaks appearing in the beamformed pattern of the
input. These estimates would be expected to be fairly good because
of the high power of the interfering sources. Using Eq. (2) and
these estimated bearing angles $\Theta_j^{\prime}$, we obtain an approximate
direction matrix **D**$^{\prime}$. If we denote the unknown complex amplitude
corresponding to the highlight of the jth interference as g_j we may
write

$$\bar{\underset{\sim}{x}} = \underset{\sim}{x} - \mathbf{D}^{\prime}\underset{\sim}{g} \tag{4}$$

where $\bar{\underset{\sim}{x}}$, given the appropriate values of the components of $\underset{\sim}{g}$,
represents the input vector essentially free of interfering

highlights. The degree to which the highlights are rejected depends on the optimization of D'.

We define the beamforming matrix K whose elements are given by

$$K_{mj} = \frac{1}{M} D'_{mj} = \frac{1}{M} \exp(i \, \frac{\omega_j}{c} \, a_m \, \sin \Theta_j) \tag{5}$$

where M is a normalizing factor corresponding to the number of array elements. The condition of null processing sets the amplitude of the field at the Θ'_j points to zero. Using the beamforming matrix, we may express this as

$$K^{\dagger} \bar{x} = \frac{1}{M} D'^{\dagger} x - \frac{1}{M} D'^{\dagger} D' g = 0 \tag{6}$$

Solving for the unknown vector g

$$g = (D'^{\dagger} D')^{-1} D'^{\dagger} x \tag{7}$$

and substituting Eq. (7) into Eq. (4) we have

$$\bar{x} = x - D' (D'^{\dagger} D')^{-1} D'^{\dagger} x \tag{8}$$

Since we define the projection operator as

$$P = D' (D'^{\dagger} D')^{-1} D'^{\dagger} \tag{9}$$

we may write

$$\bar{x} = (I - P) x \tag{10}$$

This equation represents the simultaneous rejection of strong interfering sources from the input field. To best locate the nulls in the spatial domain, the estimated values Θ' must be systematically adjusted to coincide more precisely with the actual values Θ_j. We define optimal null placement as those locations which minimize the total power of the frequency component ω integrated over all directions. We may express the total power as

$$P_T(\omega) = \langle \bar{x}(\omega) \bar{x}(\omega) \rangle$$

where $P_T(\omega)$ is the performance criterion minimized by null processing. As shown in Eq. (8) each iteration of the minimization process requires the inversion of only a JxJ matrix.

Two-Dimensional Null Processing

In the 2-DNP approach null processing is applied simultaneously to the unprocessed temporally sampled field values at each

hydrophone in both the spatial and temporal domains. The input
signal therefore becomes a matrix of dimension MxN, where M is the
number of array elements and N the number of samples taken over a
given period of time. We can express this input in a way analogous
to Eq. (1):

$$\mathbf{X}(t) = \rho_o \underset{\sim}{d}_o \underset{\sim}{t}_o{}^{\dagger} + \mathbf{DRT}^{\dagger} + \mathbf{N} \tag{11}$$

where $\mathbf{R}$ is a JxJ diagonal matrix whose elements are defined by

$$R_{ji} = \rho_{ji} \delta_{ji} \tag{12}$$

and ρ_{ji} is the complex amplitude of the jth source. The expression
$\mathbf{T}$ is an NxJ frequency matrix whose elements are given by

$$T_{nj} = \exp(i\omega_j t_n) \tag{13}$$

where ω_j is as defined above and t_n is the nth instant of the
sampling interval.

The expression $\mathbf{D}$ is an MxJ direction matrix as in Eq. (2).
The first term on the right of Eq. (11) again represents the
contribution due to the target signal where

$$\rho_o d_{mo} t_{no}{}^{\dagger} = A \exp(i \frac{\omega_o}{c} a_m \sin \Theta_o - \omega_o t_n + \phi) \tag{14}$$

and $\mathbf{N}$ is a noise matrix each element of which is a random noise
component superimposed upon every sampled input datum.

The 2-DNP method is applied much as was the ANP technique to
reduce the highlight term from the input data. A two-dimensional
Fourier transform of the input matrix $\mathbf{X}(t)$ is first carried out
(see Fig. 1). In the (Θ,ω) plane a localized source appears as a
peak. These peaks are used to estimate the frequency ω'_j as well as
the bearing angle Θ'_j of each interfering source, which lead to
approximate direction $\mathbf{D}'$ and frequency $\mathbf{T}'$ matrices. We must now
determine an unknown complex amplitude matrix G, which is not
necessarily a diagonal matrix. We begin with the following
equation:

$$\overline{\mathbf{X}} = \mathbf{X} - \mathbf{D}'\mathbf{GT}'^{\dagger} \tag{15}$$

where $\overline{\mathbf{X}}$ represents the desired interference-free matrix. Again,
interference rejection depends on choosing the proper values of
the elements of matrix G as well as on optimization of the
matrices $\mathbf{D}'$ and $\mathbf{T}'$.

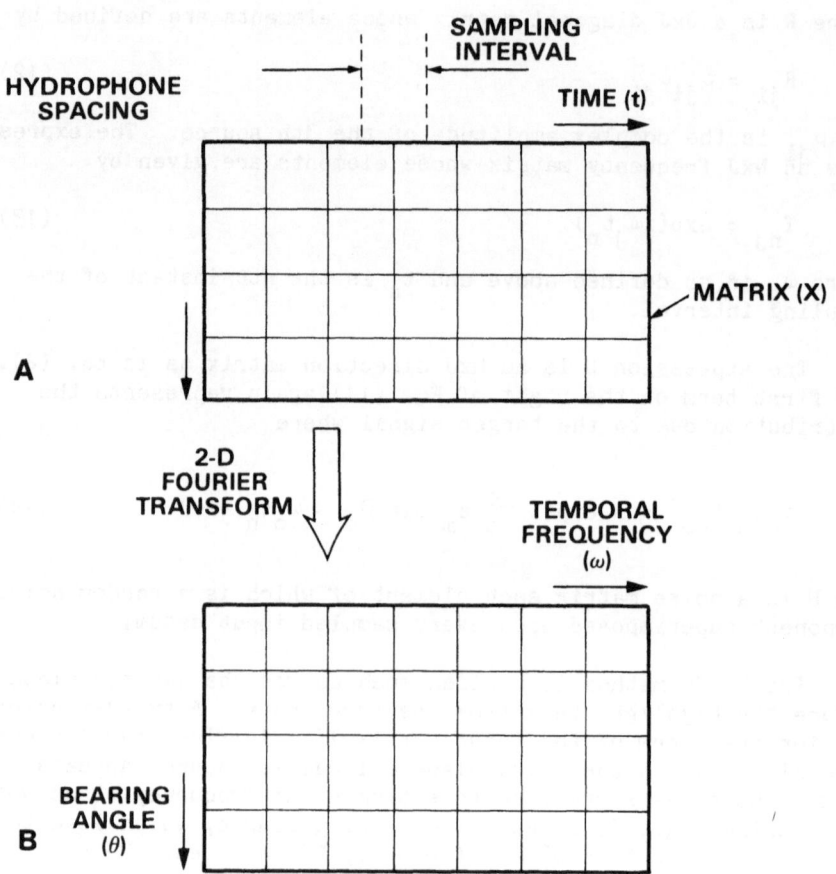

Figure 1. Input Data Matrix **X** and Its Fourier Transform
           (a) (x,t) Plane, Hydrophone-Time Domain
           (b) (Θ,ω) Plane, Bearing Angle-Frequency Domain

We define the beamforming matrix $\mathbf{K}$ as before:

$$\mathbf{K} = \frac{1}{M}\, \mathbf{D}' \tag{16}$$

We also define a spectrum-forming matrix $\mathbf{S}$ whose elements are given by

$$S_{nj} = \frac{1}{N}\, T'_{nj} = \frac{1}{N}\, \exp(i\omega'_j t_n) \tag{17}$$

where N is a normalizing factor equal to the number of signal samples. In this case we use both the beamforming and spectrum-forming matrices to express the 2-DNP condition:

$$\mathbf{K}^{\dagger}\overline{\mathbf{X}}\mathbf{S} = \left(\frac{\mathbf{D}'^{\dagger}}{M}\,\mathbf{X}\,\frac{\mathbf{T}'}{N} - \frac{\mathbf{D}'^{\dagger}}{M}\,\mathbf{D}'\mathbf{G}\mathbf{T}'^{\dagger}\,\frac{\mathbf{T}'}{N}\right) = 0 \tag{18}$$

Solving for the unknown matrix $\mathbf{G}$ we obtain

$$\mathbf{G} = (\mathbf{D}'^{\dagger}\mathbf{D}')^{-1}\mathbf{D}'^{\dagger}\mathbf{X}\,\mathbf{T}'(\mathbf{T}'^{\dagger}\mathbf{T}')^{-1} \tag{19}$$

We define the projection operators P_{Θ} and P_{ω} as

$$P_{\Theta} = \mathbf{D}'(\mathbf{D}'^{\dagger}\mathbf{D}')^{-1}\mathbf{D}'^{\dagger} \tag{20}$$

$$P_{\omega} = \mathbf{T}'(\mathbf{T}'^{\dagger}\mathbf{T}')^{-1}\mathbf{T}'^{\dagger} \tag{21}$$

Therefore, if we substitute Eq. (19) into Eq. (15) we finally obtain

$$\overline{\mathbf{X}} = \mathbf{X} - P_{\Theta}\mathbf{X}P_{\omega} \tag{22}$$

The essence of the 2-DNP technique, as expressed in Eq. (22), is the ultimate attainment of a remaining field independent of interfering sources, achieved through the optimization of the projection operators. The 2-DNP method adopts a performance criterion similar to that of adaptive null processing which minimizes the total signal power remaining after interference rejection. This performance criterion, a function of null location (Θ'_j, ω'_j), may be expressed as

$$\varepsilon(\underset{\sim}{\theta},\underset{\sim}{\omega}) = \mathrm{Tr}(\overline{\mathbf{X}}^{\dagger}\overline{\mathbf{X}}) \tag{23}$$

where $\underset{\sim}{\theta}$ and $\underset{\sim}{\omega}$ are vectors containing the estimated directions and frequencies, respectively. The simplex method is employed in minimizing ε.[7] Again, as shown in Eq. (19), this processing requires the inversion of two JxJ matrices.

The elements of the JxJ matrix formed on the left side of Eq. (18) are the values of points in the (Θ, ω) plane. The diagonal elements correspond to the values of the estimated interference locations (Θ_i', ω_i'), whereas the off-diagonal elements are the values of the points (Θ_j', ω_j'), $j \neq i$. That the values of these off-diagonal points are also reduced to zero by the null processing condition is a simplifying measure which has little or no effect on technique efficiency unless the location of the desired signal coincides with one of these points. The probability of this occurring is roughly

$$P = \frac{J(J - 1)}{MN} \tag{24}$$

Since one often deals with only a few interfering sources but a relatively large number of array elements and samples, the likelihood of problems due to this effect is small.

EFFICIENCY ANALYSIS

The scenario chosen to compare the conventional beam-forming, ANP, and 2-DNP techniques is one in which a single localized interfering source is positioned near a desired target. We evaluate the techniques' efficiency by comparing the field remaining in the target location after processing, ignoring the background white noise to simplify the analysis. The input data matrix $\mathbf{X}$ in this case is

$$\mathbf{X} = \rho_0 \underset{\sim}{\mathbf{d}}_0 \underset{\sim}{\mathbf{t}}_0{}^\dagger + \rho_1 \underset{\sim}{\mathbf{d}}_1 \underset{\sim}{\mathbf{t}}_1{}^\dagger \tag{25}$$

Conventional Beamforming (CBF) Technique

A field of the desired target in the look direction Θ_0 and in the target frequency spectrum ω_0 is obtained as follows:

$$f_{CBF} = \underset{\sim}{\mathbf{k}}_0{}^\dagger \mathbf{X} \underset{\sim}{\mathbf{s}}_0 = \rho_0 + \rho_1 d_{01} t_{01}{}^* \tag{26}$$

where $\underset{\sim}{\mathbf{k}}_0$ and $\underset{\sim}{\mathbf{s}}_0$ are related to $\underset{\sim}{\mathbf{d}}_0$ and $\underset{\sim}{\mathbf{t}}_0$ by

$$\underset{\sim}{\mathbf{k}}_0 = \frac{1}{M} \underset{\sim}{\mathbf{d}}_0 \tag{27}$$

$$\underset{\sim}{\mathbf{s}}_0 = \frac{1}{N} \underset{\sim}{\mathbf{t}}_0 \tag{28}$$

and where

$$d_{01} = \underset{\sim}{\mathbf{k}}_0{}^\dagger \underset{\sim}{\mathbf{d}}_1 \quad , \quad t_{01} = \underset{\sim}{\mathbf{t}}_1{}^\dagger \underset{\sim}{\mathbf{s}}_0 \tag{29}$$

Since

$$\overset{\dagger}{\underset{\sim}{k}}_o \underset{\sim}{d}_o = 1 \quad , \quad \overset{\dagger}{\underset{\sim}{t}}_o \underset{\sim}{s}_o = 1 \tag{30}$$

d_{o1} is the magnitude of the side lobe of the interfering high-light at the target location and t_{o1} is the leakage component of the spectrum of the interference into the target frequency band. Although both $|d_{o*}|$ and $|t_{o1}|$ are smaller than unity, the interference term $\rho_1 d_{o1} t_{o1}$ may be a large quantity since $|\rho_o| \ll |\rho_1|$ by the definition of the interference source.

Adaptive Null Processing Technique

The field of the desired target is obtained by applying the projection operator of rank 1 to the input vector and then premultiplying the modified input vector by the beamforming vector. The input vector is generated from the input matrix by postmultiplying it by the spectrum-forming vector $\underset{\sim}{s}_o$:

$$\underset{\sim}{x} = X \underset{\sim}{s}_o \tag{31}$$

The null vector is obtained by

$$\overline{\underset{\sim}{x}} = (I - \frac{\underset{\sim}{d}_1 \underset{\sim}{d}_1^{\dagger}}{M})\underset{\sim}{x} \tag{32}$$

where the projection operator P_Θ is

$$P_\Theta = \frac{1}{M} \underset{\sim}{d}_1 \underset{\sim}{d}_1^{\dagger} \tag{33}$$

The field of the desired target is derived from Eq. (25), (31), and (32):

$$f_{ANP} = \underset{\sim}{k}_o \overline{\underset{\sim}{x}} = \rho_o - \rho_o |d_{o1}|^2 \tag{34}$$

A comparison of the noise term $\rho_o |d_{o1}|^2$ with that of the CBF approach shows that the noise is significantly reduced in the ANP method.

Two-Dimensional Null Processing Technique

The null-processed input data matrix $\overline{X}$ is obtained using Eq. (20), (21), and (22):

$$\overline{X} = X - \frac{\underset{\sim}{d}_1 \underset{\sim}{d}_1^{\dagger}}{M} X \frac{\underset{\sim}{t}_1 \underset{\sim}{t}_1^{\dagger}}{N} \tag{35}$$

The desired target field is then obtained by applying the
conventional beamformer to the matrix X.

$$f_{2DNP} = \underline{k}_o \overline{X} \underline{s}_o = \rho_o - \rho_o |d_{o1}|^2 |t_{o1}|^2 \tag{36}$$

The noise term in the 2-DNP equation--$\rho_o |d_{o1}|^2 |t_{o1}|^2$ in
Eq. (36), $|t_{o1}|^2 \leqslant 1$ --is always less than that in the ANP
equation--$\rho_o |d_{o1}|^2$ in Eq. (16). The term $|t_{o1}|^2$ plays a
significant role when the desired target and the jamming
highlight are spatially very close ($|d_{o1}| \simeq 1$). In this case
the signal f_{ANP} is almost cancelled; but in the 2-DNP equation
the noise term is still small as long as $|t_{o1}| < 1$. Even when
the target and the interference are spatially overlapped, the
2-DNP technique can discriminate between the target and the
interference as long as the two are temporally separated.

Computer Simulation Results

Computer simulation tests permitted us to gauge the
performance of the 2-DNP technique, and to compare it to that of
the conventional beamforming and ANP techniques. In these
examples two localized interference sources and a weak target
signal were present. The signal-to-interference ratio was
approximately -20 db. Both the frequency and the bearing angle
of the target were unknown, but the target was assumed to be the
weakest signal.

In the first situation the two interferences bracket the
desired signal in both orientation and frequency (see Fig. 2).
The resultant signal has a major peak somewhat to the right of
the desired signal, which is identified at Θ_o. The ANP technique essentially nulls out the signal at nearly all angles; the
specific locations of the three peaks which occur in the
conventionally formed beam cause it to "misjudge" the null
positions. However, the 2-DNP process takes advantage of the
information available from the second dimension (frequency)
estimate to provide a clear and correct identification of the
target location.

In Fig. 3 the strong interferences are positioned along
virtually the same orientation but with different frequencies--a
simplified simulation of the towing vessel noise, which is
separated spatially from the target and contains many frequency
components. The 2-DNP technique again succeeds in identifying
the desired signal location.

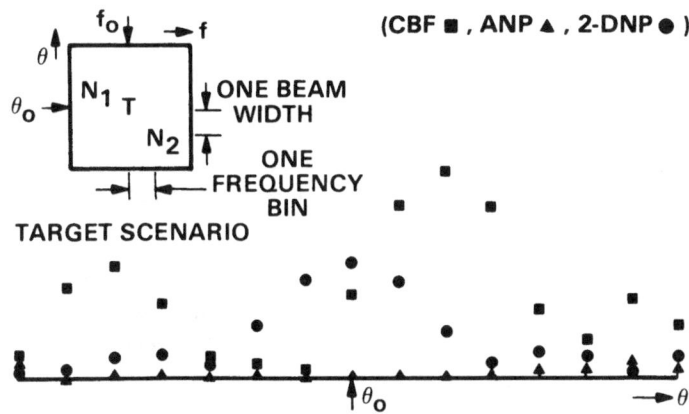

Figure 2.   Comparison of Conventional Beamforming, ANP, and
2-DNP Techniques for Scenario with Two Jammers, One
Spatially Overlapped

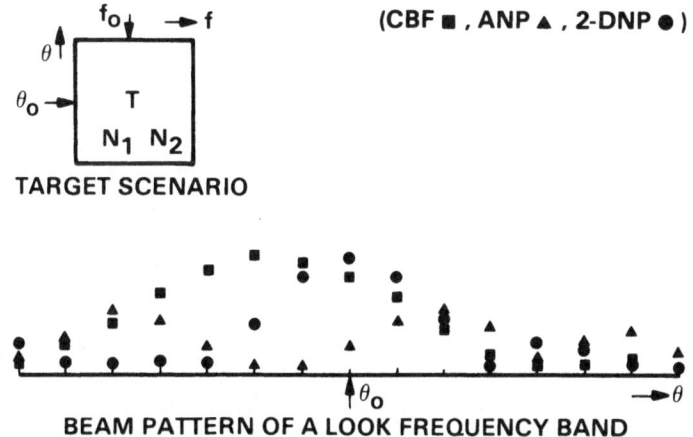

Figure 3.   Comparison of Conventional Beamforming, ANP, and
2-DNP Techniques for Scenario with Two Jammers in a
Broad Spectrum

TRACKING CAPABILITY

In any practical sonar application one can expect to
encounter targets and interfering sources whose frequency and
bearing angle change over time because of either the motion of
the array or the independent motions of the sources themselves.
These changes present any adaptive beamforming technique with a
persistent optimization problem, and the agility of the adaptive
process becomes crucial. The rapid convergence of the 2-DNP
approach made its application as a tracking filter worth
investigating.

The input matrix takes essentially the same form as in Eq.
(11), but in this case Θ_o, Θ_j, ω_o, and ω_j all become functions
of time. The elements of $\mathbf{D}$ and $\mathbf{T}$ therefore become

$$D_{mj} = \exp(i \frac{\omega_j(t)}{c} a_m \sin \Theta_j(t)) \tag{37}$$

$$T_{nj} = \exp(i\omega_j(t)t_n) \tag{38}$$

where $t = T_k + t_n$, T_k being the time at the start of the kth
sampling interval; and where

$$\omega_j(t) = \frac{c + V_a(t)}{c + V_j(t)} \omega_{jc} \tag{39}$$

In Eq. 39 $V_a(t)$ is the component of the array velocity along the
line extending from the center of the array to the jth source at
time t; $V_j(t)$ is the component of the velocity of the jth source
along the same line at time t; ω_{jc} is the rest central frequency
of the jth source; and $\Theta_j(t)$ is the bearing angle of the jth
source at time t. The velocity component is positive for
receding sources. The input matrix compiled during the kth
sampling interval is denoted $\mathbf{X}_k$.

Every input matrix will reflect some degree of spreading
due to the shift in frequency and bearing angle during a
sampling interval, but this shift is negligible for reasonable
velocities. Yet the fairly significant shifts expected to occur
between sampling intervals normally require re-optimization.
Although the 2-DNP tracking filter processes the initial input
matrix $\mathbf{X}_o$ in precisely the same way as for the static case, it
economizes its processing load in subsequent sampling intervals.

For $X_{k>0}$ the two-dimensional Fourier transform which supplied the initial estimates $(\theta, \omega)_0$ is dispensed with and the optimized values of the frequency and bearing angles found for X_{k-1} are used in the equation

$$\varepsilon(\underset{\sim}{\theta}, \underset{\sim}{\omega}) = \mathrm{Tr}(\overline{X}_k^{\dagger} \overline{X}_k) \tag{40}$$

where initially

$$\overline{X}_k = X_k - P_{\theta_{k-1}} X_k P_{\omega_{k-1}} \tag{41}$$

Thus the values of P_{θ} and P_{ω} obtained for the (k-1)th sampling interval are re-optimized using the simplex algorithm to serve as the projection operators for the kth input matrix. In this manner the 2-DNP technique can be employed to successfully track a moving target in a field which contains a number of strong moving interferences.

Figure 4 presents some indication of the 2-DNP technique's tracking ability for the scenario of the above computer simulation tests. The power spectra of three time-displaced sampling intervals and the power spectra of the processed output matrices are shown with the vertical axes of the processed power spectra expanded to retain good detail. This sequence represents a series of about 40 re-optimizations. One can readily see the shifting of all three sources in the (θ, ω) field. In each case, despite the shift in frequency and bearing angle, the interferences are successfully rejected and the target enhanced.

SUMMARY

The two-dimensional null processing technique efficiently rejects strong interferences from a field in which a weak target signal is nested. Its strength arises from its use of an additional parameter to define the location of such interferences; it is particularly apt in situations where the interferences overlap the target in the spatial dimension. In the case of a broad-spectrum, spatially localized noise source, such as a towing vessel, the 2-DNP method reduces the residual noise from the total received signal more effectively than does the conventional ANP technique.

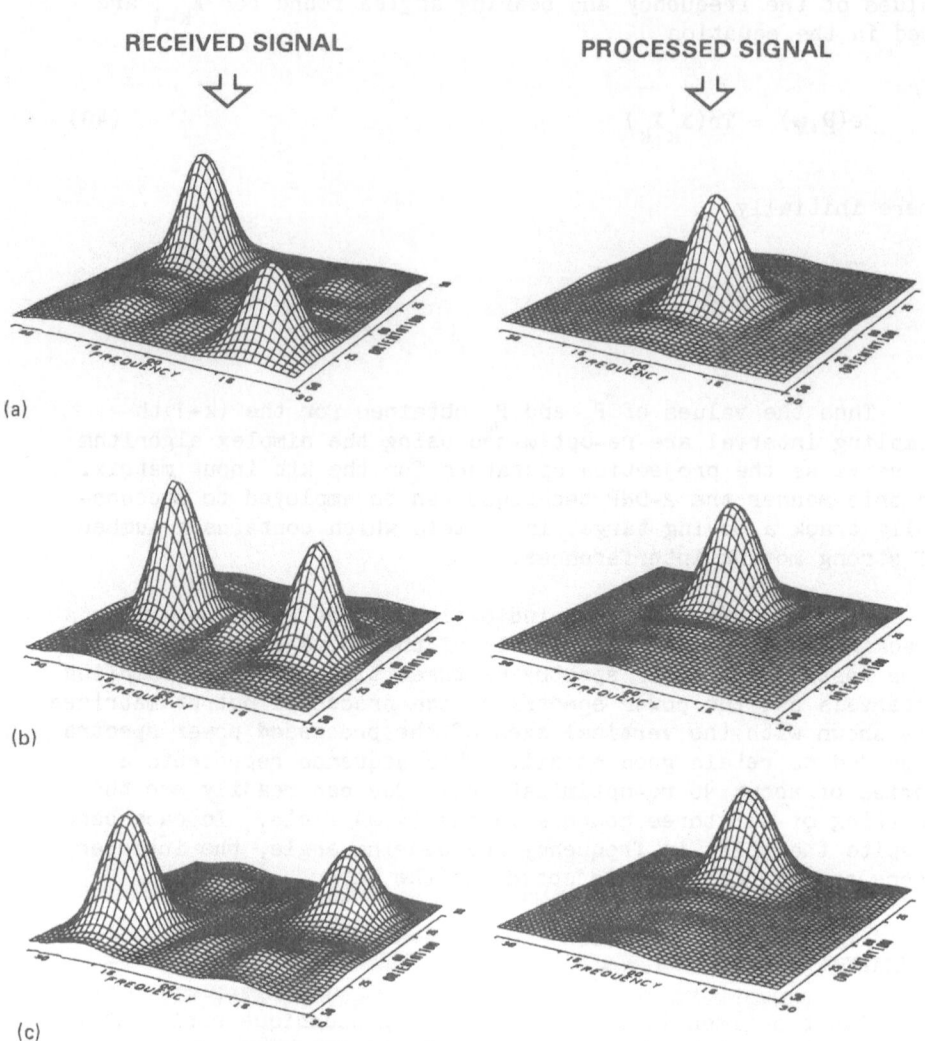

(a)

(b)

(c)

Figure 4.   Power Spectra of Input and Processed Matrices
             Representing Three Time-Displaced Sampling Intervals

             (a)   t=0
             (b)   t=1
             (c)   t=2

Other advantages of the 2-DNP technique are the size of the matrix requiring inversion and the optimization algorithm. Conventional optimal detection requires a matrix whose size equals the number of sampling points, whereas the 2-DNP technique requires a matrix whose size equals the number of jammers. The simplex technique used to obtain a minimum of the performance criterion is much more efficient than techniques such as the steepest-descent method. The 2-DNP technique also shows promise as a tracking filter.

Computer simulation results clearly indicate the ability of the 2-DNP technique to reject certain strong interferences with which conventional techniques are unable to contend. Work is presently under way to extend the input matrix to a more practical size and to include additional interferences. The tracking ability of the 2-DNP method is also undergoing more thorough tests to delineate its limitations.

REFERENCES

1. IEEE Trans. Antennas and Propagation AP-24 (Sept. 1976).

2. B. Widrow, P. E. Mantey, L. J. Griffiths, and B. B. Goode, Proc. IEEE 55:2143 (1967).

3. L. J. Griffiths, Proc. IEEE 57:1696 (1969).

4. P. N. Keating, R. F. Koppelmann, R. K. Mueller, R. F. Steinberg, and G. Zilinskas, J. Acoust. Soc. Am. 59:106 (Jan. 1976).

5. O. L. Frost III, Proc. IEEE 60:926 (1972).

6. P. N. Keating, J. Acoust. Soc. Am. 65:456 (Feb. 1979).

7. Beveridge and Schechter, "Optimization: Theory and Practice," pp. 367-383, Chemical Engineering Series, McGraw-Hill, New York (1970).

ACOUSTIC IMAGING IN MARINE SEDIMENT: A MULTIPLE MICROPROCESSOR

ARRAY PROCESSOR USING THE TRACE FUNCTION

G.L. Sackman and G.R. Vermander

Department of Electrical Engineering
Naval Postgraduate School
Monterey, CA 93940

Penetration of marine sediment requires as low a frequency as possible to reduce attenuation, whereas good resolution requires as short a pulse and as large an aperture as possible. In order to achieve a reasonable balance between these conflicting requirements, an aperture is called for with dimensions comparable to the distance to the object. Operation in such a near-field regime with extreme wavefront curvature necessitates a re-examination of the concept of beamforming. Reflected energy scattered over a wide angle is intercepted by the aperture (overcoming to some extent the usual domination of the image by specular facets). However, there remains the problem of deriving from the hydrophone signals an amplitude or intensity for each display pixel corresponding uniquely to each resolvable scattering center. One solution to this problem is to digitize the echo signal at each hydrophone, store all the data in the form of a time series for each hydrophone over a complete echo cycle and then sift through the memory contents looking for the characteristic patterns which result from every possible scattering center. An algorithm to accomplish this sifting is presented below based on the "Time Delay Trace Function" [1,2]. In order to achieve sufficiently rapid execution of the algorithm to provide near real-time image generation, it was necessary to use multiple microprocessors working together as a system [3].

TIME DELAY TRACE FUNCTION

The Time Delay Trace Function concept is philosophically related to acoustic holography (wavefront recording) but because of wavefront curvature and array geometry, it is not possible to use

Fourier techniques to simplify the image reconstruction process.
Brute force numerical and logic processing is required.

The basic procedure is outlined below, and sketched in Figure
1. Consider a single point object located in the near field of an
aperture filled with sensors numbered by an index k along with a
point source projector. Let the projector transmit an impulse
(delta function). After a single echo cycle, the time series data
of each sensor will contain an impulse at a time delay dependent
on the geometry. The pattern formed by the impulses is the Time
Delay Trace Function (TDTF) and is unique for each possible point
object location (a different aperture geometry would result in a
different trace function). The required processor is an estimator
of the object location. Assuming a particular geometry, the TDTF
can be pre-calculated for each possible object location. One estimator of the presence of a scattering center is given by over-
laying pre-computed TDTF patterns on a set of data, the best fit
indicating which object positions most probably gave rise to the
data. Estimating the goodness of fit by correlation is essentially a point-by-point multiplication and summation process, but it
can be simplified to summation if the weights are all unity.
Therefore if each impulse has unit weight and a summation is formed of all data in a TDTF pattern, the sum will equal the total
number of sensors K if the data coincides exactly with the pattern. As a result of noise and errors in geometry, the sum may be

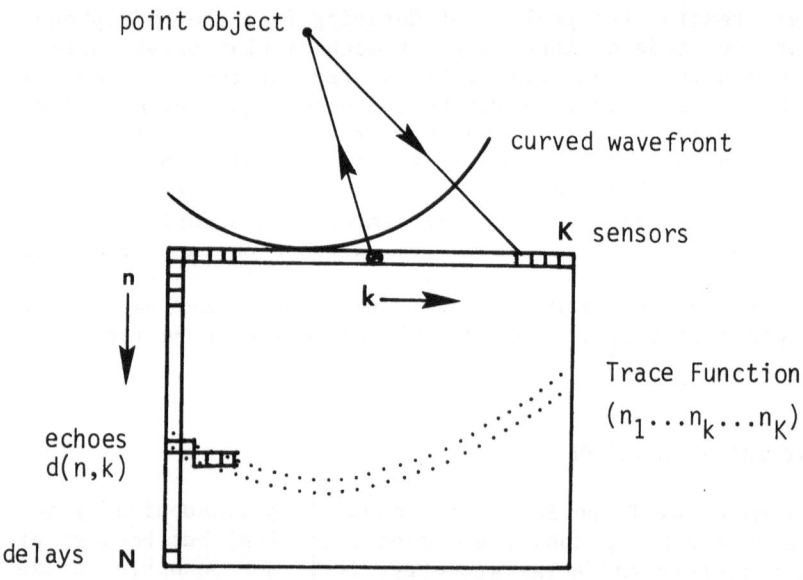

Figure 1. Echo data for point object

less than K but will still presumably have a maximum in the vicinity of the true object location.

In a practical application, this can be implemented using a
computer. The time series data from each sensor must be synchro-
nously sampled (let time be quantized by an index "n") and the amplitude quantized in an A/D converter, then stored in a read-write
memory. The data consists of a matrix of echos D with typical
matrix element d (n,k). Precomputed TDTF's are stored in read-
only-memory in the form of a list of time delay addresses $(n_1,\ldots,$
$n_k\ldots n_K)$ which can be represented by a vector $\underline{n}$ for each resolvable object location (r,θ). A list $\underline{n}(r,\theta)$ comprises a TDTF
for each resolvable object location as sketched in Figure 1.
Therefore, the processor, in "correlating" the data set to a trace
function in effect forms the sum of all data entries along the
trace function $\underline{n}(r,\theta)$ in the data matrix D.

The process can be written in matrix notation by pre-multi-
plying the data matrix D by the transpose of a matrix T_i containing the trace function of a particular scattering center. The
resulting matrix F_i.

$$F_i = T^t D$$

contains in its diagonal terms the products of the echo time series data multiplied term-by-term by the trace function "mask".
The standard matrix trace operator applied to the F_i matrix then
sums the terms on the diagonal, giving the amplitude A_i for the
pixel corresponding to the scattering center

$$A_i = Tr(F_i)$$

Note that this is a digital implementation of the classical
delay-and-sum beamformer but extended to compensate for both wavefront curvature and aperture geometry as well as maintaining a
time history of the complete echo cycle.

SYSTEM SIMULATION

A digital simulation of the processor was done first using a
Hewlett-Packard 9845T desktop minicomputer in order to explore the
effect of varying the system parameters. Typical results are presented for a 16 element linear aperture 5 meters long, pulse length
50 microseconds, and sampling with A/D conversion of the hydrophone signals at 50 microsecond intervals. The field of view was
set at $\pm$ 30°, depth of field covering the range from 5 to 15 meters
from the center element of the array. Trace functions for four

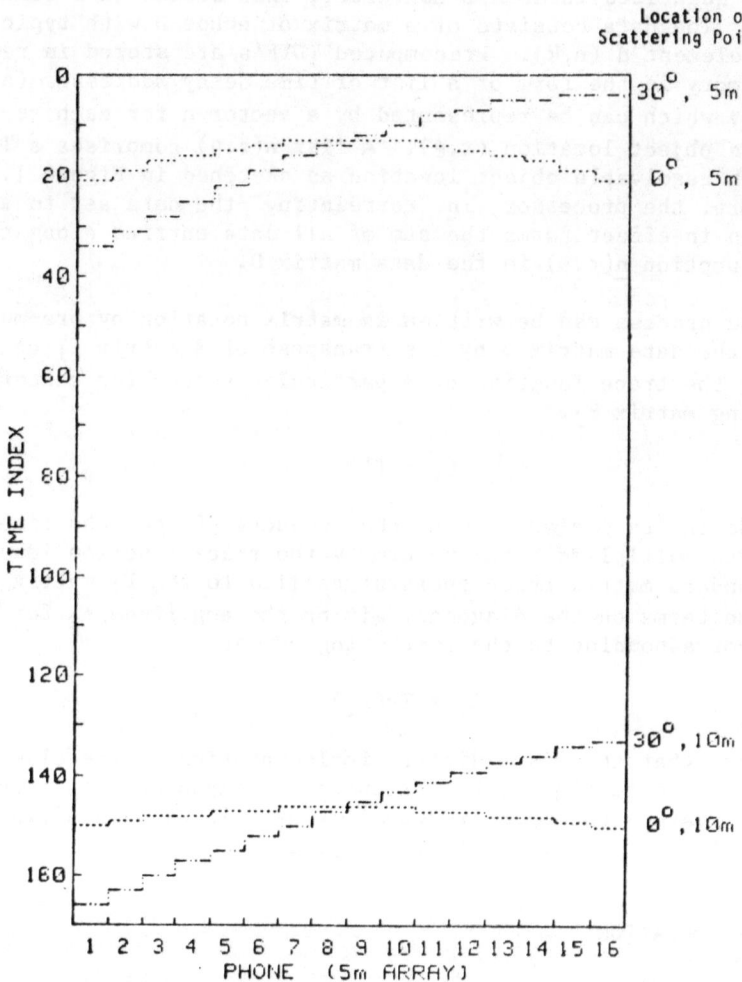

Figure 2.   Quantization of wavefront arrival times

scattering centers are shown in Figure 2. Note the effect of time quantization, which tends to limit angular resolution.

Data Structures

To visualize the signal processing operation it is useful to discuss it in terms of the data structures involved: the "Indata" array, the "Trace" matrix, and the "Amplitude" array as shown in Figure 3.

The Indata array contains the time record of the scattered acoustic field. It can be visualized as two dimensional in memory: i.e. $d(n,k)$ where n is the time index, and k is the hydrophone number. Each row would be formed by the simultaneous sampling of all hydrophones at intervals equal to the pulse width of 50 microseconds (beginning after a gate time of 6 milliseconds) and storing the results. This would continue for 167 samples until the last possible arrival time of a signal of interest (time indices 0 to 166).

The Trace matrix, $T(\theta,r,k)$ contains the traces $\underline{n}(r,\theta)$ for points in the area of interest as identified by the intersections of all range and bearing increments. It can be visualized as three dimensional where the first dimension is that of bearing, the second is that of range, and the third is the hydrophone number. For a particular range and bearing, the elements in the third dimension contain the time index sequence $\underline{n}(r,\theta)$ that is the pre-calculated trace: i.e. $n_k = T(\theta,r,k)$.

The dimensions for the Amplitude array are range and bearing. Each element, $A(r,\theta)$, is the total sum of the contents of those elements from the Indata array identified by the trace $\underline{n}(r,\theta)$:

$$A(r,\theta) = \sum_{k=1}^{K} d(n_k,k)$$

where n_k are elements of $\underline{n}(r,\theta)$

Scattering points will be evident by larger amplitude elements in $A(r,\theta)$ and the distribution of these will comprise an image of any objects present.

A decision process would then follow to determine which amplitudes represented point scatterers, as opposed to solely noise, and then these could be displayed forming the acoustical image of the insonified region.

scattering centers are shown in Figure 7. Here the effect of time quantization, which leads to limit sampling resolution.

Data Structures

To visualize the spatial processing, it is useful to consider it in terms of the basic data structures: the Indata array, the Trace matrix, and the Amplitude array as shown in Figure 3.

The Indata array contains the time record of the simulated acoustic field. It can be visualized as two images; that at lower left, Indata(i,k) where i is the time index, and k is the hydrophone number. Each row would be formed of the simultaneous sampling of all hydrophones at a time in question (a value which at 50 microseconds [sampling] rate represents on time variable and a scanning line header). Indata would contain 150 or so rows (a number possibly resulting from a sampling rate of 50 microseconds, which leads over 150 in time index values.

The Trace matrix, illustrated at upper center, is to retain in the area of interest so identified by the intersection of all range and bearing increments. There is a third dimension identified where the total resolution cell of the scanning in range and bearing exists. The cells each contain a sample for a particular voxel of space represented in the scanning system where the time when the magnitude relation that is represented exists in the actual cell.

The Amplitude array, shown at lower right, indexes the magnitude of signals at each resolution cell. The value of each element is the index of the scattering centers within that cell.

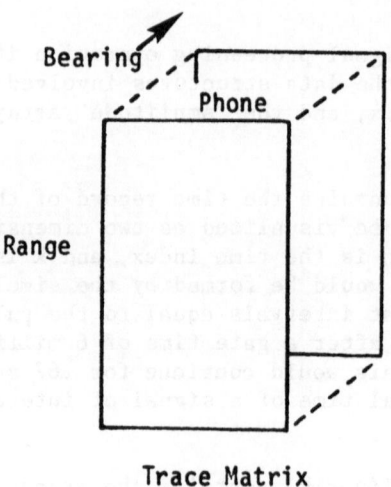

Trace Matrix

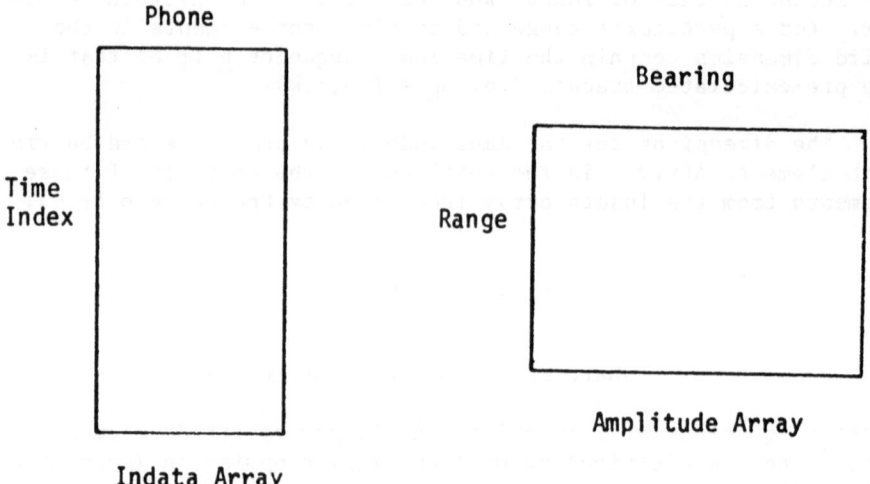

Indata Array   Amplitude Array

Figure 3. The basic data structures

scatter represented point references. The type is easily correlated and the range does to the Indata, in place the acoustical relationships are more combined.

Graphics Display

To evaluate angular resolution and imaging performance of the proposed system the graphics capability [4] of the HP9845T was utilized. It was possible to draw graphs such as shown in Figure 4 representing the received amplitudes from point scatterers and thereby obtain the system's angular resolution.

To simulate a system display, increments of 0.1 meter in range and two degree increments in bearing were calculated, a single point scatterer being represented by graphically "filling-in" an area on the screen equivalent to these dimensions and centered at the range and bearing of the point. This was termed a "target" in the developmental algorithm. A threshold was then used to decide when the magnitude of an element in the Amplitude array indicates a target present. An object was simulated by using a number of these discrete points which the graphics display then presented as a number of adjacent targets giving the image outline a stepped appearance. A rectangular grid was overlaid for geometric reference.

Simulated Images

Three large "objects", each considered as being composed of a number of point scatterers, were simulated in the field of view of the system. One was small and rounded as if it were a cross section of a long cylindrical object. The other two were long narrow objects; one was in a horizontal position, and the other at an inclined aspect angle. Figure 5 shows the positions and orientations of the simulated objects. No attempt was made to model a noise enrivonment or the reflectivity of a real buried object.

For the case of a pulse insonifying the entire sector and with clipped inputs from the receiving hydrophones, the maximum amplitude for any element in the Amplitude array was equal to the number of hydrophones. The threshold for the display of the resulting targets was varied to see the spatial distribution (in range and bearing) of the amplitudes of the image elements. Figure 6 shows the smearing of the objects resulting from the lack of angular resolution; i.e. the skirts of the amplitude distribution in bearing on each scattering point were being detected.

Similarly Figure 7 shows the results of having linear A/D conversion of inputs from the hydrophones. The largest amplitude elements were considerably larger than the nonlinear case, and the contours of the amplitude distribution in the image were much steeper. It was observed that the size, shape and aspect angle of the objects had a substantial effect on the resulting amplitude

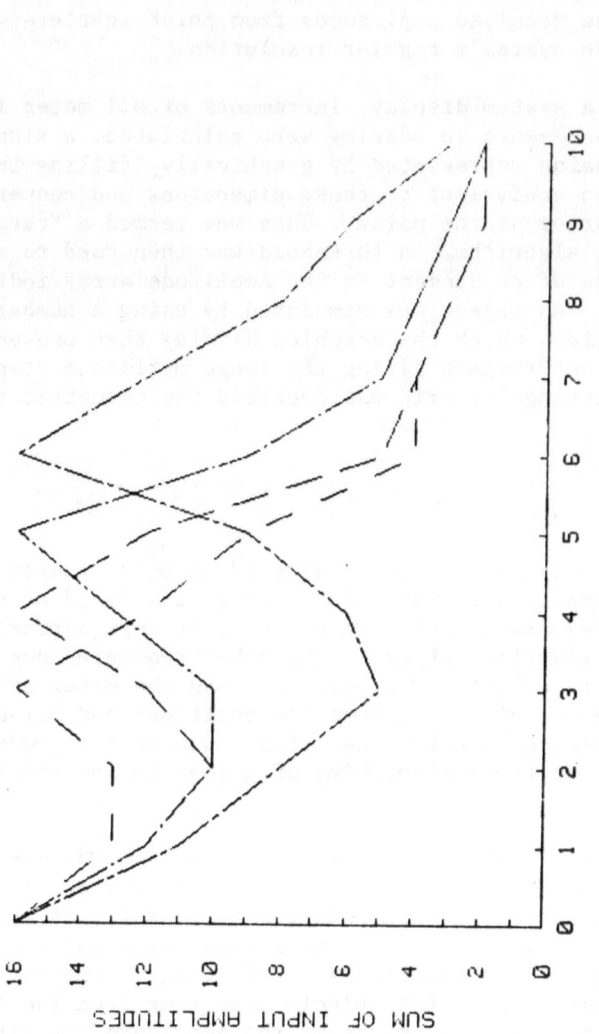

Figure 4.   The amplitude distributions for two points at a number of bearing separation angles at 10 meters range from a 5 meter line array

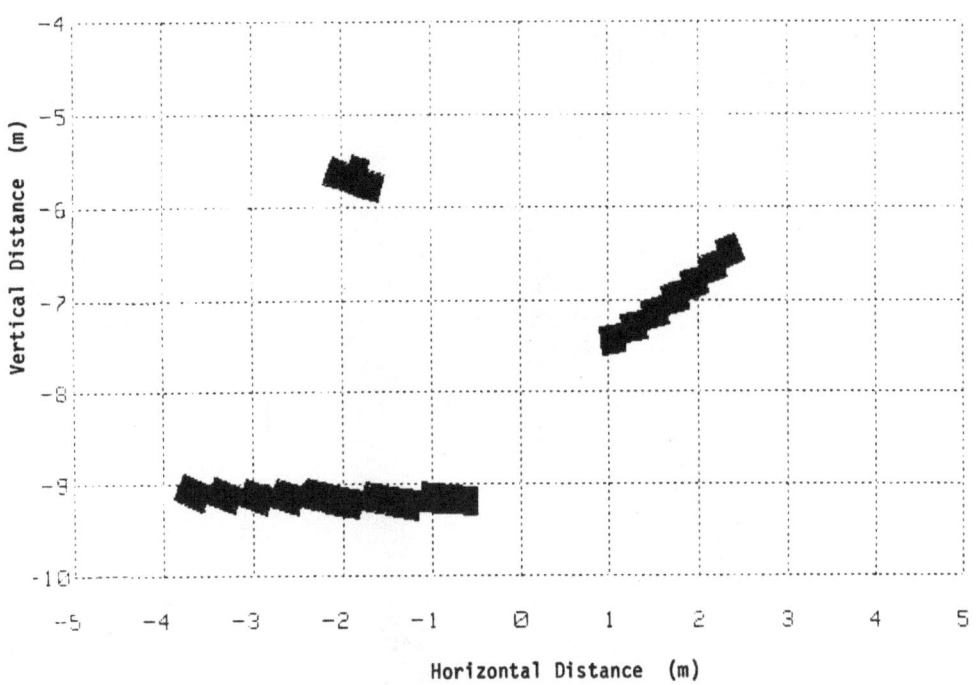

Figure 5. Simulated objects

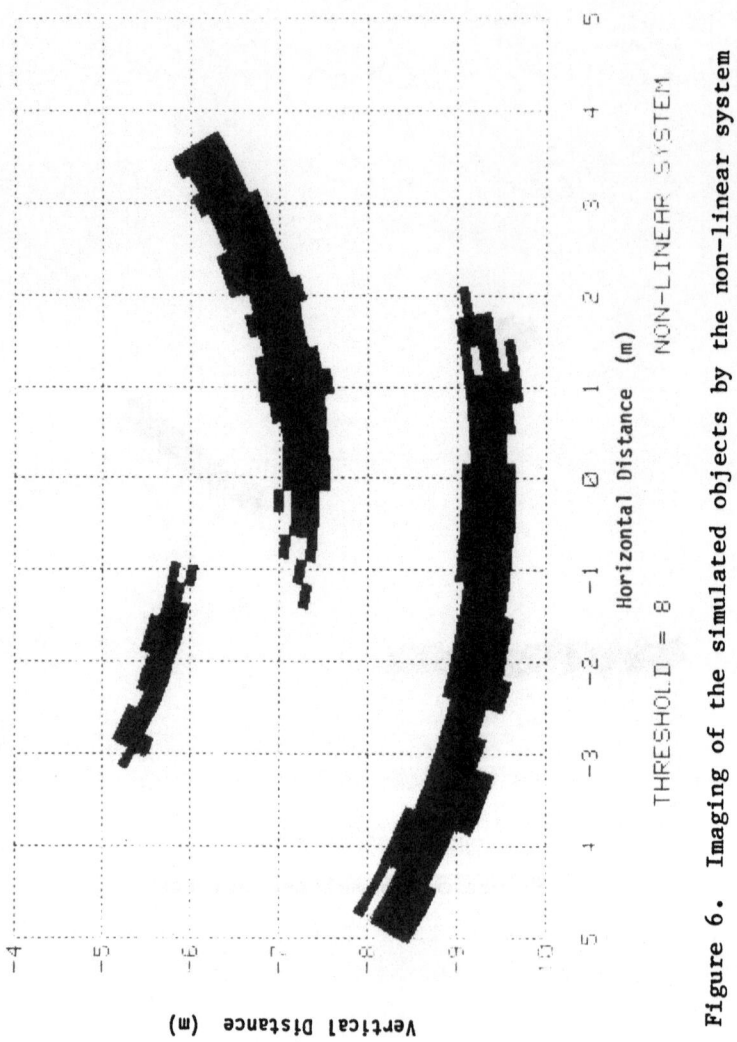

Figure 6. Imaging of the simulated objects by the non-linear system

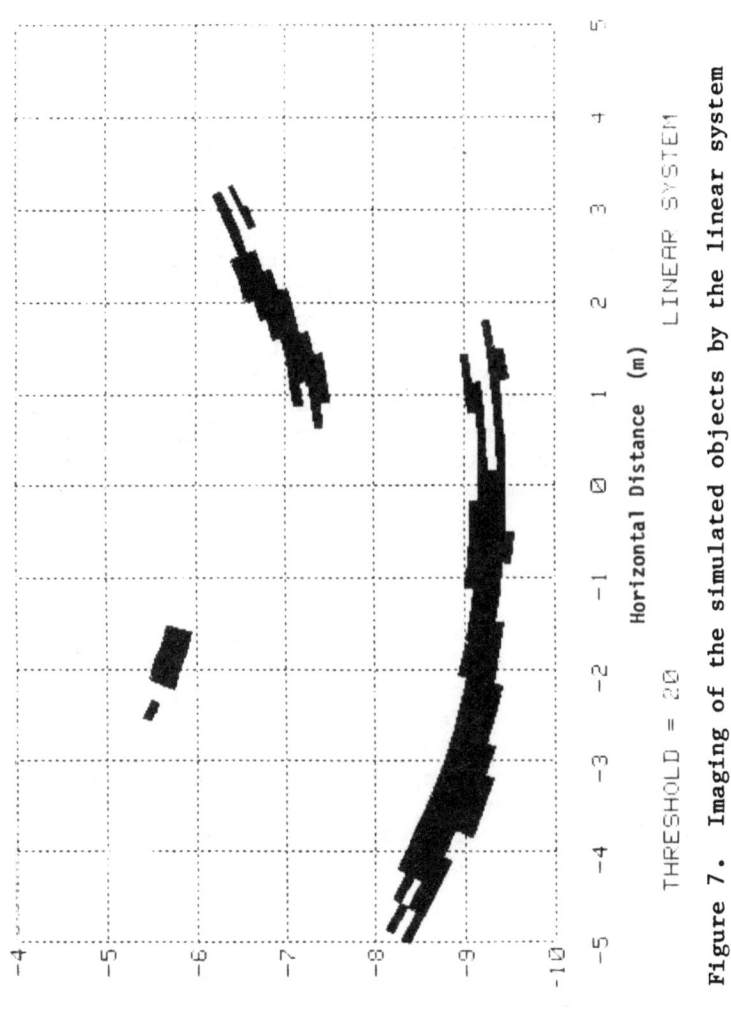

Figure 7. Imaging of the simulated objects by the linear system

distribution in the image. This would be beneficical for the
identification or classification function of an acoustic imaging
system particularly if an operator controlled threshold were
provided for the display or if pseudo color of the display
depicted the amplitudes of the image elements.

THE NEAR REAL-TIME ALGORITHM

An acoustic imaging system mounted on a maneuvering underwater vehicle would be required to provide an operator with a near
real-time display of any objects buried in the sediment as the
vehicle conducts a search pattern over the ocean floor. Although
the signal processing algorithm that was developed was basically
simple computationally, it was nevertheless very time consuming
and would have been inadequate in an operational environment.
Consequently, it was redesigned for increased execution efficiency
and implementation on multiple microprocessors while maintaining
the basic signal processing approach that has been described. The
algorithm was optimized by removing time consuming operations, together with incorporation of parallel and pipeline processes and
double buffering. Experiments were conducted using a suite of
single board computers to test and demonstrate the feasibility of
the redesigned algorithm for near real-time execution.

The Time Problem

An underwater vehicle acoustically scanning the sediment and
moving at 1 meter per second would probably require a complete
sector scan at least once per second as a minimum adequate display
rate. For example at 10 meter range, processing of the Indata array can begin approximately 15 milliseconds after pulse transmission, and would need to be completed in about 15 milliseconds for
each of the bearing increments if a scan were to be completed in
about one second. It was obvious that the developmental algorithm
would require considerable changes for an efficient microprocessor
implementation.

Algorithm Optimization

The algorithm was first examined with the objective of removing time consuming operations such as multiplication. These occurred in the addressing of the multi-dimensional arrays in the
algorithm which were stored in a row major order. Four one dimensional arrays were introduced which provided a table lookup of the
row addressing. For example, the addresses for the bearing dimension of the Trace matrix, $T(\theta, r, k)$ were stored in B_array(b), and

those for the range dimension in R_array(r). Instead of the addressing for a single element of $\overline{T(}\theta,r,k)$ requiring three additions and two multiplications, it was addressed using T(B_array(B) + R_array(r) + k); i.e., five addition operations. This was done for each of the multi-dimensional arrays, the Indata array D(n,k), the Trace matrix T(.θ,r,k), and the Amplitude array A(r,θ), thus reducing the data structures in the algorithm to strictly one dimensional arrays.

In order to test the feasibility of a microprocessor implementation of the algorithm using a number of Intel SBC 80/20 Single Board Computers (Intel 8080 CPU) [5], the algorithm was rewritten using the high level language PL/I-80 [6]. Executing the program on one SBC 80/20 it was found that after the pre-calculation of the traces in T(θ,r,k) the signal processing aspect of the algorithm required 6.45 seconds to complete one sector scan (approximately 210 milliseconds at each bearing, and not including the time required for pulse propagation and data acquisition).

Algorithm Partitioning

The structure of the signal processing algorithm was further examined in order to identify functions that could be accomplished by parallel and pipelined processes executing on a number of microprocessors for increased time efficiency. The concept of double buffering was also incorporated to reduce execution time.

Parallel Processing. The processing carried out at each of the range increments on the Indata array, for a given bearing, is independent of any of the other ranges. It was decided to partition this function into separate processes that were identical except for the scope of the range indices. These were called the Sum_amplitude processes and were executed in parallel on a number of microprocessors. Using three microprocessors for example, process S_1 performed summations along the traces for the range indices 0 to 16, process S_2 used range indices 17 to 33, and process S_3 covered range indices 34 to 50. In this manner all ranges were processing in one third the time that was required by a single process executing all range increments.

Pipeline Processing - Double Buffering. The function of data acquisition, called the Indata process, was seen to be amenable to pipeline processing with the Sum_amplitude processes. This process could control the timing of pulse transmission, sampling the hydrophone outputs, and storing the results into two buffer arrays. The Indata array of the developmental algorithm became two arrays called the A_indata array and the B_indata array. After the A_indata array was filled with the data from a particular bearing,

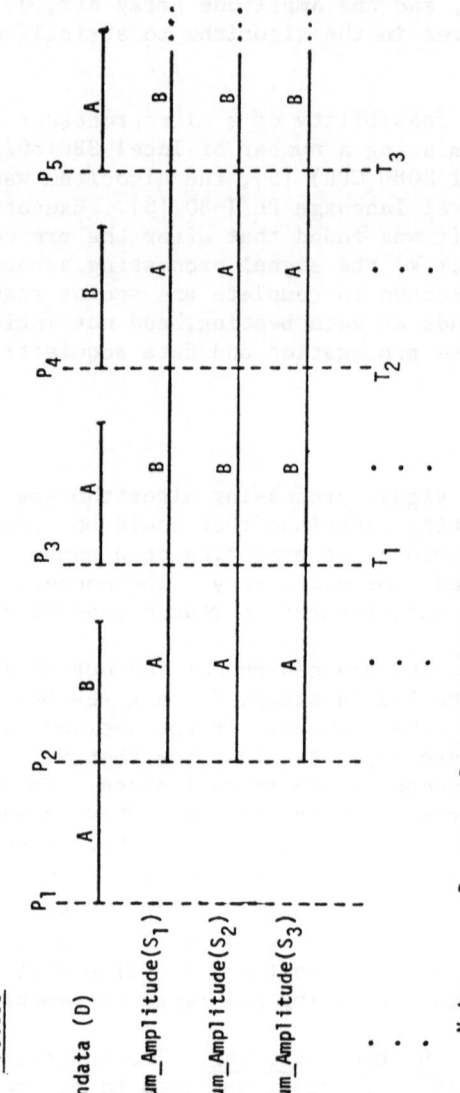

Process

Indata (D)

Sum_Amplitude(S₁)

Sum_Amplitude(S₂)

Sum_Amplitude(S₃)

Note:

P_i are pulse transmission times

T_i are times for completed processing at each
        of the bearing increments.

A   denotes operations on the A_indata buffer array.

B   denotes operations on the B_indata buffer array.

Figure 8. The relationship between the processes

the Sum_amplitude processes were permitted to begin their operations on the A_indata array. Simultaneously, the Indata process
proceeded to fill the B_indata array with the data from the next
bearing increment. When the Sum_ amplitude processes had finished
with the A_indata array, they processed the B_indata array. The
Indata process could then utilize the A_indata array for the next
bearing and so on in a pipeline fashion. Figure 8 illustrates the
relationship between the processes.

   Semaphores. Communication between the processes was accomplished with binary semaphores. The Indata process set the semaphores A_data and B_data whenever it was utilizing either the A_
indata or B_ indata array. Similarly, each of the parallel Sum_
amplitude processes set a semaphore depending on the data array it
was processing. Thus for three parallel Sum_amplitude processes
six semaphores were utilized. The Indata process waited for each
of the Sum_amplitude semaphores for one of the data buffer arrays
to be unlocked before it could proceed to fill the buffer with new
sampled data from the hydrophone outputs. Similarly, each of the
Sum_amplitude processes waited on the A_data or B_data semaphore
before it began processing the data. In this way, completely independent but synchronized operation of all the processes was
achieved.

Experimental Demonstration

   Equipment. To test and demonstrate the feasibility of near
real-time execution of the redesigned algorithm an Intellec 800
Microcomputer Development System (MDS) [7] with the CP/M monitor
control program [8] was used. The MDS is capable of being used as
a partial system simulator to check out software executing from
RAM in the Intellec MDS. A total of 64 kilobytes of memory was
available and four Intel SBC 80/20 Single Board Computers were inserted into the MDS motherboard as shown in Figure 9.

   The Processes. The contents of the Trace matrix (the pre-
calculated traces) were computed and stored in a floppy disk data
file for subsequent loading into memory. The redesigned "operational" algorithm was written as if it were to be executed on a
single processor. This was compiled and linked so the program
started at location 3000 Hex, based on the fact that addresses
3000 H to 3EFF H were available as private RAM on each of the
SBCs. All the data structures and variables such as the semaphores that were to be global to all the processes were located in
common memory above 4000 H that was accessible by all the SBCs.
An empty dummy array of the necessary size was declared to ensure
the assignment of common memory addresses to the global variables.

Figure 9.   The Microcomputer Development System with 64K of RAM and
four SBC 80/20 Single Board Computer inserted in motherboard

Using the MDS and the Dynamic Debugging Tool (DDT) program [9], the program was altered at the machine code level (hex representation) by removing the appropriate call statements and creating two versions: one which executed only the Indata procedure, and the other the Sum_amplitude procedure. In this way, the two processes stored in the private memory of separate SBCs could operate independently but in a synchronized manner using the semaphores in common memory. The data that was operated on, as well as all the data structures, were available in common memory.

The Sum_amplitude version was then amended using DDT so that all the local variables in the procedure were readdressed so they were within an SBC's private memory. This then provided a template for three subsequent copies of the Sum_amplitude process each of which was individaully altered to process a segment of the range indices.

Similarly, the addresses of variables that were local to the Indata procedure were amended so they were within an SBC's private memory. For the experimental demonstrates of the feasibility of the multiple microprocessor concept, the Indata process controlled the bearing parameter and caused the entire algorithm to scan ten complete sectors and then stop. An operational version would continue indefinitely. As the double buffering design resulted in two bearing increments being processed on each pass through the procedure, it was necessary to ensure that each sector consisted of an even number of bearing increments.

Results. Figure 11 provides a block diagram of the experimental configuration where process D is the Indata process and processes S_1 to S_3 are the Sum_amplitude processes. With the processes executing simultaneously, the time required for the amplitude summation function to be performed on ten sector scans was recorded.

Executing the algorithm with a different number of Sum_amplitude processes in parallel, while ensuring that all range indices were processed, permitted an evaluation of the effect of increasing the number of parallel processors. The resulting items for one sector scan and consequently the performance in terms of scans per second are given in Table 1.

TABLE 1. Results of the experimental demonstration of the feasibility of implementing the algorithm on multiple microprocessors

Processes	Time for one scan (seconds +0.01)	Scans per Second
D, S_1	6.45	0.155
D, S_1, S_2	3.23	0.310
D, S_1, S_2, S_3	2.15	0.465

The correct operation of the algorithm together with the performance times indicates that the design was a feasible approach to obtaining near real-time execution of the proposed signal processing algorithm. The fact that the processes were stored in the SBCs private memory and required access to the bus only when the common data structures and semaphores were referenced resulted in a minimum of bus usage. An examination of the code for the innermost loop of the Sum_amplitude process (the summation of the Indata array elements identified by the trace) indicated that the bus usage was 7.9% of the loop's execution time. The number of parallel SBCs executing simultaneously could theoretically be increased to twelve before bus contention would adversely influence the overall execution time. Table 2 clearly indicates the linear relationship between performance, in terms of scans per second, and the increasing number of parallel processors used. If the number of SBC's operating in parallel were increased to seven, plus one for the Indata process for a total of eight, then a scan rate of one second per second would be achieved.

CONCLUSIONS

Simulation first by minicomputer and then by multiple microprocessors demonstrated the feasibility of forming acoustic images in near real time using the trace function concept, for parameters typical of propagation in marine sediment. The process is easily adapted to changes in aperture geometry or other parameters by modification of software.

REFERENCES

1. G.L. Sackman & S.C. Shelef, "The Use of Time-Delay/Phase Difference Trace Functions for Bearing Estimation in Arrays", 13th Asilomar Conference on Circuits, Systems & Computers, 5-7 Nov 79, IEEE Catalog No. 79, CH 1468-8C, pp 354-358.
2. G.L. Sackman & S.C. Shelef, "The Use of Phase Difference Trace Functions for Bearing Estimation with Small Circular Arrays", IEEE Trans. Acoustics Speech & Signal Processing, Vol.ASSP-29, No. 3, part II, Jun 81
3. G.R. Vermander, "A Signal Processing Algorithm Based on Multiple Microprocessors for an Underwater Acoustic Imaging System", MS Thesis, Naval Postgraduate School, Monterey, CA, Dec 80
4. Hewlett-Packard System 45B Desktop Computer, Operating & Programming Manual, (09845B-91000), 1979, (Hewlett-Packard Desktop Computer Div., 3404 E. Harmony Rd., Ft. Collins, CO 80525)
5. SBC 80/20 and SBC 80/20-4 Single Board Computers Hardware Reference Manual, Pub. No. 98-317C, 1977, (Intel Corp., Customer Services, 3065 Bowers Ave., Santa Clara, CA 95051)

6. PL/I-80 Reference Manual, 1980, (Digital Research, PO Box 579, Pacific Grove, CA 93950)
7. Intellec 800 Microcomputer Development System (MDS) Operators Manual, Pub. No. 98-129A, 1975, (Intel Corp., Customer Services, 3065 Bowers Ave., Santa Clara, CA 95051)
8. An Introduction to CP/M Features and Facilities, 1978, (Digital Research, PO Box 579, Pacific Grove, CA 93950)
9. CP/M Dynamic Debugging Tool (DDT) Users Guide, 1978, (Digital Research, PO Box 579, Pacific Grove, CA 93950)

PHASE ERROR REDUCTION METHOD FOR A TOWED ARRAY

T. Sawatari

Bendix Advanced Technology Center
Bendix Center
Southfield, Michigan 48037

ABSTRACT

The frequency-domain adaptive interference rejection (FAIR) technique, developed to reduce interference effects, was modified to reduce the phase error effect in array data processing which results from the hydrophone position ambiguity of a towed array. The modification reduces the interference effect even when unknown phase errors exist in the input data. The new technique significantly enhances a weak signal buried by side lobes of a strong directional noise/jammers and increases target bearing estimation accuracy. In computer simulation tests the technique nearly eliminated the side lobe effect.

INTRODUCTION

It has been speculated that the position of each hydrophone in a towed array often deviates from that expected because of ocean wave and water resistance when the array is towed. A recent experimental study in which accelerometers were attached to a towed array verified that the towed array fluctuates.[1] The frequency of the fluctuation was less than 1 Hz and the amplitude of the fluctuation was less than 30 cm; no higher frequency components were observed. Although this fluctuation is small compared with the length of a towed array, significant error results in the measurement of the phase of the incident wave; e.g., if the deviation is 20 cm and the wavelength of the incident wave is 3.0 m (500 Hz), the phase error is about 24 degrees.

245

The array phase error introduces deformation in the beam
pattern of the towed array. Two factors complicate the situation: The low frequency nature of the fluctuation makes it
difficult to compensate the phase error by time-averaging of the
data, and the fluctuation may be structured (periodic and/or slowly
curved) rather than random. The periodic fluctuation gives rise to
dominant side lobes which cause more severe identification problems
if a weak desired target is located in the area.

The frequency-domain adaptive interference rejection technique
(FAIR) described in recent articles reduces the signal processing
problem to the optimization of n variables, where n is the number
of interference sources.[2] In a conventional optimal detection
scheme, such as the Wiener filter or the most likelihood filter,
the optimization is generally required with NJ degree of freedom,
where N and J are the number of array elements and frequency bins,
respectively. The optimal detection technique handles both
interference sources as well as background white noise.[3,4] The
FAIR algorithm, a suboptimal technique, only reduces the influence
of strong localized interferences.[5-7] Inexpensive and faster,
though at a sacrifice of some applicability, it is very effective
in sonar applications in which enemy jammers or surface ships act
as strong localized interference sources.

Recent work on random noise tolerance of the FAIR technique
has concluded that random phase error of less than 10 degrees does
not affect the algorithm performance.[8] This paper treats the case
where the phase error is periodic, slowly curved, or both.
Modifying the FAIR algorithm to compensate the influence of the
error results in a technique which enhances a weak desired target
and increases the accuracy of estimating the location of the weak
signal.

The following section mathematically describes the phase error
due to the periodic and slowly curved movement of a towed array,
and analyzes its effect on the beam-formed pattern. In the next
section the technique for compensating the array phase error is
described after a brief review of the FAIR algorithm. The results
of computer simulation tests which demonstrate the effect of weak
signal enhancement and improved target bearing estimation are then
presented.

ARRAY PHASE ERROR EFFECTS

Position Ambiguity and Phase Error

Figure 1 schematically shows a towed array fluctuation. A
phase ϕ of the mth hydrophone is

$$\phi = (2\pi/\lambda)(\bar{P}_m \cdot \bar{\alpha}_n) \tag{1}$$

where $\bar{P}_m$ is the three-dimensional vector of the position coordinate of the mth hydrophone, α_n is a unit vector in the same direction as the wave propagating from the nth target, and λ is the wavelength of one wave. The position vector can be divided in two components: the ideal location $\bar{P}_m^o$ of the hydrophone and the deviation $\Delta\bar{P}_m$ from the desired location. Thus

$$\bar{P}_m = \bar{P}_m^o + \Delta\bar{P}_m \tag{2}$$

Substituting this equation into (1), the phase error $\Delta\phi$ due to the position ambiguity $\Delta\bar{P}_m$ is defined as

$$\Delta\phi = (2\pi/\lambda)(\Delta\bar{P}_m \cdot \bar{\alpha}_n) \tag{3}$$

If the geometrical coordinates are chosen as in Fig. 1, where the Y axis is perpendicular to the drawing, the phase error can be represented by the major component of the dot product:

$$\Delta\phi = (2\pi/\lambda) \Delta Z_m \cos \Theta_n \tag{4}$$

where ΔZ_m is the Z component of the vector $\bar{P}_m$ and Θ_n is the direction angle component of the nth target (jammer) measured from the Z axis (normal to the array). This approximation is valid particularly when the targets are in the X plane and the angles Θ_n are small compared to $\pi/2$--the condition normally encountered in towed array operations.

Periodic Phase Error

When a towed array fluctuates in a periodic structure, the fluctuation can be expressed by an amplitude or a phase modulated sinusoidal wave. The Z component of the fluctuation may be expressed as

$$\Delta Z_m = A_m \sin (\omega_o m + \psi_m) \tag{5}$$

where $\omega = (2\pi/\Lambda)a$, Λ being the mean wavelength of the sinusoidal fluctuation and a the spacing between the hydrophones. The amplitude and phase are given by

$$A_m = A_o + A_a \sin (\omega_a m + \psi_a)$$
$$\psi_m = \psi_o + A_p \sin (\omega_p m + \psi_p) \tag{6}$$

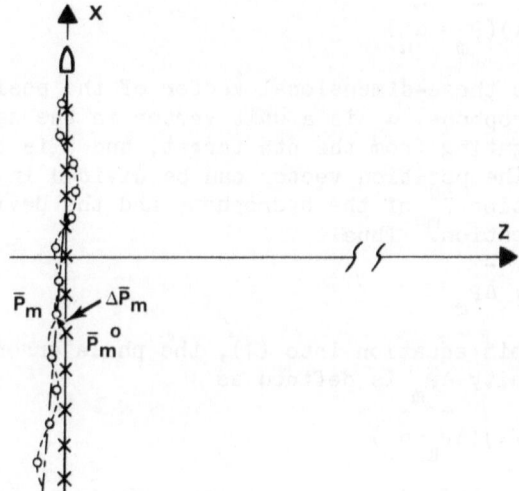

Figure 1.   Schematic of Towed Array

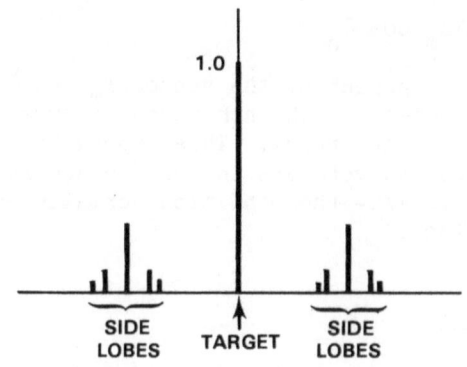

Figure 2.   Side Lobes Due to Periodic Array Phase Error

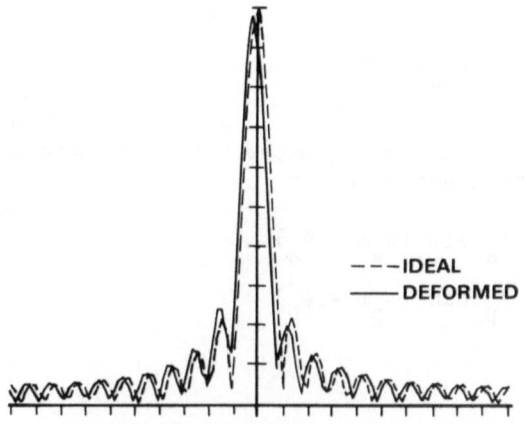

Figure 3.   Effect of Phase Error Due to Array Bending

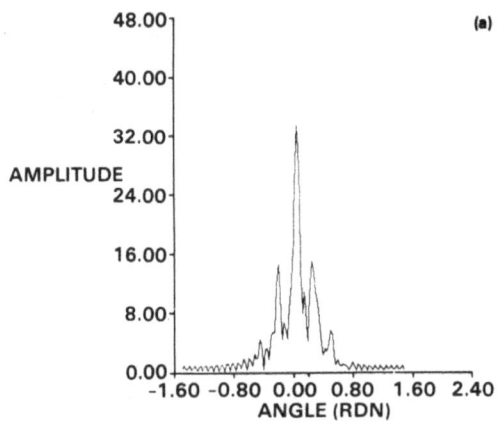

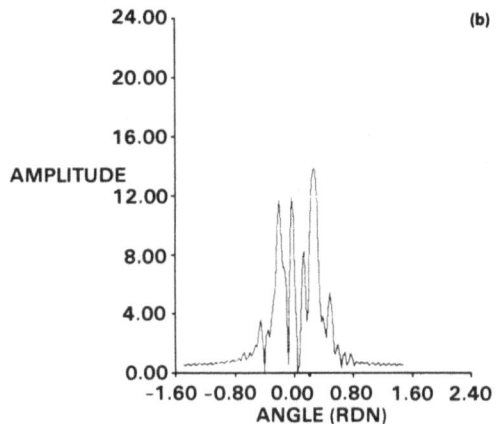

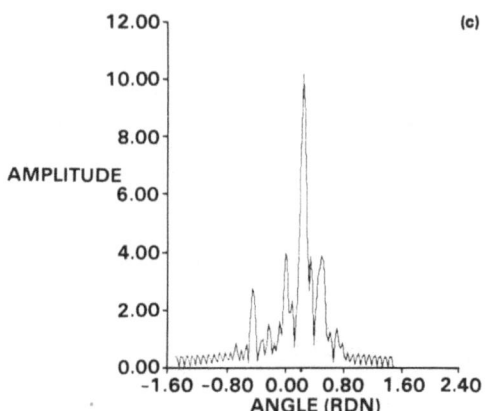

Figure 4.   Weak Signal Enhancement (amplitudes of jammer, weak
            signal, and white noise are 1.0, 0.3, and 0.0)
            (a)   Conventional Beam-Forming
            (b)   Conventional FAIR Processing
            (c)   Modified FAIR Processing

where A_a, ω_a, and ψ_a are the amplitude, the frequency, and the phase of the amplitude modulation; A_o, ω_o, and ψ_o are those of the fundamental sinusoid; and A_p, ω_p, and ψ_p are those of the phase modulation.

The side lobes that result from the phase error (Eq. (4)) due to the position ambiguity (Eq. (5)) are complicated spikes. Figure 2 shows an example in which the array size effect has been ignored to simplify the illustration. These side lobes reduce the accuracy of the target bearing estimation when one of them interferes with the desired target.

Phase Error Due to Array Bending

Another type of phase error introduced in a towed array is that due to the bending of the array. The hydrophone position ambiguity this causes (see Fig. 1) may be expressed by a poly-nomial. If the term associated with the X component is ignored because it is small and if $\Theta_n \ll 1$, then the hydrophone position ambiguity is

$$\Delta Z_m = \sum_{i=1}^{I} \beta_{i-1}(am)^i \tag{7}$$

where β represents the coefficients of each term of the polynominal.

If the bending is rather gradual, a polynomial with I less than 4 may sufficiently describe it. The first (linear) term of the polynomial, which causes a lateral shift of the image that cannot in principle be corrected without information beyond the input data, is not due to the bending per se and is thus omitted from later discussion.

Figure 3 shows the effect of the array phase error expressed by Eq. (7) on the image of a target. The peak value of the target decreases and the image shows a broadened asymmetrical feature. This deformation reduces the accuracy of the bearing estimation.

ARRAY PHASE ERROR COMPENSATION

FAIR Algorithm

The FAIR technique is implemented for two basic purposes: first, to enhance a desired target which is buried in the spread of a nearby target[9] and second, to increase the accuracy of target

bearing estimation.[5-7] The method is identical to the optimal
detection scheme when a few localized interferences are present--
the target scenario of most interest in sonar applications.

The FAIR method proceeds in three steps: (1) a projection
operator based on the direction vectors corresponding to the
roughly estimated target bearing is constructed to null all
localized targets simultaneously; (2) the total remaining power is
minimized as a function of the null location; and (3) the best
estimation of the target bearing is taken as that which minimizes
the total remaining power. These steps are explained in detail
below.

Projection Operator. The conventional beam-forming is
given by

$$\underset{\sim}{y} = F^{\dagger} \underset{\sim}{x} \tag{8}$$

where $\underset{\sim}{x}$ is the input vector, the mth component of which is the
complex amplitude of the input radiation of a particular frequency;
and $F^{\dagger}$ is the transpose of the beam-forming matrix, the column
vector of which is a beam-forming vector $\underset{\sim}{k}_j$. The beam-forming
vector $\underset{\sim}{k}_j$ is given by

$$\underset{\sim}{k}_j = [\exp(2\pi jm/K)/]M \quad ; \quad j = 1,2,\ldots,K \tag{9}$$

where M is the total number of hydrophones in the array and K is
the number of image points. The intensity distribution of the
image $|y(j)|^2$ shows high peaks when $j/K \simeq a \sin \Theta_n/\lambda$. The rough
estimation of the bearing angle Θ_n of the nth target (jammer)
is $\lambda j/aK$. The bearing angle Θ_n cannot be estimated accurately no
matter how large an integer K is chosen because of the side lobe
interference of the nearby source.

To increase the accuracy of the estimation Θ_n, a set of
direction vectors $\underset{\sim}{d}_n$ is first determined from the rough estimation
of bearing angle Θ_n:

$$\underset{\sim}{d}_n = \exp(i2\pi a\Theta_n m/\lambda) \quad n = 0,1,2,\ldots,N \tag{10}$$

where N is the total number of jammers. The direction matrix is
defined as

$$D = (\underset{\sim}{d}_0,\underset{\sim}{d}_1,\ldots,\underset{\sim}{d}_N) \tag{11}$$

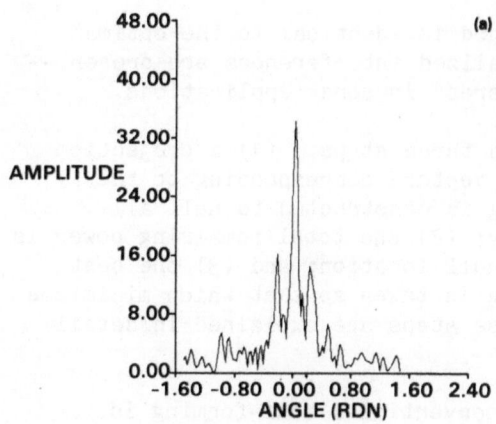

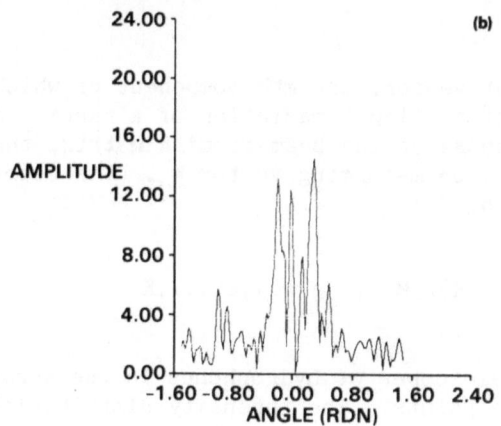

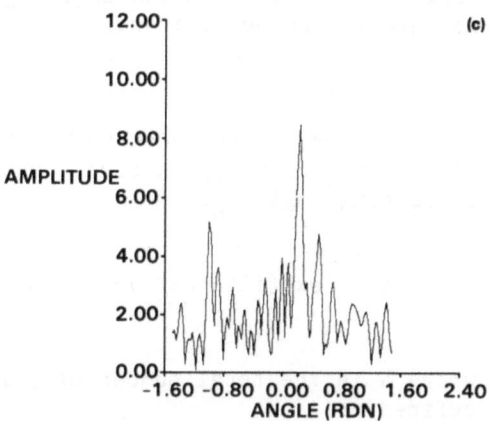

Figure 5.   Weak Signal Enhancement (amplitudes of jammer, weak
            signal, and white noise are 1.0, 0.3, and 0.3)
            (a)   Conventional Beam-Forming
            (b)   Conventional FAIR Processing
            (c)   Modified FAIR Processing

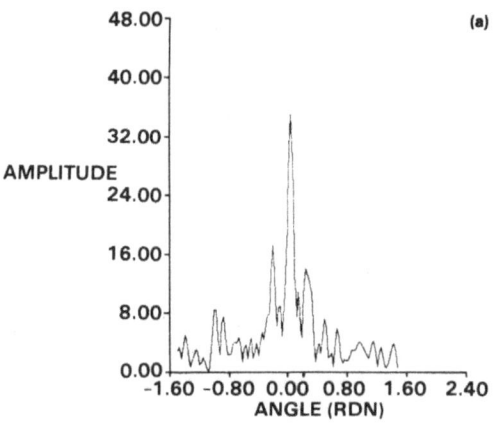

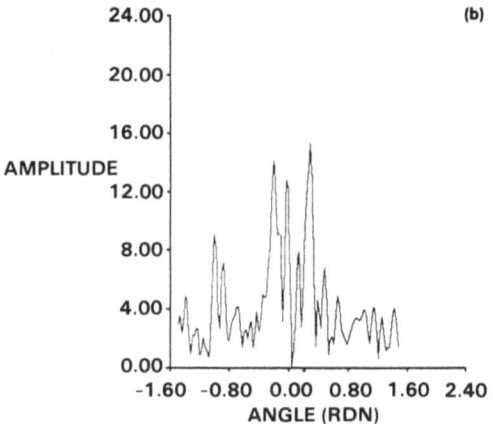

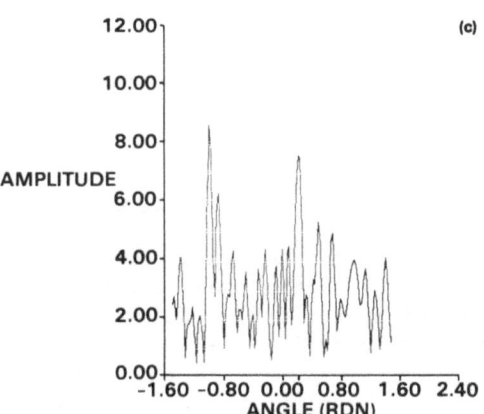

Figure 6. Weak Signal Enhancement (amplitudes of jammer, weak
           signal, and white noise are 1.0, 0.3, and 0.5)
           (a) Conventional Beam-Forming
           (b) Conventional FAIR Processing
           (c) Modified FAIR Processing

The Nth rank projection operator associated with the N
vectors $\underset{\sim}{d}_n$ can be constructed as

$$P_N(\underset{\sim}{\Theta}) = D(D^\dagger D)^{-1}D \qquad (12)$$

Then a new input vector is created as

$$\underset{\sim}{x}_N = (I - P_N(\underset{\sim}{\Theta}))\,\underset{\sim}{x} \qquad (13)$$

where I is the unit matrix. The new vector $\underset{\sim}{x}_N$ consists of the
background noise and the residue noise after the interference
sources and the desired target are nulled from the input vector.
The residue noise is due to the mismatch between the true target
bearings Θ_n^0 and the estimated bearings Θ_n.

  Minimization of Total Remaining Power. The total power
remaining after nulling the interference sources and the true
target is given by

$$Q_N(\underset{\sim}{\Theta}_j) = \underset{\sim}{x}_N^\dagger\,\underset{\sim}{x}_N \qquad (14)$$

where Q_N is a function of the estimated bearing Θ_n of the interferences. If the bearings are changed in small amounts, the amount
of the residue noise changes; therefore, the value of Q_N changes.
The best estimate of the bearings is thus defined as that which
minimizes the value $Q_N(\underset{\sim}{\Theta})$.

  Target Bearing Estimation. The problem of finding the best
estimate of the bearings is now replaced by the simple problem of
finding the minimum of $Q_N(\underset{\sim}{\Theta})$ by solving the equations

$$\underset{\sim}{\overline{V}}_\Theta(Q(\underset{\sim}{\Theta})) = 0 \qquad (15)$$

These equations can be solved numerically using approaches like the
steepest-descent method, the Newton-Raphson method, or the simplex
method.

FAIR Modification for Array Phase Error Compensation

  The FAIR technique can be modified to compensate for phase
errors by adding phase compensation variables to the direction
vector in Eq. (10). Using Eq. (4)-(7) the modified direction
vector is defined as

$$\hat{\underset{\sim}{d}}_n = \exp[i2\pi a\Theta_n m/\lambda + i\Psi_m(\underset{\sim}{\nu}) + i\Phi_m(\underset{\sim}{\xi})] \qquad (16)$$

where

$$\Psi_m(\underset{\sim}{\nu}) = [\nu_1 + \nu_2 \sin(\nu_3 m + \nu_4)]$$
$$X \sin[\nu_5 m + \nu_6 + \nu_7 \sin(\nu_8 m + \nu_9)]$$

and

$$\Phi_m(\underset{\sim}{\xi}) = \sum_{j=1}^{I} \xi_j m^j \qquad . \tag{17}$$

That is, $\underset{\sim}{\nu}$ and $\underset{\sim}{\xi}$ are the variables related to the periodic and the bending phase errors, respectively.

The projection operator P created from the direction vector $(\hat{d}_n)$ is now a function not only of $\underset{\sim}{\Theta}$ but also of $\underset{\sim}{\nu}$ and $\underset{\sim}{\xi}$. Therefore the function Q is expressed by

$$\hat{Q}(\underset{\sim}{\Theta},\underset{\sim}{\nu},\underset{\sim}{\xi}) = |I - P(\underset{\sim}{\Theta},\underset{\sim}{\nu},\underset{\sim}{\xi})\underset{\sim}{x}|^2 \tag{18}$$

The best estimate of the bearing angles $\underset{\sim}{\Theta}$ is again defined as those $\underset{\sim}{\Theta}$ which together with $\underset{\sim}{\nu}$ and $\underset{\sim}{\xi}$ minimize Q; i.e., these angles satisfy

$$\nabla_{\underset{\sim}{\Theta},\underset{\sim}{\nu},\underset{\sim}{\xi}}\hat{Q} = 0 \tag{19}$$

The numerical minimization of the function Q is not a trivial problem.

COMPUTER SIMULATION

System/Target Scenario

The simulation test used a 41-element linear uniform array. The spacing between two adjacent hydrophones was a, which was $\lambda/3$; λ was the wavelength of the radiation from the targets. The radiation from the targets, located in the far-field, was mutually coherent within a frequency band of interest. The phase errors introduced in the array system included periodic fluctuations as well as those due to gradual bending of the array. Both were assumed to be stationary during the data acquisition.

The target scenario and the array phase error were chosen such that the target bearing estimation is significantly influenced. Two targets separated by two beam-widths are present in the target plane; the frequency of the sinusoidal phase error causes one of the side lobes of one target to interfere with the other target. The magnitude of the sinusoidal errors was varied from 6 to 108 degrees with respect to the wavelength of the radiation (in a sonar towed array the expected value is about 10 degrees). The effect of background noise was studied by adding white noise to the simulated input data for phase errors of 36 degrees.

Computation Time Reduction

The rather complex function Q contains many local minima that are not the solution to the problem. A fast search method such as the steepest-descent method finds these irrelevant nearby local minima unless the initial values are close to the true value. Thus a grid search is required in which each variable is independently moved with small increments (width of grid) over its entire range and then the Q-values corresponding to each grid are compared to find the minimum. A straightforward grid search is time-consuming; e.g., even reducing the total number of variables (Θ,ν,ξ) to six will require 3,000 calculations for one input if the search range of each variable is five cells.

Taking the physical significance of the set of variables (Θ,ν,ξ) into consideration, we arrive at the following grid search procedure. The variables are divided into two groups: a set of ν and a set of Θ and ξ. We fix Θ to the initial estimate obtained in Eq. (8), choose $\xi = 0$, and then find a set of ν with the grid search which minimizes the function Q. Using this set of ν we conduct a grid search of a set of Θ and ξ which also minimizes the function Q. The obtained set of values (Θ,ν,ξ) is close enough to those values that we define as the best estimate that the steepest-descent method is now applicable. This process reduces the number of calculations in the initial value search from 3000 to 250.

RESULTS

Weak Signal Enhancement

When the modified FAIR algorithm is applied to enhance a weak signal, it is assumed that the desired target is not visible in the image plane because its weak signal is buried by a side lobe of a strong localized target which has been identified. To enhance the weak signal nulls are located on these strong interferences only. We then minimize the total remaining power using the variables Θ, ν, and ξ, and display the field remaining in the image plane. Any weak signal in the object plane will now be visible or identified as an image.

Figures 4 through 7 are typical results. In each figure the
first part (a) shows the results of conventional beam-forming, the
second part (b) the results of conventional FAIR processing, and
the third part (c) the patterns obtained with modified FAIR processing. At the array detectors the RMS amplitude of the white
noise relative to the weak signal is 0.0, 1.0, and 1.7 in Fig. 4,
5, and 6, respectively. The ratio of the amplitudes of the signal
and the interference is 0.3:1, or about -10 db.

In Fig. 4(a), 5(a), and 6(a) the weak signal, located at the
position shown by an arrow, is not seen at all. The left side lobe
of the central peak, the strong interference source, is actually
greater than the peak at the weak signal location. In Fig. 4(b),
5(b), and 6(b) the desired signal is surrounded by several high
spikes. These spikes, caused by the array phase error, have a
magnitude nearly as great as that of the weak signal. It is
difficult to determine the desired signal. In Fig. 4(c), 5(c), and
6(c) the weak signal is clearly identified, or enhanced. The high
peak on the left of the image plane in Fig. 6(c) is due to white
noise, but the weak signal is still considerably enhanced relative
to the nearby spikes.

In Fig. 7 the amplitude of the weak signal has been reduced to
0.1 (-20 db) relative to the strong interference (1.0), and the
background white noise is zero. Although the enhanced weak signal
is visible in Fig. 7(c), a level of white noise of 0.3 RMS amplitude (-9.5 db) would make the modified FAIR technique ineffective
given the small amplitude of the weak desired signal.

Accurate Bearing Estimation

The bearing angles of two targets as estimated by the conventional FAIR method and the phase error compensation method were
compared. The magnitude of the phase error or the amplitude of the
periodic error (A in Eq. (11)), and the coefficient of the array
bearing error (β_1 in Eq. (12)), were varied; the white noise level
was zero. The modified FAIR method provided estimates closer to
the true target locations, and functioned quite well up to a white
noise RMS amplitude of 0.9 relative to the unit signal amplitude.
Table 1 summarizes the errors in target bearing angle estimation
for both techniques. The results show that the phase compensation
method works very well for array phase errors between 0 to 60
degrees.

Figure 8 illustrates the various stages of the computer processing. The white noise level in this simulation was 0.1 in RMS
amplitude relative to the unit signal amplitude. In Fig. 8(a), the

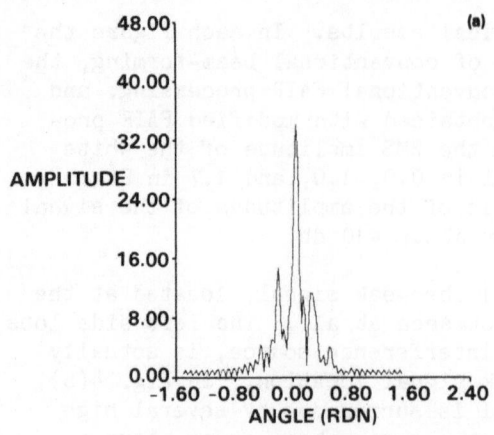

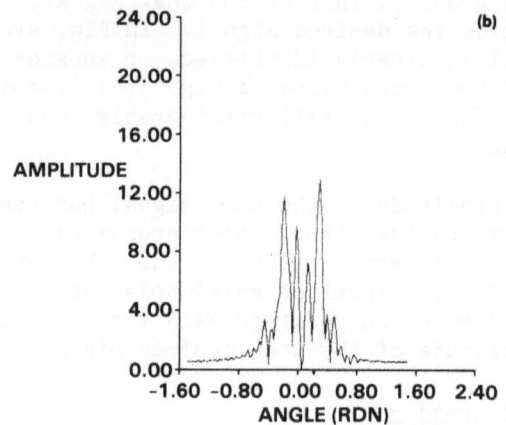

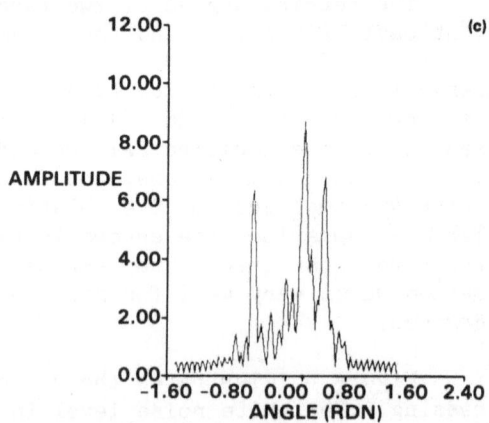

Figure 7.   Weak Signal Enhancement (amplitudes of jammer, weak
            signal, and white noise are 1.0, 0.1, and 0.0)
            (a)   Conventional Beam-Forming
            (b)   Conventional FAIR Processing
            (c)   Modified FAIR Processing

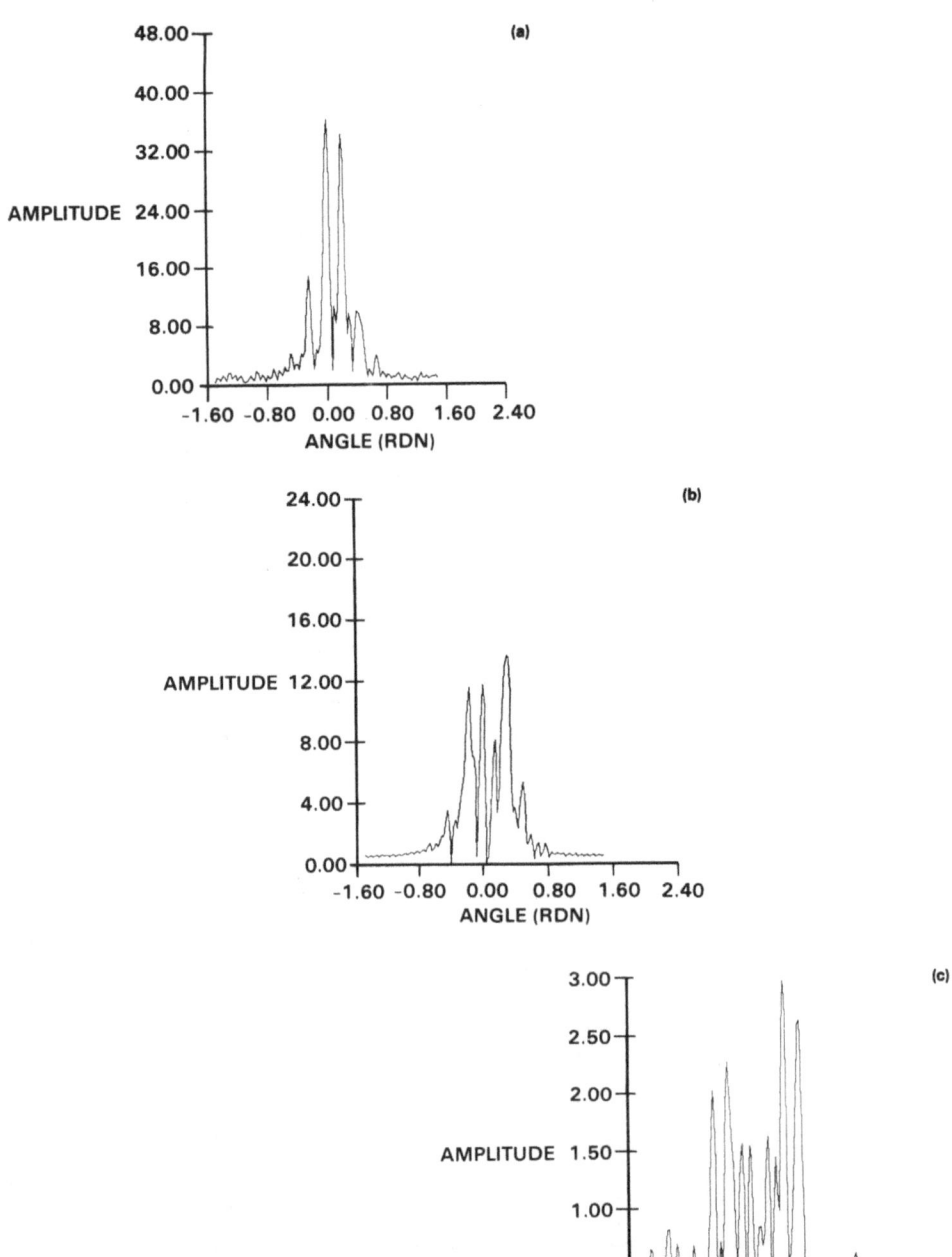

Figure 8. Target Bearing Estimation
   (a) Conventional Beam-Forming Pattern
   (b) Image Pattern After Nulling of Strongest Target
       (initial estimate of target bearings)
   (c) Field Remaining After Nulls are Optimized

beam-forming pattern of the simulated input data, the side lobes are clearly seen around the two strong targets in the center. (The higher harmonics of the side lobes are also apparent). Figure 8(b) presents the remaining field after the strongest target is nulled, with the second target and the side lobes visible. Figure 8(c) is the final remaining field after the locations of the null and the phase error compensation variables are optimized. The magnitude of the remaining spikes is very small compared with that of the side lobes in Figure 8(a).

SUMMARY

The phase error compensation technique, a modification of the FAIR algorithm, provided a fivefold enhancement of a weak signal buried by side lobes of a strong local interference and improved target bearing estimation accuracy by 30 percent at a 36 degree phase error in computer simulation tests. Close examination of the variables involved in the computation resulted in faster processing. However, since the simulation was designed only to test the feasibility of the technique, its model is rather simple if realistic. The periodic error has no phase modulation, and the simulated bending error consists of only one quadratic term. The modified FAIR technique must now be tested with real data to determine its actual performance.

Table 1. Errors in Target Bearing Estimation Measured by System Beam-Width (percent)

PHASE ERRORS / TECHNIQUES	6°	12°	36°	72°	108°
NO PHASE COMPENSATION	4	20	33	172	>200
WITH PHASE COMPENSATION	0.7	1	4	53	>200

REFERENCES

1. F. Demetz, personal communication, 1980.

2. P. N. Keating, A rapid approximation to optimal array
    processing for the case of strong localized interferences,
    J. Acoust. Soc. Am. 65:456 (1979).

3. O. L. Frost III, Proc. IEEE 60:926 (1972).

4. N. L. Owersley, Signal processing, "Proc. of the NATO Advanced
    Study Institute on Signal Processing, with Particular
    Reference to Underwater Acoustics," Loughborough, England,
    1973, J. W. R. Griffithes et al., ed., (1973).

5. R. F. Steinberg, P. N. Keating, and R. F. Koppelmann,
    Experimental implementation of advanced processing in
    acoustic holography, "Acoustical Holography," Vol. 6, N.
    Booth, ed., Plenum Press, New York (1976).

6. P. N. Keating, R. F. Koppelmann, R. K. Mueller, R. F.
    Steinberg, and G. J. Zilinskas, J. Acoust. Soc. Am. 59:106
    (1976).

7. P. N. Keating and T. Sawatari, Holographic adaptive
    processing: a comparison with LMS adaptive processing,
    "Acoustic Holography," Vol. 7, L. W. Kessler, ed., Plenum
    Press, New York (1977).

8. P. N. Keating, Effect of array errors on frequency-domain
    adaptive interference rejection, J. Acoust. Soc. Am.
    (Dec. 1980).

9. T. Sawatari, P. N. Keating, and R. E. Willey, Holographic
    adaptive interference rejection technique for low angle
    radar, IEEE Trans. Aerospace and Electronic Systems, in
    press.

DIGITAL ENHANCEMENT OF ULTRASONIC IMAGES AND ITS APPLICATION

TO NON-DESTRUCTIVE TESTING OF COMPOSITE MATERIALS[†]

P. Das[*] and R. Werner

Electrical, Computer, and Systems Engineering Department
Rensselaer Polytechnic Institute
Troy, New York 12181

ABSTRACT

An ultrasonic transmission imaging system is used for imaging composite materials. The system uses two focussed transducers with the data acquisition performed under the control of a Z-80 microprocessor based system. The images are processed by a DEANZA image processor with a PRIME 750 host computer. The computer and image processor are capable of such functions as pseudo-coloring, deconvolution to improve resolution and adaptive filtering techniques in the Fourier domain. This system has been used to study the damage caused by applied stress in composite materials. The graphite epoxy composite materials studied have fibers oriented in various directions and it is found that the damages are highly dependent on the fiber orientation.

1. INTRODUCTION

The promise of composite materials, whose development may be considered as entering the proto-type development of manufacturing phase, continues to generate strong interest. The very desirable properties of high strength, high modulus, and low density in these materials arise from the use of oriented graphite fibers embedded in a hardened epoxy resin. The fibers can be built up in layers with the laminations oriented in various directions throughout the

---

[*]On sabbatical leave at University of California, San Diego for Spring '81 semester.

[†]Partially supported by NASA under Grant No. NGL 33-018-003.

piece so that loads may be evenly transported. Realization of the
promise of fibrous composite materials depends upon the detection
of stress induced, non-catastrophic flaws.[1-4]

Stressed samples of various composite materials have been evaluated using a doubly focussed transmission imaging system. The data
gathering part of this imaging system is performed under the control
of a Z-80 microprocessor and is described in section 2. The final
images are obtained from these data after processing through the
facilities of the Image Processing Laboratory with a PRIME 750 computer and DEANZA image processing system. This is described in
section 3 which is followed by the images of the composite samples
which have been previously stressed to cause damage in a predeter-
mined way.

2. ULTRASONIC IMAGING SYSTEM

The data acquisition part of the ultrasonic imaging system is
shown in fig. 1. The data for the ultrasonic images is obtained
under the control of a Z-80 microprocessor working at a 4 MHz clock
rate. The Z-80 triggered a pulse generator which amplitude modulated the output of a signal generator oscillating at the resonant
frequency of the ultrasonic transducers. The commercially available
focussed immersion transducers had a 5 Mz resonant frequency and a
focal length of 1.5 inches. The two identical transducers were
aligned colinearly and focussed on the same plane. The imaging was
performed in a 150 gallon tank where no reflections interfered with
the received signal. The samples, lightly coated with glycerol to
stop air bubbles from clinging to the surface, were positioned in
the focal plane of the sample, the transmitted signal was received
and amplified, then sampled once during each repetition period.
The signal value at the sample time was held for analog to digital
conversion producing an eight bit number which could be stored by
the computer. On each subsequent repetition, the sample time was
incremented with respect to the trigger until 256 values had been
obtained. The sampled representation of the received waveform
could then be processed and analyzed under software control to determine the scattering caused by the sample. For the data reported
here, a simple peak detection algorithm was used. Thus, for each
element of the sample, one eight bit number representing the peak
intensity of the received signal was stored. The test sample was
scanned by attaching it to a X-Y positioning system driven by stepping motors and controlled by the Z-80. Typically a 256 by 256
array of elements is scanned for each image.

The resolution of the X-Y positioning stepping motors is 3.2μ
although it could be increased further using mechanical gears. The
image data is in general monitored by an oscilloscope while being
recorded and stored on a floppy disc. Although the system described
above uses the two focussed transducers, it is capable of handling

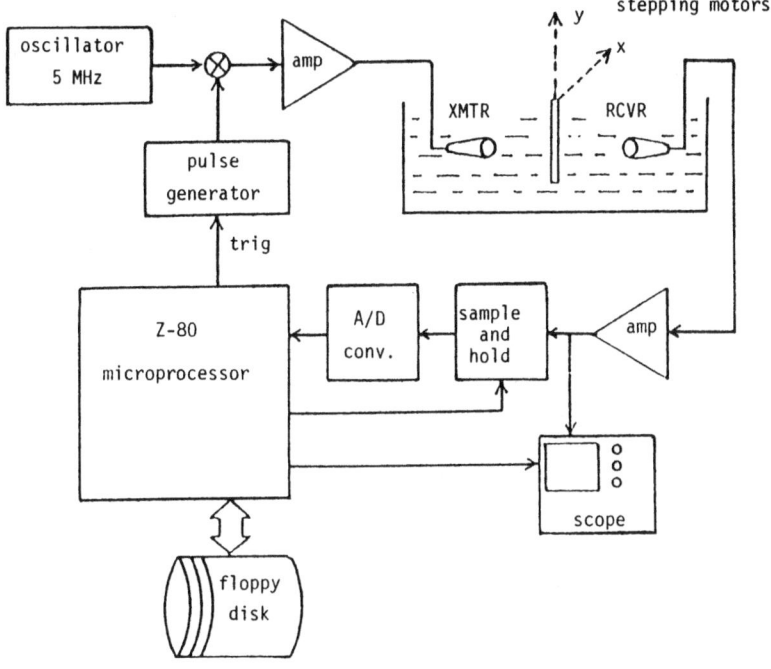

Fig. 1. Ultrasonic Imaging System

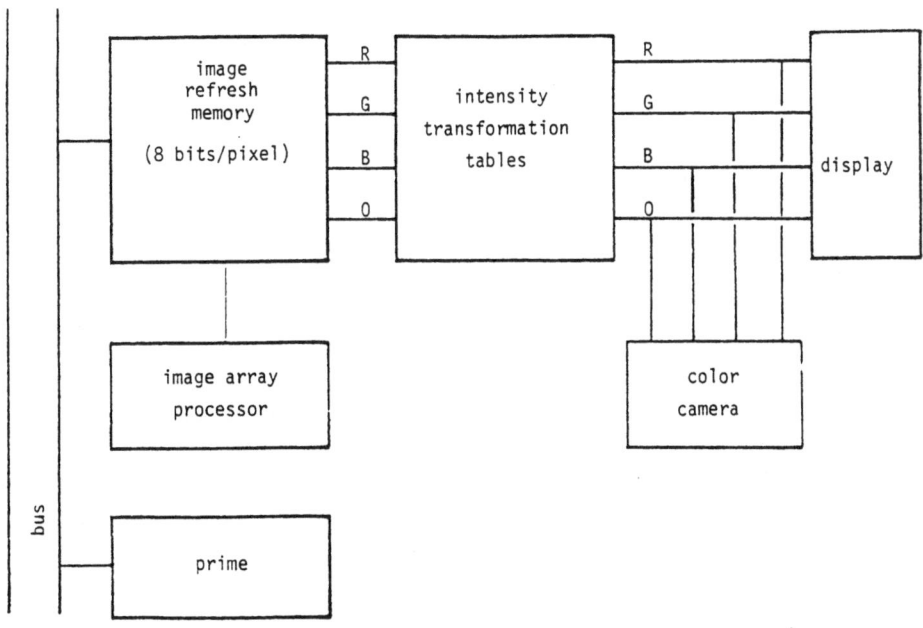

Fig. 2. Image Processing System

the trapped mosaic transducer for phased array and dynamic focussing which is also being developed.

3.   IMAGE PROCESSING SYSTEM

The block diagram of the image processing system is shown in fig. 2. The array processing DEANZA system is capable of driving a video display having a 512 x 512 array of picture elements (pixels). The data for each image is transmitted from the floppy disc storage of the Z-80 system to the PRIME 750 via a 4800 baud line.

The PRIME is the host computer of the image processor, handling data manipulation and storage. For display, the data is loaded into the image refresh memory which stores eight bits per pixel. These bits determine the intensity of the displayed pixel. The image may be displayed directly or the data may be modified by the array processor or the intensity transformation tables as described later.

The image is displayed on a color CONRAC video monitor which is connected to the system. On command this system can generate 8" x 10" Polaroid high quality color prints, 35 mm slides, or black and white hard copy of the image. All the figures of images reported in this paper were directly obtained by this system. Unfortunately, color images could not be reproduced here due to prohibitive costs, so black and white images are presented.

A typical black and white ultrasonic image as obtained is shown in fig. 3. (For a description of the object see section 4).

Fig. 4 shows a histogram of the unmodified image. The histogram, which was compiled by the array processor, shows the number of pixels (normalized to the maximum number) of each possible intensity (0-255). Recalling that each stored intensity represents the intensity of the ultrasound intensity, one sees that emphasizing the pixels of a particular value could highlight areas of damage.

A very dramatic emphasis can be achieved by pseudocoloring the image. Pseudocoloring is performed by the intensity transformation tables (ITT). There is one ITT for each of the three color channels of the display - red, green, and blue. When the tables are enabled, the eight bits of image data for each pixel (stored in the image refresh memory) are used as the address of the intensity to be displayed for each color channel. These intensities are loaded into the table by the programmer through the Prime. A graphical representation of a useful transformation scheme is shown in fig. 5. The abscissa represents the stored data (0-255) while the ordinate shows the transformed intensity displayed for each color. Here the blue and green channels have been used to show the entire sample at a low display intensity while blacking out the screen for those areas where the ultrasound has not passed through the sample. These

Fig. 3.   Black and White Ultrasonic Image

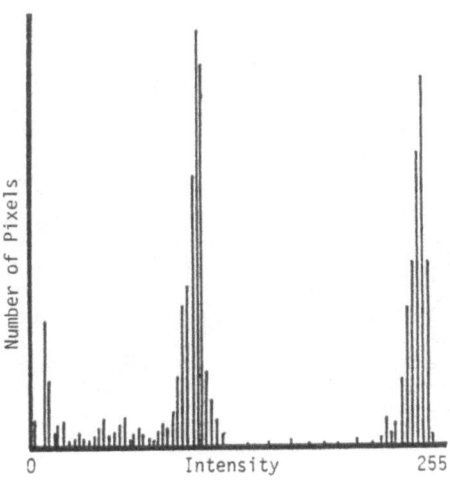

Fig. 4.   Histogram of Image in Fig. 3

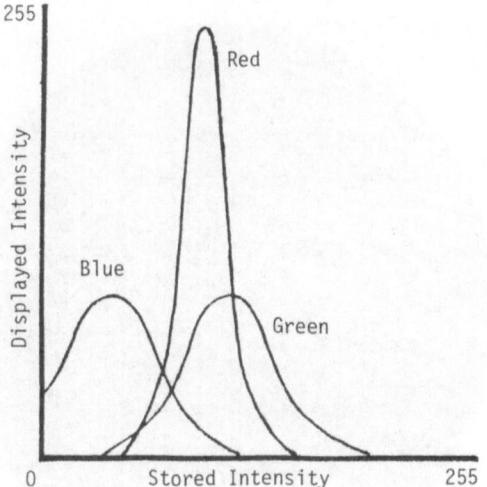

Fig. 5.   Intensity Transformation Scheme

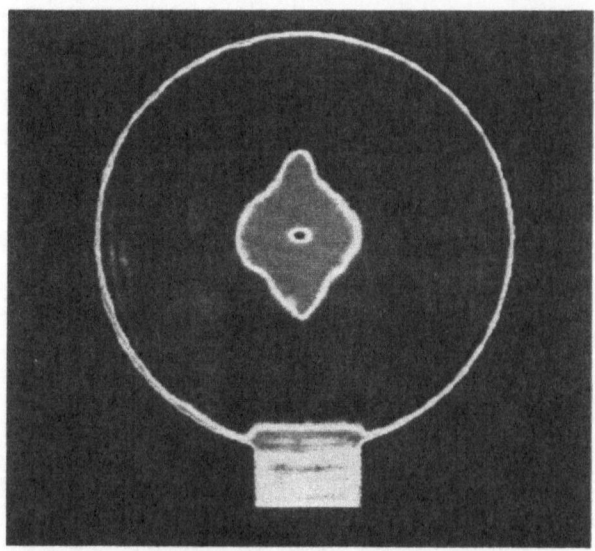

Fig. 6   Image of Fig. 3 Colored According to Intensity Transformation Scheme of Fig. 5

intensities which are to be emphasized are colored red, as indicated by the red peak in the figure. This peak can be scanned across the stored intensities to highlight the various levels of damage in the sample. Fig. 6 shows the emphasis that can be attained through this technique. (Remember that this is a black and white representation of the colored image.) The choice of the color and particular transformation can be done by the operator in an interactive fashion to obtain the optimum result. The system is capable of expanding a particular part of the image as desired. This zoom capability is shown in figs. 7-9 where in figs. 8 and 9 one can observe the individual pixels. The dimensions of each pixel in these figures were 160μ x 160μ.

It is obvious that once the digital image is stored in the computer memory, different digital signal processing functions such as deconvolution or image restoration to correct for the finite acoustic beam width or adaptive filtering such as Wiener filtering for noise reduction can be easily implemented.

4. COMPOSITE MATERIALS AND RESULTS

A Fiberite Hy-E 1048A1E pre-preg tape was used to fabricate 12 ply laminated plates with stacking sequences of $[(0/90)_3]_s$, $[0/30/60/90/120/150]_s$, and $[0/45/90/135/0/45]_s$; having a nominal fiber volume fraction of 0.6 and a nominal density of 1.54g/cc. Disks of 35.6mm diameter and 1.75mm thickness were cut from the cured plates. The disks were then tested in centro-symmetric deflection (CSD), where the disk is simply supported around the entire circumference of a hollow cylindrical die of 28.6mm inner diameter, and the load is applied perpendicularly at the center by a spherical loading nose 11.1mm in diameter. Real time traces of the applied load vs deflection of the disk center showed a distinct load drop in the positively sloped curve. Curve shapes were the same for all stacking sequences, but quantitative differences in the maximum load reached before the load drop occurred and the magnitude of the load drop were found for the different stacking sequences. The research effort associated with the CSD test and the disks examined in this paper are part of an experimental study of delamination failures produced by out of plane loading of laminated carbon-epoxy composite plates being done by Altman and Sternstein[6] for which a theoretical stress analysis is being developed by Taggart and Sternstein[7] to allow the applications of failure criteria to delamination problems. The CSD geometry is also being used to study the viscoelastic properties of composite plates.[8]

Table I shows the composite disks tested using acoustic imaging and their properties.

A comparison of figs. 10, 11, and 12 shows the striking differences between failure patterns for disks of different construction

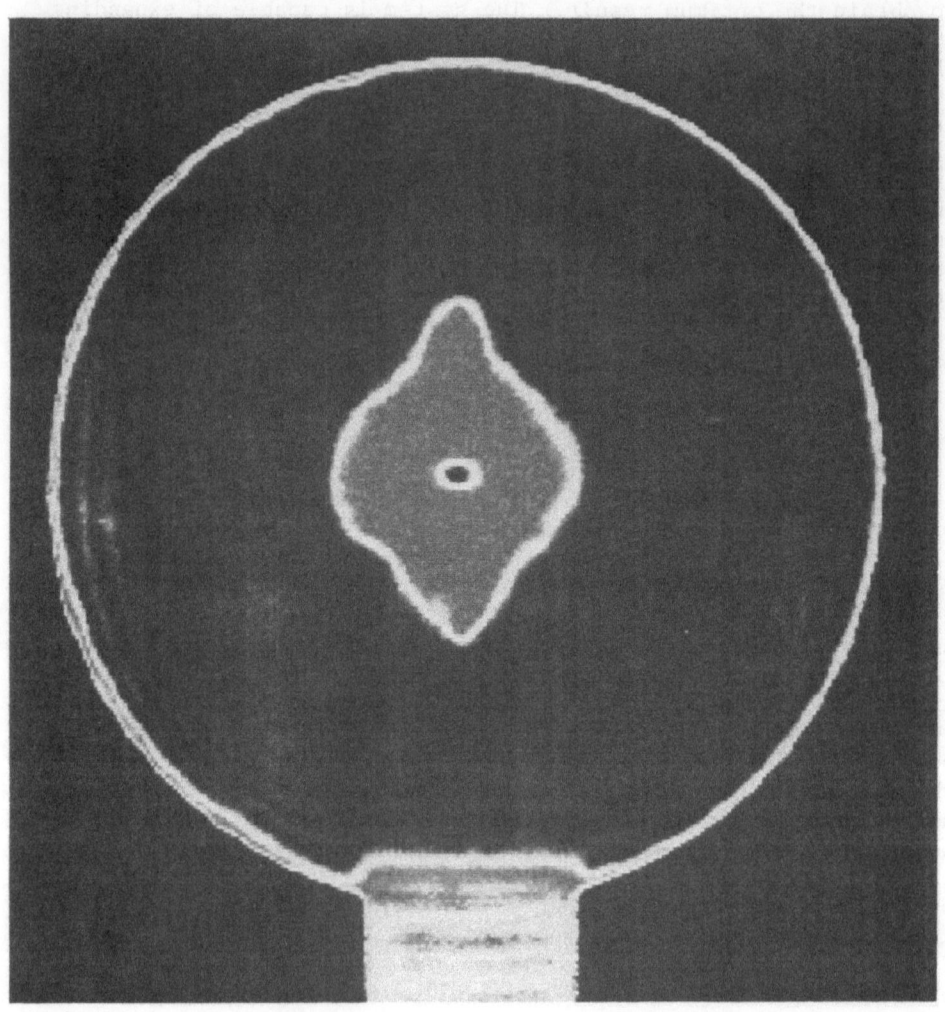

Fig. 7.   First Zoom of Fig. 6

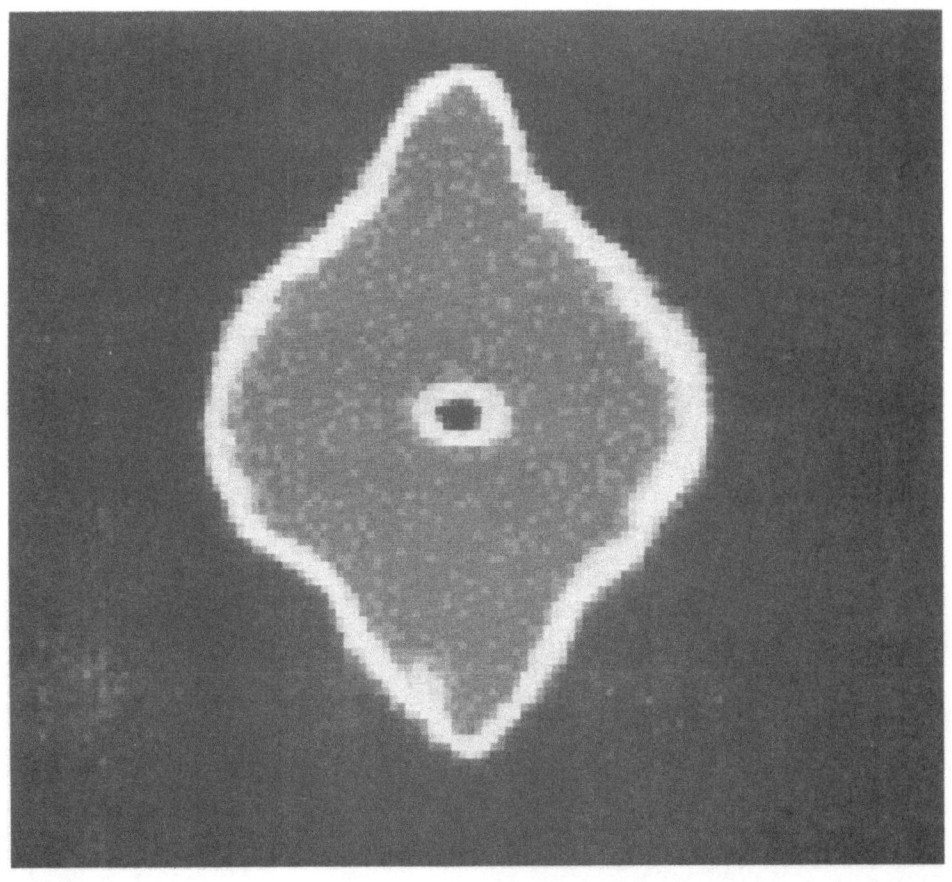

Fig. 8.   Second Zoom of Fig. 6

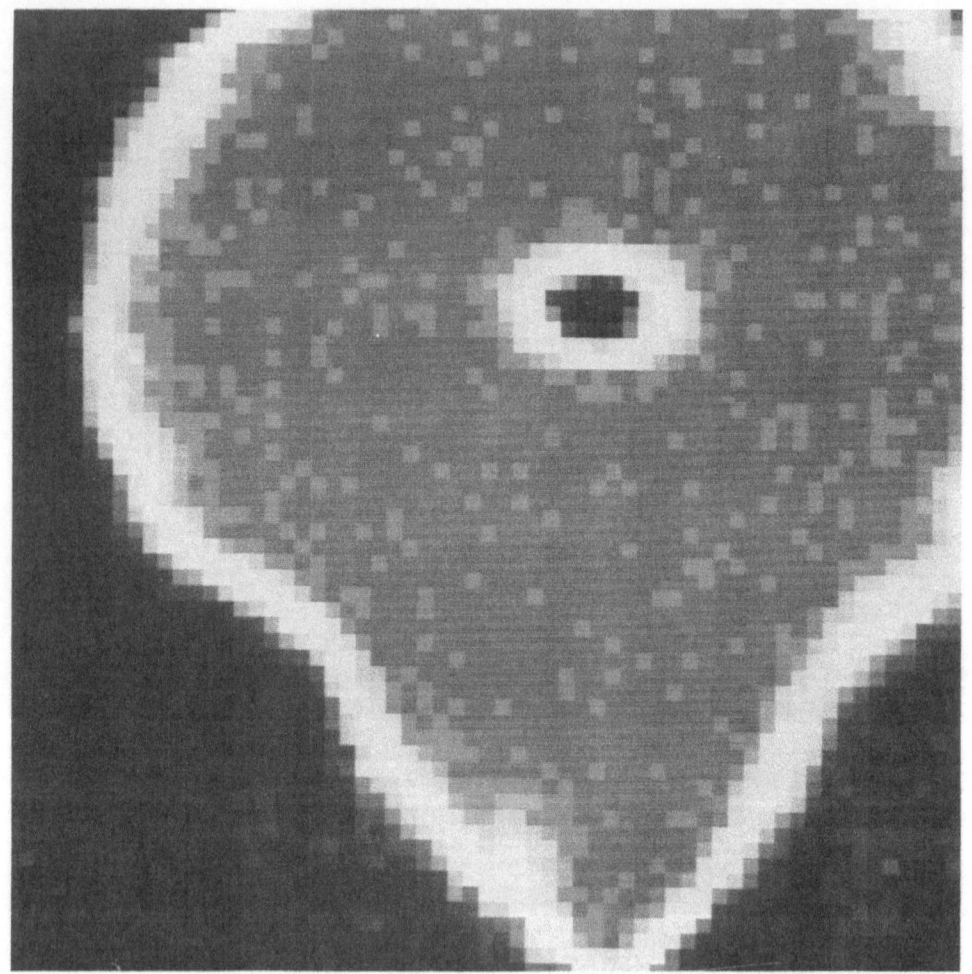

Fig. 9.   Third Zoom of Fig. 6

TABLE I

Fig. No.	Construction	Loading
10	$[(0/90)_3]_s$	first failure
11	$[0/45/90/135/0/45]_s$	first failure
12	$[0/30/60/90/120/150]_s$	0-95 kg, 1 cycle (first failure)
13	$[0/30/60/90/120/150]_s$	0-95 kg, 50 cycles
14	$[0/30/60/90/120/150]_s$	0-55 kg, 50 cycles

Furthermore, contrasting Figs. 12 and 13 show that the repeated loading has caused the damaged area to spread out from the loading point. Fig. 14 shows the pattern produced by cyclic loading with a maximum of 55 kg as opposed to 95 kg for figs. 12 and 13. Although an image is not presented here, this is especially interesting in light of the fact that the load vs. deflection curve and the same imaging techniques as described in sections 2 and 3 indicated no failure for one cycle 0-55 kg. In addition, all of the images share a common feature: the concentric rings that are visible inside the edges of each disk. From comparison with images of other materials with cut edges it is possible that these are caused by the cutting process.

CONCLUSION

Ultrasonic Imaging has been shown to be a simple and effective technique for determining damage in graphite-epoxy composite materials. In addition, the detected failure patterns for the samples tested have been shown to be highly dependent on the construction of the laminated plate. Digital enhancement of the images can improve the sensitivity of the imaging and reveal fine structure within the damaged sample. This non-destructive ultrasonic testing may be the only way these damages can be detected.

ACKNOWLEDGEMENT

It is a pleasure to acknowledge the contributions of Prof. S. S. Sternstein, Mr. C. Altman and Mr. R. Webster through many discussions and other help, especially Prof. Sternstein and Mr. Altman who fabricated and CSD tested the composites.

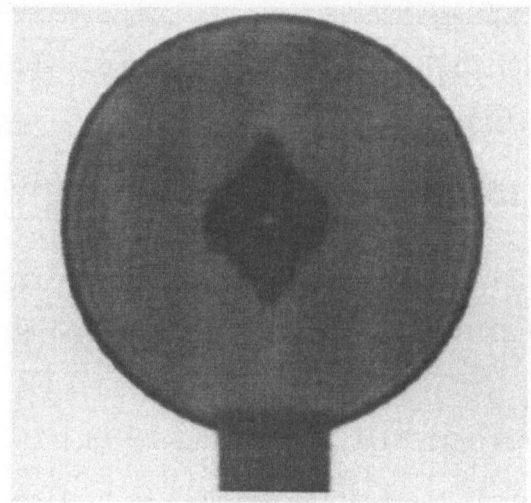

Fig. 10.   Image of $[(0/90)_3]_s$ Composite Loaded to First Failure

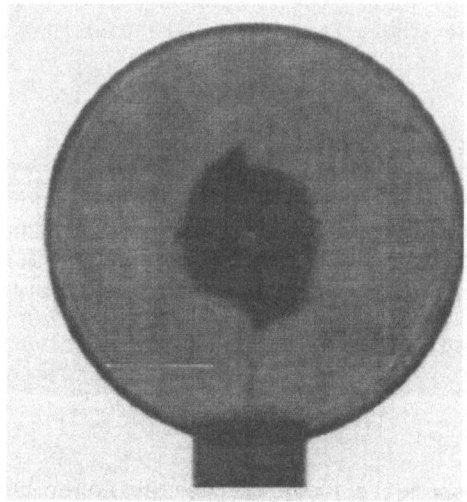

Fig. 11.   Image of $[0/45/90/135/0/45]_s$ Composite Loaded to First
           Failure

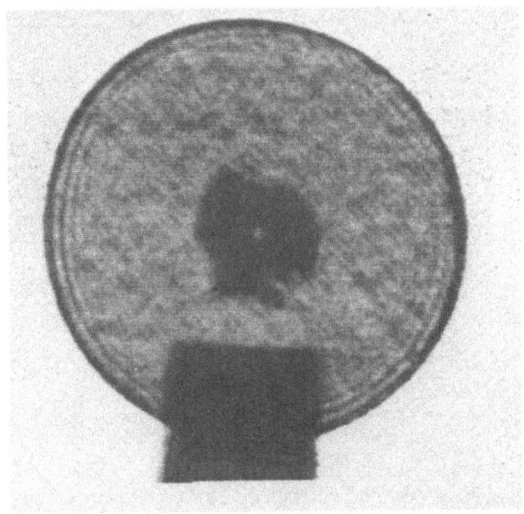

Fig. 12 Image of $[0/30/60/90/120/150]_s$ Composite Loaded to First Failure

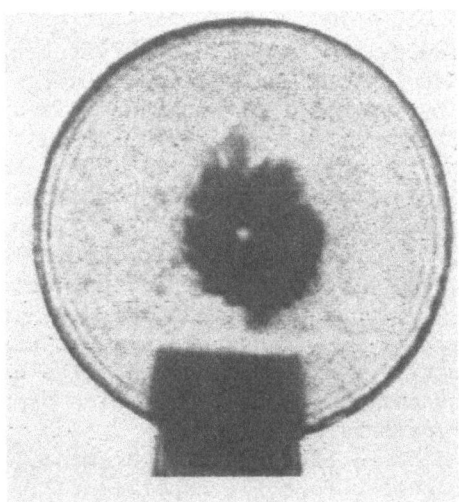

Fig. 13. Image of $[0/30/60/90/120/150]_s$ Composite, 50 cycles, 0-95 kg.

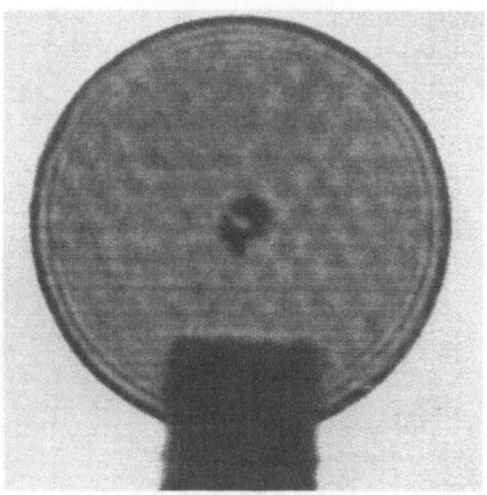

Fig. 14.   Image of $[0/30/60/90/120/150]_s$ Composite, 50 cycles, 0-55 kg.

REFERENCES

1. G. C. Knollman, D. Carver, and J. J. Hartog, "Acoustic Imaging of Composites - The Ultrasonic Test That Requires No Interpretation," Materials Evaluation, Vol. 36, No. 12, pp. 41-47, (1978).
2. P. Das, S. Talley, R. Kraft, H. F. Tiersten, and J. F. McDonald, "Ultrasonic Imaging Using Trapped Energy Mode Fresnel Lens Transducers," Acoustical Imaging, edited by K. Y. Wang, Vol. 9, Plenum Press, pp. 75092, (1980).
3. W. H. Sheldon, "Comparative Evaluation of Potential NDE Techniques for Inspection of Advanced Composite Structures," Materials Evaluation, Vol. 36, No. 2, pp. 41-46 (1978).
4. R. L. Crane, F. Chang, and S. Allinikov, "The Use of Radiographically Opaque Fibers to Aid the Inspection of Composites," Materials Evaluation, Vol. 36, No. 10, pp. 66-71 (1978).
5. R. T. Webster, P. Das, and R. Werner, "Ultrasonic Imaging for Nondestructive Evaluation of Composite Material with Digital Image Enhancement," IEEE Ultrasonics Symposium Proc., pp. 873-876, Nov. (1980).
6. C. Altman and S. S. Sternstein, "Delamination Failure in Carbon-Epoxy Laminates, Part I - Experimental," to be published.
7. D. Taggart and S. S. Sternstein, "Delamination Failure in Carbon-Epoxy Laminates, Part II - Theory," to be published.
8. P. Yang, L. Carlsson and S. S. Sternstein, "Viscoelastic Characterization of Neat Resins and Composites," appearing in American Chemical Society Polymer Preprints, August (1981).

PRELIMINARY RESULTS OF COMPUTER AIDED ACOUSTIC IMAGING

John Powers, Miltiades Economopoulos, Eugene Moon, and Harry Vasquez

Department of Electrical Engineering
Naval Postgraduate School
Monterey, California 93940

ABSTRACT

Preliminary results from a computer aided coherent acoustic imaging system are given. The system uses a raster scanned sensor to digitize samples of 1 MHz acoustic fields (both amplitude and phase) sampled at half-wavelength intervals. After transferring the data to computer memory three different types of displays have been used to display and process the data fields as images. The display devices include a black and white video system, a color graphics terminal, and a high resolution color image processing system. The advantages and disadvantages of each display system are discussed along with some of the most useful image processing operations for the acoustic images obtained. These operations include image zoom (i.e., magnification), histogram analysis, dynamic range manipulation, and other interactive capabilities.

INTRODUCTION

The Naval Postgraduate School (NPS) has been developing a general purpose non-real time test bed to investigate a wide variety of acoustic imaging techniques especially those involving computer processing and image processing of the collected data. The data acquisition portion (Fig. 1) of the system consists of a precision-screw-controlled point receiver scanned in a pattern determined by the physical setup. This receiver coherently detects the sound pattern. The most frequently used pattern to date has been a raster scan of 64 x 64 data points (sampled at

277

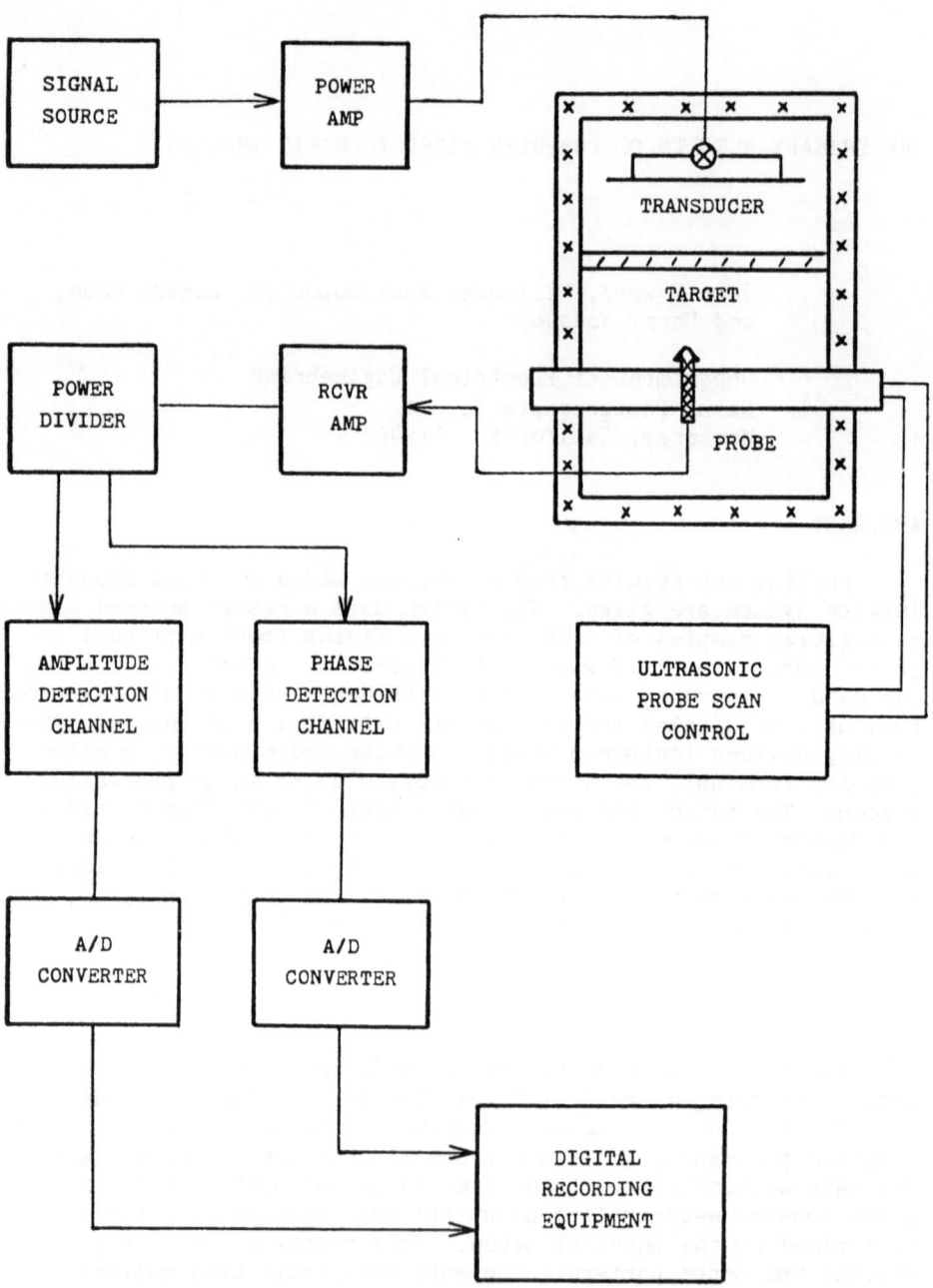

Fig. 1. Data acquisition portion of NPS acoustic imaging system.

half wavelength intervals for the 1 MHz operating frequency)
although other scans can be easily accommodated. In the present
acquisition system the scanner is controlled by a hardware
controller with the data conversion and storage (on paper tape or
cassette recorder) controlled by a microprocessor[1]. Presently a
system is being implemented with a LSI-11 microcomputer providing
the scan control and the data storage on to a floppy disk medium[2]
to allow increased ease in reconfiguring the scanning geometry.
Once the data is recorded it can be transformed to a large control
computer for processing in depth or to a smaller minicomputer that
controls the various displays used and their associated image
processing routines. Additional simple processing operations can
be done in the minicomputer but complicated algorithms
manipulating the entire complex data array would require the use
of the array processor in the minicomputer system or the larger
central computer. In this fashion a wide variety of processing
techniques can be applied to data by implementing processing
algorithms. Additionally a wide variety of image processing
algorithms are implemented in some of the display systems further
extending the flexibility.

        This paper presents representative preliminary images from
the data acquisition system and the results of some of the simple
image processing operations that can be applied to the images.
Two of the display systems can use color for pseudo-color
representation of the gray scales of the images. This allows the
operator to pick out brightness changes that are not obvious in
the black-and-white images. Experience has shown that while
arbitrary use of pseudo-color assignment has marginal value,
limited use does have a capability of presenting more pleasing
images to the operator. Due to an inability to represent color
images in this volume only black-and-white images will be
presented here.

RAMTEK GX100A SYSTEM

        The RAMTEK system is a high resolution graphics terminal
displaying 240 x 640 pixels with a wide variety of annotation
capability. The system has no image processing capability however
and was used as a beginning capability. The system has 16 levels
each of three colors allowing a wide variety of colors to be
represented. The images were primarily used to study pseudo-color
presentation of the images. Fig. 2 shows a black-and-white
representation of such an image. The object is the transmission
pattern (showing the log of the amplitude) immediately behind a
two inch aperture of an acoustic lens. Testing of such images
proved that a pseudo-color assignment similar to the visible
spectrum order of colors as the best with the cooler colors

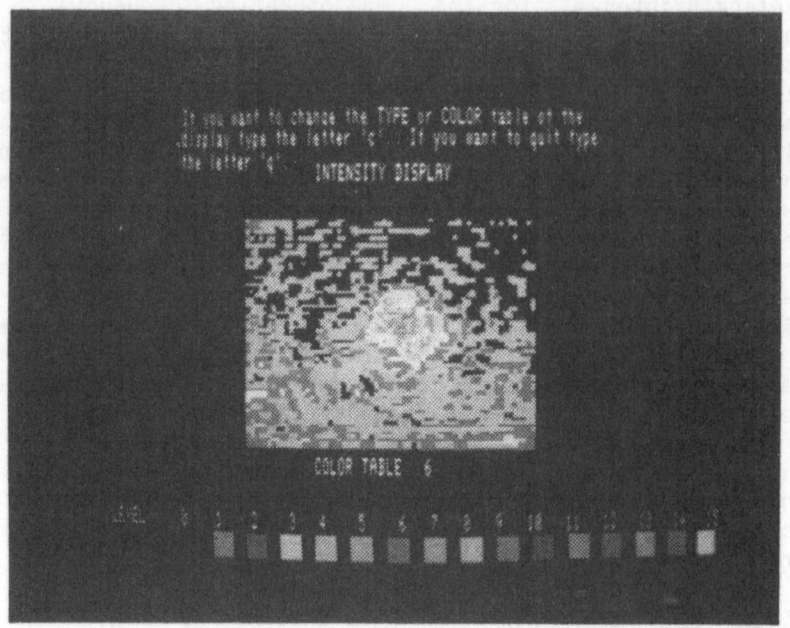

Fig. 2. B/W representation of image of acoustic lens.

representing low values of data and the hotter colors representing
the higher values.

EYECOM SYSTEM (EXPANDED)

     A system with more display and processing capability is the
Spatial Data System's EYECOM. This system consists of a video
digitizer and a display. The display of the expanded system is
capable of representing 256 gray levels in each of its 640 x 430
pixels. The system includes an annotation capability and a wide
variety of image processing functions. The most useful of these
include the capability to magnify or "zoom" an area of interest
(Fig. 3), to perform a histogram on the image (i.e., displaying
the number of pixels having a given graph scale), and to modify
the histogram of the display. This latter operation performs
contrast enhancement for some images. For example, Fig. 4
displays the same object as Fig. 3 with the gray scales between
levels 75 and 100 (i.e., where most of the data of Fig. 3 resides)
"stretched" to lie between levels 35 and 200. Values on the
original histogram are linearly interpolated to provide the new
values. The object in Figs. 3 - 9 are the transmitted sound
through a letter "N" cut out of a cork absorbing mask. As seen
the images are negative images due to the acquisition system

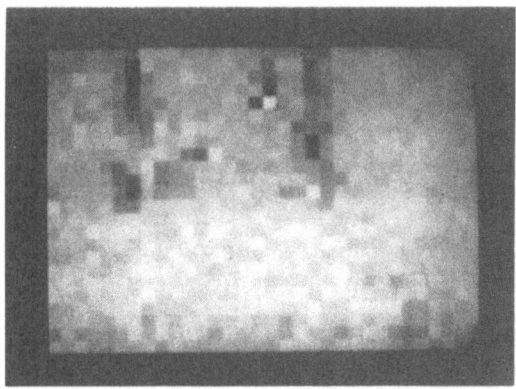

Fig. 3.   Zoomed Eyecom image of letter.

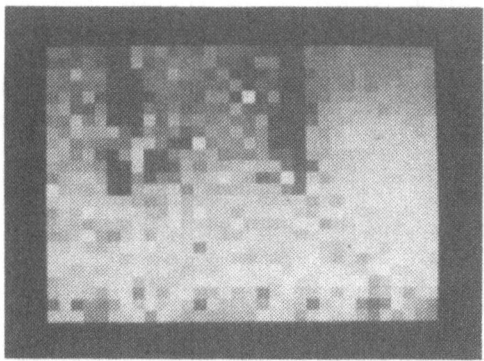

Fig. 4.   Dynamic range manipulation applied to the image of
          Fig. 3.

electronics and are reversed due to the recording geometry. The
vertical strokes of the letter are 2 2/3 acoustic wavelengths
wide. The diagonal stroke is 1 1/3 wavelengths wide. Because of
the wide dynamic range encountered in acoustic images the
amplitude data has been logarithmically compressed. Removing the
compression returns the high contrast but the compressed data
provides a more satisfactory image to the operator. Most of the
images here are presented using the compressed data unless
otherwise noted. (The phase however does not suffer from this
dynamic range problem due to its periodic nature and is recorded
linearly.)

     Averaging techniques as might be expected reduce the
graininess of image with an attendant loss in resolution. Fig. 5

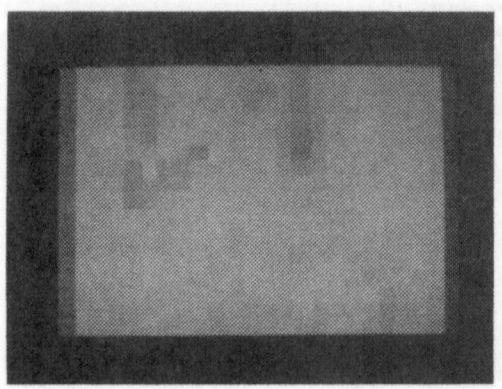

Fig. 5. Averaged image.

shows an image in which an averaging routine replaces the pixel's
value by the average value of all of the pixels in a 3 x 3 box
around the center. The dynamic range is compressed and the image
is smoother. Other routines can be used for edge enhancement.
Fig. 6 shows a processed image with a gradient operator that
approximates the function:

$$(\frac{\partial f}{\partial x})^2 + (\frac{\partial f}{\partial y})^2 \tag{1}$$

The vertical edges of N are highlighted as are the other noisy
places where the data changes significantly. A combination of
averaging to reduce the noise and then edge detection would
produce a better image. Another operator performing the Laplacian

$$\frac{\partial^2 f}{\partial x^2} + \frac{\partial^2 f}{\partial y^2} \tag{2}$$

operation is also useful in finding and accentuating edges.

COMTAL VISION/ONE SYSTEM

     The COMTAL system is a powerful image processing system with
extremely high resolution. As used in this study the basic system
had 16 levels in each of three colors or 16 gray scales over
1024 x 1024 pixels. When fully expanded the system will have 256
levels. The high frequency response of the circuit allows full
use of the dynamic range of the gray scale between adjacent pixels
producing an image extremely high fidelity even with 16 gray
scales. The system has a full complement of image processing

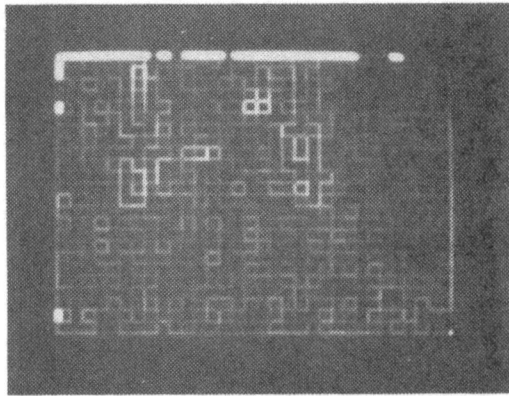

Fig. 6.   Gradient operator image.

capabilities many of which can be implemented on a frame-to-frame
basis (i.e., in 1/30 of a second). Other routines can be
implemented in software. Figures 7 and 8 show the letter "N"
amplitude data; Fig. 7 is linear amplitude data while Fig. 8 is
logarithmic data. Figure 9 shows the phase data for the letter
"N" illustrating that there is considerable object information in
the phase. No processiing has yet been attempted to the phase
data.

     Figure 10 shows the logarithmic data for a cruciform cutout
object. Each of the squares is approximately 6 2/3 wavelengths
across. The 16 level histogram of this image is shown in Fig. 11.
One of the major assets of the COMTAL system is the ability of the
operator to manipulate the histogram pattern interactively. This
manipulation can be graphical using a track ball input or analytic

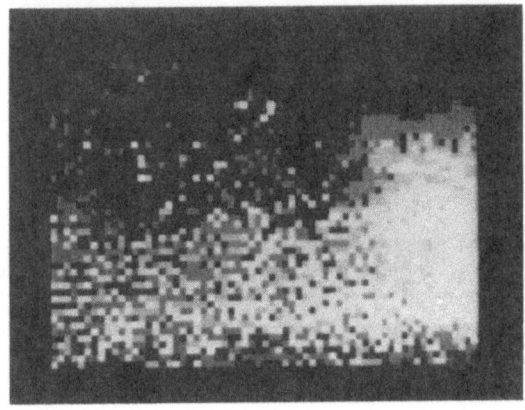

Fig. 7.   Comtal linear amplitude data.

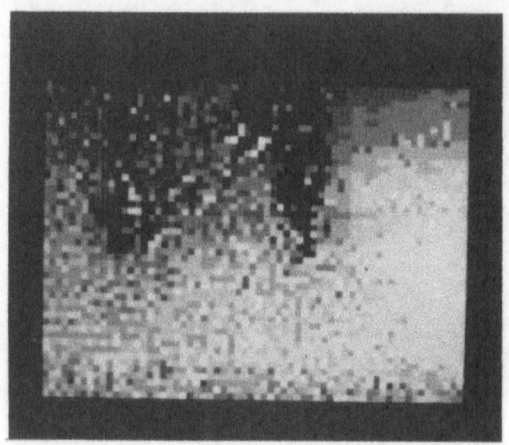

Fig. 8.   Logarithmic amplitude data.

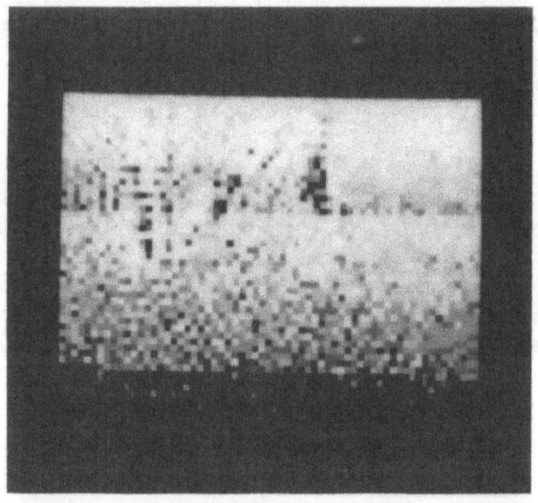

Fig. 9.   Phase data of letter object.

expressions as typed in on the keyboard. An example of the former
operation is shown in the transfer function of Figure 12. This
"hard-limiting" function will zero the intensity of any pixels
having a value greater than 5 and will produce maximum intensity
for all pixels with a value less than or equal to 5 thereby
reversing contrast and eliminating much of the noise (as well as
some of the object information) as shown in the processed image of
Fig. 13.

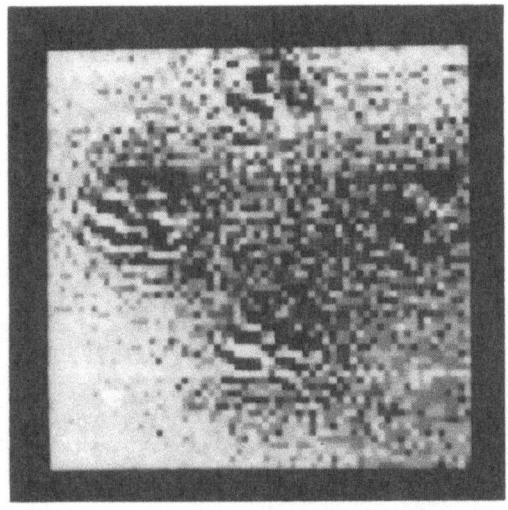

Fig. 10.   Logarithmic amplitude of cruciform.

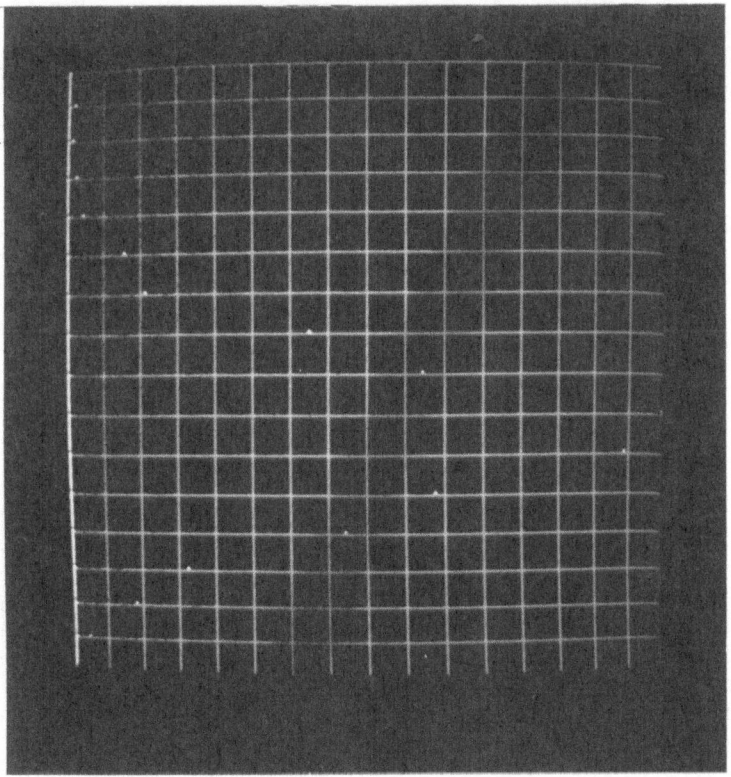

Fig. 11.   Histogram of log amplitude image.

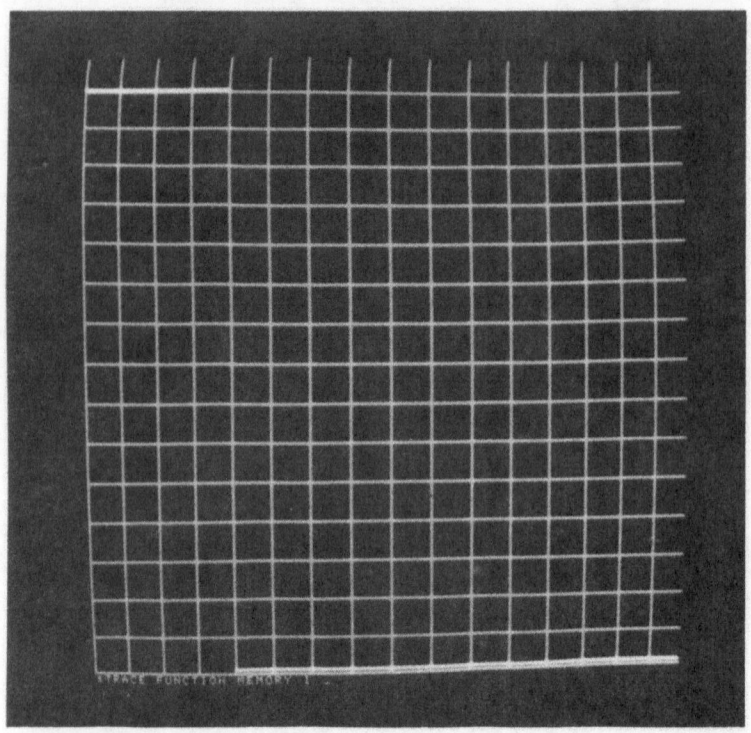

Fig. 12.   Transfer function applied to histogram. Horizontal axis
           is the gray level of the input, vertical scale is the
           gray level of the output.

SUMMARY

     This paper has presented some of the preliminary images
obtained with the NPS acoustic imaging system. Various data
display systems have been used with the newly acquired COMTAL
system showing the most promise. Application of the various image
processing techniques built into the systems demonstrate the
requirement for an interactive system since the improvement of the
images is highly subjective and the order of processing as well as
the parameters used in the processing techniques are image
dependent.

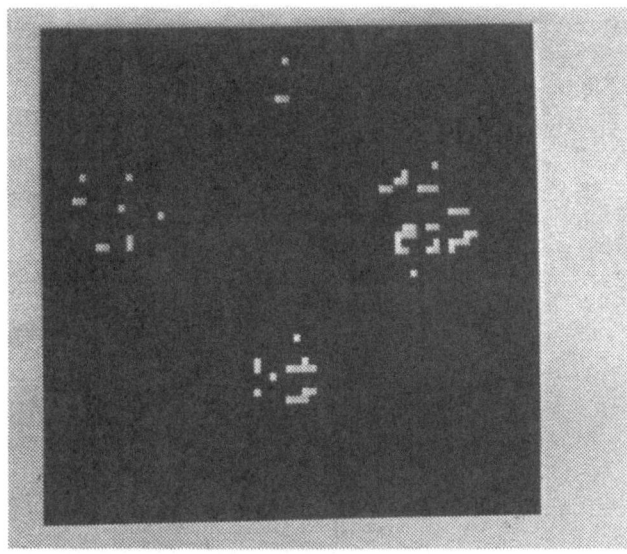

Fig. 13.   Processed image.

ACKNOWLEDGEMENT

This work was supported by the National Science Foundation under Grant No. ENG-77-21600 and by the Naval Postgraduate School Foundation Research Program.

REFERENCES

1. J. Powers, R. Carlock, and R. Colton, "Data acquisition system for computer aided acoustic imaging," Acoustical Imaging, Vol. 9, R. Wang, Ed. (Plenum Publishing Corp., New York), pp. 643-652, 1980.
2. S. E. Eriksen, "A micro-computer-based data acquisition controller for a digital acoustic imaging system," Engineer's Degree thesis, Naval Postgraduate School, Monterey, CA, 1981 (Unpublished).

AN OPTIMUM ULTRASONIC IMAGING SYSTEM USING ARMA PROCESSING

Takuso Sato, Takayoshi Yokota,
and Osamu Ikeda

The Graduate School at Nagatsuta
Tokyo Institute of Technology
4259 Nagatsuta, Midori-ku
Yokohama-shi, 227 Japan

ABSTRACT

The possibilities and conditions for the effective applications of nonlinear image reconstruction algorithms such as MEM, MLM(HRM) and ARMA for ultrasonic imaging are discussed. The discussions are focused especially on the properties of the wave field of object and the modeling of its structure. Comparison of these methods is made from the viewpoint of high fidelity of image that is reconstructed using only the data observed in a restricted finite size aperture. The special features of each method are also assured by numerical analyses. A concrete imaging method based on ARMA processing, which is selected as the most suitable means for ultrasonic imaging, is proposed. In this method the required incoherent wave field is generated actively by illuminating the object from many different directions with relatively random phases. The usefulness and effects of the proposed method are demonstrated experimentally.

1. INTRODUCTION

One of the main desires in up-to-date ultrasonic imaging systems is to get an image with as high resolution and as high fidelity as possible even from the data obtained in a restricted aperture. In the conventional coherent imaging systems the analytic or successive continuation has been applied to extrapolate the hologram data beyond the aperture plane based on the a priori information that the object area is bounded.[1,2]

On the other hand in incoherent imaging systems, several new non-linear algorithms have been developed in the field of power spectral analyses or image reconstruction processings for stellar or geophysical objects.[3]   These nonlinear methods are maximum entropy method (MEM),[4-6] maximum likelihood method (MLM),[7,8] and autoregressive-moving-average method (ARMA).[9,10]

In this paper the possibilities and conditions of the applications of these nonlinear methods to the ultrasonic objects of our interest are discussed. The following points are made clear concretely; i) what is the difference between the wave fields of stellar objects and ultrasonic objects ?, ii) can we generate the desired ultrasonic wave fields ?, iii) which algorithm is most desirable ?, and iv) what kind of effects can be expected ?

Then the special features of each method are compared numerically and experimentally. An ultrasonic imaging system is constructed based on the active generation of equivalently in-coherent wave field on the object and the ARMA image reconstruction processing. The results obtained show that the ARMA processing is most effective among them for ultrasonic objects whose structures consist of peaks and hills.

2.   PROPERTIES OF ULTRASONIC WAVE FIELD
AND CORRESPONDING IMAGE RECONSTRUCTION PROCESSINGS

2.1   Object Field

In most of the conventional ultrasonic imaging methods we have assumed implicitly that the wave field is deterministic; that is, it is coherent. In actual ultrasonic imagings, however, we often encounter the cases where the wave fields have several random natures. For instance, if there is any turbulence near the surface of an object, the wave field cannot be considered to be a coherent one, or if we illuminate the object with a set of wave fronts with relatively random phases, then the resulting field may also have random nature. In this paper we treat the cases where the object field is random and incoherent. That is, the wave fields at any two different positions on the object can be assumed to be statistically independent. In chapter 5 a concrete method to generate a quasi-incoherent wave field in an active ultrasonic imaging system will be shown.

2.2   Models of Intensity Distribution of the Object Field

Now, let us consider the image processing from the viewpoint of the parameters estimation of the intensity distribution model of the object. We consider only one dimensional case for simplicity of their treatment, but rather straightforward extension to two or three dimensional case is also possible. We put the following assumptions to proceed our discussion:

i) The object is incoherent and located in the far field.
ii) The wave field from the object is detected by a finite number of uniformly distributed array sensors.

In this case the conventional means of image reconstruction is based on the well-known van Cittert-Zernike theorem[11]; that is, the image is obtained as the inverse Fourier transform of the coherence function detected at the array. Its actual operation is carried out using discrete Fourier transform (DFT). In this method the reconstructed image is given as a Fourier series, as is shown in Fig. 1, and from the viewpoint of the modeling it consists of only zeros. So, this method may be effective for objects having relatively low order spatial frequency components. When the intensity distribution of the object field is given as a sum of peaks and hills, for example as shown in Fig. 1, however, the conventional means fail to reconstruct the peaks faithfully, resulting in the blurred image with the peaks suppressed. If we can extend the aperture infinitely physically or equivalently, the exact structure of the object may be reconstructed. Hence, the following all discussions are restricted to the problem of how we get high fidelity image even from data obtained in the limited aperture.

The maximum entropy method (MEM), which has been used extensively for the spectral estimation from such finite data, corresponds essentially to the modeling of the object field as a sum of resonance curves, hence the model is suitable for objects with only point-like components as is shown in Fig. 1. Objects of our usual interest for ultrasonic imaging, however, do not always consist of only such peak-like components, but the smooth hill-like background are also included. In this case this method is not suitable, since in this method the hills should also be represented as a sum of peaks. In the same meaning the maximum likelihood method (MLM) has almost the same property as MEM. On the other hand the model assumed in the ARMA process consists of both zeros and poles, so it may be the most desirable model for objects of our interest.

Thus, the ARMA processing is adopted as the estimation algorithm.

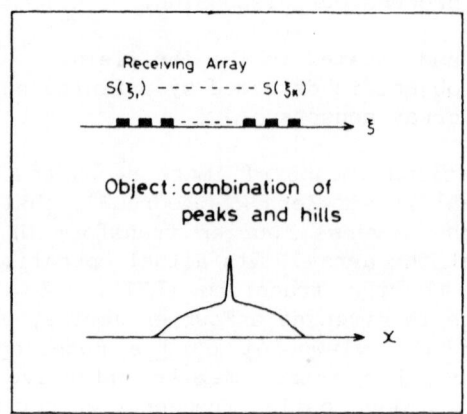

Imaging Method	Model for Object Structure	Expected Reconstructed Image
a) Back propagtion of coherence function (DFT)	$I(\Omega) = \left\| 1 + \sum_{n} b(n) e^{-jn\Omega} \right\|^2$	
b) Maximum Entropy Imaging Method (MEM)	$I(\Omega) = \left\| \dfrac{1}{1 + \sum_{m} a(m) e^{-jm\Omega}} \right\|^2$	
c) Autoregressive- Moving-Average Imaging Method (ARMA)	$I(\Omega) = \left\| \dfrac{1 + \sum_{n} b(n) e^{-jn\Omega}}{1 + \sum_{m} a(m) e^{-jm\Omega}} \right\|^2$	

where $\Omega = 2\pi x / X$

k : wave number

X : imaging area

Fig. 1.   Comparison of image reconstruction methods from the viewpoint of the model of the object's structure.

3. ARMA IMAGE RECONSTRUCTION PROCESSING

3.1 Derivation of Autoregressive (AR) and Moving-Average (MA) Coefficients

The basic equation governing the sample signal of the hetero-dyne detected complex wave field on the aperture for an object which follows the ARMA model is given by

$$\hat{s}(k) = -\sum_{m=1}^{M} a(m)s(k-m) + \sum_{n=1}^{N} b(n)w(k-n), \quad k=1,2,\ldots,K, \tag{1}$$

where $s(k)$ is the complex wave field detected by the equispaced k-th sensor on the aperture; $\hat{s}(k)$ is the predicted complex wave field; $w(k)$ is the prediction error; M and N are the orders of poles and zeros, respectively; $\{a(m)\}$ are the AR coefficients; and $\{b(n)\}$ are the MA coefficients.

The prediction error $w(k)$ is also written as follows:

$$w(k) = s(k) - \hat{s}(k)$$

$$= \sum_{m=1}^{M} a(m)s(k-m) - \sum_{n=1}^{N} b(n)w(k-n), \quad a(0)=1. \tag{2}$$

From Eqs.(1) and (2), the autocorrelation function of $s(k)$ is given by

$$R_{ss}(j) = E\{s(k)s^*(k-j)\}$$

$$= -\sum_{m=1}^{M} a(m)R_{ss}(j-m) + \sum_{n=0}^{N} b(n)E\{w(k-n)s^*(k-j)\}, \quad b(0)=1. \tag{3}$$

If we can have chosen the correct model orders, the method will be shown in detail in the next section the prediction error is reduced to a white noise. And we have the following orthogonality condition:

$$E\{w(k-n)s^*(k-j)\} = 0, \quad \text{for } j>n. \tag{4}$$

Substituting this condition in Eq.(3) the Yule-Walker equation is obtained as follows:

$$R_{ss}(j) = -\sum_{m=1}^{M} a(m)R_{ss}(j-m), \quad \text{for } j>N+1. \tag{5}$$

The AR coefficients $\{a(m)\}$ are obtained by solving a set of M equations as follows:

$$
-\begin{bmatrix} a(1) \\ a(2) \\ \cdot \\ \cdot \\ \cdot \\ a(M) \end{bmatrix} = \begin{bmatrix} R_{ss}(N) \cdots \cdots R_{ss}(N-M+1) \\ R_{ss}(N+1) \cdots \cdots R_{ss}(N-M+2) \\ \cdot \quad\quad\quad\quad\quad\quad\quad \cdot \\ \cdot \quad\quad\quad\quad\quad\quad\quad \cdot \\ \cdot \quad\quad\quad\quad\quad\quad\quad \cdot \\ R_{ss}(N+M-1) \cdots \cdots R_{ss}(N) \end{bmatrix}^{-1} \begin{bmatrix} R_{ss}(N+1) \\ R_{ss}(N+2) \\ \cdot \\ \cdot \\ \cdot \\ R_{ss}(N+M) \end{bmatrix}. \quad (6)
$$

Next, as for the MA coefficients, if we rewrite Eq.(1) as follows:

$$
y(k) = \sum_{m=0}^{M} a(m)s(k-m) \quad\quad\quad\quad\quad\quad\quad (7)
$$

$$
= \sum_{n=0}^{N} b(n)w(k-n). \quad\quad\quad\quad\quad\quad\quad (8)
$$

then, the autocorrelation function of $y(k)$ is given first from Eq.(7) as follows:

$$
R_{yy}(j) = 2\sigma_w^2 \sum_{n=0}^{N} b(n)b^*(n-j), \quad 0 \le j \le N, \quad\quad (9)
$$

and next from Eq.(8) as

$$
R_{yy}(j) = \sum_{m=0}^{M} \sum_{m'=0}^{M} a(m)a^*(m')R_{ss}(j+m'-m), \quad \text{for } j=0,1,..,N. \quad (10)
$$

where in Eq.(9) the following relation is used:

$$
E\{w(k-n)w^*(k-j-n')\} = 2\sigma_w^2 \delta_{k-n,k-j-n'}. \quad\quad (11)
$$

Hence, from Eq.(9) and Eq.(10) we have the following nonlinear relations to be solved to obtain σ_w^2 and $\{b(n)\}$:

$$
2\sigma_w^2 \sum_{n=0}^{N} b(n)b^*(n-j) = -\sum_{m=0}^{M} \sum_{m'=0}^{M} a(m)a^*(m')R_{ss}(j+m'-m),
$$

$$
\text{for } j=0,1,..,N. \quad\quad (12)
$$

The solutions exist if the process $y(k)$ satisfies the minimum phase condition.[10,12]

Finally, the image is reconstructed using the obtained coefficients $\{a(m)\}$ and $\{b(n)\}$ as follows:

$$I(x)=2\sigma_w^2\left|\frac{1+\sum\limits_{n=1}^{N} b(n)\exp(-j2\pi nx/X)}{1+\sum\limits_{m=1}^{M} a(m)\exp(-j2\pi mx/X)}\right|^2 \tag{13}$$

where x is the coordinate of the imaging area X determined by $X=\lambda Z/\Delta\xi$; λ is the wavelength of the ultrasonic waves; Z is the distance between the object and the aperture planes; and $\Delta\xi$ is the spacing between neighbouring array sensors.

3.2 Estimation of Model Orders

The estimation of the model orders from the obtained data is essential in order to apply the method effectively, since the optimality of the method is based on this estimation. So let us show a little closely the processes.

First we introduce the following two measures to estimate the orders: i) $F_a(M,N)$, to evaluate mainly the variation of the estimated AR coefficients, and ii) $F_b(N)$, to evaluate the non-whiteness of the image component due to the MA part.

From Eq.(5), $K-M-N$ sets of Yule-Walker equations are available from the observed coherence functions. That is,

$$-\begin{bmatrix} R_{ss}(N+1) & R_{ss}(N+1-1) \cdot \cdot & R_{ss}(N+1-M+1) \\ R_{ss}(N+1+1) & \cdot \cdot \cdot \cdot \cdot \cdot & R_{ss}(N+1-M+2) \\ \cdot & & \cdot \\ \cdot & \cdot & \cdot \\ \cdot & & \cdot \\ \cdot & \cdot \cdot & \\ R_{ss}(N+1+M-1) & \cdot \cdot \cdot \cdot \cdot \cdot R_{ss}(N+1) \end{bmatrix}\begin{bmatrix} a_1(1) \\ a_1(2) \\ \cdot \\ \cdot \\ \cdot \\ a_1(M) \end{bmatrix}=\begin{bmatrix} R_{ss}(N+1+1) \\ R_{ss}(N+1+2) \\ \cdot \\ \cdot \\ \cdot \\ R_{ss}(N+1+M) \end{bmatrix} .$$

$$1=0,1,2,..,K-(N+M+1) \tag{14}$$

It is clear that the solutions for every 1 are the same when the exact orders are used and there is no measurement error in $\{R_{ss}\}$. Hence, we may define the following measure for the estimation of the orders.

$$F_a(M,N) = \sum_{l=0}^{K-(N+M+1)} \sum_{m=0}^{M} \frac{\left| w_l^2 w_m^2 \, a_l(m) - \bar{a}(m) \right|^2}{\left| \bar{a}(m) \right|^2} \Bigg/ \sum_{l'=0}^{K-(N+M+1)} \sum_{m'=0}^{M} w_{l'}^2 \cdot w_{m'}^2$$

$$(15)$$

where

$$\bar{a}(m) = \sum_{l=0}^{K-(N+M+1)} w_l a_l(m) \Bigg/ \sum_{l'=0}^{K-(N+M+1)} w_{l'} \quad , \qquad (16)$$

w_l and w_m are proper real and non-negative weights required for the summation with respect to l and m, respectively. The measure in Eq.(15) actually evaluates the average of the squares of the coefficients of variation of the estimated AR coefficients. Therefore, the orders M and N which give the minimum value of $F_a(M,N)$ are chosen as the optimum model orders.

The sensibility of $F_a(M,N)$ for M is very high, as will be shown in the next chapter, while that for N is poor, since $F_a(M,N)$ doesn't evaluate MA coefficients directly.   So we adopted the following another measure $F_b(N)$ to evaluate the non-whiteness of the image component due to MA part:

$$F_b(N) = \sum_{n=0}^{N} \left| b(n) \right|^2 \quad . \qquad (17)$$

It has the following properties:

i)   When $N < N_{opt}$ the estimated AR coefficients must represent both the peaks and the hills in part. In this case the MA image tends to become that of white one with the decrease of N, hence $F_b(N)$ is reduced.   ii)   On the other hand when $N > N_{opt}$, the absolute values of the estimated AR coefficients are reduced with the increase of N due to the finiteness of the coherence functions. In this case, the AR image becomes somewhat a smoothed one, and it turns out to represent both the peaks and the hills, resulting in a flatter MA image with the increase of N.   And iii)   when $N = N_{opt}$ the effect of the MA component in the coherence functions on the estimation of the AR coefficients are eliminated, and the AR coefficients are used to represent only the peaks, while the MA coefficients are used to represent only the hills.   In other words, the MA image will have deep troughs at the places corresponding to the peaks and it tends to have a high contrast, or maximally coloured, hence $F_b(N)$ will have the largest value since it evaluates the total power of the MA coefficients.

According to these considerations we defined $F_b(N)$ in Eq. (16) as a measure to evaluate the non-whiteness of the image due to MA part.

The problem of how we choose the weight for averaging AR coefficients is left unsettled. We used tentatively a rectangular weight.

The algorithm of the image reconstruction based on ARMA processing is summerized in Fig. 2.

4. NUMERICAL ANALYSES

In this chapter, some results of the computer simulation are shown. The surface interval of the object is divided into 256 equi-segments and independent random phases are given to generate an incoherent object. The intensity distribution, that is, the object to be imaged consists of two peaks and a hill as the background as is shown in Fig. 3.

After acquiring sets of array data, say 200 in this case, first the coherence matrix is derived by averaging the data sets (ensemble average), then the coherence function is derived by spatially averaging them. By using the obtained coherence function we reconstructed images of the object by applying the several different image reconstruction algorithms. These results are shown in Fig. 3.

As for the image obtained by DFT, the peaks are blurred and fairly large sidelobes are observed. As for the images by MEM and MLM, the peaks are reconstructed more sharply than those by DFT but the resolution of the peaks is not sufficient. Moreover, the background has apparent ripples. The image by ARMA processing is the closest to the desired one.

Table I shows the values of $F_a(M,N)$ for possible pairs of M and N. From this table, we can determine easily the order M_{opt} to be 2. As for N_{opt}, however, it is rather difficult to determine its value definitely from the table. Figure 4 shows the profiles of $F_b(N)$ for the case of M=2, where w_l =1 for $0 \le l \le L$, L=2,4,6 are used for averaging the AR coefficients. From this result, we can easily determine the N_{opt} to be 4. The increase of the determined value N_{opt} from the true value N=3 may be mainly due to the incompleteness of the random phase numbers generated in the computer and to the errors of the coherence function resulting from the finiteness of data acquisition time.

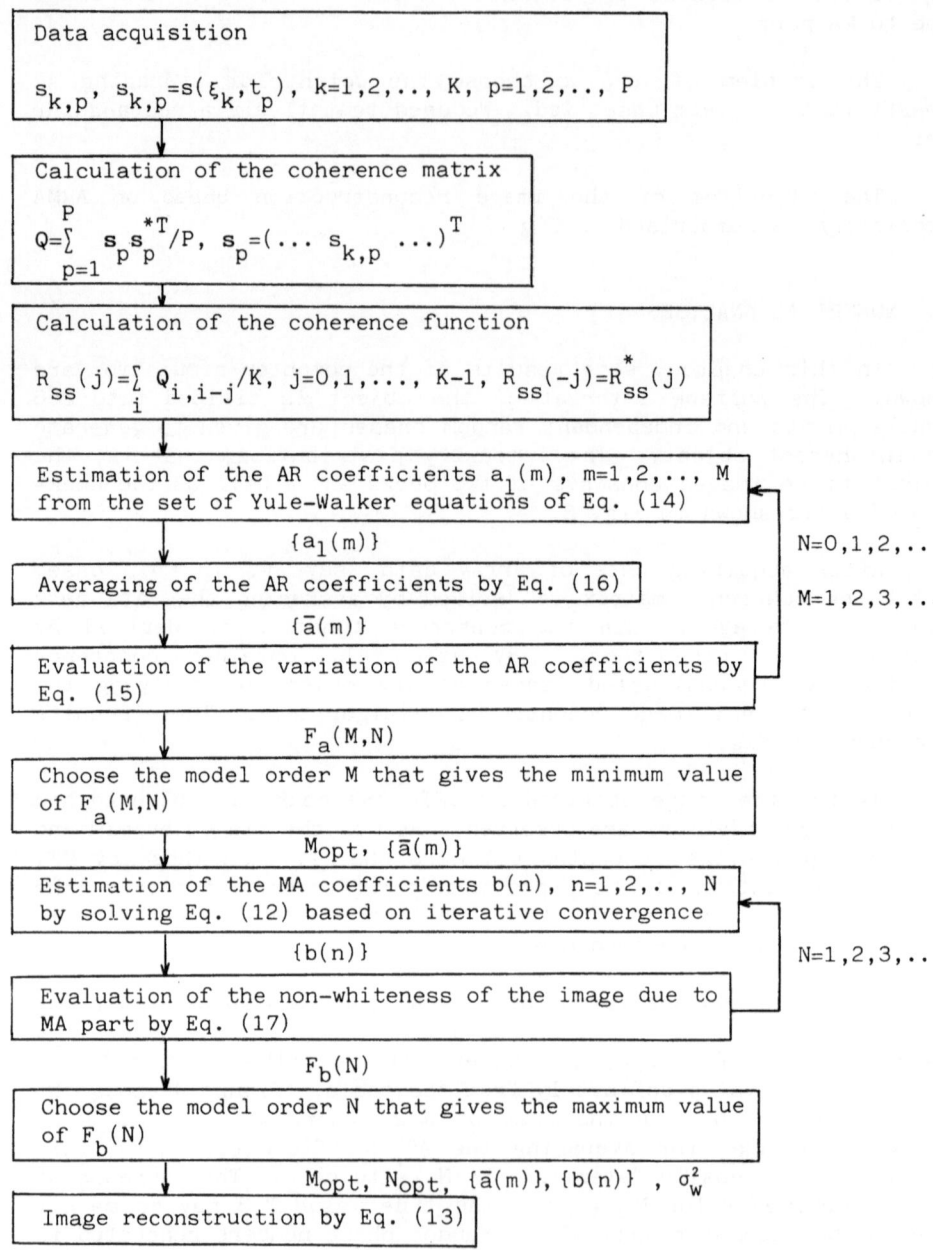

Fig. 2. Algorithm of image reconstruction based on ARMA processing.

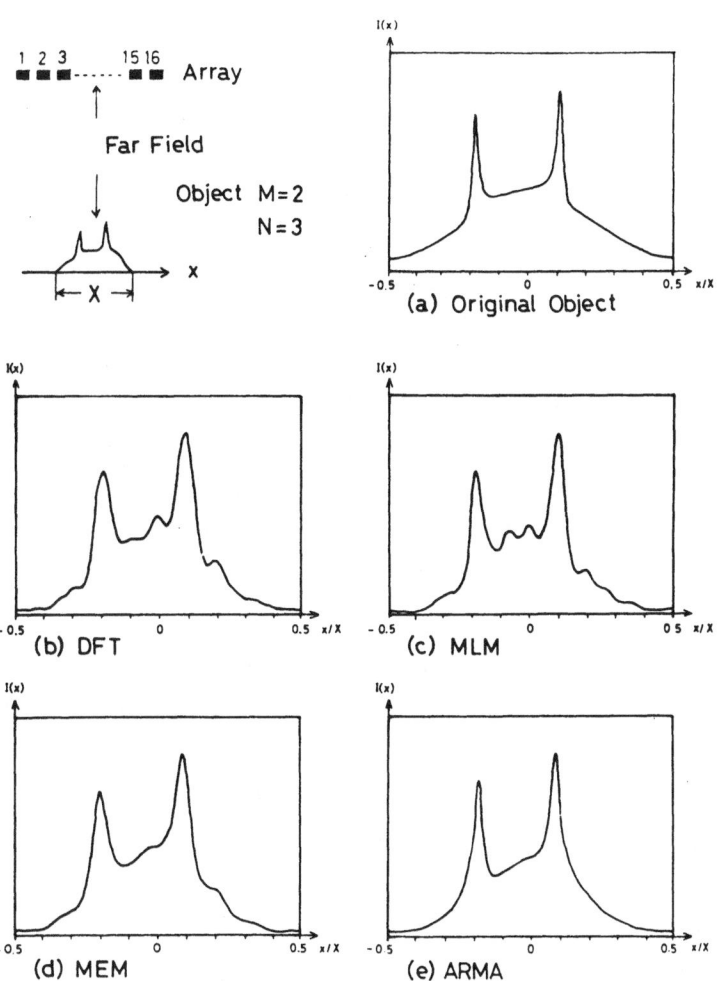

Fig. 3. Results of computer simulation (images obtained by several different image reconstructing algorithms).

Table I Table of the measure $F_a(M,N)$.

M \ N	0	1	2	3	4	5	6
1	167.88	115.21	107.80	59.61	36.66	29.78	56.09
M_{opt} → 2	0.18	0.18	0.18	0.18	0.19	0.22	0.24
3	16.34	16.66	17.82	18.88	21.33	17.66	1.22
4	56.14	54.57	65.08	25.36	21.43	29.26	92.10
5	1.74	1.60	1.65	1.67	1.82	0.76	0.71
6	2.67	2.46	2.36	0.80	0.50	0.51	0.42
7	2.99	2.35	1.45	1.00	0.63	0.27	0.19
8	8.22	9.89	6.02	3.07	1.62	1.24	1.31
9	6.50	5.27	3.82	2.05	0.87	0.50	
10	11.10	12.18	11.70	2.28	5.92		
11	8.39	6.33	3.80	0.77			
12	7.51	2.70	1.76				
13	1.33	1.83					
14	397.19						

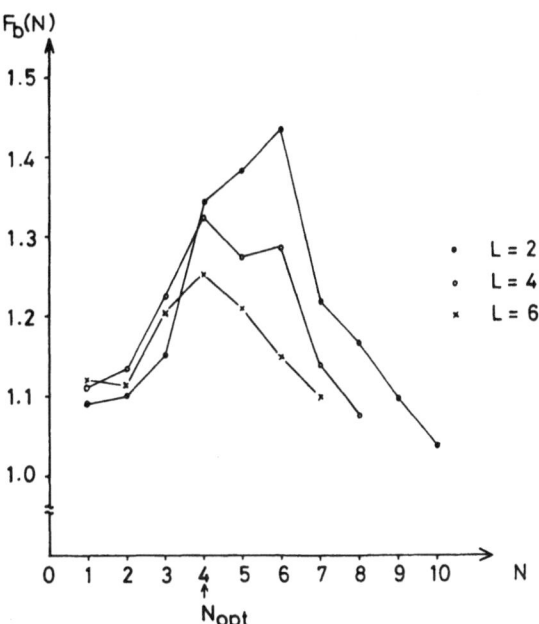

Fig. 4. Profile of the measure $F_b(N)$.

5. A MEANS OF ACTIVE GENERATION OF EQUIVALENTLY INCOHERENT
    WAVE FIELD ON THE OBJECT

    A method to generate a quasi-incoherent wave field on the
object is considered.   The geometry for this method is shown
schematically in Fig. 5.

    A set of wave fronts are transmitted from a single transducer
directly to the center of the object plane by scanning the
position of the transducer along an arc.   Actually, the transmission are carried out at $\theta_p = \{p-(P+1)/2\}\Delta\theta$, where p=1,2,..,P, P is
the total number of illuminations, and $\Delta\theta$ is the step of the
illumination angle.

    Using the Fresnel approximation, the incident wave field on
the object plane for the illumination of the wave from the
direction θ_p is given by

$$f(x,\theta_p)=\exp\{-jk(x^2-2Lx\sin\theta_p)/2L\}\exp(j\phi_p) , \qquad (18)$$

where ϕ_p is a random phase inserted in the transmission process,
and k is the wavenumber.   Then, the wave field reflected by the
object is given by

$$r(x,\theta_p)=f(x,\theta_p)g(x,\theta_p), \qquad (19)$$

where $g(x,\theta_p)$ is the reflection coefficient of the object.

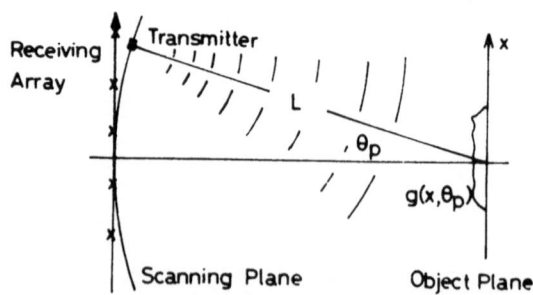

Fig. 5.   Geometry for the active generation of quasi-incoherent
           wave field on the object.

Now we impose the following assumptions on f and g:

i) ϕ_p and $\phi_{p'}$, $p \neq p'$, are independent, ii) ϕ_p distributes uniformly over $[0, 2\pi]$, iii) $g(x, \theta_p)$ and $g(x, \theta_{p'})$, $p \neq p'$ are independent. iv) The phase of $g(x, \theta_p)$ distributes uniformly over $[0, 2\pi]$ for the change of θ_p. v) The autocorrelation of $g(x, \theta_p)$ is independent of the direction, i.e., $E\{g(x, \theta_p)g^*(x', \theta_p)\} = \Gamma_g(x, x')$.

Then, the average of the wave field with respect to θ_p vanishes asymptotically for sufficiently large P as follows:

$$\frac{1}{P}\sum_{p=1}^{P} r(x, \theta_p) = \frac{1}{P}\sum_{p=1}^{P} f(x, \theta_p)g(x, \theta_p)$$

$$\simeq E\{\frac{1}{P}\sum_{p=1}^{P} f(x, \theta_p)\}E\{g(x, \theta_p)\}$$

$$= 0. \tag{20}$$

And, the autocorrelation of the wave field has the following asymptotical form:

$$\frac{1}{P}\sum_{p=1}^{P} r(x, \theta_p)r^*(x', \theta_p) \simeq E\{\frac{1}{P}\sum_{p=1}^{P} f(x, \theta_p)f^*(x', \theta_p)\}$$

$$\times E\{g(x, \theta_p)g^*(x', \theta_p)\}$$

$$= \Gamma_f(x, x')\Gamma_g(x, x'), \tag{21}$$

where $\Gamma_f(x, x')$ is given from Eq. (18) as follows:

$$\Gamma_f(x, x') = E\{\frac{1}{P}\sum_{p=1}^{P} f(x, \theta_p)f^*(x', \theta_p)\}$$

$$= \exp\{-jk(x^2 - x'^2)/2L\}\frac{1}{P}\sum_{p=-P_0}^{P_0} \{\sum_{n=-\infty}^{\infty} J_n(A)\exp(jnp\Delta\theta)\}$$

$$= \exp\{-jk(x^2 - x'^2)/2L\}$$

$$\times \{J_0(A) + 2\sum_{m=1}^{\infty} J_{2m}(A)\sin(Pm\Delta\theta)/P\sin(m\Delta\theta)\}, \tag{22}$$

where $A = k(x - x')$, and $P_0 = (P-1)/2$.

The width of the peaks of the periodic function $\sin(Pm\Delta\theta)/$ $P\sin(m\Delta\theta)$, $m=1,2,..,\infty$ in Eq. (22) are reduced according to the increase of $P\Delta\theta$ (the total illumination angle). The values of m which give the peaks of the $\sin(Pm\Delta\theta)/P\sin(m\Delta\theta)$ are given approximately by

$$m_s = [l\pi/\Delta\theta], \quad l=1,2,\ldots,\infty , \qquad (23)$$

where $[\ ]$ denotes Gauss operator. If we take a sufficiently large P so that $P\Delta\theta=\pi$ holds, then Eq. (22) is reduced to

$$\Gamma_f(x,x')=\exp\{-jk(x^2-x'^2)/2L\}$$

$$\times\{J_0(A)+2\sum_{m_s=1}^{\infty} J_{2m_s}(A)\}. \qquad (24)$$

By the way, $J_n(x)$ has significant values only in the range of $n<|x|<2n$, $n>0$, so $J_{2m_s}(A)$ in Eq. (24) has significant values for $|A|>2[l\pi/\Delta\theta]$. Hence, if we take the object's region of our interest within the value $\lambda/\Delta\theta$, then we obtain

$$\Gamma_f(x,x')=J_0(A)\exp\{-jk(x^2-x'^2)/2L\}. \qquad (25)$$

The result implies that the effective correlation width of Γ_f is of the order of the wavelength of the ultrasonic waves. Therefore, we can realize a quasi-incoherent wave field on the object regardless of the correlation length of Γ_g.

This method may be regarded as a version of the multiple-aperture imaging method which is used in the conventional coherent imaging system for the purpose of the speckle noise reduction.

6. EXPERIMENTAL RESULTS

Figure 6 shows the experimental set-up used for the generation of a quasi-incoherent wave field on the objects.

The object consists of two steel rods mounted on a sponge with random rough surface, and the object was illuminated from 16 different directions over 32 degrees.

The far field approximation was applied with the proper compensations.

Figure 7 shows a typical experimental result. The algorithm based on the ARMA model gives the desired sharply resolved image with reduced sidelobes and the speckle noises suppressed. In this case the estimated orders were M=2 and N=2.

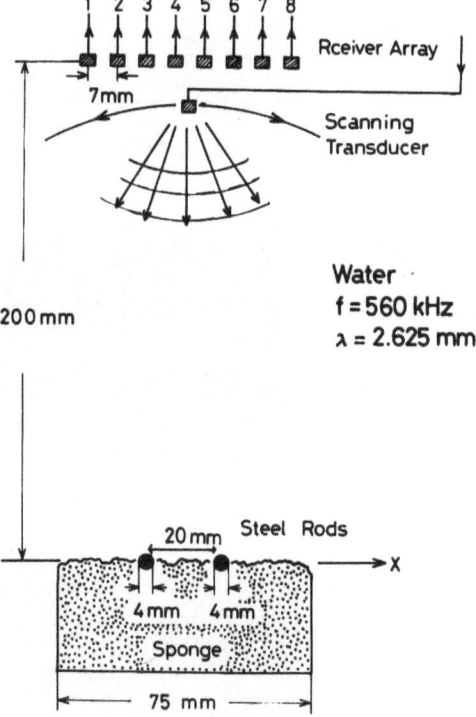

Fig. 6. Experimental set-up.

a) 16 Conventional Images
   Obtained by Wave Field
   Back Propagation for
   Changing the Angles
   of Illumination

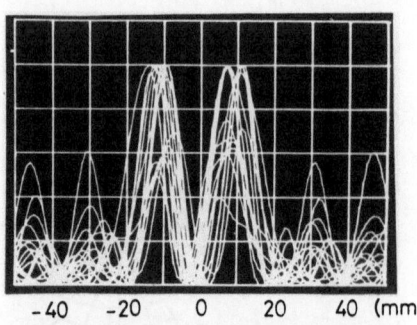

b) Image obtained by
   Using DFT of the
   Coherence Function

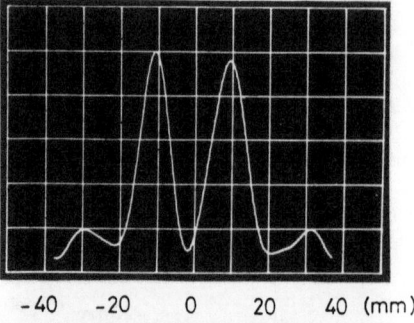

c) Image Obtained by
   ARMA Processing

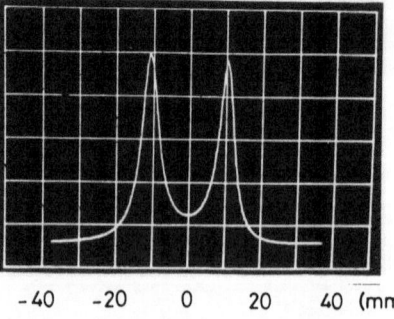

Fig. 7. Experimental results: comparison of images obtained by
         ARMA and conventional methods.

7. DISCUSSIONS AND CONCLUSION

In this paper we have discussed the possibilities and conditions for the effective applications of nonlinear processings such as MEM, MLM(HRM), and ARMA for ultrasonic imaging to get high resolusion and high fidelity images. Concretely the following points were made clear; i) if the wave field is incoherent, the method based on the ARMA processing is the most desirable one to get high fidelity images for usual ultrasonic objects of our interest, ii) the generation of equivalently incoherent wave field on the object is possible if we use appropriate illumination methods, iii) the ARMA-based incoherent imaging method promises not only the reduction of speckle noises but also the suppression of the sidelobe levels, and superresolved images can be expected, iv) estimation of the model orders can be carried out by using proper measure functions.

One of the problems left unsettled is that our algorithm based on ARMA does not always guarantee the minimum phase condition for the calculated spectrum (image), although all examples in our study did not show this problem. Another problem is on the choice of the optimum weight for averaging the AR coefficients.

Rather straightforward extension of this method to the case of two-dimensional or three-dimensional object is possible. More effective means are under study.

REFFERENCES

1. Barnes, C. W., Object Restoration in a Diffraction-Limited Imaging System, J. Opt. Soc. Am., 56:575 (1966).
2. Takuso Sato, Kazuho Uemura, and Kimio Sasaki, Super-Resolution Acoustical Passive Imaging System Using Algebraic Reconstruction, J. Acoust. Soc. Am., 67:1802 (1980).
3. "Nonlinear Methods of Spectral Analyses", S. Haykin, ed., Springer-Verlag (1979).
4. J. P. Burg, Maximum Entropy Spectral Analyses, 37th Int. SEG Meeting (1967).
5. T. J. Ulrich and T. N. Bishop, Maximum Entropy Spectral Analysis and Autoregressive Decomposition, Rev. Geophys. and Space Phys., 13:183 (1975).
6. S. J. Werneck and L. R. D'Addario, Maximum Entropy Image Reconstruction, IEEE Trans. Comput., C-26:351 (1977).
7. J. Capon, High-Resolution Frequency-Wavenumber Spectrum Analysis, Proc. IEEE, 57:1408 (1969).
8. G. L. Duckworth, Adaptive Array Processing For Acoustic Imaging, Acoustical Imaging, vol. 9, Prenum Press, New York, (1980).

9. G. E. Box and G. M. Jenkins, "Time Series Analysis Forecasting and Control", Holdan-Day, San Francisco, Calif., (1970).
10. P. R. Gutowski, E. A. Robinson, and S. Treitel, Spectral Estimation: Fact or Fiction, IEEE Trans. Geosci. Electron., GE-16:80 (1978).
11. M. Born and E. Wolf, Chapter X, Interference and Diffraction with Partially Coherent Light, in: "Priciple of Optics," Pergamon Press, Oxford (1970).
12. S. Brussone and M. Kaveh, On some Suboptimum ARMA Spectral Estimators, IEEE Trans. ASSP., ASSP-28:753 (1980).

ULTRASONIC PLANAR SCANNED TOMOGRAPHY

Hua Lee, Carl Schueler[†], Gail Flesher, and Glen Wade

Department of Electrical & Computer Engineering

University of California, Santa Barbara, California 93106

Abstract

Linear x-ray tomography has been used for decades as a technique that trades depth information for quantity of resolved volume. In linear focal-plane x-ray tomography, a synchronous opposing motion of x-ray source and film during exposure results in an image which has the midplane of the object in sharp focus with all other planes overlaid with various amounts of linear blurring. Nevertheless, the in-focus plane stands out well enough that x-ray diagnosticians still make extensive use of the technique.

We present a novel ultrasonic analog to the x-ray focal-plane technique that seeks to eliminate the out-of-focus planes. As an ultrasonic wave propagates through an object, it is attenuated and phase shifted. The transmitted magnitude and phase of the continuous wave may be measured to reconstruct tomograms on a complex number basis, if desired. The data taking requirements are presented. Two independent computer reconstruction algorithms are described. One offers the advantage of allowing non-linearities in the system, but retains the standard x-ray assumption of straight-ray propagation. The other allows spreading, but is restricted to a linear system. Either method allows reconstruction of attenuation and refractive index variation in the object, and either method allows independent reconstruction of any desired horizontal plane of the object. Finally, the limitations of each algorithm are discussed.

[†] Advanced Applications Department, Santa Barbara Research Center, Goleta, California 93117.

309

INTRODUCTION

The advent of axially scanned, computer-reconstructive x-ray tomography has caused a revolution in medical diagnostics. [1-3] Modern CT scanners produce images with very good resolution, due in part to the fact that they construct an effectively infinite aperture. Their excellent performance is due in part also to the fact that, unlike the focal-plane x-ray technique, images of all parts of the volume under scan are equally in focus, and no image overlapping occurs. In spite of the imaging superiority of CT scanners, x-ray focal-plane tomography, suffering from poorer resolution and overlapping image blur, still remains a popular medical diagnostic tool. This is because focal-plane tomography is simple and cheap compared to the CT systems.

Ultrasonic tomography is under study as a possible competitor for medical diagnostic use. [2-5] Ultrasound has the advantage of causing little or no tissue damage, but suffers from poorer resolution than x-ray techniques due in part to the much longer ultrasound wavelengths. In addition, imaging is made more difficult by the fact that ultrasound refracts and diffracts as it propagates through tissue. In principle, this does not impair resolution, but it does complicate the image reconstruction algorithms required to image the tissues of interest. The algorithms required in the x-ray case are simpler because straight ray propagation is a more accurate approximation for x-rays than for ultrasound.

Ultrasonic tomography algorithms have been developed that could be implemented in a 180° axially scanned system. [4,5] Such a device would probably be comparable in complexity and cost to the x-ray CT scanners that are presently used in some hospitals. The question naturally arises as to whether a simpler, planar ultrasonic imager, analogous to the focal-plane x-ray device, might offer advantages over the full axial scan implementation in terms of complexity and cost. To help answer that question, we have developed algorithms useful to test such an ultrasonic analog to the focal-plane x-ray technique. Our algorithms demonstrate that not only is such an analog feasible, but that one of the chief disadvantages of the x-ray focal-plane technique can be overcome.

The focal-plane x-ray tomogram includes a sharp image of one object plane. Superimposed on this image are blurred images of the other object planes that degrade the quality of the in-focus image. In the analogous ultrasonic planar-scanned tomogram, computer-processing of the measured data, under certain approximations, allows imaging of any plane individually, with no superimposed blurred images. The availability of such a calculated image would exist throughout the volume of the object, thus allowing computer reconstruction of a two- or three-dimensional image from a single data-taking scan. Therefore, we demonstrate how one may: (1) improve the x-ray focal plane technique by computer-processing, and (2) develop an analogous ultrasonic technique. Furthermore, we have developed two algorithms. One is suitable for implementation not only with ultrasound, but also in principle with x-rays. The other is suitable for implementation with ultrasound only.

THE DATA ACQUISITION SYSTEM

Fig. 1 shows the general structure of the proposed data acquisition system. On the top, an ultrasonic source is scanned in the horizontal source plane. The function $s(x,y,t)$ represents the source wavefield distribution over the source plane at $z=0$. The ultrasonic signals are received in a plane parallel to the source plane at $z=z_o$. The source function may vary with time, so that both the source wavefield $s(x,y,t)$ and the detected signal $r(x,y,t)$ are, in general, functions of space and time.

The unknown object distribution is located between these two planes. While the source scans above the object, the ultrasonic waves travel through the object and reach the receiving plane, where both the amplitude and the phase of the received signal are measured and recorded for processing. In practice, both the source and detection planes may be one- or two-dimensional. For simplicity, we develop the algorithms for the one-dimensional case. Consequently, the one-dimensional case has a source scanning along a straight line in the x-direction at height $z=0$. The receiver measures the signal along a line parallel to the source scan at depth $z=z_0$. The results of the scan are presented on a set of coordinate axes (x,t) called the "history" plane for the received signal. For this case, the acquired data would be sufficient to reconstruct a single vertical tomographic slice of the object, as in axially-scanned CT tomography. Extension to two horizontal spatial dimensions will be covered in the discussion, and for that case, the acquired data would allow reconstruction of a three-dimensional image of the entire volume under scan.

Two novel algorithms to reconstruct images from the data are presented. Each has a limitation which the other does not have, and we may compare one to the other. The first algorithm is reminiscent of x-ray focal plane tomography, and might be applied with x-rays if one used an array of x-ray detectors and a digital computer for data processing, rather than photographic film, as in the usual x-ray case. This first algorithm requires the assumption of straight-line propagation of ultrasound (or x-rays). However, the algorithm allows for non-linear acoustic interaction along the line of propagation. The second algorithm

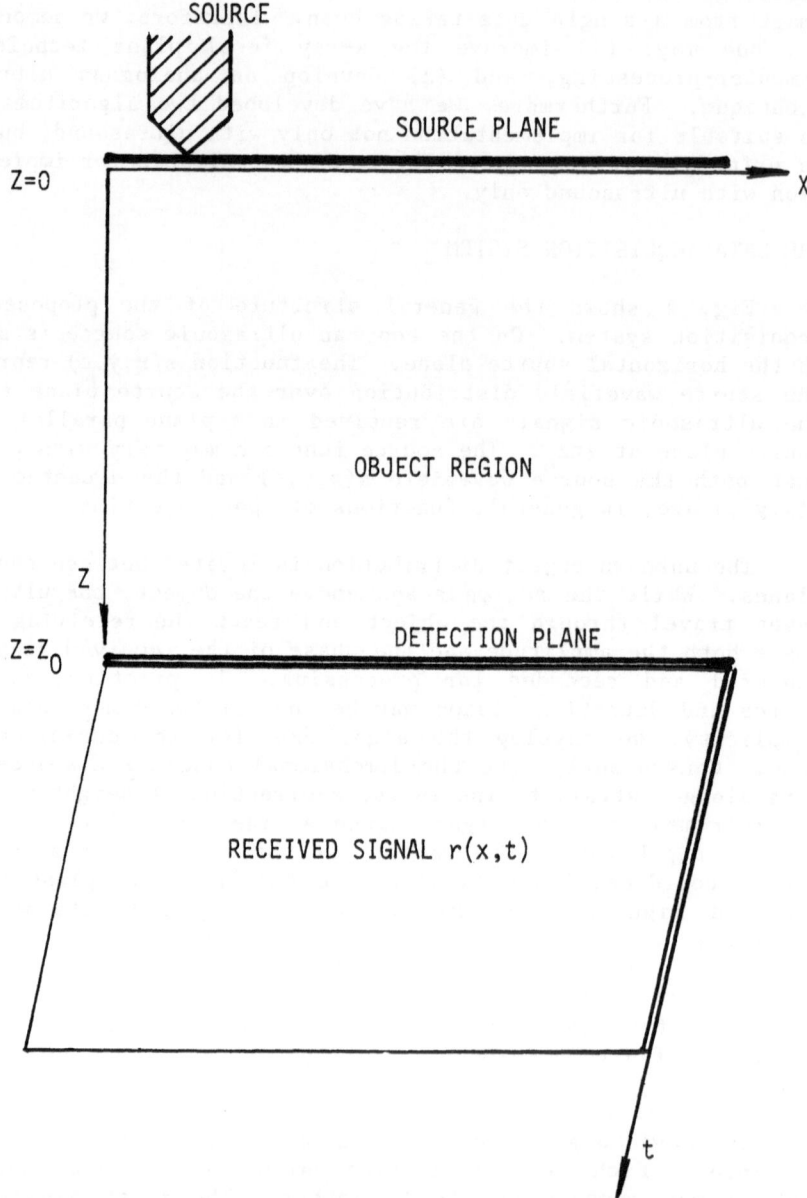

Figure 1. Data Acquisiton system for Planar-Scanned Tomography.
Shown in a representation for the "one-dimensional" case
discussed in the text. For the "two-dimensional" case,
the source would scan the (x,y) plane normal to the paper.
The received signal in the "history" plane, (x,t), would
then occupy a "history" volume, (x,y,t).

employs a holographic approach, and although spreading of ultrasound through the medium of interest is allowed, linear superposition of system effects along the z-axis is assumed. Therefore,
each technique has a restriction not imposed by the other. This
provides a choice according to the physical situation of interest.

ALGORITHM 1: COMPLEX LINE-INTEGRATION

The first approach is similar to focal-plane tomography
employing x-rays. Two simplifying assumptions are made: (1) We
assume that the ultrasonic wave propagates in a straight-line and
diffraction is negligible. (This assumption is often closely
approximated in practical medical applications of ultrasonic
tomography. [3]) (2) The distribution $s(x,t)$ is assumed to be a
point source moving linearly in a horizontal direction with a
constant speed. (The assumption of constant speed will be removed in the discussion.) The point source has unit amplitude
and radiates uniformly to the object region in all directions.

The complex object function $f(x,z)$ can be described as

$$f(x,z) = \exp\{\alpha(x,z) + j\beta(x,z)\}$$

$$= \exp\{\alpha\}\exp\{j\beta\} \qquad (1\text{-}1)$$

where α and β are real functions of position in the object.
$\exp\{\alpha\}$ is the amplitude of the object function, and represents
the attenuation due to the object. The term $\exp\{j\beta\}$ indicates
the phase delay factor caused during propagation through the
object.

A source point $s(x_s,t)$ and a detection point $r(x_r,t)$ define
a straight line. $r(x_r,t)$ is given by

$$r(x_r,t) = s(x_s,t) \, n(x_r,x_s) \, \exp\{\int(\alpha+j\beta)d\ell\} \qquad (1\text{-}2)$$

where the directional normalizing factor $n(x_r,x_s)$ is a function
of the angle of incidence, and the line integral is integrated
along the line joining x_s and x_r. Hence the line integral is

$$r_n(x_r,t) = \int(\alpha+j\beta)d\ell = \ln \frac{r(x_r,t)}{s(x_s,t) \, n(x_r,x_s)} \qquad (1\text{-}3)$$

Corresponding to the point source $s(x,t)$ at x and time t, a
point in the object region will create a shadow on the receiving
line. When the point source moves, the shadow also moves along
the receiving line. During the scan, a point on the line at
height $z=z_1$ in the object region traces out a straight-line

shadow pattern in the history plane (x,t). This line has a slope which is different for each height in the object. The slope is a function of the ratio $z_1/(z_9-z_1)$. After the normalization and logarithm operations in (1-3), $r_n(x_r,t)$ has a constant value along the straight line shadow pattern caused by a single object point. It can be seen that an object point at any height produces a straight line in the history plane, and points at different heights generate straight lines with different slopes. Finally, different points at the same object height generate a set of parallel, non-colinear lines in the history plane (x,t).

The received signal $r(x,t)$ has contributions from object points at all heights of the object region. Hence the normalized, received signal $r_n(x,t)$ is a linear superposition of many sets of parallel, non-colinear lines. Each set has different slope, and each slope corresponds to a unique height in the object.

Consider the two-dimensional received signal $r(x,t)$ for the case when all object points lie at a single height, z_1. For each object point, the normalized logarithmic signal $r_n(x,t)$ is constant along the history plane direction defined by the slope corresponding to the height z_1. If we take a two-dimensional Fourier transformation in x and t following the logarithm operator, we will obtain a single line of spectral values passing through the origin and normal to the slope direction in $r_n(x,t)$. [6] This line contains all the spectral information about the signal variation normal to the set of parallel, noncolinear lines in the (x,t) plane.

In the case for which object points lie at more than one height, $r_n(x,t)$ contains sets of parallel, non-colinear lines. Each set has a slope corresponding to a particular object point height. Each set also gives rise to a single spectrum line with a unique slope in the transform plane. Consequently, we can separate the information from each height in the object region by filtering the spectrum-line corresponding to that height from the spectral plane. We can filter one line-spectrum from all the rest because the line spectra for different object planes intersect only at zero frequency. Then merely a one-dimensional Fourier transform is needed, for every object height, to reconstruct the spatial-temporal variation in the original data $r_n(x,t)$ corresponding to the object point distribution at a single height. Finally, a simple scaling operation yields the spatial object point distribution at each height.

We can summarize the steps of this proposed complex line-integration algorithm as follows:

1)   Calculate the normalization factor.

$$n(x,x_s) = [1 + (\frac{x-x_s}{z_o})^2]^{-\frac{1}{2}}$$

2)   Normalize the received signal.

$$r'(x,t) = \frac{r(x,t)}{s(x_s,t)\ n(x,x_s)}$$

3)   Perform the logarithm operation on the two-dimensional normalized data.

$$r_n(x,t) = \log r'(x,t)$$

4)   Perform a two-dimensional Fourier transformation.

$$R(f_x,f_t) = F_{x,t}\ \{r_n(x,t)\}$$

5)   Separate the components by using line-spectrum selection for the object heights.

$$R_z(f_r) = R(f_x,f_y)$$

where $\dfrac{f_x}{f_y} = \dfrac{z}{z_o-z}$ and $f_r = (f_x^2+f_y^2)^{\frac{1}{2}}$

6)   Perform a one-dimensional Fourier transformation for each line-spectrum.

$$f(x;z) = F^{-1}\{R_z(f_r)\}$$

7)   Perform spatial scaling for each object height.

$$\tilde{f}(x;z) = f(cx;z)$$

where

$$c = \frac{z_o}{[(z_o-z)^2 + z^2]^{\frac{1}{2}}}$$

8)   Perform the exponent operation to obtain the resultant image at height z.

$$\hat{f}(x;z) = \exp\{\tilde{f}(x;z)\}$$

9)   Repeat steps 5-8 for each object height of interest

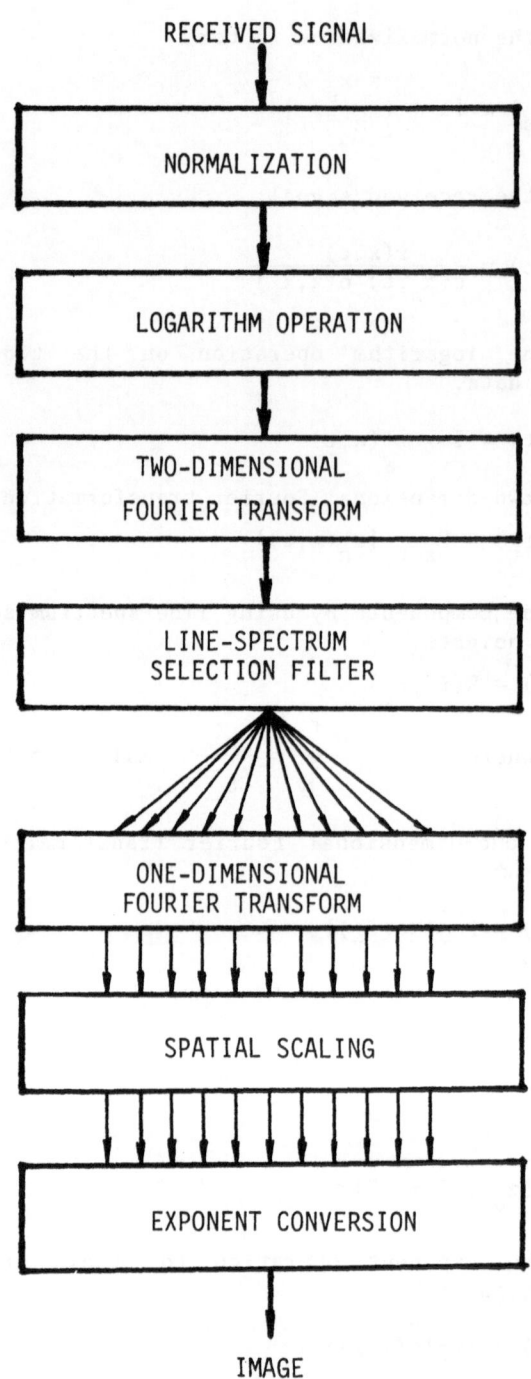

Figure 2. Flow diagram for the Complex-Line-Integration Alogrithm.

ALGORITHM TWO: "BACK AND FORTH" WAVE CORRELATION

The second algorithm is based on a holographic approach using both backward and forward wave propagation. [7] Diffraction and refraction effects on the wave are approximated as the wave passes through the object, so that no straight-line propagation assumption need be made. The accuracy will be greatest for the case of objects with lowest scattering strength. Secondary scattering is assumed small.

The physical arrangement for data-taking is identical to that for the first algorithm. While the imaging procedure can again be two-dimensional (planar) at each height z, for simplicity the following explanation is for the one-dimensional case. The source field $s(x,t)$ at height $z=0$, and the resulting received signal $r(x,t)$ at height $z=z_0$, are functions of space and time.

First, we consider only the contribution to the received signal of a thin object layer at $z=z_1$. The spatial frequency content of the source is found by taking the spatial Fourier transform of $s(x,t)$ at $z=0$.

$$S(f_x;t) = F_x\{s(x,t)\} \qquad (2\text{-}1)$$

Assuming that all the scattering points in the object are located at height z_1, the sound field at z_1 would be found by forward-propagating the source field to z_1. [8]

$$U_1(f_x;z_1,t) = S(f_x,t) \exp\{j2\pi z_1(\lambda^{-2}-f_x^2)^{\frac{1}{2}}\} \qquad (2\text{-}2)$$

where λ is the wavelength in the ambient medium. The spatial sound field is found by taking the inverse spatial Fourier transform of (2-2).

$$u_1(x;z_1,t) = F_x^{-1} [S(f_x,t) \exp\{j2\pi z_1(\lambda^{-2}-f_x^2)^{\frac{1}{2}}\}] \qquad (2\text{-}3)$$

The object distribution $f(x;z_1)$ at level z_1 modifies the ultrasonic field amplitude at each point, giving the transmitted field u_2 as follows.

$$u_2(x;z_1,t) = f(x;z_1)u_1(x;z_1,t) \qquad (2\text{-}4)$$

The spatial frequency content is

$$U_2(f_x;z_1,t) = F_x\{f(x;z_1)u_1(x;z_1,t)\} \qquad (2\text{-}5)$$

To first order, this field propagates without further modification to the receiver at $z=z_0$. (Successive scattering interaction is entirely neglected at this point in the development.) The received signal spectrum at z_0 due to the object at z_1 is

$$R(f_x,t) = U_2(f_x;z_1,t) \exp\{j2\pi(z_0-z_1) (\lambda^{-2}-f_x^2)^{\frac{1}{2}}\} \quad (2-6)$$

The receiver measures a contribution from each level, z, of the object. Each received signal $r(x;z,t)$ has a spatial Fourier transform $R(f_x,t)$ in the form of (2-6).

If the received field is back-propagated to height $z=z_1$ as in holographic reconstruction, only the effect of $f(x;z_1)$ in (2-4) will give the desired field $u_2(x;z_1,t)$. Scattering due to interaction with object points at other heights than z_1 will give rise to interference, which will be a function of both space and time. This interference will cause the received signal to be such that the results of back propagation to z_1 will differ from u_2 as given in (2-4). We can denote this actual value of the backpropagated field as $u_2'(x;z_1,t)$. Then $u_2'(x;z_1,t)$ is the sum of u_2, given in (2-4), and a complex interference term $v(x,t)$:

$$u_2'(x;z_1,t) = u_2(x;z_1,t) + v(x,t)$$
$$= f(x;z_1)u_1(x;z_1,t) + v(x,t) \quad (2-7)$$

As t varies, $v(x,t)$ will vary in both magnitude and phase. If we regard $v(x,t)$ as a noise term with zero mean, we can obtain a least square estimate of the object function at $z=z_1$: [9]

$$\hat{f}(x;z_1) = \frac{\displaystyle\int_0^T u_1^*(x;z_1,t)u_2'(x;z_1,t)dt}{\displaystyle\int_0^T |u_1(x;z_1,t)|^2 dt} \quad (2-8)$$

where the time T of integration is the length of time that data was taken.

In (2-8), u_2' is the back propagated actual received field, at z_0, and is obtained as illustrated in the next two equations.

$$U_2'(f_x;z_1,t) = R(f_x,t) \exp\{-j2\pi(z_0-z_1)(\lambda^{-2}-f_x^2)^{\frac{1}{2}}\} \quad (2-9)$$

and

$$u_2'(x;z_1,t) = F_x^{-1}\{U_2'(f_x;z_1,t)\} \quad (2-10)$$

This last step is the back-propagation part of the algorithm. The forward-propagation is indicated by equation (2-2). The forward and backward-propagated signals are correlated by equation (2-8) to estimate the object distribution at an arbitrary height z.

We can summarize the back-and-forth wave correlation algorithm as follows:

1) Perform spatial Fourier transformations on the source sound field s(x,t) and the received signal r(x,t).

$$S(f_x,t) = F_x\{s(x,t)\}$$

$$R(f_x,t) = F_x\{r(x,t)\}$$

2) Backward-propagate the received sound field and forward-propagate the source sound field to an object height z.

$$U_2'(f_x;z,t) = R(f_x,t) \exp\{-j2\pi(z_0-z)(\lambda^{-2}-f_x^2)^{\frac{1}{2}}\} \qquad \text{"Back"}$$

and

$$U_1(f_x;z,t) = S(f_x,t) \exp\{j2\pi z(\lambda^{-2}-f_x^2)^{\frac{1}{2}}\} \qquad \text{"Forth"}$$

3) Inverse Fourier transform $U_1(f_x;z,t)$ and $U_2'(f_x;z,t)$ to obtain the estimated sound field distributions before and after the modulation by the object function at height z.

$$u_1(x;z,t) = F_x^{-1}\{U_1(f_x;z,t)\}$$

$$u_2'(x;z,t) = F_x^{-1}\{U_2'(f_x;z,t)\}$$

4) Obtain the least square estimate of the object function at z.

$$\hat{f}(x;z) = \frac{\int_0^T u_1^*(x;z,t)\, u_2'(x;z,t)dt}{\int_0^T |u_1(x;z,t)|^2 dt}$$

5) Repeat steps 2 to 4 for each desired object height.

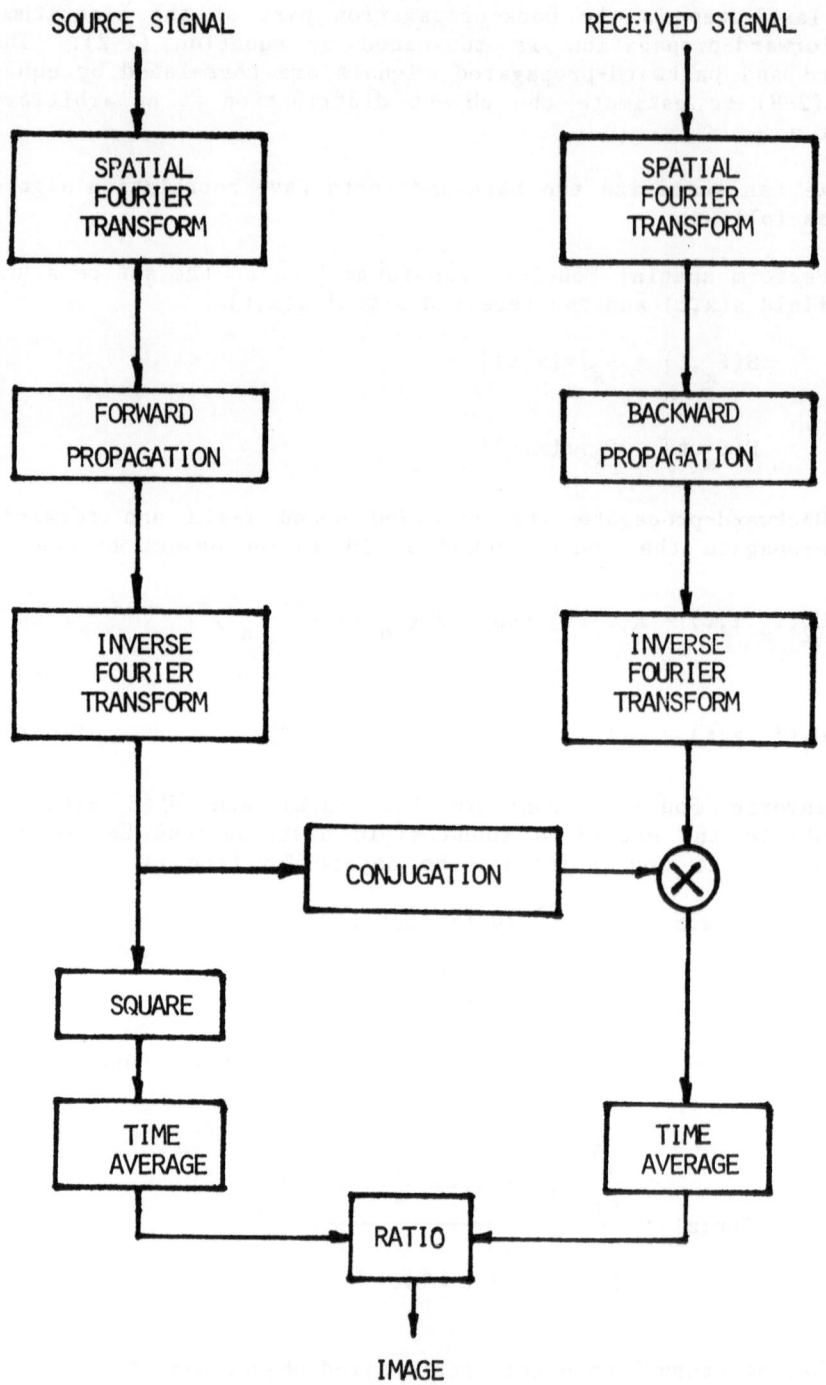

Figure 3. Flow diagram for the "Back-and-Forth" Wave Correlation
          alogrithm.

DISCUSSION

A.   Complex-Line Integration

Complex line integration requires a point source moving horizontally along the source height z=0. If the point source scan speed varies, proper rescaling can correct the time-coordinate of the received signal r(x,t). Because straight-line propagation is the major assumption for this method, the processing algorithm requires only a small modification to extend the treatment to two spatial dimensions for each height. For example, the same algorithm could be repeated sequentially for each of many vertical, two-dimensional slices through the object. Finally, the entire three-dimensional object distribution could be reconstructed. (A similar process is used to obtain three-dimensional reconstruction using axially scanned x-ray CT tomography.)

During the spectrum-selection for different object heights, the zero frequency component is common to every object height. Due to the exponent operator at the end of the algorithm, the zero-frequency component becomes a multiplicative common factor for the resultant image. Therefore, the zero frequency components should be set to zero to reduce the complexity. In addition, the practical problem of spectrum selection is more difficult at lower spatial frequency. This is because at low frequencies there is less space between line spectra than at higher frequencies, and a digital implementation of the algorithm would not have perfect resolution.

B.   "Back-and-Forth" Wave Correlation

Back-and-forth wave correlation does not require a point source, nor does the source have to be linearly scanned. Therefore, one is less restricted in the detailed design of the data acquisition system for back-and-forth wave correlation than for complex line integration. Also, because diffraction is included, no normalization process is necessary for back-and-forth wave correlation. However, the source sound field must be coherent, due to the use of forward and backward propagation operators.

The back-and-forth correlation algorithm requires more computation capacity than complex line integration because Fourier transformations are needed for both the source sound field and the received data. In addition, inverse transformations are needed for every object height. However, Fourier transformations operate on only the space variables. Therefore, when the space coordinate is extended to two horizontal dimensions for each value of z, the only modification to the algorithm is to use the two-dimensional spatial Fourier transformations.

C.   General Considerations

For both algorithms, the resolution of the resultant image is limited by the aperture size and time duration. Diffraction effects degrade the resolution when complex-line integration is used and the presence of secondary scattering reduces the accuracy for back-and-forth wave propagation.

CONCLUSION

Two algorithms have been presented for acoustic imaging of objects. Both rely on measurement of wave fields that have been transmitted through the object. The first algorithm for complex-line integration tomography considers the effect of the attenuation and phase shift that a wave experiences when traversing the object, but does not include the effect of diffraction and refraction. The second algorithm for "Back and Forth Correlation" acoustic imaging considers the first order diffractive and refractive effects of wave propagation through the object, but does not include the higher order diffractive and refractive effects on the net attenuation and phase shift. Both algorithms would be most accurate for an object with mild diffractive and refractive properties.

ACKNOWLEDGEMENT

The authors would like to thank Tracy Hamilton of the Electrical and Computer Engineering Department at the University of California at Santa Barbara for her expert and expeditious preparation of this manuscript.

REFERENCES

1.   G. Wade, R.K. Mueller, and M. Kaveh, "A Survey of Techniques for Ultrasonic Tomography," Proceedings of IFIP TC-4 Working Conference Computer-Aided Tomography and Ultrasonics in Medicine, J. Raviv, Editor, Amsterdam, The Netherlands, North Holland, 1978.

2.   R.K. Mueller, M. Kaveh, and G. Wade, "Reconstructive Tomography and Applications to Ultrasonics," Proceedings of the IEEE, Vol. 67, No. 4, April 1979, pp. 567-587.

3.   G. Wade, "Ultrasonic Imaging by Reconstructive Tomography," Acoustical Imaging, Vol. 9, K. Wang, Editor, Plenum Press, New York, 1980, pp. 379-432.

4.   J.F. Greenleaf, S.A. Johnson, S.L. Lee, G.T. Herman, and E.H. Wood, "Algebraic Reconstruction of Spatial Distributions of Acoustic Absorption Within Tissue from their Two-

Dimensional Acoustic Projections," Acoustical Holography, Vol. 5, P.S. Green, Editor, Plenum Press, New York, 1974, pp. 591-603.

5.   G. Wade, S. Elliott, I. Khogeer, G. Flesher, J. Eisler, D. Menza, N.S. Ramesh, and G. Heidbreder, "Acoustic Echo Computer Tomography," Acoustical Imaging, Vol. 8, A.F. Metherell, Editor, Plenum Press, New York, 1979, pp. 565-576.

6.   R. Bracewell, The Fourier Transform and Its Applications, McGraw-Hill, New York, 1965, Chapter 12.

7.   C.F. Schueler, G. Wade, and J. Fontana, "Ultrasonic Underwater Imaging System with Computer Image Processing and Reconstruction," Proceedings of Ultrasonics International 1979, IPC Science and Technology Press, Ltd., 1979, pp. 372-377.

8.   J.W. Goodman, Introduction to Fourier Optics, McGraw-Hill, 1968, pp. 54.

9.   A.D. Whalen, Detection of Signals in Noise, Academic Press, 1971, pp. 335-336.

A COMPARISON OF BORN AND RYTOV APPROXIMATIONS

IN ACOUSTIC TOMOGRAPHY

M. Kaveh, M. Soumekh and R. K. Mueller

Department of Electrical Engineering
University of Minnesota
Minneapolis, MN 55455

ABSTRACT

This paper gives a comparison of acoustic tomograms based on
the wave equations using the Born and Rytov approximations. The
comparisons are based on the distortions in the reconstructions
using these two methods. It is shown that when a large forward
scattering component exists that tomograms based on Born's
approximation are greatly in error. Rytov's method has a wider
range of validity, in practice, but suffers from the problems
associated with the determination of the phase of the scattered
wave.

INTRODUCTION

The objective of diffraction tomography is to reconstruct,
from the observed scattered field, the spatial distribution of
appropriate parameters in a test object. The data is normally
obtained from measurements of the scattered field from a number
of object orientations. The mathematical nature of the reconstruction is then one of solving an inverse problem of the
appropriate equation of motion. The challenge, of course, is in
obtaining a solution that is computationally efficient and subject
to reasonable interpretation. This in general is not possible.
In cases of interest to acoustic transmission tomography, however,
the relatively small perturbation of object parameters compared
to the surrounding medium give the possibility of using approximations to the equation of motion based on perturbation techniques.
These techniques substantially simplify the inversion problem and
lend themselves to computational advantages of fast Fourier
transform processing.

325

Previous work in this area has included the theoretical formulation of the problem and solution based on the Born and Rytov approximations[1,2,3,4]. The experimental results given in reference 3 were based on the Born approximation and no comparison of the two methods were given. In this paper, the inversion algorithms based on Born and Rytov approximations are revisited for a relatively simple medium with small perturbations in the velocity of propagation and small attenuation. We will specifically concentrate on the nonlinear terms that are neglected in these two approximations. These will be used as indicators of error in the reconstructed images to show the performance capabilities of the two techniques.

The paper is organized as follows. The wave equation and its perturbation approximations are first given. The function that is imaged in our reconstruction algorithm is then related to the assumptions made in the Born and Rytov approximations. Finally, some practical issues concerning the algorithm based on the Rytov approximation are discussed and simulation and experimental results are presented.

THE APPROXIMATE WAVE EQUATIONS

A general equation of motion for the propagation of a sound field through biological tissue is given in reference 4. Here we neglect density variations in the object and assume two-dimensional propagation. Then, using a single frequency, ω_o, plane wave excitation we represent the wave propagation by a complex wavenumber, k, given by[5]

$$k^2(x,y) = k_o^2[1 + f(x,y)] \tag{1}$$

where

$$k_o = \frac{\omega_o}{C_o} \text{ and } f(x,y) = f_r(x,y) + if_i(x,y)$$

with $|f(x,y)| \ll 1$, $f_i(x,y)$ representing the small attenuation of the medium and C_o the uniform velocity of propagation. The wave equation, omitting the (x,y) notation, is given by:

$$\nabla^2\psi + k^2\psi = 0 \tag{2}$$

The aim is to estimate f from measurements of ψ for different orientations of the object. An approach for $|f| \ll 1$ is to linearize equation (2) to obtain computationally efficient (e.g., Fourier domain) solutions. The two approximations that have been used are the Born and Rytov approximations[1-4]. In this paper we re-examine these two methods, concentrating on the reconstructed functions in a comparative fashion.

Tomograms Based on Born's Approximation

The wave function is expressed in terms of the incident and scattered waves ψ_o and ψ_s. Then, the wave equation is:

$$\nabla^2(\psi_o+\psi_s) + k_o^2(1+f)(\psi_o+\psi_s) = 0 \tag{3}$$

or

$$\nabla^2\psi_s + k_o^2\psi_s = -\psi_o k_o^2 f - \left[\psi_s k_o^2 f\right] \tag{4}$$

In Born's approximation the term in [ ] in equation (4) is neglected and the resulting linear differential equation is inverted to obtain an estimate of f^{1-3}. The problem is that most practical objects are large enough to produce $|\psi_s|$ on the order of $|\psi_o|$. Therefore, Born's inversion results in a reconstruction of a function h given by

$$h = (1 + \frac{\psi_s}{\psi_o})f = f + g_B \tag{5}$$

g_B is then the nonlinear distortion that appears in the images based on Born's approximation. This term is dominated by the relatively uniform average characteristics of f that cause a significant amount of forward scattering. A consequence of this distortion is, for example, a scrambling of the reconstruction of the real and imaginary parts of f. That is:

$$h_r = f_r + g_{Br} \text{ and } h_i = f_i + g_{Bi}$$

However, both g_{Br} and g_{Bi} contain the real and imaginary components of f. Examples of these effects will be given in the section on simulation and experimental results.

Tomograms Based on Rytov's Approximation

In Rytov's approximation, one defines a function χ as:

$$\chi = \ln\psi \text{ or } \psi = e^\chi \tag{6}$$

Substituting for ψ in (2) results in an alternative differential equation[2] given by

$$\nabla^2\chi + \nabla\chi\cdot\nabla\chi + k_o^2(1+f) = 0 \tag{7}$$

Letting

$$\chi = \chi_o + \chi_s \text{ and } \chi_1 = \chi_s\psi_o$$

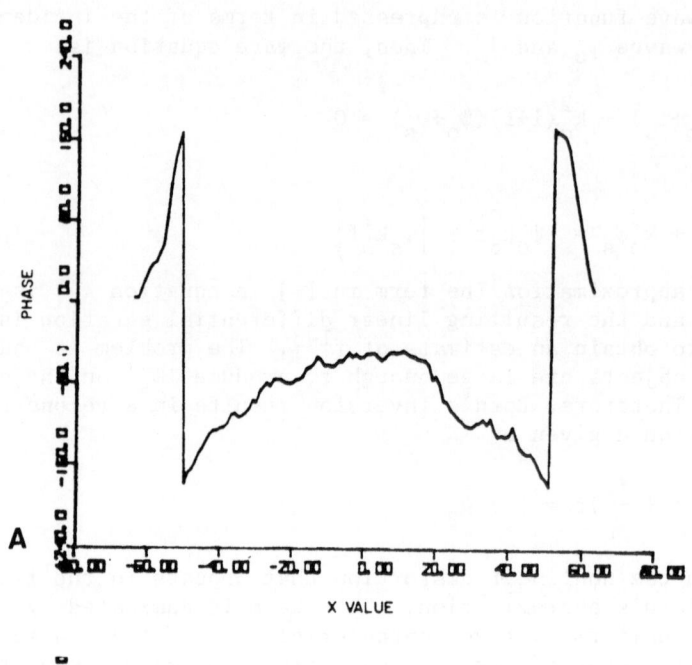

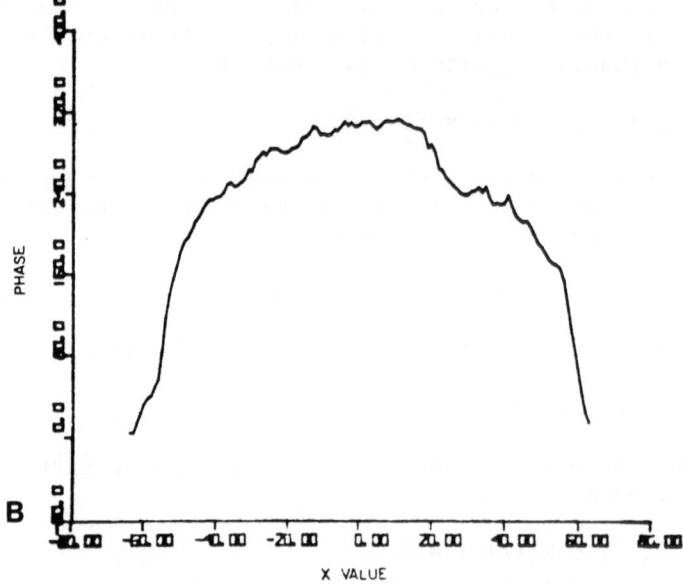

Fig. 1.   Phase profile of 120λ diameter gelatine cylinder.
          (a) measured phase, (b) unwrapped phase.

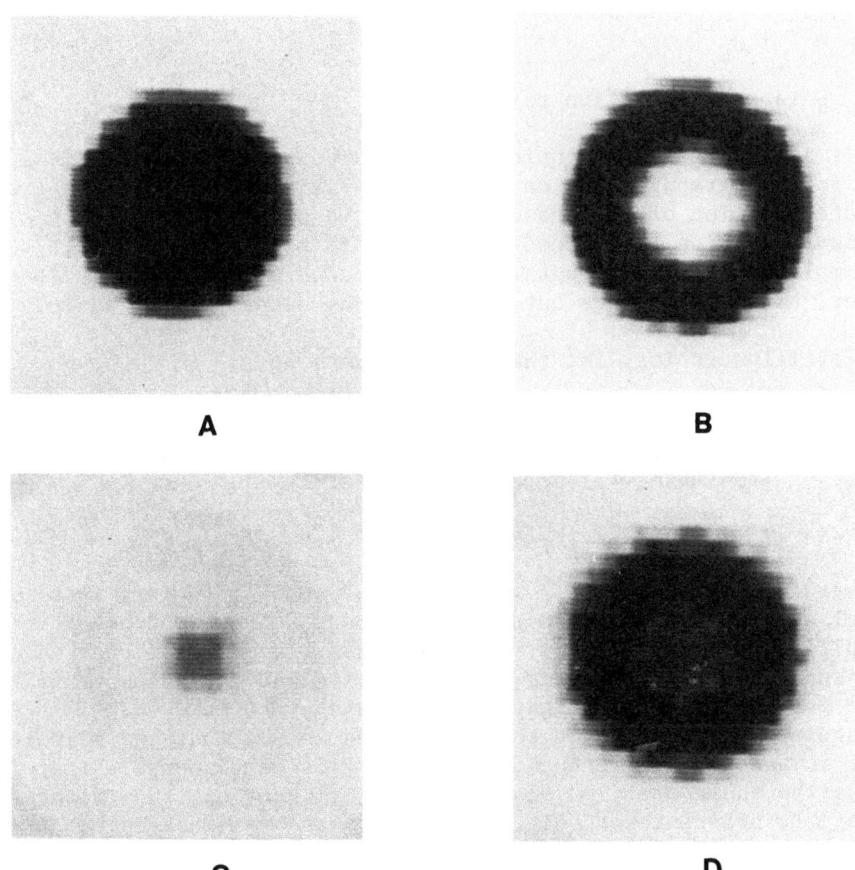

A B

C D

Fig. 2. Complex annular cylinder. $f_r = -0.1$, $r \leq 2\lambda$.
$f_i = 0.19$ $\lambda \leq r \leq 2\lambda$, numerical simulations,
Rytov: (a) $\bar{f}_r$, (b) f_i; Born: (c) f_r, (d) f_i.

where χ_o is the homogenous solution to (7), one obtains

$$\nabla^2 \chi_1 + k_o^2 \chi_1 = -\psi_o k_o^2 f - \left(k_o^2 \psi_o \nabla \chi_s \cdot \nabla \chi_s \right) \tag{8}$$

In Rytov's approximation the term in [ ] is neglected and the resulting linear differential equation is inverted[1]. Therefore, the actual tomogram is now given by

$$\ell = f + g_R \tag{9}$$

where g_R is the distortion term in (8).

It is important to consider, at this point, a very practical issue for tomography based on Rytov's approximation. This concerns the determination of the phase of χ. It is obvious that in the Born approximation the needed phase of ψ is restricted to $[-\pi,\pi]$. The complex logarithmic operation used in the Rytov approximation, however, naturally brings out the ambiguous nature of the phase.

For illustration, let the measured wave be

$$\psi(x_o,y) = A(y)e^{iP(y)} \tag{10}$$

Taking the logarithm of (10) gives

$$\chi_2(x_o,y) = \ell n A(y) + iP_1(y) \ , \ \left| P_1(y) \right| \leq \pi \tag{11}$$

Reconstruction based on (11) is, of course, unsatisfactory in general and would lead to an erroneous image. Thus, one needs to carefully determine an exact relative phase $P(y)$/profile that is commensurate with the physics of the problem and the assumptions on the medium under investigation. We make use of the weak inhomogeneity assumption on the medium and assume that the measured signal is sampled fine enough so that $P(y)$ is continuous. Figure 1 shows the phase profiles of a gelatine cylinder of 120 λ diameter with a 12 λ off-centered hole. Both $P_1(y)$ and the "unwrapped" phase $P(y)$ are shown. A tomogram based on $P_1(y)$ was found to be completely in error.

PERFORMANCE COMPARISONS

Conditions for the validity of Born and Rytov approximations have been considered before[1,4]. In this paper, however, we compare the performances of the two methods based on the reconstructed distortion functions g_B and g_R. We have already noted the direct dependence of g_B on the desired function f. This is indeed a very disturbing effect and can lead to totally misleading information.

As an example, if f is purely real and $\mathrm{Re}\left(\dfrac{\psi_s}{\psi_o}\right) \simeq -1$ then $h_r \simeq 0$

and h_i is proportional to f.

On the other hand, the distortion in Rytov's method is not directly related to f and depends on the <u>rate</u> <u>of</u> <u>change</u> of the smoothed scattered wave. Thus one expects distortion possibly at the edges of sharp discontinuities, without a direct mixing of the real and imaginary parts of the disturbance. The latter is a pleasing property. In the following a comparison of tomograms obtained by the two methods is given to demonstrate the above effects. The results are based on both numerically generated as well as experimentally obtained scattered waves.

i) Complex Annular Cylinder. This example is numerically generated as follows:

$$f(r) = \left\{ \begin{array}{ll} -0.1 + j0.19 & \lambda < r \leq 2\lambda \\[2ex] -0.1 & r \leq \lambda \end{array} \right.$$

Therefore in the absence of any distortion the tomogram should indicate a real disk of magnitude 0.1 and radius 2λ and an imaginary ring of width λ. The simulated data was obtained from a sampling spacing of $\lambda/4$ with 32 samples/profiles and 128 profiles. Figure 2 shows the Born and Rytov reconstructions for this simulation. The distortion terms discussed earlier are prominent here. The pictures are all normalized to the true value of the disturbance. Figures 2(a) and 2(b) (Rytov reconstruction) show good separation of the real and imaginary parts with quantitative fidelity. Figures 2(c) and 2(d) show the effect of g_R. The real part, h_r, of the tomogram in this case is "washed out" as g_r-f_r, whereas h_i contains a mixture of f_r and f_i. Thus, there is basically no quantitative information in this reconstruction.

ii) Gelatine Cylinder with Holes. This phantom is a 60λ diameter cylinder with two 8λ holes in water. The measurement setup and data acquisition were described previously[3]. 64 samples/profiles at 2λ spacing and 64 profiles were used. The disturbance imaged by the Rytov method (Figures 3(a) and (b)) is mainly real, indicating the difference in the velocity of propagation. The tomogram based on Born's method (Figures 3(c) and (d)) is again in error, with the real part washed out and the imaginary part incorrectly showing an outline of the object. We suspect some degradation in these experimental results due to deviations of the insonifying wave from a plane wave. This

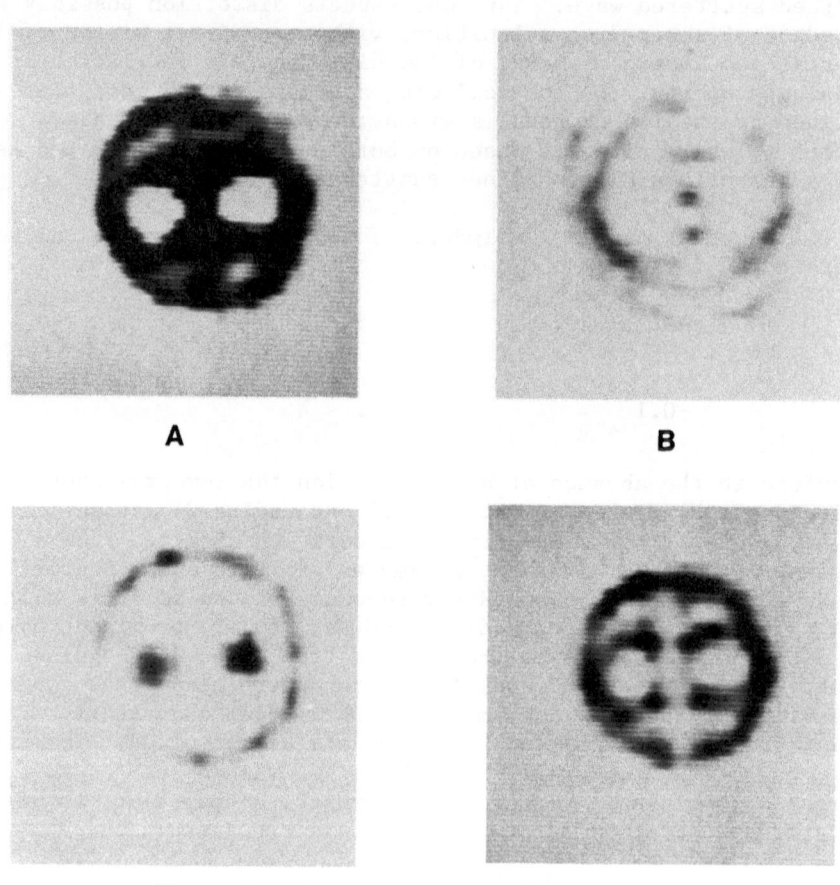

Fig. 3.   60λ diameter gelatine cylinder with 8λ holes.
          Rytov: (a) f_r, (b) f_i; Born: (c) f_r, (d) f_i.

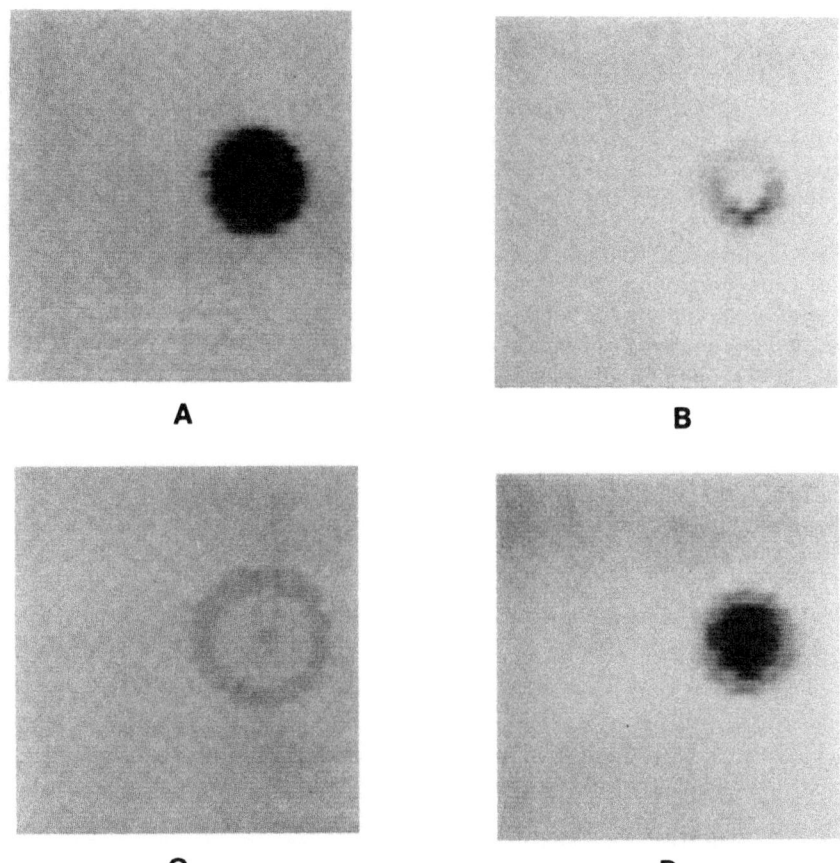

A   **B**

C   **D**

Fig. 4. 5% saline solution cylinder. Rytov: (a) f_r, (b) f_i; Born: (c) f_r, (d) f_i.

example is also a correction to the explanation of the quality of
the tomograms in reference 3 based on Born's approximation. In
reference 3 the magnitude of the reconstructions were shown.

iii) Off-Centered Saline Phantom. Figure 5 shows Rytov and
Born reconstructions of a cylindrical 5% saline solution in water.
Rytov's method gave the correct quantitative result for f_r. The
tomograms based on Born's approximation are again greatly in error.

CONCLUSIONS

We have compared the qualities of diffraction tomograms
based on the Born and Rytov methods. The comparison was based on
the severity of the degradation in the reconstructed images due
to the neglected terms in the two approximations. It was shown
that in media with appreciable forward scattering that the Born's
approximation yields useless images. On the other hand reconstructions based on Rytov's method are reasonable if the phases of
the profiles can be determined accurately. The performance of
this method on more complex media such as tissue, however, remains
to be demonstrated in practice.

ACKNOWLEDGEMENT

The authors acknowledge helpful discussions with
Professors James Greenleaf, Steve Johnson and Mr. David Fishbaine.
The saline phantom profiles were measured at the Mayo Clinic by
Dr. Greenleaf. This work was supported by the National Science
Foundation under Grants ENG76-84521 and ECS-7926008.

REFERENCES

1. M. Kaveh, R. K. Mueller and R. D. Iverson, Ultrasonic
    tomography based on perturbation solutions of the wave
    equation, Computer Graphics and Image Processing,
    9:105 (1979).
2. R. K. Mueller, M. Kaveh and G. Wade, Reconstructive tomography
    and applications to ultrasonics, IEEE Proceedings,
    67:567 (1979).
3. M. Kaveh, R. K. Mueller, R. Rylander, T. R. Coulter and
    M. Soumekh, Experimental results in ultrasonic diffraction
    tomography, in: "Acoustical Imaging, Vol. 9," K. Wang, ed.,
    Plenum, New York (1980).

4. R. K. Mueller, Diffraction tomography I: the wave equation,
    Ultrasonic Imaging, 2:213 (1980).

5. P. M. Morse and K. U. Ingard, "Theoretical Acoustics",
    McGraw-Hill, New York (1968).

TOMOGRAPHIC EVALUATION OF SOUND FIELDS

FROM ACOUSTO-OPTIC DATA

Bill D. Cook* and John F. Laflin*

Department of Mechanical Engineering
University of Houston, Houston, Texas 77004

Charles F. Gaumond and Henry D. Dardy

U.S. Naval Research Laboratory
Washington, D.C. 20375

ABSTRACT

The principles of computerized transverse tomography can be
applied to the acousto-optic reconstruction of the local sound
pressure of an ultrasonic field. For sufficiently narrow beams
of ultrasound in the low megahertz region, the total optical
phase retardation of an interrogating light beam can be considered as a projection of the sound field pressure. (Fourier
techniques for the numerical reconstruction of the pressure field
yield as intermediate steps a Fourier domain associated with the
angular spectrum of plane waves comprising the sound field.) Consequently the sound field can be reconstructed in other regions
than the plane of interrogation. In this work we discuss two alternative methods for acquiring data. One method builds the
Fourier domain along radial spokes which is inconvenient for
numerical processing by DFFT algorithms. The other procedure builds the Fourier domain in a nearly rectangular format compatible
with two-dimensional DFFT algorithms. With this latter method,
it is possible to evaluate the pressure along a line with limited
data and a one-dimensional DFFT.

* Work supported in part by Physical Acoustics Center Program at
the Naval Research Laboratory, Code 5130, Washington, D.C.
20375.

1.0   INTRODUCTION

A sound field of low ultrasonic power, low ultrasonic frequency, and narrow beam width behaves as an optical phase grating. Collimated light passing through such a sound field experiences an optical phase retardation proportional to the local
sound pressure, integrated over the light path. This constitutes
a "projection" of the sound field and numerical methods of computerized transverse tomography can be applied to estimate the
local sound pressure.

Numerical techniques using Fourier transforms are useful in
pressure field evaluation since an intermediate step yields the
Fourier domain associated with the angular spectrum of plane
waves comprising the sound field. Moreover, by modification of
the phase terms of each plane wave of the angular spectrum, it is
possible to construct the sound field at different planes. In
other words, an estimate of most of the sound field can be computed from a set of acousto-optic data taken over a single
transverse plane. The total field, however, cannot be constructed everywhere since evanescent waves near the sound source are
not accounted for. This total field concept is valid when the
sound field can be described by the Helmholtz equation, thus eliminating application to non-linear or highly attenuated sound
fields.

In a series of papers[1-3] directed toward transducer calibration Cook and Berlinghieri have described one method of data
collection which we will call Method A. Acousto-optic data is
collected at the terminus of the light paths as shown in Figure
1. These light paths are parallel to each other and are in a
plane parallel to the surface of the transducer. Sufficient data
can be acquired from interrogation of the field in one direction
if the field is symmetric. If the field is not symmetric, Method
A involves rotation of the transducer about an axis normal to the
transducer surface, such as CC'.

Here, we present an alternative method of data collection
which we will call Method B. In Method B the transducer is rotated about a line AA' parallel to the transducer surface and located in the plane of the light paths. The line AA' is also perpendicular to the light paths.

We will demonstrate how both methods allow the generation of
data in the angular spectrum (plane wave decomposition) domain
with both phase and amplitude information to allow evaluation of
the pressure field at the plane of measurement using Fourier
transforms.

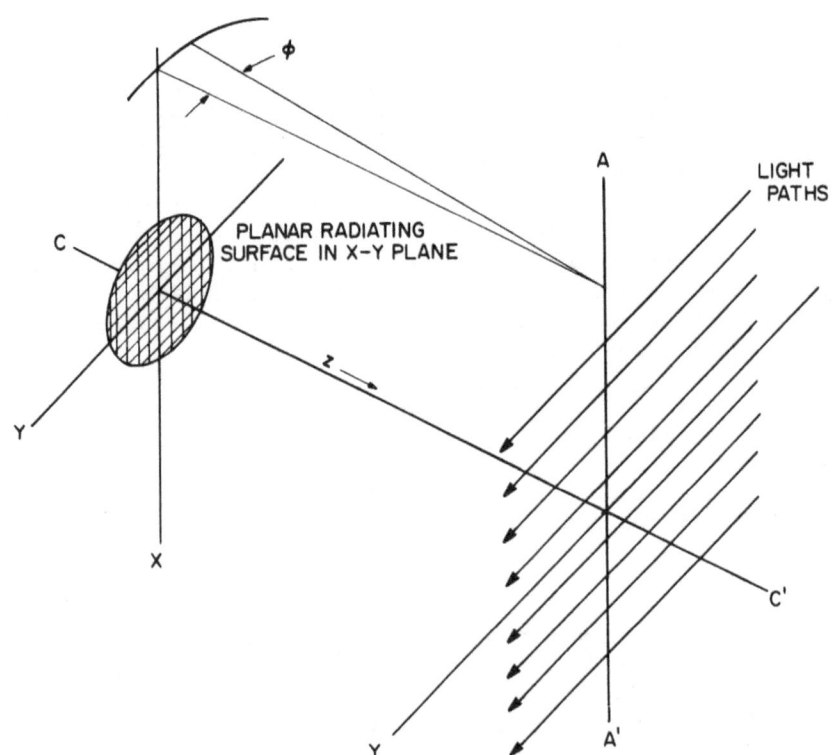

Fig. 1. Light path geometry.

Data acquired using Method A builds the angular spectrum domain in a polar format through a series of one-dimensional Fourier transforms. The pressure field at the plane of measurement
can then be obtained by a two- dimensional inverse Fourier transform. DFFT algorithms, however, require data to be in a rectangular format. Two alternatives are to interpolate the rectangular data from the polar data or to perform a Hankel
(Fourier-Bessel) transform. The polar data becomes less dense
away from the origin so interpolation becomes questionalble
there. On the other hand, development of efficient algorithms
for Hankel transforms is now an active area of research.[4-6]

Method B exhibits three attractive features. The first is
that through a series of one-dimensional transforms the angular
spectrum domain can be built in a format which closely approximates a rectangular raster. The second feature is that the data
collected by this method lies midway between the angular spectrum
domain and the time-space domain. Evaluation of the pressure
field at the plane of measurement, therefore, requires only a
series of inverse, one-dimensional transforms. A third feature
is that acoustic pressure can be computed along a line transverse
to the direction of sound propagation by a single inverse
one-dimensional transform.

2.0   THEORY OF METHOD A

Consider a harmonic sound field being produced by a planar
transducer. Let the pressure at a distance z from the transducer
be expressed as

$$p(x,y,z,t) = \tilde{p}(x,y,z)\exp(-j\omega t) \tag{1}$$

In the following discussion the time variance will be dropped for
convenience.

Line integrals across the pressure field at $z+z_0$ can be
written

$$\tilde{p}(x,z_0) = \int \tilde{p}(x,y,z_0)dy \tag{2}$$

where the limits of this integral and others are taken from minus
infinity to plus infinity. $\tilde{p}(x,z_0)$ can be seen as a "projection"
of the pressure field $\tilde{p}(x,y,z_0)$.

In the design of this experiment, $\tilde{p}(x,z_0)$ is obtained from measurement of the Raman-Nath parameter V defined as

$$V(x,z_0) = [2\pi k/\lambda] \; \tilde{p}(x,z_0) \qquad (3)$$

where λ is the optical wavelength in vacuum and k is the medium piezo-optic coefficient which relates the index of refraction to changes in acoustic pressure. This parameter V is a measure of the optical phase- retardation induced by the sound field. It is a common paramter used in most theories and can be inferred from acousto-optic measurements. This parameter, in our case, is to be measured as a phasor. Various techniques for acquiring the necessary phase and amplitude information can be found in the literature[7-8].

We will show the relation between the projected pressure and the Fourier domain assuming $p(x,z_0)$ to be a measurable quantity. Substitution of a two-dimensional transform expression into the integral of Equation (2) gives

$$\tilde{p}(x,z_0) = \iiint \tilde{p}(k_x,k_y;z_0)\exp[j(k_x x+k_y y)]dydk_x dk_y \qquad (4)$$

where k_x and k_y are components of the acoustic wave vector k . The integration of the y-variable can be completed yielding the Dirac-δ function $2\pi\delta$ (k). The sifting properties of the δ function upon integration over k produce the desired result

$$\tilde{p}(x,z_0) = 1/2\pi \int p(k_x,0;z_0)\exp(jk_x x)dk_x \; . \qquad (5)$$

This result, sometimes referred to as the "Fourier projection-slice theorem," states that the Fourier transform of a projection is a slice of the Fourier transform of the projected function. In other words, the one-dimensional transform of $\tilde{p}(x,z_0)$ produces a single line in the Fourier domain. This line lies perpendicular to the direction of the light paths, that is the line of interrogation. Consequently, if one rotates the transducer as specified for Method A in equal angular increments (essentially around the sound field axis), the projections obtained are related to values in the Fourier domain located along radial lines. In other words, taking a discrete one-dimensional transform of the projections $\tilde{p}$ (x,z_0) yields values shown as circles in Figure 2. This result is not restricted to sound fields and is a general result of projection theory.

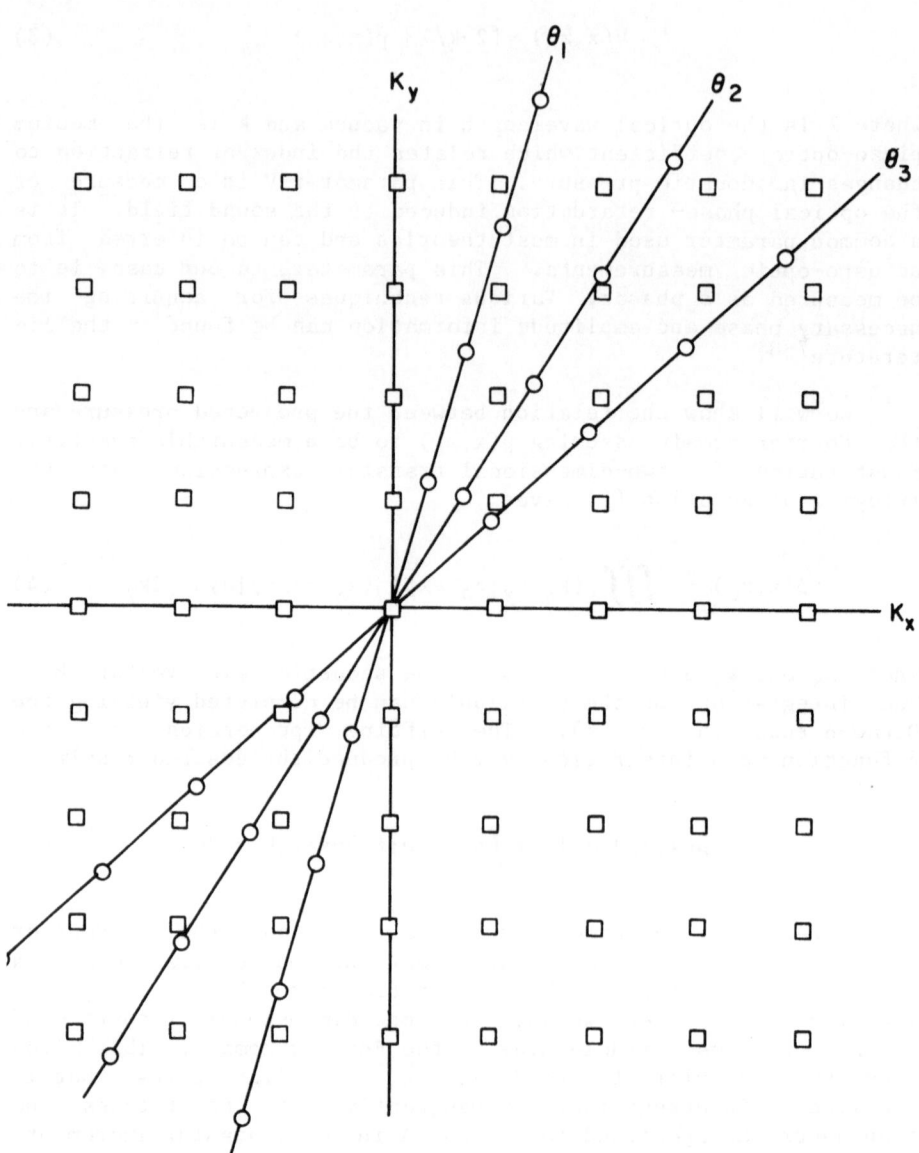

Fig. 2.   Fourier domain, Method A.

The previously stated difficulty is in the inversion of data in the Fourier domain to the time-space domain from the polar format. In addition to interpolation to a rectangular format, Mersereau and Oppenheim suggest other schemes to circumvent this problem.[9]

3.0 THEORY OF METHOD B

Consider the origin of the (x,y,z) coordinate axes to be the point of intersection of the acoustic axis and the plane of the light slice. Again, z is the acoustic axis. The light path slice will now be at an angle with the y-axis. We can write the projection of the pressure field with the notation changed to account for the angle as

$$\tilde{p}(x,\phi,z) = \int \tilde{p}(x,y,z)dy' \tag{6}$$

where dy' is along the light paths at an angle ϕ with the y-axis.

The Fourier domain pressure can be expressed as

$$\tilde{p}(k_x,k_y,z) = \tilde{p}_0(k_x,k_y)\exp(jk_z z) \tag{7}$$

where $\tilde{p}_0(k_x,k_y)$ is the Fourier component at $z = 0$ and k_z is the z-component of the acoustic wave vector. We again substitute the Fourier description of the pressure field in Equation (6) and incorporate Equation (7) to give

$$\tilde{p}(x,\phi,z) = \iiint \tilde{p}_0(k_x,k_y)\exp[j(k_x x+k_y y+k_z z)]dy'dk_x dk_y \tag{8}$$

The light path slice and the x-axis defines a new coordinate system (x,y',z'). This is related to the (x,y,z) coordinate system by the transformation

$$\begin{aligned} y &= y'\cos\phi + z'\sin\phi \\ z &= -y'\sin\phi + z'\cos\phi \end{aligned} \tag{9}$$

Substituting this expression into Equation (8) and collecting terms of integration of dy', we find the integral

$$\int \exp[j(k_y\cos\phi - k_z\sin\phi)y']dy' = 2\pi\delta(k_y\cos\phi - k_z\sin\phi) \tag{10}$$

If we define

$$k_y' = k_z \tan\phi \qquad (11)$$

we can write the Dirac-δ function of Equation (10) as $2\pi\delta(k_y-k_y')$. Equation (8) can now be written as

$$\tilde{p}(x,\phi,z) = 1/2\pi \int \tilde{p}_o(k_x,k_y)\delta(k_y-k_y') \; X$$

$$\exp[j(k_x x + k_y z' \sin\phi + k_z z' \cos\phi)]dk_x dk_y \qquad (12)$$

The integral over dy' can be evaluated using the sifting properties of the δ function.

$$\tilde{p}(x,\phi,z) = 1/2\pi \int \tilde{p}_o(k_x,k_y') \; X$$

$$\exp[j(k_x x + k_y' z' \sin\phi + k_z z' \cos\phi)]dk_x \qquad (13)$$

Since the origin of the coordinate system is in the plane of the light path slice, we have $z' = z_0 = 0$. Equation (13) now becomes

$$\tilde{p}(x,\phi,z_0) = 1/2\pi \int \tilde{p}_o(k_x,k_y')\exp(jk_x x)dk_x \qquad (14)$$

Equation (11) can be restated as

$$k_y' = (k^2 - k_x^2)\sin\phi. \qquad (15a)$$

If k_x is small compared to k, then the following approximation holds.

$$k_y' = k\sin\phi \qquad (15b)$$

When this approximation is substituted into Equation (14), $\tilde{p}(x,\phi,z_0)$ becomes $\tilde{p}(x,k_y',z_0)$

$$\tilde{p}(x,k_y',z_0) = 1/2\pi \int \tilde{p}_o(k_x,k_y')\exp(jk_x x)dk_x \qquad (16)$$

which is the main result of Method B.

It is important to note that $\tilde{p}(x,k_y',z_0)$ lies midway between the Fourier domain and the time-space domain. If we take the inverse Fourier transform of $\tilde{p}(x,k_y',z_0)$ with respect to the variable k_y', we obtain

$$1/2\pi \int \tilde{p}(x,k_y',z_0)\exp(jk_y'y)dk_y' =$$

$$(1/2\pi)^2 \iint \tilde{p}_0(k_x,k_y',z_0)\exp[j(k_xx+k_y'y)]dk_xdk_y' \qquad (17)$$

The term on the right hand side can be recognized as $p(x,y,z_0)$. Thus from a set of measurement taken at a given elevation (x fixed) and varying angle, we can apply a one-dimensional Fourier transform to obtain the pressure along the y-axis for that value of x. The pressure over a specified x-y plane can also be obtained by a series of such transforms taken at equally spaced values of x.

If we take the Fourier transform of $\tilde{p}(x,k_y',z_0)$ with respect to the variable x, we obtain

$$\tilde{p}(k_x,k_y',z_0) = \int \tilde{p}(x,k_y',z_0)[\exp(-jk_xx)]dk_x \qquad (18)$$

which is the pressure field in the Fourier domain.

Implementation of Method B using DFFT algorithms requires the approximation that k_y' does not depend on k_x. This approximation is valid when the ultrasound is confined to a narrow beam as with sound field produced by most medical and NDE transducers.

To illustrate this approximation we show in Figure 3 the nearly rectangular format for data obtained from Equation (16) using an DFFT. The error incurred by assuming this format to be rectangular will be small if most of the radiated energy is near the origin in the Fourier domain. The illustration in Figure 3 is for a sound field produced by a circular transducer of radius $a = 10\lambda$ where λ is the acoustic wavelength. Each of the larger concentric circles has associated with it a percentage of total radiated energy contained within the circle. The percentages were calculated using an Airy pattern to approximate the sound field. Figure 3 shows the format to be essentially rectangular for 98% of the radiated energy in this case. Figure 4 shows the general curve for the fraction of total energy radiated as a function of $ka(\sin\phi)$ for the Airy function.

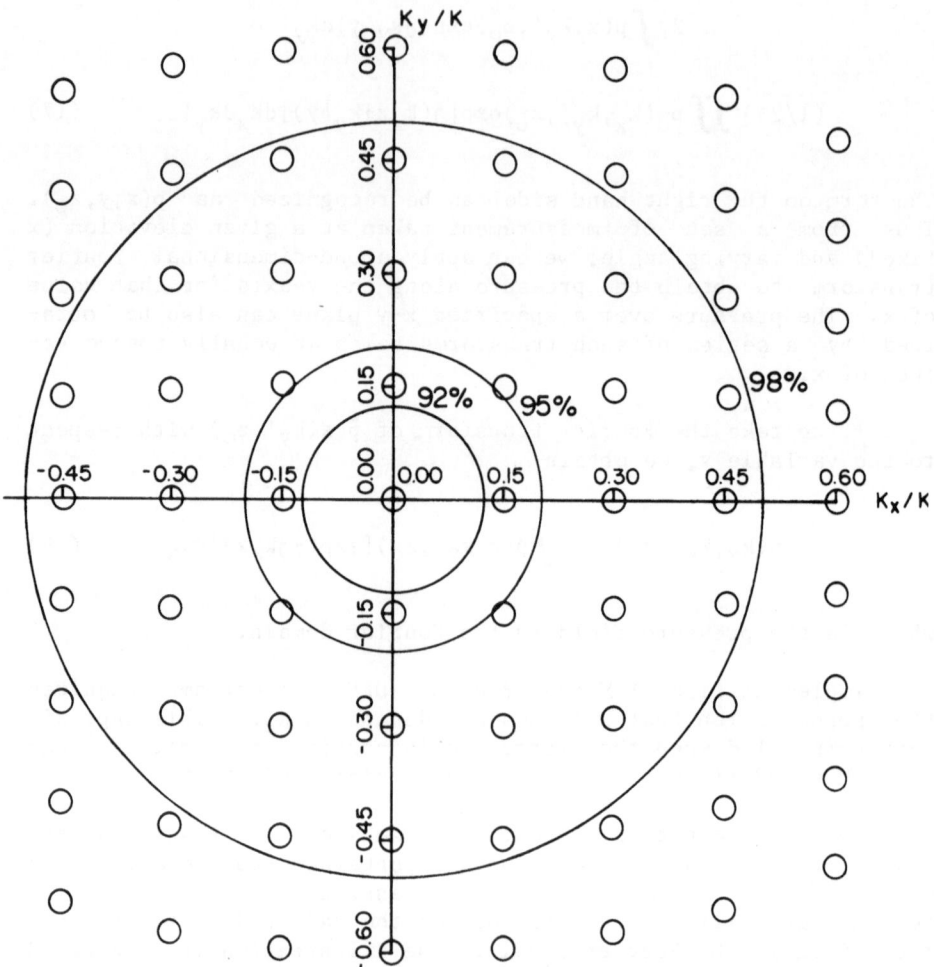

Fig. 3. Fourier domain, Method B.

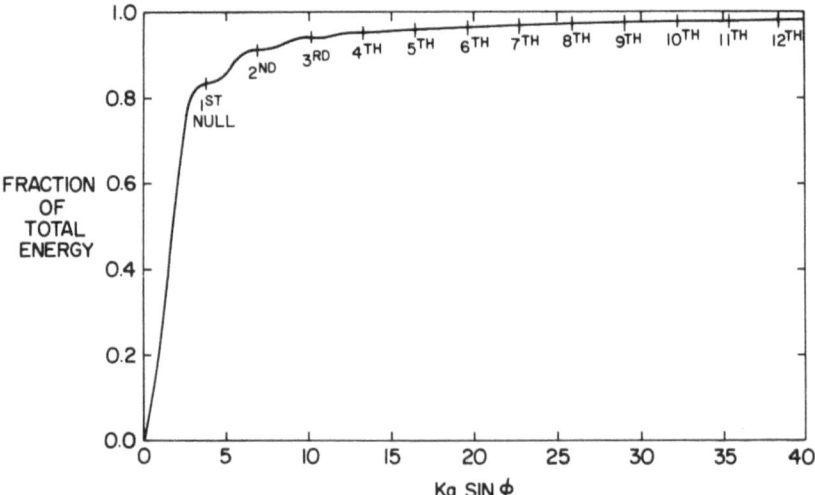

Fig. 4.   Distribution of radiated energy.

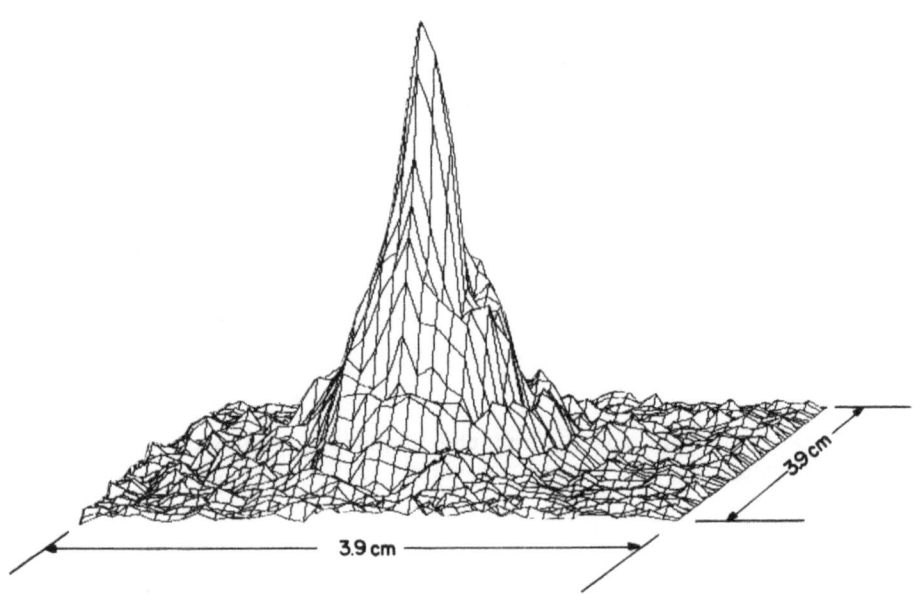

Fig. 5.   Reconstructed image.

4.0   DATA ACQUISITION USING METHOD B

Equations (15a) and (15b) indicate that the angle ϕ should change by equal increments of $\sin\phi$, or that

$$\sin\phi \;=\; +/- \; n\alpha \qquad\qquad (19)$$

where $n = 0,1,2,\ldots,N$ with $2N+1$ = number of experimental points and α = specified increment for $\sin\phi$.

The maximum value of k_y observed in the data is

$$(k_y)_{max} = N\,k\,\alpha \qquad\qquad (20)$$

From Equation (15b), we also know

$$(k_y)_{max} = k(\sin\phi_{max}) \qquad\qquad (21)$$

Combining Equations (19), (20) and (21) yields

$$\Delta k_y \;=\; k\alpha \qquad\qquad (22)$$

The increments of y in the transformed data returned by the DFFT are $y = m\Delta y$ where $m = 0,1,2,\ldots,M$ and M is the number of points used in the DFFT. From the periodicity of the DFFT, we find

$$M\Delta k_y\Delta y \;=\; 2\pi \qquad\qquad (23)$$

or

$$\Delta y \;=\; c/(M\alpha f) \qquad\qquad (24)$$

where c is the sound speed, and the acoustic frequency $f = c/\lambda$. The total range on the y-axis is then

$$M\Delta y \;=\; c/(\alpha f) \;. \qquad\qquad (25)$$

5.0 EXPERIMENTAL RESULTS

Cook and Berlinghieri have experimentally demonstrated the validity of Method A.[2] They generated a rectangular array of points in the Fourier domain by rotating the transducer through the proper angles and sampling the linear scan at proper intervals. They then used a two-dimensional DFFT to reconstruct the acoustic field in the plane of interrogation and in nearby planes.

Method B was used to calculate the pressure distribution over a transverse plane for a 1.25 cm. diameter PZT transducer submersed in water. The transducer was operating at 2.2 Mhz (a = 9.3 λ). A total of 41 x 41 data points using a value of ϕ = sin(1 degree) were used to measure the pressure in a plane 5 cm. from the transducer face. This data was used to construct the pressure field over an area 3.9 x 3.9 cm. The results are shown in Figure 5. The effects of the approximation made in Equation (15) are noticeable in the reconstruction. The reconstructed field is slightly elliptical rather than circular due to the outward displacement of the k wave-vector components in the reconstruction.

6.0 REFERENCES

1. J. Berlinghieri and B. Cook, J. Acoust. Soc. Am. 58, 823-827 (1975).

2. B. Cook and J. Berlinghieri, Proc. IEEE Ultrason. Symp. 75, CHO 944-SU, 133-135 (1975).

3. B. Cook and J. Berlinghieri, J. Acoust. Soc. Am. 61, 147-1480 (1977).

4. J. P. Clero and C. H. Durney, Proc. IEEE 67, 1463 (1979).

5. Alan V. Oppenheim, G. V. Frisk, and D. R. Martinez, Proc. IEEE 66, 264-265 (1978).

6. A. J. Jerri, Applicable Analysis 7, 97-109 (1978).

7. Ward A. Riley, J. Acoust. Soc. Am. 65, 82-85 (1979).

8. B. Cook, J. Acoust. Soc. Am. 60, 95-99 (1976).

9. R. Mersereau and A. Oppenheim, Proc. IEEE 62, 1319-1338 (1974).

EFFECTS OF DIFFRACTION ON ULTRASONIC COMPUTER-ASSISTED TOMOGRAPHY

J. F. Greenleaf*, P. J. Thomas*, and
B. Rajagopalan[†]

*Biodynamics Research Unit, Mayo Clinic/
 Foundation, Rochester, MN 55901

[†]Department of Radiology, University of
 Texas, San Antonio, TX 78284

INTRODUCTION

Ultrasonic computer-assisted tomography has been developed as a analog to the x-ray computer-assisted tomography techniques[1] and has been used for measuring acoustic speed and attenuation in tissues[2,3,4,5]. The advantages of computer-assisted tomography are that quantitative images representing parameters such as attenuation or acoustic speed, can be obtained relating values represented by an image to fundamental properties of the tissue. The disadvantage of computer-assisted tomography is that a large plcurality of angles of view are required to obtain the necessary data[6]. This results in the technique being applied primarily to imaging accessible organs such as the breast around which 360° of views can be obtained using specially designed ultrasound scanners[7]. The most commonly used form of ultrasonic transmission tomography algorithm is that in which the ultrasonic energy is assumed to travel in a straight line[8], although more advanced techniques have been proposed by ourselves and others such as Mueller, et al.[9], Stenger and Johnson[10] and Dines and Kak[11] and Schomberg[12].

The use of straight line assumptions for computer-assisted tomography has the advantage that algorithms

351

used for x-ray tomography can be easily implemented for
ultrasound, but of course the disadvantage is that aberrations occur due to the fact that ultrasonic energy
does not travel in a straight lines through tissues.
These aberrations are caused by modes of energy deflection such as refraction and diffraction that cause the
energy to travel in curved lines, the loci of which are
unknown, obviating the use of classic straight line
reconstruction algorithms such as convolution backprojection[13] and the Algebraic Reconstruction Technique
(ART)[14]. The effects of refraction are well known and
can apparently be corrected by ray tracing using theories
of geometric optics[15]. However, deflection of energy
caused by diffraction cannot be corrected using methods
of geometric optics[16].

The purposes of this paper are 1) to describe the
character of diffraction in simple objects having
geometries and refraction indexes of the order of those
found in breast tissues, 2) to correct with computer
simulations the effects of diffraction, at least in
the simple cases of the models studied here, and 3) to
describe a transducer design which corrects for most
of the diffraction found in the simple models.

METHODS

To carefully study the effects of diffraction,
simple objects such as finger cots filled with saline or
alcohol/water mixtures were scanned in a general-purpose
experimental water filled scanner[7]. The transmitter
consisted of a polyvinylfluoride membrane[17], 30 cm long,
6 cm high, and 30 microns thick. This membrane was configured on a concave surface of a cylinder focusing on
a line 20 cm in front of the transducer and parallel
to its long axis. The receiver was a 2 mm x 5 mm element of PZT resonant at about 5 MHz in the thickness
mode. The receiver was scanned in a linear fashion
using stepper motors under the control of a Z80 microprocessor which in turn was controlled by a 7/32 Perkin
Elmer computer. The signals from the 5 MHz PZT receiver
were amplified with a wide band pre-amplifier having a
gain of approximately ten and then A/D converted at
either twenty or fifty megasamples per second using a
Biomation Transient Recorder Model 8100. The resulting
data were displayed on an eight-bit gray scale display
device having 512 x 512 pixel resolution. Photographs
were obtained by taking Polaroid photographs of the face
of a high resolution monitor. Variations in refractive

indexes of the media in the finger cots were obtained by varying the concentration of alcohol in water or of salt in water[18].

RESULTS

In order to describe pure diffraction effects at medical frequencies, we scanned a finger cot filled with air so that no acoustic energy traversed through the finger cot but diffraction occurred at its edges. Figure 1 illustrates an image of the resulting data. One can see that diffraction from the surface of the air-filled finger cot describes a hyperbola through the data space whose characteristics depend not on the contents of the finger cot but on the geometry of the scanning situation.

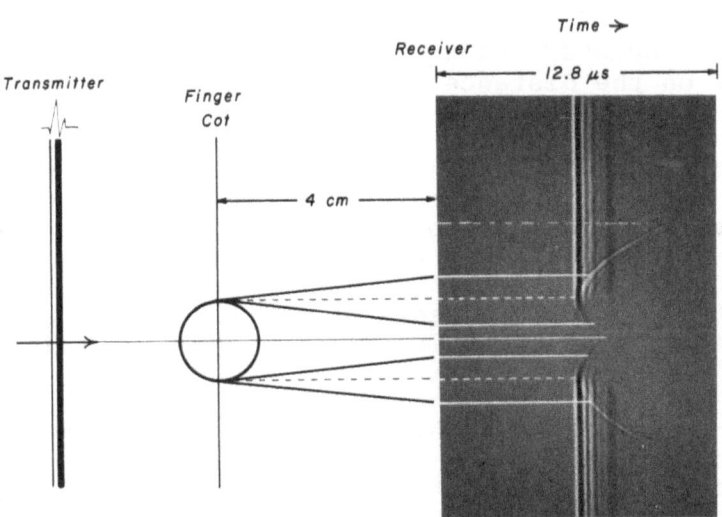

Figure 1   Digitized signal scattered from an air-
            filled finger cot. Note the hyperbolas
            at each of the horizontally projected
            edges which represent the arrival of cy-
            lindrical wave originating at the edge
            of the finger cot. These hyperbolas are
            examples of diffraction and are indepen-
            dent of the contents of the finger cot.

A model of the measured diffracted pulse pressure is

$$P(t,y) \cong A(t - \sqrt{\frac{L^2+Y^2+L}{C}})$$ Eq. 1

where L = Diffraction source to scan line distance.
      Y is the vertical distance of the receiver from
         the diffraction source, A(t) is the impulse
         response of the system and
      C is the speed of sound in the medium.

That is, the source of diffraction "transmits" a signal which is a cylindrical wave and is detected at the receiver as a cylindrical wave.

This is due to the fact that a simple model of diffraction considers the diffraction edge to act like a point source or in more complex geometries a distribution of point sources[16] resulting in the Kirkhof integral[15].

The curvature of the hyperbola in the signal space depends on the distance L of the scanning locus from the center of the finger cot. To relate relative contributions of refraction and diffraction to acoustic energy as it traverses a simple cylindrical shape we scanned a finger cot having high refractive index containing pure ethyl alcohol, n = 1.29 shown in Figure 2, and a finger cot containing fluid of a low refraction index, saline, n = .90 shown in Figure 3. One can see that even though the effects of the contents of the finger cot are very different, the diffraction effects are virtually identical.

In general, measurements of the time of flight between the transmitter and receiver which are required to reconstruct distributions of acoustic speed[8] are obtained by measurng the time from the transmitter trigger to the trigger obtained by thresholding on the earliest arrival time of the signal. This technique decreases the effects of signals arriving from multiple paths which may affect later peaks or zero crossings in the signals. One can see that if the first arrival time techniques are used to measure the profile in Figure 2 or in Figure 3, errors will occur in the resulting image reconstructed using straight line methods causing a cylinder of low refraction index to be reconstructed with too large a diameter while an image of a high

(Isopropyl Alcohol,   22°, 1 cm Radius)

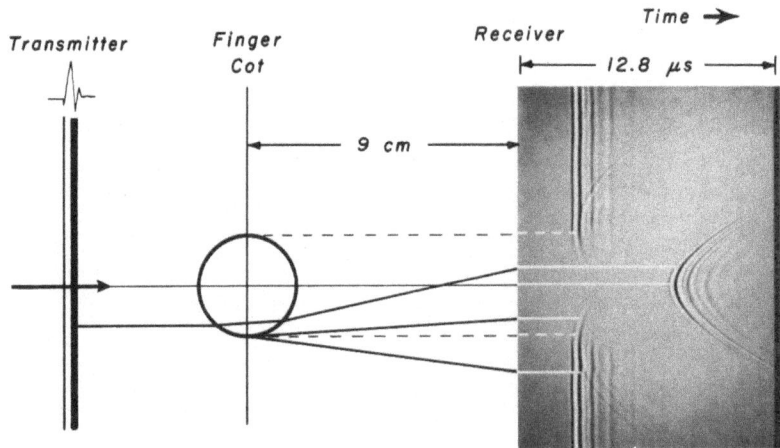

Figure 2   Digitized signal scattered from a finger
            cot filled with alcohol. Note the dif-
            fraction hyperbolas which are similar to
            the ones in Figure 1. The part of the
            wave scattered by refraction arrives much
            later in some regions than that due to
            diffraction.

refraction index cylinder will appear too small as can
be appreciated from Figure 4.

It is clear that if the diffraction occurring as
the wave propagates between the finger cot and the receiver can be reversed, i.e., if the wave can be backward propagated in some way, the diffraction hyperbolas
could be "focused" nearly to points or at least greatly
minimized by making L very small in Equation 1.

A computer program was written which back propagates the received signal much the same way as is done
in seismic migration[19]. The computer program reconstructs the signal which would have been received at each
point along a line through the center of the finger cot
L = o by summing all the received signals at the original

(5% Saline, 22°, ~1cm Radius)

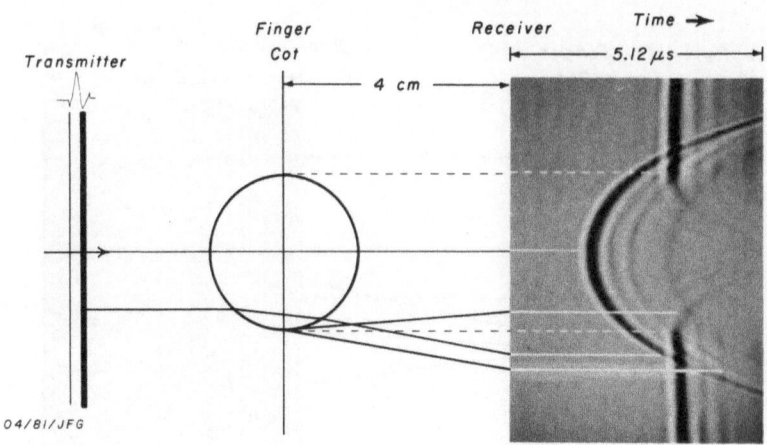

Figure 3   Digitized signal scattered from a finger
            cot filled with saline. The diffraction
            hyperbolas are the same as those in Figures
            1 and 2. However, the high acoustic speed
            of the saline results in early arrival of
            the refracted portion of the signal.

locus after shifting each measured signal by an amount
depending on its position relative to the new point.

    If P(x,y,t) is the scattered pressure wave measured
for insonification by a plane wave traveling in the plus
x direction, then with the spatial origin at the center
of the finger cot the measured signal is P(L,y,t) and
the backward propagated signal we want is

$$P(o,y,t) = \int_{y-D}^{y+D} P(L,s,t + \frac{\sqrt{L^2+(y-s)^2}-L}{C})\, ds, \quad \text{Eq. 2}$$

where 2D is an aperture distance selected to be
large enough for the problem.

    The experimentally received signal shown in the
right panel of Figure 5 was backward propagated by computer using Equation 2 and the result is shown in the
left panel. One can see that the resulting profile

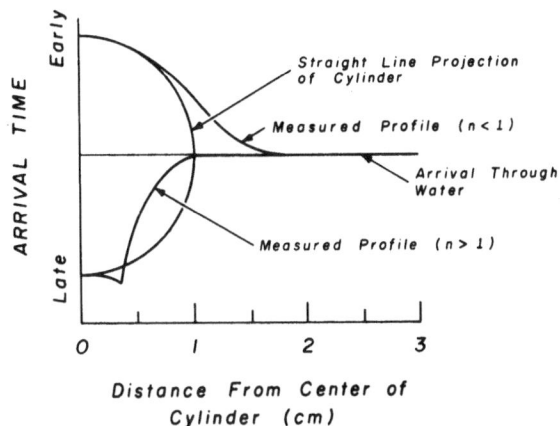

Figure 4   Schematic diagram of profiles which would
            be measured from earliest arrival method
            of detecting signal and the hypothetical
            true profiles. The profile for n < 1 ap-
            pears oversized and the profile for n > 1
            is undersized and complex. The exact
            character of the profile for n > 1 depends
            on whether the scattered signal is mea-
            sured nearer or farther away than the
            focal region of the cylinder.

relates qualitatively very well to the expected profile
of the high refraction index material which is shown in
Figure 4.

    We noticed that such a backward propagation pro-
cedure could be accomplished using a cylindrical transducer. A cylindrical transducer "delays" and sums each
pressure in a manner identical to Equation 2 since its
surface is curved in a cylindrical fashion. In other
words, the cylindrical transducer/receiver focuses on a
line at the origin of the radius of curvature of the
cylinder accomplishing the delay and summation which
are required by the computer simulation of backward
propagation or migration (Equation 2).

(*Isopropyl*,   *22° C*,   *2 cm   dia.*)

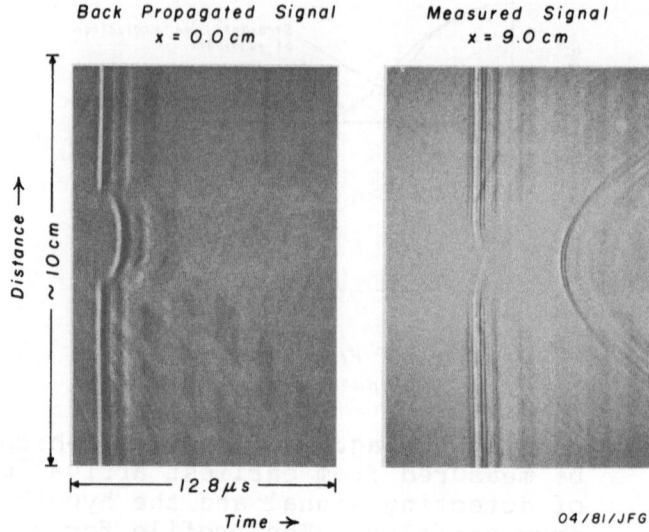

Figure 5   Computer calculated backward propagation
            of a digitized signal scattered from a cy-
            linder insonified with a plane wave. The
            profile in the left panel was calculated
            from the received signal on the right
            using Equation 2 in the text.

        To test this hypothesis we constructed cylindrical
transducer having a radius of twelve cm, a height of 2
cm and a length of material of 16 cm using polyvinylfluoride mounted on a lucite form. This transducer was
used to scan finger cots containing a low and a high
refractive index material as shown in Figure 6 and
Figure 7, respectively. One can see that when the received wave was back propagated to the center of the
finger cot, that is, when the cylindrical receiver was
focused at the center of the finger cot, the measured
signal resulted in a pressure profile very close to that
obtained by computer simulation and resulted in a quantitatively correct profile to be used for straight line
reconstruction.

        In order to test whether the cylindrical trans-
ducer could be used to obtain data for accurately reconstructing the distribution of high refraction index (the
most difficult case) in the finger cot, the cylindrical

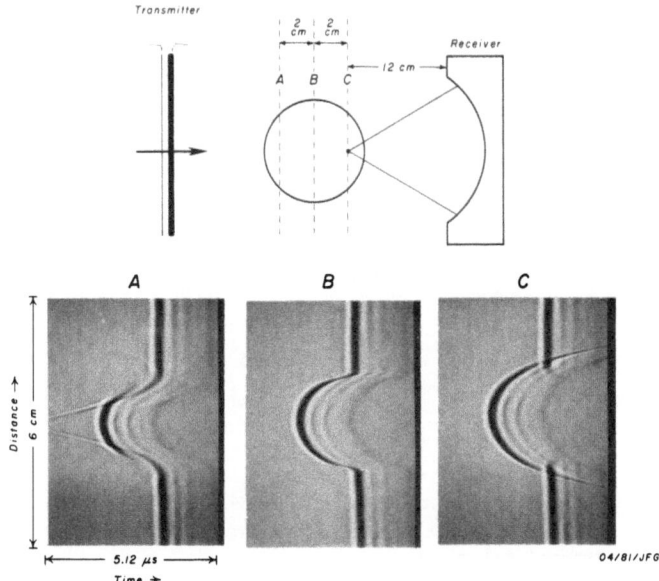

(5% Saline, 22°, ~1 cm Radius)

Figure 6   Digitized signal scattered from a finger
            cot measured with a cylindrical receiver
            used to accomplish backpropagation described
            in the text n = .90.

transducer was used to receive signals which were then
thresholded to obtain the time-of-flight profiles[8] required for straight line reconstruction. The resulting
reconstruction is shown in Figure 8. This reconstruction was obtained by measuring time of flight of the received signal at 400 points in space separated by .254
mm for each of 60 views spaced equally around 180°. It
can be seen that when the transducer focuses precisely
at the center of rotation of the finger cot, the reconstruction is virtually perfect having a very flat interior of the correct speed.

DISCUSSION

    The cylindrical transducer simulates a small dif-
fraction limited receiver aperture at its center of
focus. This simulated receiver element can be scanned
as close to the object as possible, or through the

(50% Ethanol, 27°C, ~1 cm Radius, n = 1.078)

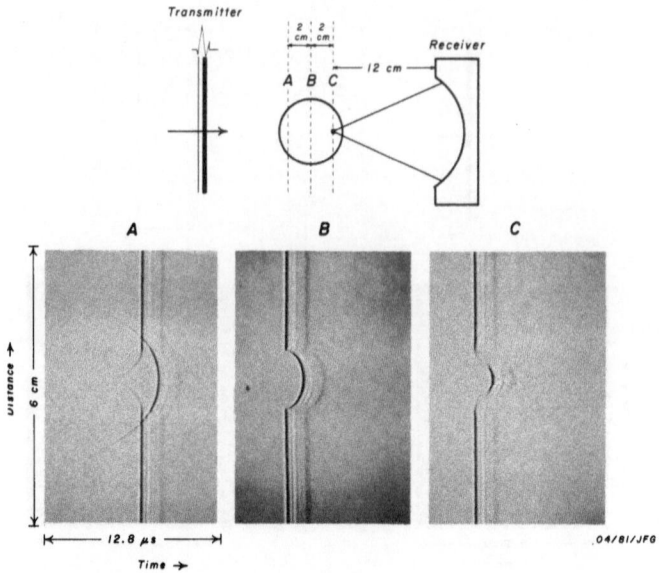

Figure 7   Same format as Figure 6 but with n = 1.08.

object if desired. The small effective f number
(f ≅ 1.0) of the cylindrical transducer, results in a
very shallow depth of field (Figures 2 and 3). This
apparently results in high sensitivity of the shapes
of the profiles to small variations in distance from
the receiver to the object. We could not test whether
the method would work in tissues because of the low
sensitivity of the transmitter/receiver combination.

     It is clear that diffraction is a large component
of the scattered signal at medical frequencies. Therefore, methods of correcting transmission tomography
images using geometric optics, thus ignoring diffraction,
will not adequately correct the aberrations since important information concerning refraction occurs later in
the signal than the first arriving signals. Careful
designs of transducer and scanning geometries are also
necessary for reducing the effects of diffraction.

(26°C, 2 cm Radius, n = 1.29)

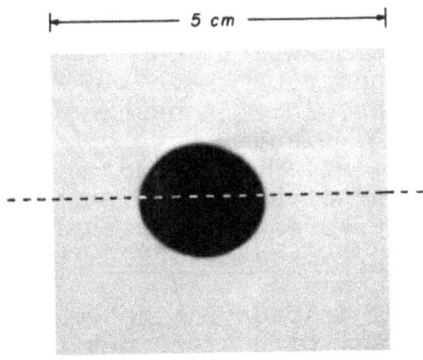

Central Line Through Image

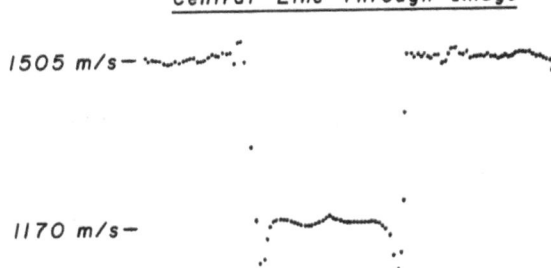

1505 m/s —

1170 m/s —

Figure 8   Reconstruction of highly diffracting and
            refracting cylinder (n = 1.29) obtained
            using first arrival measurements from
            signal received with a cylindrical trans-
            ducer focused at the center of the finger
            cot. Central region of finger cot appears
            flat with a nominal speed of about 1171.6
            + 1.08 meters/second. Literature value
            for ethanol is 1170 m/s[18].

ACKNOWLEDGMENT

        The authors greatly appreciate the secretarial
assistance of Ms. Elaine Quarve and the graphical
assistance of Ms. Julie Lauer and Mr. Steve Richardson.

        This work was supported in part by Grants CA 24085
from NCI, GM 24994 from NIH and ECS 7296008 from NSF.

REFERENCES

1. J. F. Greenleaf, S. A. Johnson, S. L. Lee, G. T. Herman, and E. H. Wood, Algebraic reconstruction of spatial distributions of acoustic absorption with tissues from their two-dimensional acoustic projections, in: "Acoustical Holography," P. S. Green, ed., Plenum, New York (1974), pp. 591-603.

2. J. F. Greenleaf, S. A. Johnson, and A. H. Lent, Measurement of spatial distribution of refractive index in tissues by ultrasonic computer assisted tomography, Ultrasound Med. Biol. 3:327 (1978).

3. G. H. Glover and J. L. Sharp, Reconstruction of ultrasound propagation speed distribution in soft tissue: time of flight tomography. IEEE Trans on Sonics and Ultrasonics SU-24(4):229-234 (1977).

4. P. L. Carson, T. V. Oughton, W. R. Hendee, and A. S. Ahuja, Imaging soft tissue through bone with ultrasound transmission tomography be reconstruction, Med. Phys. 4(4):302 (1977).

5. M. O'Donnell, J. W. Mimbs, B. E. Sobel, and J. B. Miller, Ultrasonic attenuation in normal and ischemic myocardium, in: "Proc. Nat. Bureau Standards 2nd Int. Symp. Ultrasonic Characterization," Gaithersburg, MD, June 13-15, 1977, p. 63.

6. R. A. Brooks and G. Di Chiro, Principles of computer assisted tomography (CAT) in radiographic and radioisotopic imaging, Phys. Med. Biol. 21:689 (1976).

7. J. F. Greenleaf, and R. C. Bahn, Signal processing methods for transmission ultrasonic computerized tomography, 1980 Ultrasonics Symposium Proc. (1980) pp. 966-972.

8. J. F. Greenleaf, Computerized transmission tomography, in: "Methods of Experimental Physics-Ultrasound," P. D. Edmonds, ed., Academic, New York (1981).

9. R. K. Mueller, M. Kaveh, and G. Wade, Acoustical reconstructive tomography and applications to ultrasonics, Proc. IEEE 67:567 (1979).

10. F. Stenger and S. A. Johnson, Ultrasonic transmission tomography based on the inversion of the Helmholtz wave equation for plane and spherical wave insonification, Appl. Math. Notes 4:102 (1979).

11. K. A. Dines and A. C. Kak, Measurement and reconstruction of ultrasonic parameters for diagnostic imaging, Purdue University, Lafayette, IN, Rep. TR-EE-77-4 (Dec. 1976).

12. Schomberg, H., An improved approach to reconstructive ultrasound tomography, J Appl D: Appl Physics 11:L181 (1978).

13. G. N. Ramachandran and A. V. Lakshminarayanan, Three-dimensional reconstruction from radiographs and electron micrographs: applications of convolutions instead of Fourier transforms, Proc. Nat. Acad. Sci, U.S. 68(9):2236 (1971).

14. G. T. Herman, A relaxation method for reconstructing objects from noisy x-rays. Math. Programming 8:1 (1975).

15. M. Born and E. Wolf, "Principles of Optics," Pergamon, (1964) pp. 133.

16. J. B. Keller, Geometrical theory of diffraction, J. Acoust. Soc. Am. 52(2):116 (1962).

17. KYNAR PVDF, Trademark of Pennwalt Corporation, King of Prussia, PA 19406.

18. F. J. Millero and T. Kubinski, Speed of sound in sea water as a function of temperature and Salinity at 1 atm, J. Acoust. Soc Am. 57(2):312 (1975).

19. E. A. Robinson and S. Treitel, "Geophysical Signal Analysis," Prentice Hall Signet Processing Series, (1980), pp. 374.

AN EXACT THEORY FOR COHERENT ACOUSTIC PROBING

W. Ross Stone

IRT Corporation
1446 Vista Claridad
La Jolla, CA   92037

1. INTRODUCTION AND SYNOPSIS

In an inverse scattering problem, the fields in the inhomogeneous wave equation are known, and it is desired to solve for the source term. N. N. Bojarski has derived an Exact Inverse Scattering Theory for such "inverse source" problems. The problem of determining the generalized refractive index (i.e., the complex permeability and dielectric constant for an electromagnetic problem, or the velocity and absorption for an acoustic problem) distribution of an inhomogeneous medium from measurements of the fields scattered by the medium can also be treated using this theory. This solution is applicable to all remote probing problems, including radar, sonar, "profiling" of inhomogeneous propagation media, nondestructive evaluation, and seismic exploration.

The whole field of synthesis problems--in which it is desired to determine the set of currents and charges, and/or the set of refracting, reflecting, or diffracting elements which will produce a given set of fields or a given reception pattern--can also be treated by the Exact Theory. An exact closed form, analytic solution for such problems is presented below. This solution applies to optical, microwave, and acoustic lens design, antenna and antenna array synthesis, acoustic transducer design, and many related problems.

Of greatest importance are two very recent results which are presented in this paper. A closed form analytic solution for the total field, including the field in the source region, has been obtained by Bojarski. This holds the same significance for the inverse scattering problem as the Kirchoff and Stratton-Chu integrals hold for the direct scattering problem. A closed form analytic solution for the source term in the inverse scattering problem has subsequently been obtained by Stone.   These solutions have provided important new insights into the fundamental physics of inverse scattering.

This paper reviews the Exact Inverse Scattering Theory. The Theory results in an integral equation which can be solved numerically in an efficient, well posed manner. This integral equation is derived in Section 2, and the uniqueness of the solution is discussed in Section 3. This numerical solution and its characteristics with regard to incomplete, sampled, and noisy data (i.e. real-world measurement conditions) are discussed in Sections 4, 5, and 6. The relationship between holography and inverse scattering is discussed in a companion paper in this volume. Sections 7 and 8 present an exact closed form solution to the synthesis problem. Section 9 presents a closed form solution to the remote probing problem which, although not an exact solution, should often be quite useful. Section 10 places these synthesis results in perspective with respect to more traditional approaches. Section 11 presents the exact, closed-form analytic solution for the fields, and Section 12 presents the exact, closed-form analytic solution for the sources in the inverse scattering problem.

Although this paper uses scalar notation, all of the results have been shown to apply to the general, full vector field and tensor medium quantities. The equations applicable to the electromagnetic cases are used; however, the theory and results apply equally well to the acoustic equations.

This paper often refers to the problem of identifying a target from its scattered field as a remote probing (or inverse medium) problem. This terminology is used because, from a physical standpoint, a target can be considered known (although not necessarily recognized or identified) if its three dimensional refractive index distribution is known. The Exact Theory provides this information. Traditionally only a subset of this information--the boundaries between the target and its surrounding medium, usually termed the target's shape--is employed.

2. THE BOJARSKI EXACT INVERSE SCATTERING THEORY

This section presents a derivation of N. N. Bojarski's (Refs. 1-4) "Exact Inverse Scattering Theory." The theory is derived in the form leading to an integral equation which can be solved numerically in a stable and efficient manner. Sections 11 and 12 present derivations which parallel that of this section, but which result in exact, closed-form analytic solutions for the total field and the source term. This order of presentation has been chosen because it has proven to be the simplest for newcomers to the field of inverse scattering to understand. It is important to note that although consideration of the integral equation is the simplest means of introducing the Exact Theory, the analytic solutions of Sections 11 and 12 are by far the most significant results presented in this paper. Consider a source $\rho(\underline{x})$ in a domain D bounded by a surface S. Then the time harmonic field, $\phi(\underline{x})$, due to $\rho(\underline{x})$ is the solution to the inhomogeneous wave equation

$$\nabla^2 \phi(\underline{x}) + k^2 \phi(\underline{x}) = -\rho(\underline{x}), \quad \underline{x} \in D \tag{1}$$

where $k = 2\pi/\lambda$.

A direct scattering problem is one in which $\rho(\underline{x})$ is known or specified, and a solution for $\phi(\underline{x})$ is sought. The inverse scattering problem is one in which $\phi(\underline{x})$ is known, and $\rho(\underline{x})$ is sought. For the inverse source problem, $\phi(\underline{x})$ is measured over some surface, and the object is to determine $\rho(\underline{x})$. In general, $\rho(\underline{x}) = \rho_m(\underline{x}) + \rho_s(\underline{x})$, where ρ_m is due to interaction with the medium, and ρ_s is due to actual sources. If $n(\underline{x})$ is the complex refractive index of the medium, then

$$\rho_m(\underline{x}) = k^2 \left[n^2(\underline{x}) - 1 \right] \phi(\underline{x}) \tag{2}$$

In most remote probing problems, $\rho_s(\underline{x})$ is known, and $\rho_m(\underline{x})$ is sought to yield $n(x)$. This is termed the inverse medium problem. If $\phi(\underline{x})$ is specified (as the desired field) and $\rho(\underline{x})$ or $n(\underline{x})$ is sought so as to produce that $\phi(\underline{x})$, the problem is termed an inverse synthesis problem.

For the acoustic case, Equation 1 can be written as

$$\nabla^2 \phi(\underline{x}, \omega) + k_m^2 \phi(\underline{x}, \omega) = 0 \tag{1a}$$

where

$$k_m = \omega/v_m(\underline{x},\omega)$$

and v_m is the velocity of the medium. ω is the radian frequency. Note that the variation of the velocity with the spatial coordinates and with frequency is explicitly included at this point. Equation 2 can be written as

$$\rho(\underline{x}, \omega) = \left[(\omega^2/c^2) - (\omega^2/v_m^2) \right] \phi(\underline{x}, \omega) \tag{2a}$$

where c is the homogeneous medium propagation velocity (or an arbitrarily chosen reference velocity). Substitution of Equation 2a into Equation 1a yields Equation 1.

Let the following field quantity, $\phi_H(\underline{x})$, be defined:

$$\phi_H(\underline{x}) = \oint \left[g^*(\underline{x}-\underline{x}')\nabla\phi(\underline{x}') - \phi(\underline{x}')\nabla g^*(\underline{x}-\underline{x}') \right] d\underline{S}' \qquad (3)$$

where $g(\underline{x})$ is the free space Green's function and the asterisk denotes complex conjugation. g satisfies Equation 1 with $\rho(\underline{x}) = \delta(\underline{x})$. ϕ_H is in the form of the Kirchoff integral with g complex conjugated. Note that if the Kirchoff integral is applied to the field $\phi(\underline{x})$ on S and evaluated at any point $\underline{x}$ inside D, it is identically zero: The Kirchoff integral is nonzero only for points <u>outside</u> D. Conversely, $\phi_H(x)$ is nonzero only for points <u>inside</u> D. Points outside of D are associated with the direct scattering problem; points inside D are of interest for the inverse scattering problem. This topological difference is the reason why direct scattering solutions are mathematically ill-posed when applied to the inverse scattering problem.

It should also be noted that ϕ_H is the mathematical expression for the reconstruction obtained from a hologram (ϕ in Equation 3) recorded on S. The relationship between holography and inverse scattering, along with an analysis of the consequences for remote probing and coherent imaging applications, has been presented by Stone (Ref. 5). ϕ_H is, in general, known for inverse problems, since ϕ is known over S. ϕ is measured over S for the inverse source and medium problems, or specified over S for the inverse synthesis problem.

Applying Gauss' theorem to Equation 3 converts the surface integral into a volume integral:

$$\phi_H = \int dV \left(g^*\nabla^2\phi - \phi\nabla^2 g^* \right) \qquad (4)$$

From Equation 1,

$$\nabla^2\phi = -k^2\phi - \rho \qquad (5)$$

and, by complex conjugation of Equation 1 for g,

$$\nabla^2 g^* = -k^2 g^* - \delta \qquad (6)$$

Substitution of Equations 5 and 6 into Equation 4 gives

$$\phi_H = \int dV \left[g^*(-k^2\phi - \rho) - \phi(-k^2 g^* - \delta) \right]$$

$$= \int dV \left(\phi\delta - g^*\rho \right) \qquad (7)$$

and, carrying out the integration over the delta function,

$$\phi_H = \phi - \int dV \ g^*\rho \tag{8}$$

Direct scattering theory gives the result that

$$\phi = \int dV \ g \ \rho + \oint dS(g \nabla \phi - \phi \nabla g)$$

$$= \int dV \ g \ \rho + \phi_i \tag{9}$$

In Equation 8, the first integral is just the superposition integral over the sources. The second term is the Kirchoff integral, and is asssociated with the incident field, ϕ_i. For the inverse scattering problem, ϕ_i can be assumed to be known without loss of generality (e.g., it is the known probing field for the inverse medium case, or the specified incident field in the inverse synthesis case).

Equations 8 and 9 are two independent simultaneous equations in two unknowns, ϕ and ρ. Substitution of Equation 9 into Equation 8 yields

$$\phi_H = \int dV \ g \ \rho - \int dV \ g^*\rho + \phi_i$$

$$= \int dV \ (g - g^*)\rho + \phi_i \tag{10}$$

or

$$\phi_H(\underline{x}) = 2i \int dV' \ \text{Im} g(\underline{x} - \underline{x}')\rho(\underline{x}') + \phi_i(\underline{x}) \tag{11}$$

where Im denotes the imaginary part. Equation 11 is the basic equation of the Exact Inverse Scattering Theory. It is an integral, convolution equation for the single unknown, $\rho(\underline{x})$. It can be solved by standard deconvolution techniques. Quite recently, Stone has presented a closed-form solution to Equation 11, based on a closed form solution by Bojarski for the total field, ϕ, in Equation 8 (see Sections 11 and 12).

3. UNIQUENESS OF THE SOLUTION

The uniqueness of the solution to Equation 11 was first deduced by Bojarski (Refs. 2,4) and later proven more rigorously by Bleistein and Cohen (Ref. 7). A simpler and more physically understandable proof was presented by Stone (Ref. 8). The result is that the solution to Equation 11 for the source, $\rho(\underline{x})$ is unique if $\rho(\underline{x})$ is identically zero outside some finite domain (i.e., is of bounded support), has finite energy, and does not contain any nonradiating components. A nonradiating source is a source component which produces a field which is identically zero outside a finite region. Although the "nonuniqueness" associated with nonradiating sources has been somewhat troublesome from a mathematical standpoint, it does not affect the uniqueness of results for practical applications (Bojarski Ref. 9). Quite recently, Stone (Ref. 10) has proven that a conjecture by Bleistein and Bojarski (Ref. 4) that nonradiating sources are nonphysical is true.

4. INCOMPLETE KNOWLEDGE OF $\emptyset(\mathbf{x})$, AND RESOLUTION

Practical inverse scattering problem measurements almost always involve discrete measurements over a limited aperture, as opposed to the continuous measurements over a closed surface used in the above theory. Mager and Bleistein (Ref. 11) have shown that, in the physical optics limit, the spatial bandwidth over which measured data is known is the spatial bandwidth over which the source term can be determined. A similar but more general result follows directly from analysis of the three dimensional spatial Fourier transform of Equations 3 and 11. Let $\underline{\nu}$ be the spatial frequency variable, and let capital letters denote the transformed functions. For the general case, $\Phi_H(\underline{\nu})$ is known over the whole surface S, and thus for all $0 \leq \underline{\nu} \leq \underline{\nu}_0$ (the upper bound is $\underline{\nu}_0$ rather than ∞, since D, and thus S, are of finite size). Since the left side of the transformed version of Equation 11 is known for all $0 \leq \underline{\nu} \leq \underline{\nu}_0$, it follows that $P(\underline{\nu})$ is determined over this range. Now let $\phi(\underline{x})$ be measured at discrete points over a limited aperture. Then $\Phi(\underline{\nu})$ is determined for $\underline{\nu}_1 \leq \underline{\nu} \leq \underline{\nu}_2$, where these spatial bandlimits are determined by the aperture size and sample spacing. It follows from the transform of Equation 3 that $\Phi_H(\underline{\nu})$ is similarly band-limited, and from the transform of Equation 11 that $P(\underline{\nu})$ can be determined over this band of spatial frequencies.

A related but more general result on the resolution obtainable in a coherent imaging process has been obtained by Stone (Ref 12). The lateral resolution for a coherent process is $\Delta x = (\lambda z / D)(\Delta \phi / 2\pi)$, where λ is the wavelength, z is the distance from the aperture to the target, and D is the size of the aperture over which the data is recorded. $\Delta \phi$ is the accuracy with which phase can be measured across the aperture. Note that as $\Delta \phi \rightarrow 2\pi$, the resolution goes to the classical (Rayleigh) result for incoherent imaging. The second term, involving $\Delta \phi$, represents the gain in resolution as a result of the added information present in the phase recorded in a coherent process. A similar result is obtained for the longitudinal resolution: $\Delta z = (\lambda z^2 / D)(\Delta \phi / 2\pi)$. The accuracy with which phase can be measured is determined by the power signal-to-noise ratio of the measurement, S/N: $\Delta \phi = (2S/N)^{-\frac{1}{2}}$. These theoretical results have been verified by numerical experiments (Refs 12,13) and by results obtained with the Holographic Radio Camera (Ref 33).

5. THE EFFECTS OF NOISY MEASUREMENTS (REF. 13)

The effects of measurement noise on the reconstructed refractive index can be seen by writing $\phi_H(\underline{x}) = \phi_H(\underline{x}) + \phi_N(\underline{x})$, where $\phi_N(\underline{x})$ contains a contribution due to noise. It follows, using Equation 11, that this is equivalent to a source term $\rho(\underline{x}) + \rho_N(\underline{x})$, where $\rho(\underline{x})$ is the true source and $\rho_N(\underline{x})$ contains the effect of the noise. From this it can be seen that the signal-to-noise ratio of the solution is the signal-to-noise ratio of $\phi_H(\underline{x})$. Since $\phi_H(\underline{x})$ depends on the integral over the measurement surface of the measured field values (Equation 3), the signal-to-noise ratio of $\phi_H(\underline{x})$ [and thus of $\rho(\underline{x})$] is not less than (and may be greater than) the signal-to-noise ratio of the measured data. This has been confirmed by numerical experiments (Ref 13). Recently, Stone (Ref 29) has proven analytically that the signal-to-noise ratio of $\rho(\underline{x})$ is greater than or equal to the signal-to-noise ratio of the measured data. It should also be mentioned that in addition to being numerically well-posed, the solution can be implemented in an extremely efficient form. This permits addressing problems heretofore impractical because of computational effort or storage limitations.

6. THE INVERSE MEDIUM PROBLEM

Based on the above theory, the inverse medium problem can be solved by the following steps:

A. Compute $\phi_H(\underline{x})$, using the measured field values in Equation 3 (note that the surface of integration, S, is the measurement surface).

B. Solve Equation 11 for $\rho(\underline{x})$, using $\phi_H(\underline{x})$ from A and the known incident field, $\phi_i(\underline{x})$.

C. Compute the total field, $\phi(\underline{x})$, from the direct scattering result, Equation 9, using $\rho(\underline{x})$ from B.

D. Solve Equation 2 for the desired complex refractive index, $n(\underline{x})$, using $\rho(\underline{x})$ from B and $\phi(\underline{x})$ from C.

Note that for the inverse source problem, only steps A and B are required. However, for the inverse medium problem it is necessary to carry out steps C and D in addition. The solution to the direct scattering problem (step C) is a necessary step in solving the inverse medium problem. It is important to emphasize that the computations involved in steps A through C are all convolution integrals: They can be carried out using fast Fourier transform techniques. As a result, computation time and storage requirements are proportional to $N \log_2 N$, where N is the number of data points. Step D is an algebraic operation.

The closed form, analytic solutions of Sections 11 and 12 permit a closed form, analytic solution to the inverse medium problem.

7. THE SYNTHESIS PROBLEM

There is very close relationship between the synthesis problem and the inverse medium problem. In the inverse medium problem a known probing (incident) field is used, and the scattered field is <u>measured</u>. This data is sufficient to obtain a unique solution for n(x), using the four steps in Section 6. In the synthesis problem, a <u>specified</u> incident field and a <u>desired</u> scattered field are chosen, and the n(x) required to produce this scattered field is sought. If there are no constraints (other than physical realizability) on the desired n(x), the same four steps in Section 6 will solve the synthesis problem. If there are constraints (e.g., a desired range of values for n(x), etc.), it is necessary to regularize the solution to these constraints.

8. A CLOSED-FORM SOLUTION TO THE SYNTHESIS PROBLEM

The author has carried the solution to the synthesis problem one step further. The result is a closed-form, exact solution for the desired refractive index distribution. Let T denote the desired relationship between the incident field, $\phi_i(x)$, and the scattered field, $\phi_s(x)$. T can be a function, an operator, or, in the most general case, any desired algorithm. The only requirement is that the operation of T on ϕ_i (denoted $T\phi_i$) result in a field which is a valid solution of the inhomogeneous wave equation. Thus,

$$\phi_s = T\phi_i \tag{12}$$

By definition, the total field, ϕ, is the sum of the incident and scattered fields:

$$\phi = \phi_i + \phi_s \tag{13}$$

Substituting Equation 12 into Equation 13,

$$\phi = (1 + T)\,\phi_i \tag{14}$$

Using convolution notation, Equation 9, the direct scattering result, gives

$$\phi = \rho * g + \phi_i \tag{15}$$

The source term, ρ, is related to ϕ and n(x) by the constitutive Equation 2. Substituting Equation 2 for ρ and Equation 14 for ϕ into Equation 15 yields

$$(1 + T)\phi_i = \left[k^2(n^2-1)(1+T)\phi_i \right] * g + \phi_i \tag{16}$$

After some algebra, Equation 16 can be solved for n:

$$n^2(\underline{x}) = \left\{ k^2 \left[1+T\right] \phi_i \right\}^{-1} \left[\mathcal{F}^{-1}\left\{\widetilde{T}\phi_i/\widetilde{g}\right\}\right] + 1 \tag{17}$$

where the tilde indicates the three dimensional spatial Fourier transform, and $\mathcal{F}^{-1}$ denotes the inverse transform.

Equation 17 is a closed-form solution to the synthesis problem. Furthermore, it is very attractive from the system designer's standpoint. The designer need only specify the desired input field to output field transformation, T, and Equation 17 provides the complex refractive index distribution which will produce that transformation. From the Exact Inverse Scattering Theory it can readily be shown that Equation 17 is numerically stable. Furthermore, it can be evaluated with great efficiency using fast Fourier transform techniques. It is also readily ameanable to regularization for the purpose of incorporating design constraints. Finally, the solution of Equation 17 is unique, and has the same behavior with respect to noise and incomplete measurements as discussed in the sections above.

In the form given in Equation 17, the solution is most suited to lens design (acoustic, microwave and optical), randome design, and synthesis problems in which a refractive or diffractive medium is employed. It is well suited to synthesizing holographic optical elements. However, the same solution can also be applied to the design of tranducers and transducer arrays. In such problems it is desired to determine either the distribution of acoustic impedances which will produce a desired radiated field (or, equivalently, a desired receiving pattern), or the distribution of acoustic sources which will produce the desired field. If it is the distribution of acoustic refractive index which is sought, then it is only necessary to replace $T\phi_i$ in Equation 17 with ϕ_s, which becomes the (specified) radiated field. $n(\underline{x})$ then gives the required arrangement of refractive index. To obtain the source distribution the closed form analytic solution presented in Section 12 can be used.

9. A CLOSED-FORM SOLUTION TO THE REMOTE PROBING PROBLEM

Let ϕ_i be the known incident field in a remote probing problem, and let ϕ_s be the measured, scattered field. Then the operator T, defined in Equation 17, can be determined, and Equation 17 is a closed-form solution to the remote probing problem.

Unfortunately, this involves a hidden approximation. The field measured is usually the total field, ϕ, _ not just the scattered field, ϕ_s _ over some surface. Obtaining ϕ_s throughout the volume from ϕ over a surface can be as complex as solving the inverse medium problem. However, under certain conditions, the approximation of ϕ_s throughout the volume by ϕ_H, as determined by Equation 3, may be adequate. Where such an approximation is good, Equation 17 provides a closed-form solution to the remote probing problem. A discussion of the conditions and implications associated with such an approximation has been given by Stone (Ref. 5; see also companion paper by Stone in this volume). Note that no approximation is involved in the closed-form solution of the above section for the synthesis problem: A designer has the freedom to specify ϕ_s. Indeed, this is usually the desired quantity for specification.

10. SOME COMMENTS ABOUT "STANDARD" APPROACHES TO THE SYNTHESIS PROBLEM

The synthesis problem is usually approached using direct scattering techniques. As discussed in the second section, this approach is inherently ill-posed. An initial guess at the solution $[n(\underline{x})]$ is made, a direct scattering analysis is carried out to obtain the scattered field, and this sequence is iterated, changing the $n(\underline{x})$ in an attempt to minimize the difference between the computed and desired scattered fields. Ray tracing is the most common direct scattering technique employed. There are many very important reasons for not using such synthesis methods: They are iterative, with no guarantee of the nature or rate of convergence; they are mathematically and numerically ill-posed; they require the designer to specify an optimization criterion which is usually not related to desired design requirements; and (in the case of ray tracing) they are only applicable where the geometrical optics approximation is valid. The previous two sections present two solutions which eliminate all of these objections. However, it is also important to realize that the inverse scattering approaches are many powers of 10 more efficient than standard techniques. One example from optical system synthesis will suffice to demonstrate this. Using the inverse scattering techniques, computation of $n(\underline{x})$ for 200,000 complex values requires of the order of one second using a 10 year old minicomputer with an FFT processing board. A state-of-the-art ray trace design program can, at best, compute 4,000 field values through one element of a guessed $n(\underline{x})$ per iteration in one second, using state-of-the-art, special purpose hardware (faster than a CDC 7600 or an IBM 360/195) and several thousand iterations are commonly required.

11. AN EXACT CLOSED-FORM ANALYTIC SOLUTION FOR THE TOTAL FIELD (REFS 16,24)

In the derivation of Section 1, Equations 8 and 9 were arrived at as two simultaneous equations in two unknowns: The sources, ρ, and the total field, ϕ. It was chosen to eliminate ϕ and solve for ρ. Bojarski has shown (Ref 24) that by starting with a slightly different form of the initial equations and eliminating ρ instead of ϕ, it is possible to obtain a closed form, analytic solution for the total field, ϕ.

The derivation begins with a somewhat more general form of the wave equation.

$$\nabla^2 \phi(\underline{x}, \omega) + \left[\omega^2/c^2(\underline{x}, \omega) \right] \phi(\underline{x}, \omega) = -\rho(\underline{x}, \omega) \tag{18}$$

where ω is the temporal frequency (which will often be suppressed in this section), and c is a velocity which varies as a function of both $\underline{x}$ and ω. If the constitutive equation is written as

$$\rho_t(\underline{x}, \omega, v) = \left\{ \left[\omega^2/c^2(\underline{x}, \omega) \right] - \omega^2/v^2 \right\} \phi + \rho(\underline{x}, \omega) \tag{19}$$

where v is an arbitrarily chosen constant "free space" reference velocity, then Equation 18 can be written as

$$\nabla^2 \phi(\underline{x}, \omega) + \left[\omega^2/v^2 \right] \phi(\underline{x}, \omega) = -\rho_t(\underline{x}, \omega, v) \tag{20}$$

ρ_t is introduced to permit writing Equation 20 with a constant coefficient in front of ϕ. If G is the Green's function associated with the constant reference velocity v, then G satisfies Equation 20 with a delta function source. Note that G satisfies Equation 20, and not Equation 18.

Let a quantity θ be defined as

$$\theta \equiv \oint d\underline{s} \cdot (G_r \nabla \phi - \phi \nabla G_r) \tag{21}$$

where $G_r = \text{Re} G$, which also satisfies Equation 20 with a delta function source. θ is analogous to ϕ_H of Equation 3, and has many of the same properties. In particular, θ depends only on the measured data, and is thus known. If the same steps that were carried out in going from Equation 3 to Equation 8 are applied to Equation 21, the following integral equation is obtained.

$$\theta = \phi - \int dV \, G_r \rho_t \tag{22}$$

Equations 8 and 22 are quite similar.

Let a new source term ρ_γ be introduced which has support only over the volume of integration V:

$$\rho_\gamma(\underline{x}, \omega, v) \equiv \begin{matrix} \rho_t(\underline{x}, \omega, v) & , & \underline{x} \in V \\ 0 & , & \underline{x} \notin V \end{matrix} \tag{23}$$

Then Equation 22 can be written as

$$\theta(\underline{x}, \omega, v) = \phi(\underline{x}, \omega) - \int_{-\infty}^{\infty} G_r(\underline{x}|\underline{x}', \omega, v) \rho_\gamma(\underline{x}', \omega, v) \, d^n x' \tag{24}$$

In cartesian coordinates G_r is a difference kernal and the integral in Equation 24 becomes a convolution. Using superscript tildes to denote the n-dimensional spatial Fourier transform of quantities, the spatial Fourier transform of Equation 24 becomes

$$\widetilde{\theta}(\underline{k}, \omega, v) = \widetilde{\phi}(\underline{k}, \omega) - \widetilde{G}_r(\underline{k}, \omega, v) \widetilde{\rho}_\gamma(\underline{k}, \omega, v) \tag{25}$$

where $\underline{k}$ is the transform variable. Note that ϕ and its spatial Fourier transform cannot depend on the arbitrary reference velocity v.

The spatial Fourier transforms of the real and imaginary part of the Green's function are

$$\widetilde{G}_r(\underline{k}, \omega, v) = P \, 1/(k^2 - \omega^2/v^2)$$

$$= \begin{cases} 1/(k^2 - \omega^2/v^2) \,, & k^2 \neq \omega^2/v^2 \\ 0 & , & k^2 = \omega^2/v^2 \end{cases} \tag{26}$$

$$\widetilde{G}_i(\underline{k}, \omega v) = (\pi/2k) \left[\delta(k - \tfrac{\omega}{v}) - \delta(k + \tfrac{\omega}{v}) \right]$$

where $k = |\underline{k}|$ and P denotes the principal value. Note that the support of $\widetilde{G}_r$ is everywhere except on the sphere $k^2 = \omega^2/v^2$, and the support of $\widetilde{G}_i$ is only on the sphere $k^2 = \omega^2/v^2$. This sphere is termed the Ewald sphere. Note also that the spatial Fourier transform of G is invariant with respect to the dimensionality of the space. Because the support of $\widetilde{G}_r$ is everywhere except on the Ewald sphere,

$$\widetilde{G}_r(\underline{k}, \omega, v)\Big|_{v = \omega/k} = \widetilde{G}_r(\underline{k}, \omega, \omega/k) = 0 \tag{27}$$

If Equation 25 is evaluated for $v^2 = \omega^2/k^2$ (on the Ewald sphere), then, by Equation 27,

$$\widetilde{\theta}(\underline{k}, \omega, \omega/k) = \widetilde{\phi}(\underline{k}, \omega) \tag{28}$$

Taking the inverse spatial Fourier transform of Equation 28 yields

$$\phi\,(\underline{x},\,\omega) = (1/2\pi)^n \int\limits_{-\infty}^{\infty} e^{-i\underline{k}\cdot\underline{x}}\,\widetilde{\theta}\,(\underline{k},\,\omega,\,\omega/k)\,d^n k \qquad (29)$$

This is an exact, closed form analytic solution for the total field, ϕ. Note that ϕ does not depend on the arbitrary reference velocity, v, used in the Green's function, since ϕ must satisfy Equation 18, which is independent of v.

Previous solutions to the inverse scattering problem have yielded expressions for the source which were dependent on the velocity used in the Green's function. Without additional information, determination of the source off the Ewald sphere surface is precluded. Stone (Ref 30) has shown that the requirement that the source be of compact (finite) support is both necessary and sufficient. However, the solution given by Equation 29 achieves the same effect in an elegant fashion. As the free parameter v is varied, the Ewald sphere on which $\widetilde{\theta}$ is evaluated sweeps throughout spatial Fourier transform space to obtain the complete spatial reconstruction of $\phi(\underline{x})$.

A more complete derivation of these results is given in Bojarski (Ref 24).

12. AN EXACT CLOSED-FORM ANALYTIC SOLUTION FOR THE SOURCE TERM

A major advantage of solving the inverse scattering problem using numerical solution of the integral equation, Equation 11, as discussed in Section 2, is that fast Fourier transform (FFT) techniques can be employed. This makes treatment of three-dimensional, real world problems well within standard computer capabilities. However, numerical evaluation of Equation 29 is not necessarily as efficient, because of the problems associated with evaluating $\widetilde{\theta}$. Furthermore, although Bojarski (Ref 24) has shown that by evaluating the Laplacean of Equation 29 in the Fourier domain (using the Fourier differentiation theorem) an expression for the source can be obtained, it would be preferable to have an expression for the source directly in terms of the measured data.

The author has derived two results presented here. First, an expression for $\widetilde{\theta}$ in terms of Fourier transform operations on the measured fields is presented. This makes numerical evaluation of the solution for the fields using FFT's possible. Second, an expression for the source in terms only of $\widetilde{\theta}$ (and thus the measured data) is obtained.

Bojarski (Ref 24) has pointed out that since on the Ewald sphere shell $G_r = 0$, for this case

$$\theta = -\oint d\underline{s}\,\nabla G_r\,\phi \qquad (30)$$

This avoids the requirement to evaluate $\nabla\phi$ over the measurement surface, which is quite desirable when the measured field contains noise. Consider the case where the measurement surface is the x-y plane (more general, nonplanar surfaces can be accomodated using the same formulation by adjusting the phase at each measurement point such that the measured data is the same as if it had been measured over a plane). Then Equation 30 becomes (ω and v are temporarily suppressed)

$$\theta(x,y,z) = -\int dx'dy'\,\phi(x',y',0)\ \partial/\partial z\ G_r\ (x-x',\ y-y',\ z) \tag{31}$$

Evaluation of Equation 31 using numerical integration requires spatially sampling ϕ on the measurement surface at an interval small compared to oscillations in G_r, and thus small compared to a wavelength. This is impossible in most real world situations; numerical integration is also slow. These limitations can be avoided by evaluating ϕ in the Fourier transform domain, using FFTs. This requires an analytical expression for the two dimensional spatial Fourier transform of the normal derivative of G_r. Noting that $2G_r = G + G^*$, this reduces to evaluating $\frac{1}{2}\,\partial/\partial z\ (G + G^*)$. Harrington (Ref 31) derives the following expression for G^*:

$$\frac{e^{-iqr}}{r} = (-i/2\pi)\int\!\!\!\int_{-\infty}^{\infty} e^{i\,2\pi(\nu_x x + \nu_y y)}$$

$$\cdot\ \frac{e^{-iz\sqrt{q^2-(2\pi\nu_x)^2-(2\pi\nu_y)^2}}}{\sqrt{q^2-(2\pi\nu_x)^2-(2\pi\nu_y)^2}}\ d(2\pi\nu_x)d(2\pi\nu_y) \tag{32}$$

where $q \equiv \omega/v$ and $r = (x^2 + y^2 + z^2)^{1/2}$. Taking the derivatives of both sides of Equation 32 with respect to z and applying the definition of the Fourier transform yields

$$\mathcal{F}_{xy}\left\{\partial/\partial z\ G^*\right\} = \mathcal{F}_{xy}\left\{\partial/\partial z\left(\frac{e^{-iqr}}{4\pi r}\right)\right\}$$

$$= (-1/2\pi)e^{-iz\sqrt{q^2-(2\pi\nu_x)^2-(2\pi\nu_y)^2}} \tag{33}$$

Sherman (Ref 32) gives

$$\mathcal{F}_{xy}\left\{\partial/\partial z\ G\right\} = \mathcal{F}_{xy}\left\{\partial/\partial z\left(\frac{e^{iqr}}{4\pi r}\right)\right\}$$

$$= (-1/2\pi)e^{iz\sqrt{q^2-(2\pi\nu_x)^2-(2\pi\nu_y)^2}} \tag{34}$$

$\mathcal{F}_{xy}$ denotes the two dimensional spatial Fourier transform from x,y to ν_x, ν_y.

Equation 31 is a convolution in x and y. It follows, using Equations 33 and 34 and the convolution theorem that

$$\theta(x,y,z,\omega,v) = \mathcal{F}_{xy}^{-1}\left\{\mathcal{F}_{xy}\left\{\phi(x,y,0,\omega)\right\}\ P(\nu_x,\ \nu_y,\ z,\omega,v)\right\} \tag{35}$$

where

$$P(\nu_y, \nu_y, z, \omega, v) = (1/4\pi) \; e^{-iz \sqrt{q^2 - (2\pi \nu_x)^2 - (2\pi \nu_y)^2}}$$

$$+ \; e^{iz \sqrt{q^2 - (2\pi \nu_x)^2 - (2\pi \nu_y)^2}} \qquad (36)$$

This expression is strictly valid for $v = \omega/k$. A similar expression, involving an additional term of the same form as in Equation 35, has been derived for $v \neq \omega/k$.

The importance of Equation 35 is that it permits evaluating θ (and thus $\tilde{\theta}$) using FFTs. Indeed, the algorithm is quite simple:

1. Compute the two dimensional spatial Fourier transform of the measured field data, $\phi(x,y,0)$, using an FFT.

2. Multiply, in the spatial frequency domain, by the "propagator" function, P, given in Equation 36.

3. Compute the two dimensional inverse spatial Fourier transform of the product, using an FFT.

This method of evaluating $\tilde{\theta}$ is analogous to the Fourier domain method often used for reconstructing holograms (c.f., Section 15), and for evaluating the Kirchoff integral (Ref 32).

It should also be pointed out that Equations 35 and 36 can be viewed as a Fourier domain, monochromatic analogue to the time domain, effectal field representation for ϕ_H discussed in Reference 16.

Given an efficient means of computing θ, the evaluation of ϕ becomes equally efficient. To obtain an equally efficient expression for the source term, ρ_γ, substitute Equation 26 into Equation 25. Requiring that $k \neq \omega/v$, this yields

$$\tilde{\theta}(\underline{k}, \omega, v) = \tilde{\phi}(\underline{k}, \omega) - (k^2 - \omega^2/v^2)^{-1} \tilde{\rho}_\gamma(\underline{k}, \omega, v) \qquad (37)$$

Substituting Equation 28 into Equation 37 and carrying out some algebra gives

$$\tilde{\rho}_\gamma(\underline{k}, \omega, v) = [\tilde{\theta}(\underline{k}, \omega, \omega/k) - \tilde{\theta}(\underline{k}, \omega, v)] \; (k^2 - \omega^2/v^2) \qquad (38)$$

Equation 38 is an exact, closed form analytic solution for the source term in the inverse scattering problem. Physically, it says that ρ_γ can be written as the difference between the θ field on the Ewald sphere and the θ field off the Ewald sphere. Although they will not be repeated here, all of the results in the preceeding sections concerning the nonexistence of nonradiating sources, the effects of incomplete measurements and noise, the well-posedness of the solution, etc. can be deduced from this analytic solution.

13. CONCLUSIONS

The following conclusions can be drawn from this work:

A. Inverse scattering problems fall into three classes: Inverse source, inverse medium, and synthesis problems.

B. The Bojarski Exact Inverse Scattering Theory provides solutions to all three of these problems, and in particular, to the inverse medium and synthesis problems.

C. An exact, closed-form solution to the synthesis problem has been presented in this paper.

D. As shown elsewhere, the solutions of B, C, G, and H are unique, well-posed, insensitive to noisy measurements, applicable with incomplete data, and computationally efficient.

E. In addition, the closed-form solution to the synthesis problem, presented in this paper, provides a closed-form solution to the remote probing (inverse medium) problem. However, the application to the remote probing problem involves an approximation.

F. The approaches to the synthesis problem presented here are both exact and many powers of 10 more efficient than standard ray tracing techniques.

G. An exact analytic closed form solution for the fields in the inverse scattering problem has been obtained.

H. An exact analytic closed form solution for the source term has also been obtained, and a method derived to permit its numerical evaluation in a stable, efficient manner.

REFERENCES

1. N. N. Bojarski, "Inverse Scattering," Naval Air Systems Command, third quarterly report to Contract N00019-73-C-0312 (NASC-C2-Q3), October 1973.

2. N. N. Bojarski, "Inverse Scattering," Naval Air Systems Command, final report on Contract N00019-73-C-0312, February 1974.

3. N. N. Bojarski, "Exact Inverse Scattering," presented at the Annual Meeting of USNC/URSI, October 20-23, 1975, Boulder, Colorado.

4. N. Bleistein and N. N. Bojarski, "Recently Developed Formulations of the Inverse Problem in Acoustics and Electromagnetics," report MS-R-7501, (AD/A-003 588) Department of Mathematics, Denver Research Institute, University of Denver, Colorado, 1974.

5. W. R. Stone, "Holographic Reconstruction is Usually a Poor Solution to the Inverse Scattering Problem: A Comparison Between the Bojarski Exact Inverse Scattering Theory and Holography as Applied to the Holographic Radio Camera," presented at the 1980 International Optical Computing Conference, April 8-11, 1980, Washington, DC; to appear in SPIE Proceedings, Vol. 231.

6. N. N. Bojarski, "Exact Inverse Scattering," presented at the 1980 DARPA/AF Review of Quantitative NDE, July 14-18, 1980, La Jolla, CA; to appear in conference proceedings.

7. N. Bleistein and J. Cohen, "Nonuniqueness in the Inverse Source Problem in Acoustics and Electromagnetics," Journal of Mathematical Physics, 18: 194-201, 1977.

8. W. R. Stone, "A Uniqueness Proof for the Bojarski Exact Inverse Scattering Theory, and Its Consequences for the Holographic Radio Camera," presented at the URSI National Radio Science Meeting, Boulder, Colorado, January 9-13, 1978.

9. N. N. Bojarski, "A Wave Equation for Radiating Source Distributions," presented at the URSI National Radio Science Meeting, Boulder, Colorado, November 6-9, 1979.

10. W. R. Stone, "The Nonexistence of Nonradiating Sources and the Uniqueness of the Solution to the Inverse Scattering Problem," presented at the North America Radio Science meeting, June 2-6, 1980, Quebec, Canada.

11. R. D. Mager and N. Bleistein, "An Examination of the Limited Aperture Problem of Physical Optics Inverse Scattering," IEEE Trans. Ant. Prop. AP-26: 695-699, 1978.

12.   W. R. Stone, "The Resolution of an Aperture for Coherent Imaging," presented at the Optical Society of America Annual Meeting, San Francisco, California, October 23-27, 1978.

13.   W. R. Stone, "Numerical Studies of the Effects of Noise and Spatial Bandlimiting on Source Reconstructions Obtained using the Bojarski Exact Inverse Scattering Theory," presented at the URSI spring meeting, College Park, Maryland, May 15-19, 1978.

14.   W. M. Lewis, "Physical Optics Inverse Scattering", IEEE Trans. Ant. Prop., AP-17, 308-314, 1969.

15.   N. N. Bojarski, "Exact Inverse Scattering Theory", report on contract N00014-76-C-0082, March 1980.

16.   N. N. Bojarski, "Passive Seismic Artillery Location by Exact Inverse Scattering", report on contract N00014-79-C-0189, July 1980.

17.   Y. T. Lo, "A Mathematical Theory of Antenna Arrays with Randomly Spaced Elements, "IRE Transactions on Antennas and Propagation, AP-12, May 1964.

18.   Bernard D. Steinberg, "The Peak Sidelobe of the Phased Array Having Randomly Located Elements," IEEE Transactions on Antennas and Propagation, AP-20, March 1972.

19.   Bernard D. Steinberg, Principles of Aperture and Array System Design, John Wiley & Sons, New York, 1976, Chapters 8 and 9.

20.   Earl N. Powers and Bernard D. Steinberg, "Valley Forge Research Center Adaptive Array," USNC/URSI Annual Meeting, Boulder, CO, October 1975.

21.   N. N. Bojarski, "Three-Dimensional Electromagnetic Short-Pulse Inverse Scattering," Report SPL 67-3, Syracuse University Research Corporation, Syracuse, New York, February 1967.

22.   N. N. Bojarski, "Electromagnetic Inverse Scattering Theory," Syracuse University Research Corporation, Syracuse, New York, December 1968.

23.   N. N. Bojarski, "Electromagnetic Inverse Scattering," Naval Air Systems Command, June 1972.

24.   N. N. Bojarski, "Inverse Scattering Inverse Source Theory," Office of Naval Research, Contract N00014-76-C-0082, November 1980.

25.   N. N. Bojarski, "Low Frequency Inverse Scattering," Office of Naval Research, Contract N00014-76-C-0082, July 25, 1980.

26. N. Bleistein and J. K. Cohen, "A Survey of Recent Progress on Inverse Problems," University of Denver Department of Mathematics Report MS-R- 7806, 1978.

27. R. J. Collier, C. B. Burkhart, and L. H. Lin, Optical Holography, New York: Academic Press, 1971, Section 13.3.

28. B. P. Hildebrand and B. B. Brenden, An Introduction to Acoustical Holography, New York: Plenum Press, 1972, Chapter 4.

29. W. R. Stone, "Numerical Considerations in the Application of the Exact Inverse Scattering Theory," presented at the 17th Annual Meeting of the Society of Engineering Science, Atlanta, GA, December 15-17, 1980.

30. W. R. Stone, "How to Get Off the Fourier Domain ("Ewald") Sphere in the Inverse Scattering Problem," presented at the National Radio Science Meeting of USNC/URSI, Boulder, CO, January 12-16, 1981.

31. R. F. Harrington, Time Harmonic Electromagnetic Fields, New York, McGraw-Hill Book Company, 1961, Equations 4 through 124.

32. G. C. Sherman, "Application of the Convolution Theorem to Rayleigh's Integral Formulas," J. Opt. Soc. Am., 57, 546-547, 1967.

33. W. R. Stone, "The Concept, Design, and Operation of a Demonstration Holographic Radio Camera," Ph.D. dissertation, Applied Physics and Information Sciences, University of California, San Diego, 1978.

ACOUSTICAL HOLOGRAPHY IS, AT BEST, ONLY A PARTIAL

SOLUTION TO THE INVERSE SCATTERING PROBLEM

W. Ross Stone

IRT Corporation
1446 Vista Claridad
La Jolla, CA 92037

Coherent acoustical imaging and acoustical holography are rarely employed solely for the purpose of recording and reconstructing a coherent acoustical field. Usually, the ultimate information desired concerns the characteristics of an acoustic source, a scatterer, or the inhomogeneity in a medium. The image or holographic reconstruction is an intermediate step in extracting this information (often by human visual recognition). The determination of such characteristics from measurements at a distance of emitted or scattered acoustic fields is, in fact, an inverse scattering problem. In a companion paper by the author in this volume, (hereafter referred to as "Stone-C"), the Exact Inverse Scattering Theory is reviewed as it applies to the acoustic probing problem. The purpose of this paper is to quantify the relationships among imaging, holography, and inverse scattering. It is shown that a holographic reconstruction is only a partial solution to the inverse scattering problem at best. Furthermore, it is shown that holographic data is sufficient to provide an exact solution to the inverse scattering problem, using the Exact Theory. Finally, a common but rarely acknowledged approximation associated with the interpretation of a holographic reconstruction is clarified. It is often assumed and/or stated that the field reconstructed from a hologram is the field which existed in the region of the scatterer (in optical terms, in "object space") when the hologram was recorded. The nature of this approximation is quantified, and the conditions under which it becomes a good approximation are derived.

Throughout this paper scalar notation is employed in presenting the mathematics. It is very important to note that all of the relations discussed, especially with regard to the Bojarski Exact Inverse Scattering Theory, have been shown to apply to the full vector electromagnetic (and acoustic) equations, in anisotropic media. The theories on which the results of this paper are based are completely general. Scalar notation is employed only to simplify the presentation.

1. THE BOJARSKI EXACT INVERSE SCATTERING THEORY

In order to show the relationship between holography and inverse scattering, it is first necessary to provide a mathematical basis. This section presents a derivation of N. N. Bojarski's (Refs 1-4) "Exact Inverse Scattering Theory." This theory is discussed in more detail in Stone-C." The theory is derived in the form leading to an integral equation which can be solved numerically in a stable and efficient manner. In Stone-C derivations are presented which parallel that of this section, but which result in exact, closed-form analytic solutions for the total field and the source term.

Consider a source $\rho(\underline{x})$ in a domain D bounded by a surface S. Then the time harmonic field, $\phi(\underline{x})$, due to $\rho(\underline{x})$ is the solution to the inhomogeneous wave equation

$$\nabla^2 \phi(\underline{x}) + k^2 \phi(\underline{x}) = -\rho(\underline{x}), \quad \underline{x} \in D \tag{1}$$

where $k = 2\pi/\lambda$.

A direct scattering problem is one in which $\rho(\underline{x})$ is known or specified, and a solution for $\phi(\underline{x})$ is sought. The inverse scattering problem is one in which $\phi(\underline{x})$ is known, and $\rho(\underline{x})$ is sought. For the inverse source problem, $\phi(\underline{x})$ is measured over some surface, and the object is to determine $\rho(\underline{x})$. In general, $\rho(\underline{x}) = \rho_m(\underline{x}) + \rho_s(\underline{x})$, where ρ_m is due to interaction with the medium, and ρ_s is due to actual sources. If $n(\underline{x})$ is the complex refractive index of the medium, then

$$\rho_m(\underline{x}) = k^2 \left[n^2(\underline{x}) - 1 \right] \phi(\underline{x}) \tag{2}$$

In most remote probing problems, $\rho_s(\underline{x})$ is known, and $\rho_m(\underline{x})$ is sought to yield $n(x)$. This is termed the inverse medium problem. If $\phi(\underline{x})$ is specified (as the desired field) and $\rho(\underline{x})$ or $n(\underline{x})$ is sought so as to produce that $\phi(\underline{x})$, the problem is termed an inverse synthesis problem.

For the acoustic case, Equation 1 can be written as

$$\nabla^2 \phi(\underline{x}, \omega) + k_m^2 \phi(\underline{x}, \omega) = 0 \tag{1a}$$

where

$$k_m = \omega / v_m(\underline{x}, \omega)$$

and v_m is the velocity of the medium. ω is the radian frequency. Note that the variation of the velocity with the spatial coordinates and with frequency is explicitly included at this point. Equation 2 can be written as

$$\rho(\underline{x}, \omega) = \left[(\omega^2/c^2) - (\omega^2/v_m^2) \right] \phi(\underline{x}, \omega) \tag{2a}$$

where c is the homogeneous medium propagation velocity (or an arbitrarily chosen reference velocity). Substitution of Equation 2a into Equation la yields Equation 1.

Let the following field quantity, $\phi_H(\underline{x})$, be defined:

$$\phi_H(\underline{x}) = \oint \left[g^*(\underline{x}-\underline{x}')\nabla\phi(\underline{x}') - \phi(\underline{x}')\nabla g^*(\underline{x}-\underline{x}') \right] d\underline{S}' \tag{3}$$

where $g(\underline{x})$ is the free space Green's function and the asterisk denotes complex conjugation. g satisfies Equation 1 with $\rho(\underline{x}) = \delta(\underline{x})$. ϕ_H is in the form of the Kirchoff integral with g complex conjugated. Note that if the Kirchoff integral is applied to the field $\phi(\underline{x})$ on S and evaluated at any point $\underline{x}$ inside D, it is identically zero: The Kirchoff integral is nonzero only for points outside D. Conversely, $\phi_H(\underline{x})$ is nonzero only for points inside D. Points outside of D are associated with the direct scattering problem; points inside D are of interest for the inverse scattering problem. This topological difference is the reason why direct scattering solutions are mathematically ill-posed when applied to the inverse scattering problem.

It should also be noted that ϕ_H is the mathematical expression for the reconstruction obtained from a hologram (ϕ in Equation 3) recorded on S. Indeed, the spatial Fourier transform of Equation 3 results in the "plane wave spectrum" formulation for computed holographic reconstruction (Refs 9,10). . ϕ_H is, in general, known for inverse problems, since ϕ is known over S. ϕ is measured over S for the inverse source and medium problems, or specified over S for the inverse synthesis problem.

Applying Gauss' theorem to Equation 3 converts the surface integral into a volume integral:

$$\phi_H = \int dV \left(g^* \nabla^2 \phi - \phi \nabla^2 g^* \right) \tag{4}$$

From Equation 1,

$$\nabla^2 \phi = -k^2 \phi - \rho \tag{5}$$

and, by complex conjugation of Equation 1 for g,

$$\nabla^2 g^* = -k^2 g^* - \delta \tag{6}$$

Substitution of Equations 5 and 6 into Equation 4 gives

$$\phi_H = \int dV \left[g^*(-k^2\phi - \rho) - \phi(-k^2 g^* - \delta) \right]$$
$$= \int dV \ (\phi\delta - g^* \rho) \tag{7}$$

and, carrying out the integration over the delta function,

$$\phi_H = \phi - \int dV \ g^* \rho \tag{8}$$

Direct scattering theory gives the result that

$$\phi = \int dV \ g \rho + \oint dS(g \nabla \phi - \phi \nabla g)$$
$$= \int dV \ g \rho + \phi_i \tag{9}$$

In Equation 8, the first integral is just the superposition integral over the sources. The second term is the Kirchoff integral, and is asssociated with the incident field, ϕ_i. For the inverse scattering problem, ϕ_i can be assumed to be known without loss of generality (e.g., it is the known probing field for the inverse medium case, or the specified incident field in the inverse synthesis case).

Equations 8 and 9 are two independent simultaneous equations in two unknowns, ϕ and ρ. Substitution of Equation 9 into Equation 8 yields

$$\phi_H = \int dV \ g \rho - \int dV \ g^* \rho + \phi_i$$
$$= \int dV \ (g - g^*) \rho + \phi_i \tag{10}$$

or

$$\phi_H(\underline{x}) = 2i \int dV' \, \text{Img}(\underline{x}-\underline{x}') \, \rho(\underline{x}') + \phi_i(\underline{x}) \tag{11}$$

where Im denotes the imaginary part. Equation 11 is the basic equation of the Exact Inverse Scattering Theory. It is an integral, convolution equation for the single unknown, $\rho(\underline{x})$. It can be solved by standard deconvolution techniques. Quite recently, Stone has presented a closed-form solution to Equation 11, based on a closed form solution by Bojarski for the total field, ϕ, in Equation 8 (see Stone-C).

2. HOLOGRAPHY, INVERSE SCATTERING, AND IMAGING

Imaging is an inverse scattering problem: Based on observed field data, information about the scatterer (contained in $\rho(\underline{x})$ in the above formulation) is sought. If only the shape of the scatterer is sought, then it can be shown that a physical optics far-field approximation to the exact inverse scattering solution is often sufficient (Refs 4,5). However, if any information other than the shape is desired (e.g., the complex refractive index), then the full exact solution is required.

It was noted above that ϕ_H is the field obtained from a holographic reconstruction. From Equation 8,

$$\phi_H = \phi - \int dV \, g^* \rho \tag{8}$$

Several conclusions follow from this equation:

1. ϕ_H is a field. It is not the source term, ρ. The source term, ρ, contains all the information about the source. Simple examination of the holographic reconstruction, ϕ_H, cannot give ρ. To obtain ρ, Equation 11 must be solved.

2.   Holographers commonly make the statement, "The holographic reconstruction is the field which existed in object [source] space." This is an approximation which, from Equation 8, is given by

$$\phi_H \approx \phi \tag{12}$$

3.   If, and only if, the scatterer is of such a form, and if and only if the wavelength and all other dimensions in the problem are such, that the field scattered by the scatterer has a perturbation introduced into it which is the shape of the scatterer, then, to the degree that Equation 12 holds, it is possible to deduce the shape of the scatterer from a holographic

reconstruction. In other words, the shape can be deduced when the field "looks like" the scatterer. This is often true for objects with sharp edges imaged using optical holography; it <u>may</u> be true for certain acoustic scatterers imaged using acoustical holography.

4. The approximation in Equation 12 can only be true when the second term of Equation 8 is 0. The kernel, g*, is of the form $[\cos(kr) - i \sin(kr)]/4\pi r$. The integral will thus be small as $r \to \infty$. Thus, the statements of paragraphs 2 and 3 hold only in the far field. Note that this requires that reconstructions, "near" enough the position of the scatterer so that the field "looks like" the scatterer, must still be in the far field in terms of wavelengths. One would not expect this condition to be easily satisfied.

The statement, that the requirement that the scattered field "look like" the scatterers, is often true for objects with edges that are "sharp" and can be expressed in a more quantitative manner. Consider the kernel of the integral in Equation 11. In three dimensions, Im g is given by

$$\text{Im } g = \sin(kr)/4\pi r$$

Using convolution notation, Equation 11 can be written as

$$\phi_H = 2i\rho \; * \; \sin(kr)/4\pi r$$

Now consider the short wavelength limit. As $\lambda \to 0$, $\sin(kr)/4\pi r \to \delta(r)/2$ (where δ denotes a three-dimensional delta distribution), and

$$\lim_{\lambda \to 0} \phi_H = 2i \; \rho * \delta(r)/2$$

$$\lim_{\lambda \to 0} \phi_H = i\rho$$

Physically, this says that when the wavelength is small enough compared to any structure in the scatterer (i.e. the edges of the scatterer are "sharp"), the convolution with Im g acts like a convolution with a delta distribution, replicating the scatterer in the field ϕ_H. Note that if the intensity distribution $\phi_H \phi_H^*$ is viewed in this limit, this image will be equal to the magnitude of the source term.

To summarize, holography is not a solution to the inverse scattering problem. It may give an adequate approximation to the shape of the object under certain conditions. However, holographic data <u>is</u> all that is required to solve the inverse scattering problem uniquely, using the Exact Scattering Theory.

3. AN EFFICIENT METHOD OF VISUALIZING SCATTERERS: THE EFFECTAL FIELD

From Equation 10, the holographic field, ϕ_H, is given by (the incident field can be taken to be zero for a holographic reconstruction)

$$\phi_H = \int dV(g-g^*)\rho \tag{13}$$

Letting double arrows denote Fourier transformation with respect to time,

$$G(r, t-r/c) \longleftrightarrow g(r, \omega r/c) \tag{14}$$

Since the Green's function is causal, it follows that

$$G(r, t+r/c) \longleftrightarrow g^*(r, \omega r/c) \tag{15}$$

Thus, the temporal inverse Fourier transform of Equation 13 is

$$\phi_H(\underline{x}, t) = \int dV \left[G(r, t-r/c) - G(r, t+r/c) \right] \star \rho(\underline{x}, t) \tag{16}$$

where the $\star$ denotes convolution with respect to time. The first convolution on the right hand side of Equation 16 is clearly the real, physical time retarded causal field radiated by the sources $\rho(\underline{x},t)$. The second convolution on the right hand side of Equation 16 is a fictitious time advanced anti-causal field "radiated" by the sources, $\rho(\underline{x},t)$. The difference of these two, $\phi_H(\underline{x},t)$, is, in the time domain, termed the effectal field (Ref 6). The effectal field is thus due to the destructive interference between the time retarded causal field and the fictitious time advanced anti-causal field.

It is clear from Equation 16 that if the sources contain components that are spatially and temporally impulsive, then the effectal field will also contain impulsive components at the same temporal and spatial locations as the impulsive source components. A graphical space-time display of the effectal field thus yields these correct spatial and temporal locations. Specifically, it will clearly yield the spatial locations of discontinuities in the refractive index: It shows where the boundaries of cracks, flaws, sonar targets, etc. are located. If only these locations are needed, such a graphical display (which is also subject to computer quantification) can obviate the need for a solution to the inverse scattering integral Equation 11.

The time domain effectal field also adds important insight into the relationship between holography and inverse scattering, discussed, in the frequency domain, in the previous section. It was shown in the last section that the holographic reconstruction contains information about the shape and location of refractive index discontinuities when the field, ϕ_H, has imposed upon it a perturbation in the shape of the scatterer. It was also shown that the approximation of Equation 12 is good in the short wavelength, far field regime. Equation 16 shows that, in the time domain, the field ϕ_H becomes an exact picture of the shape and location of the scatterer when the scatterer is impulsive. The discussion of the previous section shows what can be accomplished with a single temporal frequency component; this section shows what is obtainable with the full spectrum. Note that the wavelength and far field conditions under which the frequency domain holographic reconstruction contains the most information about the shape and location of the scatterer are those that make the scatterer appear most "impulsive."

An example of the information present in a graphical display of the effectal field is shown on the following four pages. These plots, taken from Bojarski (Ref 6), show the reconstruction of an acoustic point source in two dimensions. An analytic expression for the field radiated by the source was used to compute data values on the boundary of a 128 x 128 point grid. The holographic field, Equation 13, was computed for 256 frequencies at all points in the grid, and a fast Fourier transform was used to compute the effectal field, Equation 16, in the time domain. A plot of every fifth point for t = -5 to t = 70 of the effectal field is shown on the following two pages (Figure 1). The subsequent two pages (Figure 2) show the same computation, with random noise added to the data prior to computation of the holographic field. The signal to noise ratio is O dB (a 1 to 1 signal-to-noise ratio)!

4. CONCLUSIONS

A coherent acoustical image, and in particular, the reconstruction obtained from an acoustical hologram, is a representation of a field. All of the information about the source, object, or medium being imaged is contained in the source term of the wave equation. Simple examination of the field will not normally provide information about the source. If, and only if the scatterer is of such a form, and if and only if the wavelength and all other dimensions in the problem are such, that the field scattered by the scatterer has a perturbation introduced into it which is the shape of the scatterer, then it may be possible to deduce the shape of the scatterer from a holographic reconstruction. The shape can be deduced when the field "looks like" the scatterer.

Holographers commonly make the statement, "The holographic reconstruction is the field which existed in object (source) space." This is an approximation, the nature of which was quantified in Section 2. The conditions for this approximation to be "good" are that reconstructions, "near" enough the position of the scatter so that the field "looks like" the scatterer, must still be in the far field of the scatterer in terms of wavelengths. Under certain conditions, in the short wavelength limit, the image obtained with a holographic reconstruction can approach being equal to the magnitude of the source term.

An efficient method of visualizing scatterers when wide bandwidth, or time domain, holographic data is available was presented.

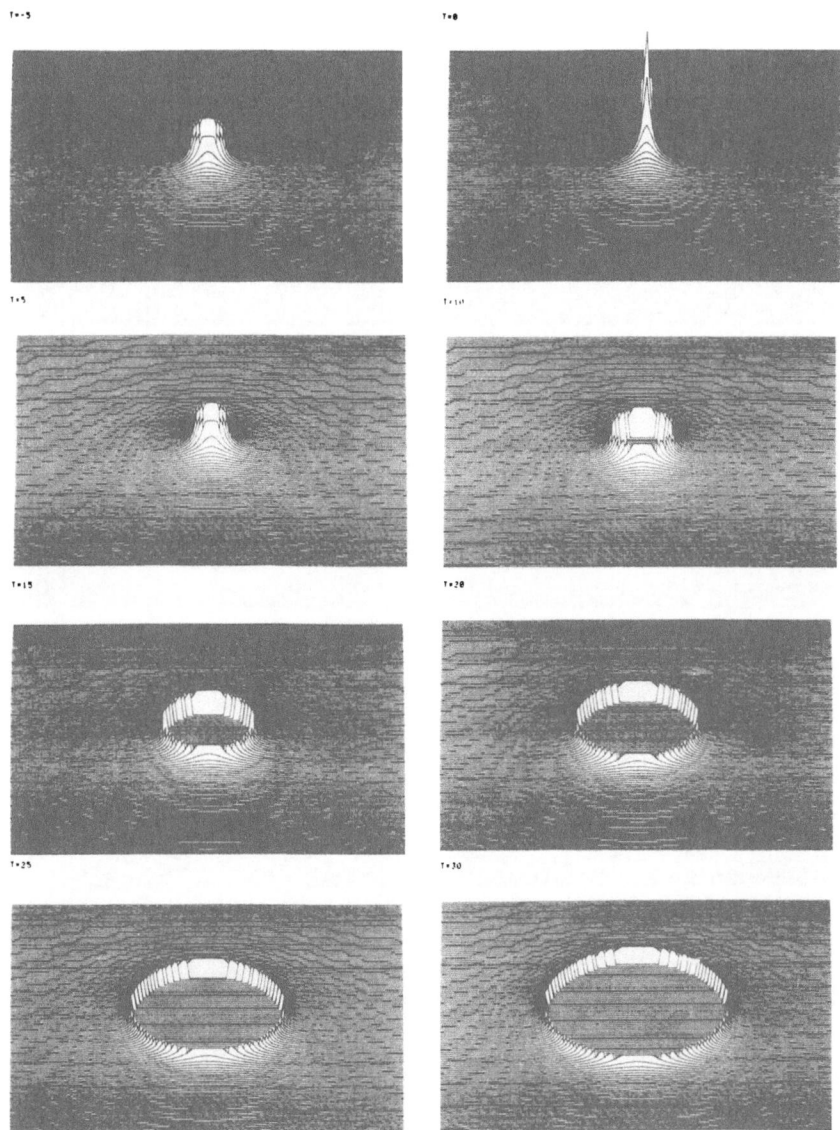

Figure 1. Reconstruction of a point source using the effectal field space/time display (Sheet 1 of 2 Sheets) (Ref 6)

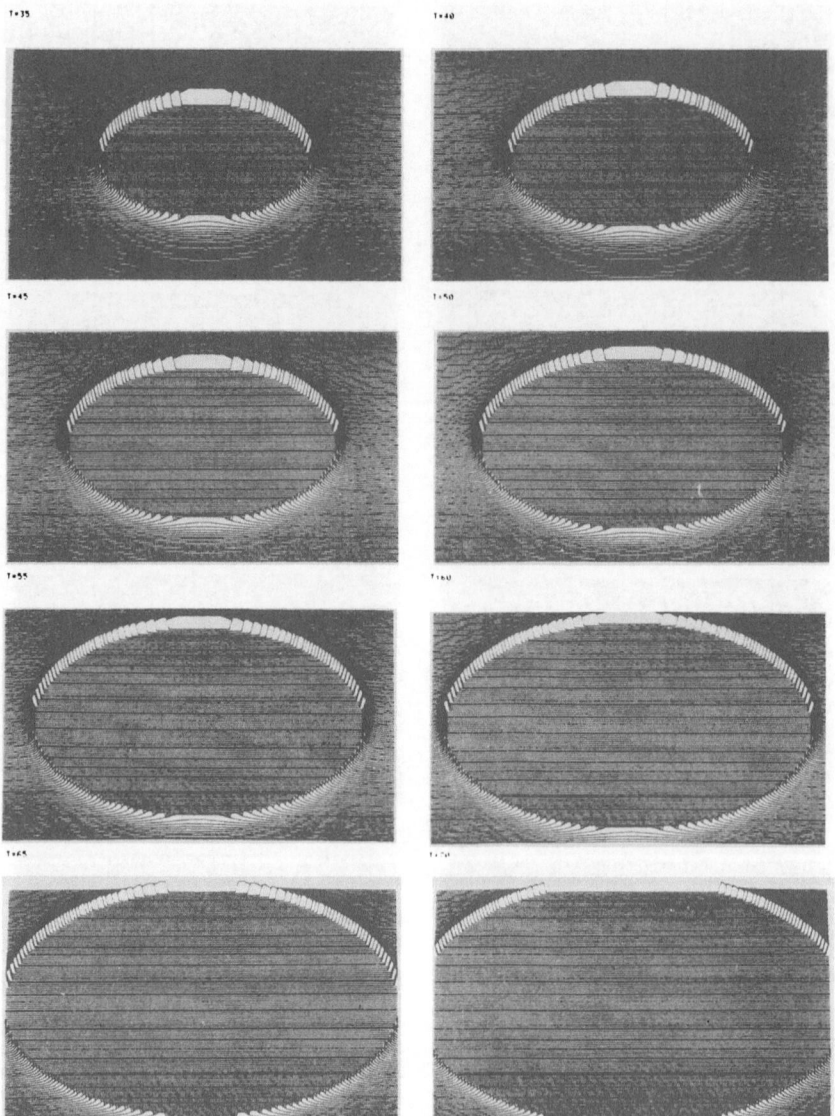

Figure 1. Reconstruction of a point source using the effectal field
space/time display (Sheet 2 of 2 Sheets) (Ref 6)

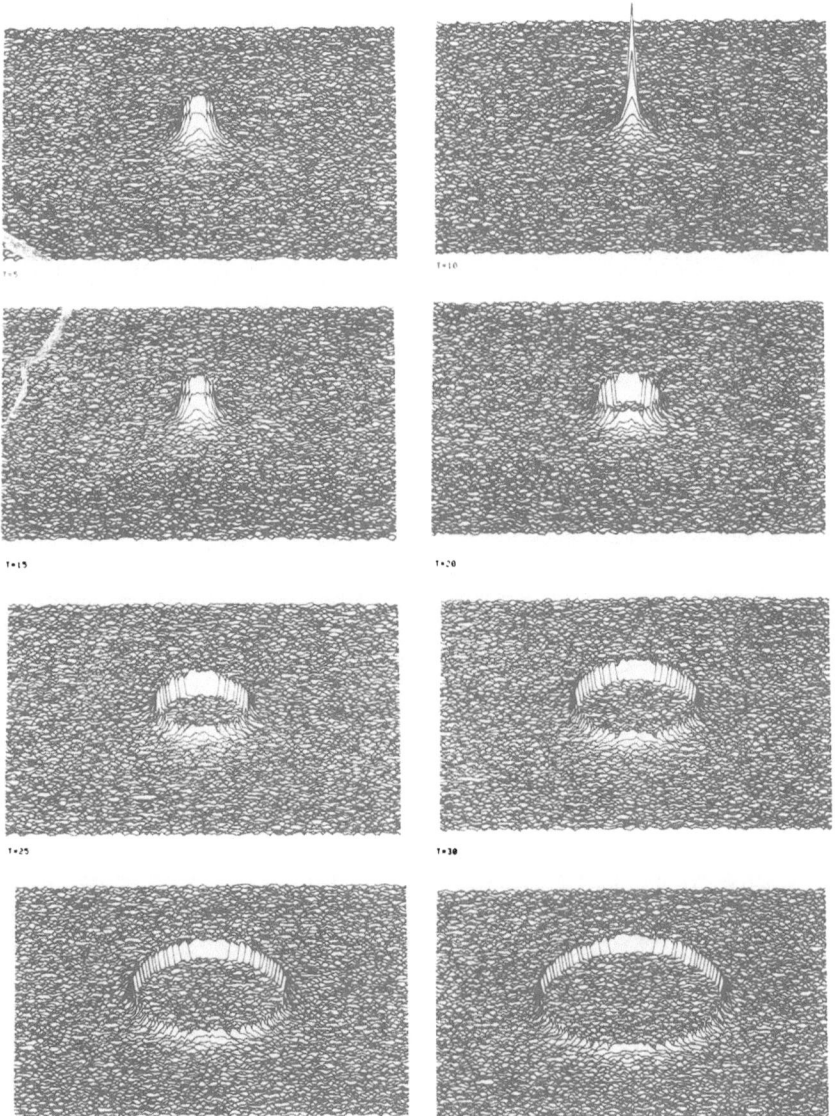

Figure 2. Reconstruction of the point source in Figure 1 with noise added
to the original data. The signal-to-noise ratio for the added
noise is 0 dB (Sheet 1 of 2 Sheets) (Ref 6)

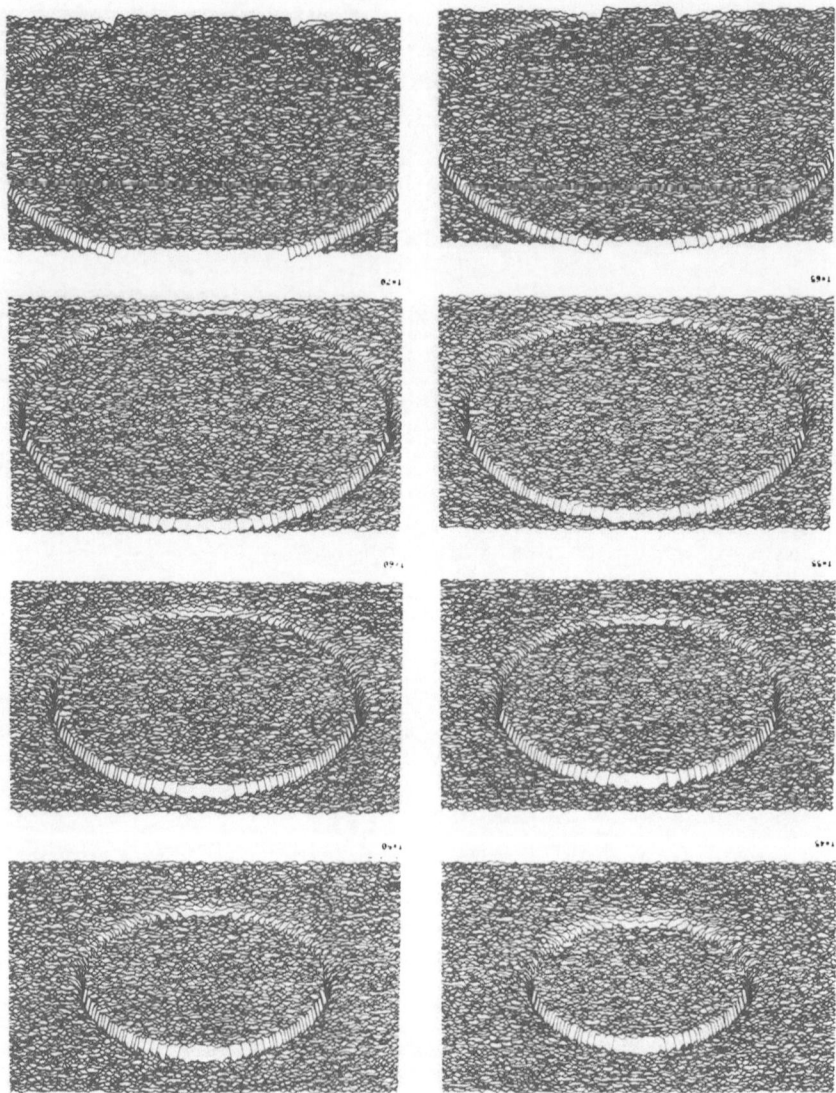

Figure 2. Reconstruction of the point source in Figure 1 with noise added
to the original data. The signal-to-noise ratio for the added
noise is 0 dB (Sheet 2 of 2 Sheets) (Ref 6)

In summary, the information sought in most imaging applications represents the solution to the inverse scattering problem. In general, a holographic reconstruction is not a solution to the inverse scattering problem. It may give an adequate approximation to the shape of a scatterer under certain conditions, discussed above. However, holographic data is all that is required to solve the inverse scattering problem uniquely, using the Exact Inverse Scattering Theory.

Many of the results obtained above have been experimentally verified by the results obtained with the Holographic Radio Camera, in an experiment which obtained the first three dimensional images of the ionosphere (Refs 7,8).

REFERENCES

1. N. N. Bojarski, "Inverse Scattering," Naval Air Systems Command, third quarterly report to Contract N00019-73-C-0312 (NASC-C2-Q3), October 1973.

2. N. N. Bojarski, "Inverse Scattering," Naval Air Systems Command, final report on Contract N00019-73-C-0312, February 1974.

3. N. N. Bojarski, "Exact Inverse Scattering," presented at the Annual Meeting of USNC/URSI, October 20-23, 1975, Boulder, Colorado.

4. N. Bleistein and N. N. Bojarski, "Recently Developed Formulations of the Inverse Problem in Acoustics and Electromagnetics," report MS-R-7501, (AD/A-003 588) Department of Mathematics, Denver Research Institute, University of Denver, Colorado, 1974.

5. W. M. Lewis, "Physical Optics Inverse Scattering", IEEE Trans. Ant. Prop., AP-17, 308-314, 1969.

6. N. N. Bojarski, "Passive Seismic Artillery Location by Exact Inverse Scattering", report on contract N00014-79-C-0189, July 1980.

7. W. R. Stone, "A Holographic Radio Camera Technique for Three-Dimensional Reconstruction of Ionospheric Inhomogeneities," Journal of Atmospheric and Terrestrial Physics, 38, 583-592, 1976.

8. W. R. Stone, "The Concept, Design, and Operation of a Demonstration Holographic Radio Camera," Ph.D. dissertation, Applied Physics and Information Sciences, University of California, San Diego, 1978.

9.   J. W. Goodman, <u>Introduction to Fourier Optics</u> (McGraw-Hill, San Francisco, 1968) Sections 3 through 7.

10. R. J. Collier, C. B. Burckhardt, and L. H. Lin, <u>Optical Holography</u> (Academic Press, New York, 1971) Appendix I and Section 5.4.

INVERSE SCATTERING, INVERSE FIELD, AND

INVERSE SOURCE THEORY

Norbert N. Bojarski

16 Pine Valley Lane
Newport Beach, California 92660
(714)-640-7900

Abstract. Treated is the inverse scattering inverse source problem associated with the inhomogeneous Helmholtz wave equation, the (special case) Sturm-Liouiville (acoustic wave) equation, and the time-independent Schrodinger equation. To this end, the concepts of a reference wave velocity, and an associated free reference space Green's function spectrum, are introduced. A modified Kirchhoff surface integral, containing only the gradient of the real part of this free reference space Green's function spectrum and the fields on a measurement surface, is formulated, yielding an integral equation for the unknown fields and sources in the interior of the closed surface on which the (remotely sensed) fields are known. A well-posed, analytic closed form solution of this integral equation for the unknown fields and their Laplacean is obtained with the aid of a (modified) spatial Fourier transform in which the reference velocity is continually varied in such a fashion that the Ewald sphere shell sweeps to fill the entire transform space. The unknown potential or medium properties, and the unknown sources, are then determined algebraically for the inverse scattering and inverse source problems, respectively. The effects of finite sampling density and incomplete observation domain are discussed briefly.

1. Introduction. Presented is a unified formulation and solution to the inverse scattering inverse source problems for the time-independent Schrodinger equation

$$\frac{\hbar^2}{2m} \nabla^2 \phi + (E-V)\phi = 0 \quad , \tag{1}$$

the (special case) Sturm-Liouiville (acoustic wave) equation

$$\nabla^2 \phi(\mathbf{X}, \omega) + \frac{\omega^2}{c^2(\mathbf{X}, \omega)} \phi(\mathbf{X}, \omega) = 0 \quad , \tag{2}$$

and the inhomogeneous Helmholtz wave equation

$$\nabla^2 \phi(\mathbf{X}, \omega) + \frac{\omega^2}{c^2} \phi(\mathbf{X}, \omega) = -\rho(\mathbf{X}, \omega) \quad , \tag{3}$$

subject to the constitutive equation

$$\rho(\mathbf{X}, \omega) = V(\mathbf{X}, \omega) \, \phi(\mathbf{X}, \omega) \quad . \tag{4}$$

To this end, the single mixed wave equation

$$\nabla^2 \phi(\mathbf{X}, \omega) + \frac{\omega^2}{c^2(\mathbf{X}, \omega)} \phi(\mathbf{X}, \omega) = -\rho(\mathbf{X}, \omega) \tag{5}$$

is introduced, (still subject to the constitutive equation (4)), which reduces to (1), (2), or (3), depending on the choice of c and ρ in (5); i.e., (5) reduces to (3) if c is a known constant, (5) reduces to (2) if $\rho=0$, and (5) reduces to (1) if $\omega^2/c^2 = 2m/h^2$ E and ρ is given by the constitutive equation (4).

It is argued that the inverse scattering inverse source solution presented is an alternative (to the direct 1859 Kirchhoff) integration of the wave equation. It is thus appropriate to review this direct integration of (5), as well as some of the properties of this direct integration.

The direct integration of (5) is accomplished in the following fashion: Let a reference potential V_{r}, a reference source distribution ρ_{r}, and a total source distribution ρ_{t} be defined respectively as

$$V_{r} \equiv \left(\frac{\omega^2}{c^2(\mathbf{X}, \omega)} - \frac{\omega^2}{v^2} \right) \tag{6}$$

$$\rho_{r} \equiv V_{r} \, \phi \tag{7}$$

$$\rho_{t} \equiv \rho + \rho_{r} \quad , \tag{8}$$

where v is an arbitrarily chosen constant "free space" reference velocity. With the aid of (6-8), (5) can be rewritten as

$$\nabla^2 \phi(\mathbf{X}, \omega) + \frac{\omega^2}{v^2} \phi(\mathbf{X}, \omega) = -\rho_{t}(\mathbf{X}, \omega) \quad . \tag{9}$$

If G is chosen as the free space Green's function associated with the constant reference velocity ν, then Kirchhoff's direct integration of (9) is

$$\int_v dv \; G \; \rho_t + \oint_s d\mathbf{s} \cdot (G \; \nabla\phi - \phi \; \nabla G) = \begin{cases} \phi \; , & \forall \; \mathbf{x} \in v \\ \\ 0 \; , & \forall \; \mathbf{x} \notin v \end{cases} \qquad (10)$$

which is an equivalent INTEGRAL REPRESENTATION of the partial differential equation (9).

If the free space Green's function G satisfies the Sommerfeld radiation condition at infinity, then the Kirchhoff surface integral in (10) can be recognized as the incident field (the field in v due to all the sources not in v); i.e.,

$$\oint_s d\mathbf{s} \cdot (G \; \nabla\phi - \phi \; \nabla G) = \phi_i \quad . \qquad (11)$$

If the total source distribution ρ_t is related to the field ϕ by the constitutive equation

$$\rho_t = V_t \; \phi \quad , \qquad (12)$$

then (10), (11), and (12) can be combined to yield the direct scattering Lipman-Schwinger INTEGRAL EQUATION

$$\phi - \int_v dv \; G \; V_t \; \phi = \phi_i \quad . \qquad (13)$$

A brief review of some of the properties of the Kirchhoff surface integral

$$\oint_s d\mathbf{s} \cdot (G \; \nabla\phi - \phi \; \nabla G) \qquad (14)$$

is now in order. Specifically, this Kirchhoff surface integral is an equivalence statement relating the field at a field point on one side of the closed surface produced by all the sources on the other side of the closed surface, via the fields produced by these sources on this closed surface (an equivalence statement that permitted the identification of the incident field (11) and the formulation of the Lipman-Schwinger direct scattering integral equation (13)). The inverse scattering inverse source problem is, however, characterized by both the field point for the unknown fields as well as all the unknown sources (that produce these fields) being on the same side of the closed surface (on which the remote sensing is accomplished), for which situation the Kirchhoff surface integral vanishes, thus rendering this Kirchhoff surface integral as useless for the inverse scattering inverse source problem. A modified Kirchhoff surface integral, which does not suffer from this pathology, is introduced next.

2. <u>The Inverse Scattering Inverse Source Integral Equation.</u> Let G be the free reference space Green's function satisfying the inhomogeneous Helmholtz wave equation

$$\nabla^2 G + \frac{\omega^2}{v^2} G = - \delta \tag{15}$$

and the Sommerfeld radiation condition at infinity, where v is any arbitrarily chosen reference velocity.

Next, let an effectal field θ be defined as

$$\theta \equiv \oint_S d\mathbf{S} \cdot (G_r \nabla \phi - \phi \nabla G_r) \quad , \tag{16}$$

where

$$G_r \equiv \mathcal{R}e \, G \quad , \tag{17}$$

which, by (15), also satisfies the inhomogeneous Helmholtz wave equation

$$\nabla^2 G_r + \frac{\omega^2}{v^2} G_r = - \delta \quad . \tag{18}$$

It should be noted that, by (16), the effectal field $\theta(x)$ can be computed for $\forall x \in v$ from mere knowledge of the field $\phi(x)$ for $\forall x \in s$.

By Green's theorem, (16) reduces to

$$\theta = \int_V dv \, (G_r \nabla^2 \phi - \phi \nabla^2 G_r) \quad , \tag{19}$$

which, by (5) and (18), further reduces to

$$\theta = \int_V dv \left[G_r \left(- \frac{\omega^2}{c^2} - \rho \right) - \phi \left(- \frac{\omega^2}{v^2} - \delta \right) \right] \tag{20}$$

$$= \int_V dv \, \delta \, \phi - \int_V dv \, G_r \left[\left(\frac{\omega^2}{c^2} - \frac{\omega^2}{v^2} \right) \phi + \rho \right] \quad , \tag{21}$$

where it should be noted that $c = c(\mathbf{x}, \omega)$.

By the very definition of the Dirac Delta function

$$\int_v dv' \; \delta(\mathbf{x}-\mathbf{x}') \; \phi(\mathbf{x}') = \begin{cases} \phi(\mathbf{x}) & , \quad \forall \; \mathbf{x} \in v \\ \\ 0 & , \quad \forall \; \mathbf{x} \notin v \quad . \end{cases} \tag{22}$$

With the aid of (6-8)

$$\left(\frac{\omega^2}{c^2} - \frac{\omega^2}{v^2} \right) \phi + \rho = \rho_t \quad . \tag{23}$$

With the aid of (22) and (23), (21) reduces to

$$\theta = \begin{cases} \phi - \int_v dv \; G_r \; \rho_t & , \quad \forall \; \mathbf{x} \in v \\ \\ - \int_v dv \; G_r \; \rho_t & , \quad \forall \; \mathbf{x} \notin v \quad , \end{cases} \tag{24}$$

which is a proper (i.e., $\mathbf{x},\mathbf{x}' \in v$) inverse scattering inverse source integration equation.

With the aid of the direct integration (10-11), this inverse scattering inverse source integral equation (24) reduces to the less general 1973 inverse scattering inverse source integral equation of this author [1], which was studied extensively by Bleistein and this author [2], Bleistein and Cohen [3], and others.

3. Solution of the Inverse Scattering Inverse Source Integral Equation. For the purpose of solving the inverse scattering inverse source integral equation (24), it becomes convenient to introduce the characteristic function $\gamma(\mathbf{x})$ for the volume of integration v, defined by

$$\gamma(\mathbf{x}) \equiv \begin{cases} 1 & , \quad \forall \; \mathbf{x} \in v \\ \\ 0 & , \quad \forall \; \mathbf{x} \notin v \quad . \end{cases} \tag{25}$$

With the aid of (25), (24) can be written as

$$\omega,v) = \gamma(\mathbf{x}) \; \phi(\mathbf{x},\omega) - \int_{-\infty}^{\infty} G_r(\mathbf{x}|\mathbf{x}',\omega,v) \; \gamma(\mathbf{x}') \; \rho_t(\mathbf{x}',\omega,v) \; d^n x' \quad . \tag{26}$$

Introducing the characteristic source distribution $\rho_\gamma(\mathbf{x},\omega,v)$

$$\rho_\gamma(\mathbf{x},\omega,v) \equiv \gamma(\mathbf{x}) \; \rho_t(\mathbf{x},\omega,v) \quad , \tag{27}$$

permits the rewriting of (26) as

$$\theta(\mathbf{x},\omega,\upsilon) = \gamma(\mathbf{x}) \; \phi(\mathbf{x},\omega) - \int_{-\infty}^{\infty} G_r(\mathbf{x}|\mathbf{x}',\omega,\upsilon)\rho_\gamma(\mathbf{x}',\omega,\upsilon)d^n\mathbf{x}', \quad (28)$$

which, in cartesian coordinates, can be rewritten as

$$\theta(\mathbf{x},\omega,\upsilon) = \gamma(\mathbf{x})\phi(\mathbf{x},\omega) - \int_{-\infty}^{\infty} G_r(\mathbf{x}-\mathbf{x}',\omega,\upsilon) \; \rho_\gamma(\mathbf{x}',\omega,\upsilon) \; d^n\mathbf{x}'. \quad (29)$$

Since the Green's function G_r in (29) is a spatial difference kernal in cartesian coordinates, (29) can be further rewritten as the n-dimensional spatial convolution

$$\theta(\mathbf{x},\omega,\upsilon) = \gamma(\mathbf{x}) \; \phi(\mathbf{x},\omega) - G_r(\mathbf{x},\omega,\upsilon) * \rho_\gamma(\mathbf{x},\omega,\upsilon) \quad . \quad (30)$$

Taking the n-dimensional spatial Fourier transform of (29) thus yields the algebraic product equation

$$\tilde{\theta}(\mathbf{k},\omega,\upsilon) = \tilde{\gamma}(\mathbf{k}) * \tilde{\phi}(\mathbf{k},\omega) - \tilde{G}_r(\mathbf{k},\omega,\upsilon) \; \tilde{\rho}_\gamma(\mathbf{k},\omega,\upsilon) \quad . \quad (31)$$

It should be noted here that the field ϕ, and its spatial Fourier transform $\tilde{\phi}$, does not depend on the arbitrarily chosen reference velocity . This is mathematically self-evident since the wave equation (5) does not contain this arbitrarily chosen reference velocity υ, and physically self-evident since the physical field and its spatial Fourier transform cannot depend on the arbitrary choice of the reference velocity υ.

A digression examining some of the properties of the spatial Fourier transform of the Green's function is now in order.

The spatial Fourier transform of the Green's function is

$$\tilde{G}(\mathbf{k},\omega,\upsilon) = P \; \frac{1}{k^2 - \frac{\omega^2}{\upsilon^2}} + \frac{i\pi}{2k} \left[\delta(k - \frac{\omega}{\upsilon}) - \delta(k + \frac{\omega}{\upsilon}) \right] \quad , \quad (32)$$

where P denotes the principal value; i.e.,

$$P \; \frac{1}{k^2 - \frac{\omega^2}{\upsilon^2}} \equiv \begin{cases} \frac{1}{k^2 - \frac{\omega^2}{\upsilon^2}} \; , & \forall \; k^2 \neq \frac{\omega^2}{\upsilon^2} \\ \\ 0 \; , & \forall \; k^2 = \frac{\omega^2}{\upsilon^2} \end{cases} \quad . \quad (33)$$

It should be noted that the functional form of the spatial Fourier transform of the Green's function in k-space is invariant to the dimensionality of the space, which is not the case for the Green's function in x-space.

Furthermore, the spatial Fourier transform of the real and imaginary parts of the Green's function are respectively

$$\tilde{G}_r(\mathbf{k},\omega,v) = P \frac{1}{k^2 - \frac{\omega^2}{v^2}} \tag{34}$$

$$\tilde{G}_i(\mathbf{k},\omega,v) = \frac{\pi}{2k} \left[\delta(k - \frac{\omega}{v}) - \delta(k + \frac{\omega}{v}) \right] . \tag{35}$$

(The notation used here for the imaginary part of the Green's function and its spatial Fourier transform is consistent with the notation used for the real part of the Green's function and its spatial Fourier transform; i.e., $ReG = G_r \leftrightarrow \tilde{G}_r$ and $ImG = G_i \leftrightarrow \tilde{G}_i$.)

Next, the support of the spatial Fourier transform of the real and imaginary parts of the Green's function is examined. By (32-35)

$$Sup\ \tilde{G}_r(\mathbf{k},\omega,v) \in \forall \quad k^2 \neq \frac{\omega^2}{v^2} \tag{36}$$

$$Sup\ \tilde{G}_i(\mathbf{k},\omega,v) \in \forall \quad k^2 = \frac{\omega^2}{v^2} \tag{37}$$

i.e., $\tilde{G}_r(\mathbf{k},\omega,v)$ is nonzero everywhere except on the Ewald sphere shell $k^2 = \frac{\omega^2}{v^2}$, and $\tilde{G}_i(\mathbf{k},\omega,v)$ is nonzero only on the Ewald sphere shell $k^2 = \frac{\omega^2}{v^2}$; and conversely, $\tilde{G}_r(\mathbf{k},\omega,v)$ is zero only on the Ewald sphere shell $k^2 = \frac{\omega^2}{v^2}$, and $\tilde{G}_i(\mathbf{k},\omega,v)$ is zero everywhere except on the Ewald sphere shell $k^2 = \frac{\omega^2}{v^2}$.

Thus (on the Ewald sphere shell)

$$\tilde{G}_r(\mathbf{k},\omega,v) \Big|_{v=\frac{\omega}{k}} = \tilde{G}_r(\mathbf{k},\omega,\frac{\omega}{k}) = 0 \quad , \tag{38}$$

where $k = |\mathbf{k}|$.

Equation (31), with the aid of (38), thus yields (for on the Ewald sphere shell)

$$\tilde{\theta}(\mathbf{k},\omega,\tfrac{\omega}{k}) = \tilde{\gamma}(\mathbf{k}) * \tilde{\phi}(\mathbf{k},\omega) \qquad . \qquad (39)$$

It is noteworthy that for far-fields ϕ in (16), for which the volume v is infinite and $\gamma(\mathbf{k}) * \delta(\mathbf{k})$, on the Ewald sphere shell in the spatial Fourier transform space, the known effectal field is equal to the unknown field.

(Earlier attempts at solving the inverse scattering inverse source problem have yielded somewhat similar results; i.e., a solution for the characteristic source distribution on the Ewald sphere shell in the Fourier transform space. The difficulty with such solutions, however, is that the characteristic source distribution depends on the arbitrarily chosen reference velocity, which precluded the determination of the characteristic source distribution off the Ewald sphere shell, whereas the field, as used in this solution, does not depend on this arbitrarily chosen reference velocity.) It is thus possible to vary the arbitrary reference velocity in such a fashion that the Ewald sphere shell sweeps to fill the entire Fourier transform space. This is accomplished simply as follows.

Taking the spatial inverse Fourier transform of (39), and recalling the definition (25), thus yields the desired solution for the field ϕ in the volume v; i.e.,

$$\frac{1}{(2\pi)^n} \int_{-\infty}^{\infty} e^{-i\mathbf{k}\cdot\mathbf{x}} \, \tilde{\theta}(\mathbf{k},\omega,\tfrac{\omega}{k}) \, d^n k = \begin{cases} \phi(\mathbf{x},\omega) & , \ \forall \ \mathbf{x} \in v \\ \\ 0 & , \ \forall \ \mathbf{x} \notin v \end{cases} \qquad (40)$$

and, by the Fourier transform differentiation rule, the solution for the Laplacean of the field $\nabla^2 \phi$ in the volume v; i.e.,

$$-\frac{1}{(2\pi)^n} \int_{-\infty}^{\infty} e^{-i\mathbf{k}\cdot\mathbf{x}} \, \tilde{\theta}(\mathbf{k},\omega,\tfrac{\omega}{k}) \, k^2 \, d^n k = \begin{cases} \nabla^2 \phi(\mathbf{x},\omega) & , \ \forall \mathbf{x} \in v \\ \\ 0 & , \ \forall \mathbf{x} \notin v \end{cases} \qquad (41)$$

This solution (40-41) thus has the (previously mentioned) desired properties converse to the properties of the direct Kirchhoff integration; i.e., for the case of the (unknown) sources being in the volume v, a solution is obtained for the fields if the field point is also in this volume v, and zero is obtained if the field point is not in this volume v; or, more generally, a solution for the fields is obtained if the sources and the field point are on the same side of the closed surface s, and zero is obtained if the sources and the field point are on different sides of the closed surface s.

The above solution (40-41) depends only on $\tilde{\theta}(\mathbf{k},\omega,\tfrac{\omega}{k})$, i.e., only on the values of the effectal field in the spatial Fourier transform space which are on the Ewald sphere shell $\quad k^2 = \frac{\omega^2}{v^2}\quad$. Examination of (16) which defines this effectal field   in real space shows it to consist of two terms: The first term depends only on the real part of the Green's function, and the second term depends only on the gradient of the real part of the Green's function.   Thus, in the spatial Fourier transform space, this effectal field consists also of two terms: The first depends only on the spatial Fourier transform of the real part of the Green's function, and the second term depends only on the spatial Fourier transform of the gradient of the real part of the Green's function.   However, on the Ewald sphere shell $k^2 = \frac{\omega^2}{v^2}$ in the spatial Fourier transform space this first term vanishes by (38). The second term, however, does not vanish on the Ewald sphere shell $\quad k^2 = \frac{\omega^2}{v^2}$ in the spatial Fourier transform space, since its support in k-space behaves like the support of the spatial Fourier transform of the imaginary part of the Green's function (see (32) et seq.). It thus follows that this first term contributes nothing to the solution, which depends only on this second term. For the purpose of the solution (40-41), (16) can thus be redefined as consisting only of this second term, i.e.,

$$\theta \equiv - \oint_S d\mathbf{s} \cdot \nabla G_r \, \phi \quad . \tag{42}$$

This is a particularly gratifying result since it obviates the need to evaluate (or measure) the gradient of the measured field, which in practice is difficult to accomplish accurately.

It should be noted that solution (40-41) cannot be implemented for the D.C. case (i.e., $\omega = 0$), since $\omega = 0$ precludes the required sweeping of the Ewald sphere shell $\quad k^2 = \frac{\omega^2}{v^2}\quad$ over all values of k by varying v, and locks the Ewald sphere shell to the null sphere shell k=0.

It should further be noted that because of the functional form of the reference potential V_h which goes to ∞ as v goes to zero (see (6) and (21) et seq.) and the functional form of the argument $\omega r/v$ of the Green's function G_r, in practice, $\tilde{\theta}(\mathbf{k} = \omega, v)$ cannot be evaluated numerically for $v = 0$, and hence not for $k = \infty$. It thus follows (from the spatial Fourier transform relationships involved) that infinite spatial resolution for the fields and their Laplacean cannot be achieved.

The entire solution (16-41) contains the spatial Fourier transforms of the field and the Green's function in such a fashion that the effects of finite sampling density (of ϕ in (16)) and incomplete observation domain (partial, open, not closed, surface of integration in (16)) results in solution (40-41) yielding a degraded resolution of the fields and their Laplacean. The degree of this degraded resolution is determined by the spatial Fourier transform uncertainty rules.

Once the field ϕ and its Laplacean $\nabla^2\phi$ has been determined by (40-41), the unknown potential $V(\mathbf{x},\omega)$ or the unknown medium propagation velocity $c(\mathbf{x},\omega)$ can be evaluated algebraically by the appropriate wave equation and constitutive equation (1-4) for the inverse scattering case; and similarly, the unknown source distribution $\rho(\mathbf{x},\omega)$ for the inverse source case.

4. References

1   Bojarski, N. N., "Inverse Scattering," Sect. II, October 1973, Third Quarterly Company Report, Naval Air Systems Command Contract N00019-73-C-0312.

2   Bleistein, N., and Bojarski, N. N., "Recently Developed Formulations of the Inverse Problem in Acoustics and Electromagnetics," Denver Research Institute Report MS-R-7501, NTIS #AD/A-003 588 (1974).

3   Bleistein, N., and Cohen, J. K., "Non-Uniqueness in the Inverse Source Problems in Acoustics and Electromagnetics," Journal of Mathematical Physics, Volume 18, No. 2, pp 194-201, February 1977.

WAVE EQUATIONS AND INVERSE SOLUTIONS FOR SOFT TISSUE

Steven A. Johnson†, Frank Stenger‡, Calvin
Wilcox‡, James Ball*, and Michael J. Berggren†

Departments of †Bioengineering, ‡Mathematics, and
*Physics
University of Utah
Salt Lake City, Utah 84112

INTRODUCTION

Ultrasound has been used in the echo mode as a diagnostic tool
for many years. The images produced by the echo mode are not qualitative and do not measure absolute tissue properties. Rather,
echo-mode images display the changes in acoustic impedance. Thus,
the boundaries of tissues are imaged. Such images are of value for
study of anatomy and tissue morphology. Tumors are detected by
their shape and not by their tissue type. Thus, classification of
malignant or benign tumors is difficult by use of echo-mode imaging
alone. The recently developed straight-line ultrasound transmission tomographic methods provide images which are qualitative and
absolute (not relative), but have inferior resolution compared to
the echo methods [Greenleaf, et al., 1978].

Straight-line time of flight and amplitude ultrasound tomography has a resolution of only about 10 wavelengths. The value of
this method lies in its ability to discriminate cancer from normal
tissue by the greater acoustic velocity of cancerous tissue [Green-
leaf and Bahn, 1981].

An important improvement in diagnostic capability would acrue
if an imaging method could be found which is both qualitative and
has high spatial resolution. A method for achieving this end has
been proposed based upon inversion of the Helmholtz wave equation
[Mueller, et al., 1979]. The Helmholtz wave equation may be written

$$\nabla^2 p(\vec{r}) + k^2 p(\vec{r}) = 0 \tag{1}$$

$$k = \frac{\omega}{c(\vec{r})} \left(1 + i\,\alpha_{loss}\right) \tag{2}$$

Here p = acoustic pressure, $c(\vec{r})$ = speed of sound, k = complex wave number, and α_{loss} = attenuation. This equation is usually inverted by applying a perturbation expansion to p and k^2, [Mueller, et al., 1979]. The inversion provides a separate image of $c(\vec{r})$ and $\alpha(\vec{r})$.

This equation has been inverted by Kaveh, et al. for small objects with low attenuation using the Rytov approximation [Kaveh, et al., 1980]. For large objects with large attenuation, the perturbation approaches do not produce good results.

IMPROVED WAVE EQUATIONS FOR SOFT TISSUE

A basic limitation with equation (1) is its failure to describe reflection of ultrasound. It is well known that reflection of ultrasound is determined by the difference in acoustic impedance at a boundary. For oblique incidence on a plane boundary between media A and media B, the reflection coefficient is given by

$$R = \frac{Z_A \cos\theta_A - Z_B \cos\theta_B}{Z_A \cos\theta_A + Z_B \cos\theta_B} \;,\quad Z_i = c_i \rho_i \tag{3}$$

where R = reflection coefficient, Z_A and Z_B are the acoustic impedances of media A and media B, respectively, c_i = speed of sound in media i, and ρ_i = density in media i. Also θ_A and θ_B are angles of incidence and transmission from the normal to the surface for matrials A and B, respectively.

The correct reflection coefficient given by equation (3) is predicted by a wave equation which includes a term in density. We have recently proven this result, which is included in Appendix 1, using the following wave equation:

$$\nabla^2 p + k^2 p - \nabla \ln\rho \cdot \nabla p = 0 \tag{4}$$

Here the first two terms are identical to those in equation (1) and k is given by equation (2). A derivation of equation (4) is found in Appendix 1. Note the third term contains density information. By the substitution,

$$p = \sqrt{\rho}\, p_1 \tag{5}$$

note that equation (4) may be written

$$\nabla^2 p_1 + \left[k^2 - \sqrt{\rho}\, \nabla^2\, \sqrt{1/\rho}\right] p_1 = 0 \tag{6a}$$

This equation may also be written in the form

$$\nabla^2 p_1 + \left[k^2 + \frac{1}{2\rho} \nabla^2 \rho + \frac{3}{4\rho^2} \nabla\rho \cdot \nabla\rho\right] p_1 = 0 \tag{6b}$$

In these forms, the scattering from ρ is seen to occur on surfaces normal to which ρ or gradient ρ change abruptly. The substitution,

$$c^2 = 1/(\rho\kappa) \tag{7}$$

where κ is compressibility, transforms equation (4) into the form

$$\nabla^2 p + k_o^2 p = -k_o^2 \gamma_\kappa p - \nabla \cdot \left[\gamma_\rho \nabla p\right] \tag{8}$$

where

$$k_o^2 = \omega^2 \rho_o \kappa_o, \quad \gamma_\kappa = \frac{\kappa - \kappa_o}{\kappa_o}, \quad \text{and} \quad \gamma_\rho = \frac{\rho - \rho_o}{\rho} \tag{9}$$

and where κ_o and ρ_o are constant κ and ρ, respectively, outside the body in a water bath. A general form of this derivation is found in Appendix 2. In this form, it can be shown that monopole scattering arises from γ_κ and dipole scattering from γ_ρ. Equation (8) is identical to equation (8.1.12) from Morse and Ingard [Morse and Ingard, 1968]. Equation (8) has the advantage that the left-hand side is homogeneous with constant coefficients, while tissue parameters are introduced in the right-hand side.

Equation (8) may also be used to describe attenuation of acoustic pressure if γ_κ is replaced by γ_κ' where

$$\gamma_\kappa' = \gamma_\kappa - \frac{i\sigma\rho_o c_o^2}{\omega} \tag{10}$$

Here γ_κ is defined by equation (9), ρ_o is the background density, c_o is the background speed of sound, and σ is a factor which is proportional to absorption of ultrasound. The derivation of this result is found in Appendix 2.

Finally we present an equation for propagation of plane waves in solids, which describes the effect of shear wave losses, which has been derived by Mueller [Mueller, 1980].

$$\left(\nabla^2 + 2ik \cdot \nabla\right)\psi = k^2(\alpha + \beta + \gamma) + \nabla^2\left(2\alpha - \frac{\gamma}{2}\right) + \frac{1}{2k^2}\nabla^4\alpha \qquad (11)$$

where

$$\psi = \left\{ \begin{matrix} d_1/d_o \\ u_s \end{matrix} \right\} + 2\alpha + \beta - \frac{\gamma}{2} + \left(\nabla^2 - 2i\vec{k} \cdot \nabla\right)\frac{\alpha}{2} \qquad (12)$$

Here

$$\alpha = \theta(\mu - \mu_o)/(\lambda_o + 2\mu_o)$$

$$\beta = \theta(\lambda - \lambda_o)/(\lambda_o + 2\mu_o)$$

$$\gamma = \theta(\rho - \rho_o)/\rho_o \qquad (13)$$

are normalized elastic parameters.

The qualities μ, λ are the Lame coefficients defined by

$$\sigma_{ik} = 2\mu \, \epsilon_{ik} + \lambda \, \epsilon_{ii}\delta_{ik} \qquad (14)$$

where σ_{ik} is the stress tensor and ϵ_{ik} is the strain tensor for an elastic medium. The qualities μ_o and λ_o are the values of μ and λ in a homogeneous fluid which contains the object to be imaged. In equation (13), ρ is density and ρ_o is the density in the homogeneous fluid which contains the object to be imaged.

In equation (13), θ is a function which is 1 inside the object to be imaged and is zero in the fluid in which the object is immersed.

In equation (12), (d_1/d_o) is the ratio of the dilatation with the object present to the dilatation with only the fluid present and is used for the Born approximation. The dilatation is proportional to the signal on the transducer receiver. In equation (12), u_s is the quantity used in the Rytov approximation and is given by

$$u_s = \ln \left(\frac{d}{d_o} \right) \tag{15}$$

In equations (11) and (12), the elastic constants α, β, the density γ, and the constant k are to be reconstructed. If k is taken to be complex, as in equation (2), then attenuating, density, and the Lame numbers μ and λ are parameters to be reconstructed. Alternately, attenuation, speed of sound, density, and one of the Lame numbers could be reconstructed. Such reconstructions may require use of multiple frequencies.

Also note that equation (11) assumes an incident plane wave, while equations (1) and (4) are for arbitrary waves.

A wave equation which is isomorphic to equation (11) has been derived by C. H. Wilcox using a viscous, compressible, heat conducting fluid as a model [Wilcox, 1981].

DISCUSSION

We have presented wave equations from three levels of sophistication for describing the propagation of acousic energy in tissues. In the first class as represented by equation (1), only speed of sound and attenuation are modeled.

In the second class, speed of sound and density are modeled as represented by equations (4), (6a), (6b), or (8). For complex k, equations (4), (6a), or (6b) also can describe attenuation through the definition of losses contained in equation (2). We show in Appendix 2 that equation (8) with complex γ_κ can also describe attenuation.

In the third class of wave equations as represented by equations (10), (11), and (12), speed of sound, density, and attenuation are represented by a more sophisticated model equation. Accordingly, attenuation is derived from several mechanisms including viscosity, relaxation, or shear wave losses. The inclusion of shear wave propagation results in a vector wave equation. However, in tissues, shear waves are highly attenuated and thus one wave equation containing the longitudinal component of velocity, or dilatation, both scalars, is sufficient.

In any application, the question arises which wave equation to use. The first class is a special case of the second class. The second class is derivable from a subset of the assumptions used to derive the third class. Thus the third class is descriptive for all applications, but their complexity may preclude application in many cases. We believe that the second class of equations is adequate to model propagation of acoustic energy through tissues and may be used as the basis for producing images of tissue parameters. One general approach which we propose is the solving of the inverse scattering problem. In this approach, either a differential equation or an integral equation derived from the differential equation is solved for the distribution of tissue parameters. Equation (8) may be transformed into an integral equation which contains the necessary conditions on the boundary. Such an equation is isomorphic to equation 8.1.13 on page 410 of Morse and Ingard, and is given below [Morse and Ingard, 1968].

$$p_\omega(\vec{r}) = p_i(\vec{r}) + \iiint \left(k_o^2 \gamma_\kappa' p_\omega g_\omega + \gamma_\rho \nabla_o p_\omega \cdot \nabla_o g_\omega \right) dv_o$$

where

$$g_\omega(\vec{r}|\vec{r}_o) = \frac{1}{4\pi |\vec{r} - \vec{r}_o|} \exp\left(ik_o |\vec{r} - \vec{r}_o| \right) \tag{16}$$

Here $k_o = \omega_o / c_o$, γ_κ' is defined by equation (10), and γ_ρ is defined by equation (9). p_ω is the total scattered field and $p_i(\vec{r})$ is the incident field.

In summary we predict that equation (16), or its equivalent of equation (8) plus boundary conditions, will be an accurate model of ultrasound interaction in soft tissues. This prediction is strengthened by the correct treatment of reflection which equation (8) gives, as is described in Appendix 1. We therefore plan as part of our strategy to develop imaging algorithms based upon the inversion of equations (8) or (16).

APPENDIX 1

REFLECTION OF SOUND AT A PLANE INTERFACE

Field Equations

    Inhomogeneous lossless fluids are considered with:

$c(\vec{x}\,)$ = sound speed at point $\vec{x}$

$\rho(\vec{x})$ = equilibrium density at $\vec{x}$

The basic field equations are

$$\frac{\partial \vec{v}}{\partial t} + \frac{1}{\rho}\, \nabla p = \vec{0} \ , \quad \text{(conservation of momentum)}$$

$$\frac{\partial p}{\partial t} + c^2 \rho \nabla \cdot \vec{v} = 0 \ , \quad \text{(conservation of mass)}$$

(1.1)

where

$\vec{v} = \vec{v}(t,\ \vec{x})$ = acoustic velocity field

$p = p(t,\ \vec{x})$ = acoustic pressure field

Eliminating $\vec{v}$ between these equations gives the wave equation

$$\frac{\partial^2 p}{\partial t^2} - c^2 \rho \nabla \cdot \left(\frac{1}{\rho}\, \nabla p \right) = 0 \tag{1.2}$$

Steady-state sound fields (time-dependence $e^{-i\omega t}$) satisfy

$$i\omega \vec{v} = \frac{1}{\rho}\, \nabla p \tag{1.3a}$$

and

$$i\omega p = c^2 \rho \nabla \cdot \vec{v} \tag{1.3b}$$

and thus on elimination of $\vec{v}$

$$c^2 \rho \nabla \cdot \left(\frac{1}{\rho} \nabla p \right) + \omega^2 p = 0 \qquad (1.4)$$

which is equivalent to equation (4).

Jump Conditions at an Interface

If $c(x)$ and $\rho(x)$ are piece-wise continuous with discontinuities on a fixed surface S, then the stability of the interface requires

p continuous across S   (1.5a)

$\vec{n} \cdot \vec{v}$ continuous across S   (1.5b)

where $\vec{n}$ is a unit normal vector on S. Equations (1.3) imply that (1.5b) can be replaced by

$$\frac{1}{\rho} \vec{n} \cdot \nabla p = \frac{1}{\rho} \frac{\partial p}{\partial n} \text{ continuous across S} \qquad (1.5c)$$

Reflection of Plane Waves at a Plane Interface

Assume

$$c(x) = \begin{cases} c_1, & x_1 < 0 \\ c_2, & x_1 > 0 \end{cases}$$

$$\rho(x) = \begin{cases} \rho_1, & x_1 < 0 \\ \rho_2, & x_1 > 0 \end{cases}$$

Consider a plane wave incident on the interface at $x_1 = 0$:

$$P_{inc}(x) = e^{i\vec{k}\cdot\vec{x}} \quad , \quad k = (k_1, k_2, k_3), \quad k_1 > 0$$

A total acoustic field of the following form is assumed:

$$p(x) = \begin{cases} P_{inc}(x) + P_{reft}(x) \quad , \quad x_1 < 0 \\[2em] P_{trans}(x) \qquad\quad , \quad x_1 > 0 \end{cases}$$

where

$$P_{reft}(x) = R\, e^{ik^{(r)}\cdot\vec{x}} \quad , \quad k^{(r)} = \left(k_1^{(r)}, k_2^{(r)}, k_3^{(r)}\right)$$

$$P_{trans}(x) = T\, e^{ik^{(t)}\cdot\vec{x}} \quad , \quad k^{(t)} = \left(k_1^{(t)}, k_2^{(t)}, k_3^{(t)}\right)$$

Condition (1.5a) implies that (since $x_1 = 0$ at interface)

$$e^{i\left(k_2 x_2 + k_3 x_3\right)} + R\, e^{i\left(k_2^{(r)} x_2 + k_3^{(r)} x_3\right)} = T\, e^{i\left(k_2^{(t)} x_2 + k_3^{(t)} x_3\right)}$$

for all x_2, x_3. It follows that

$$k_2 = k_2^{(r)} = k_2^{(t)}, \qquad k_3 = k_3^{(r)} = k_3^{(t)} \tag{1.6}$$

The validity of the wave equation (1.4) in $x_1 < 0$ and $x_1 > 0$ implies that

$$|k|^2 = |k^{(r)}|^2 = \frac{\omega^2}{c_1^{\,2}}, \qquad |k^{(t)}|^2 = \frac{\omega^2}{c_2^{\,2}} \tag{1.7}$$

whence

$$k_1^2 = k_1^{(r)2} \text{ and therefore } k_1^{(r)} = -k_1 \tag{1.8}$$

This is the usual law of reflection.

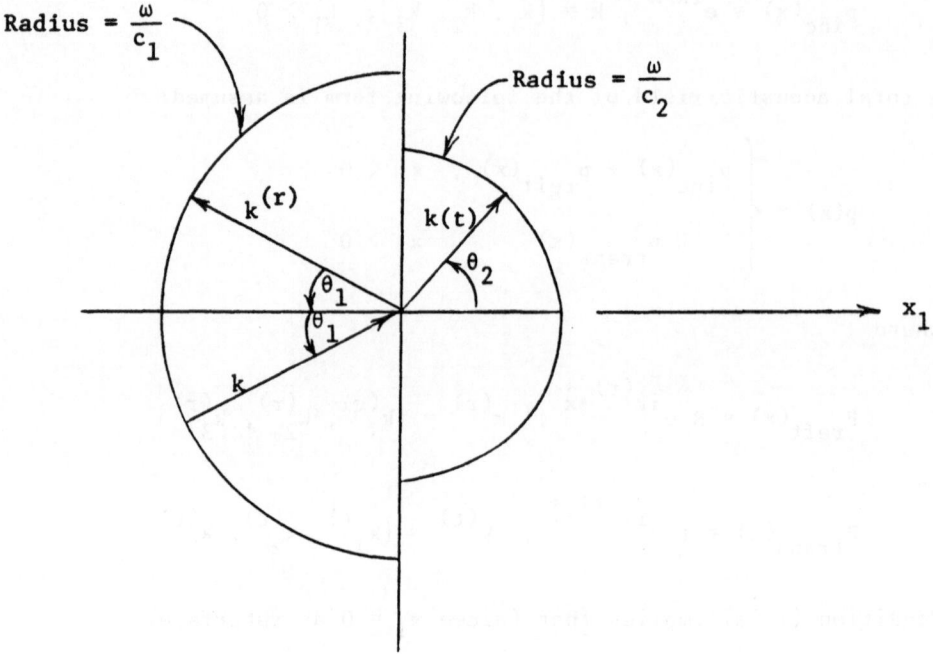

Conditions (1.6) imply

$$k_2^2 + k_3^2 = k_2^{(t)2} + k_3^{(t)2} \text{ or}$$

$$\frac{\sin \theta_1}{c_1} = \frac{\sin \theta_2}{c_2} \tag{1.9}$$

which is Snell's law.   The case where

$$\frac{c_2}{c_1} \sin \theta_1 > 1 \tag{1.10}$$

corresponds to <u>total reflection</u>.   In this case,

$$k_1^{(t)} = i\mu \text{ (pure imaginary)},$$

μ = coefficient of falloff of reactive energy.

Calculation of R and T

Equations (1.5a), (1.5c), and (1.6) give

$$P_{inc}(0, x_2, x_3) + P_{reft}(0, x_2, x_3) = P_{trans}(0, x_2, x_3)$$

$$\frac{1}{\rho_1} \frac{\partial P_{inc}}{\partial x_1}(0, x_2, x_3) + \frac{1}{\rho_1} \frac{\partial P_{reft}}{\partial x_1}(0, x_2, x_3) = \frac{1}{\rho_2} \frac{\partial P_{trans}}{\partial x_1}(0, x_2, x_3)$$

or

$$1 + R = T$$

$$\frac{k_1}{\rho_1} - \frac{k_1}{\rho_1} R = \frac{k_1^{(t)}}{\rho_2} T$$

where

$$k_1 = \frac{\omega}{c_1} \cos \theta_1$$

$$k_1^{(t)} = \begin{cases} \dfrac{\omega}{c_2} \cos \theta_2 & \text{if} \quad \dfrac{c_2}{c_1} \sin \theta_1 \leqslant 1 \\[4ex] -i\mu & \text{if} \quad \dfrac{c_2}{c_1} \sin \theta_1 > 1 \end{cases}$$

Solving for R and T gives

$$R = \frac{\dfrac{k_1}{\rho_1} - \dfrac{k_1^{(t)}}{\rho_2}}{\dfrac{k_1}{\rho_1} + \dfrac{k_1^{(t)}}{\rho_2}} \quad , \quad T = \frac{2 \dfrac{k_1}{\rho_1}}{\dfrac{k_1}{\rho_1} + \dfrac{k_1^{(t)}}{\rho_2}}$$

Case 1

$$\frac{c_2}{c_1} \sin \theta_1 < 1$$

$$R = \frac{\dfrac{\cos \theta_1}{\rho_1 c_1} - \dfrac{\cos \theta_2}{\rho_2 c_2}}{\dfrac{\cos \theta_1}{\rho_1 c_1} + \dfrac{\cos \theta_2}{\rho_2 c_2}} = \frac{\rho_2 c_2 \cos \theta_1 - \rho_1 c_1 \cos \theta_2}{\rho_2 c_2 \cos \theta_1 + \rho_1 c_1 \cos \theta_2} \quad , \quad (\text{real})$$

$$T = \frac{2 \dfrac{\cos \theta_1}{\rho_1 c_1}}{\dfrac{\cos \theta_1}{\rho_1 c_1} + \dfrac{\cos \theta_2}{\rho_2 c_2}} = \frac{2 \rho_2 c_2 \cos \theta_1}{\rho_2 c_2 \cos \theta_1 + \rho_1 c_1 \cos \theta_2} \quad , \quad (\text{real})$$

Squaring gives

$$\frac{\cos \theta_1}{\rho_1 c_1} R^2 + \frac{\cos \theta_2}{\rho_2 c_2} T^2 = \frac{\cos \theta_1}{\rho_1 c_1}$$

Case 2

$$\frac{c_2}{c_1} \sin \theta_1 > 1, \text{ for total internal reflection}$$

$$R = \frac{\dfrac{k_1}{\rho_1} - \dfrac{i\mu}{\rho_2}}{\dfrac{k_1}{\rho_1} + \dfrac{i\mu}{\rho_2}} \quad , \quad \text{(complex)}$$

In this case,

$$|R|^2 = 1$$

and all the incident energy is reflected. The transmitted wave is

$$P_{trans}(x) = T\, e^{-\mu x_1}\, e^{i\left(k_2 x_2 + k_3 x_3\right)}$$

APPENDIX 2

DERIVATION OF MEANING OF COMPLEX FRACTIONAL COMPRESSIBILITY

As a differential equation which models ultrasound propagation through tissues, we propose

$$\nabla \cdot \left(\frac{1}{\rho} \nabla p \right) - \frac{1}{c^2} \frac{\partial^2 p}{\partial t^2} + \sigma \frac{\partial p}{\partial t} = 0 \tag{2.1}$$

The time dependence is removed by the substituion

$$p = p_\omega(\vec{r}) e^{-i\omega t} \tag{2.2}$$

Then equation (2.1) becomes

$$\nabla^2 p_\omega - \nabla \ln \rho \cdot \nabla p_\omega + \left(\frac{\omega^2}{c^2} + i\omega\rho\sigma \right) p_\omega = 0$$

Set

$$k^2 = k_R^2 + i\omega\sigma\rho, \quad k_R^2 = \omega^2/c^2 \tag{2.3}$$

Then for a plane wave,

$$p = p_\omega \, e^{-i\omega t} \, e^{ik\hat{k}\cdot\vec{r}} \tag{2.4}$$

where $\hat{k}$ is a unit vector normal to the plane waves in the direction of propagation and where k is complex, and is given by

$$k = k_R \left(1 - \frac{\omega^2 \sigma^2 \rho^2}{k_R^4} \right)^{1/4} \exp \left[\frac{i}{2} \arctan \left(\frac{\omega\sigma\rho}{k_R^2} \right) \right] \tag{2.5}$$

Let ρ_o and κ be the constant density and compressibility of water, respectively, outside of the body. Then multiply equation (2.1) by ρ and add $\frac{\rho}{\rho_o} \nabla^2 p - \nabla^2 p = 0$. Let $\gamma_\rho = (\rho - \rho_o)/\rho$. Then equation (2.1) becomes

$$-\nabla \cdot \left(\gamma_\rho \nabla p\right) - \kappa\rho_o \frac{\partial^2 p}{\partial t^2} + \sigma\rho \frac{\partial p}{\partial t} = -\nabla^2 p \qquad (2.6)$$

Noting that

$$\kappa\rho_o = \frac{\kappa}{c_o^2 \kappa_o} \;,\; \text{adding } \frac{1}{c_o^2} \frac{\kappa_o}{\kappa_o} \frac{\partial^2 p}{\partial t^2} - \frac{1}{c_o^2} \frac{\partial^2 p}{\partial t^2} = 0$$

and defining $\gamma_\kappa = \left(\kappa - \kappa_o\right)/\kappa_o$, equation (2.6) becomes

$$-\nabla \cdot \left(\gamma_\rho \nabla p\right) - \frac{1}{c_o^2} \gamma_\kappa \frac{\partial^2 p}{\partial t^2} + \sigma\rho \frac{\partial^2 p}{\partial t^2} = -\nabla^2 p + \frac{1}{c_o^2} \frac{\partial^2 p}{\partial t^2} \qquad (2.7)$$

on substituion of equation (2.2) into equation (2.7), we obtain

$$\nabla^2 p_\omega + \frac{\omega^2}{c_o^2} p_\omega = -k_o^2 \gamma_\kappa' p_\omega + \nabla \cdot \left[\gamma_\rho \nabla p_\omega\right] \qquad (2.8)$$

where

$$k_o^2 = \omega^2/c_o^2 \text{ and } \gamma_\kappa' = \gamma_\kappa - i\sigma\rho_o c_o^2/\omega \qquad (2.9)$$

Thus γ_κ', the complex fractional compressibility, is seen to be composed of a real part which is the real fractional compressibility and an imaginary part which contains an asorption parameter σ to the first power.

ACKNOWLEDGMENTS

     We appreciate the helpful discussions with Professor Douglas
A. Christensen.   We also appreciate helpful discussions with R. K.
Mueller of the University of Minnesota and J. F. Greenleaf of Mayo
Clinic.   We appreciate the secretarial help of Marian Swenson and
Ruth Eichers.   This work was supported in part by NCI grant 5 R01
CA23430 and American Cancer Society grant PDT-110A, and by U.S.
Army contract DAAG29-80-K-0089.

REFERENCES

Greenleaf, J. F., and R. C. Bahn, "Clinical Imaging with Transmis-
sive Ultrasonic Computerized Tomography", IEEE Transactions on
Biomedical Engineering, Vol. BME-28, No. 2, February 1981.

Greenleaf, J. F., S. A. Johnson, and A. H. Lent, "Measurement of
Spatial Distribution of Refractive Index in Tissues by Ultrasonic
Computer Assisted Tomography", Ultrasound in Medicine and Biology,
Vol. 3, pp. 327-339, 1978.

Kaveh, M., R. K. Mueller, R. Rylander, T. R. Coulter, and M. Soumekh, "Experimental Results in Ultrasonic Diffraction Tomography",
Acoustical Imaging, Vol. 9 (in press).

Morse, P. M., and K. U. Ingard, Theoretical Acoustics, McGraw Hill
Book Company, New York, 1968.

Mueller, R. K., M. Kaveh, and G. Wade, "Reconstructive Tomography
and Applications to Ultrasonics", Proceedings of the IEEE, Vol. 67,
No. 4, April pp. 567-587, April 1979.

Mueller, R. K., "Diffraction Tomography I: The Wave Equation",
Ultrasonic Imgaing, Vol. 2, No. 3, July 1980, pp. 213-222.

Wilcox, C. W., The wave equation for a viscous, compressible, heat
conducting fluid was derived in April 1981 and is the subject of a
paper planned for future publication.

ASYMPTOTIC ULTRASONIC INVERSION BASED ON USING MORE THAN ONE FREQUENCY*

Frank Stenger

Department of Mathematics
University of Utah
Salt Lake City, UT 84112

ABSTRACT

Let the spacial sound pressure u in a body B be governed by the equation $\nabla^2 u + \{\omega/c(\overline{r})\}^2 u = 0$ where $c(\overline{r})$ denotes the speed of sound at the point $\overline{r}$ in $\mathbb{R}^3$. Given either plane wave excitation $\exp\{i\overline{k}\cdot\overline{r} - i\omega t\}$ or spherical wave excitation $\exp\{ik|\overline{r}-\overline{r}_s|\}/\{4\pi|\overline{r}-\overline{r}_s|\}$, if $k = \omega/c_0$ is large, where c_0 denotes the speed of sound in the medium surrounding B , Rytov's approximation yields a simple expression for u in terms of f where $f(\overline{r}) = c_0^2/c^2(\overline{r}) - 1$. While the reconstruction of f based on varying the direction of $\overline{k}$ (or varying the source point $\overline{r}_s$ for the case of a spherical wave source) and reading u or surfaces that enclose B is a three dimensional problem, this three dimensional problem is reduced to a one-dimensional problem

*Research supported by U. S. Army Research Contract No. DAAG 29-80-K-0089.

425

(i.e., "x-ray inversion") using more than one frequency. That is,
the diffraction effects may be eliminated using more than one
frequency.

1.   INTRODUCTION AND SUMMARY

Let the sound pressure $p = p(\bar{r}, t)$ in $\mathbb{R}^3 \times (0, \infty)$ be
governed by the equation

$$(1.1) \qquad [\nabla^2 - \frac{1}{c^2(\bar{r})} \frac{\partial^2}{\partial t^2}]p = 0$$

where ∇ denotes the gradient operator, t denotes time, and
$c(\bar{r})$ is the speed of sound at the point $\bar{r}$ in $\mathbb{R}^3$.

For the case of sonic plane wave excitation

$$(1.2) \qquad p_0 = u_0(\bar{r})e^{-i\omega t} \; ; \; u_0(\bar{r}) = e^{i\bar{k}\cdot\bar{r}}$$

or sonic spherical wave excitation emanating at $\bar{r} = \bar{r}_s$,

$$(1.3) \qquad p_0 = u_0(\bar{r})e^{-i\omega t} \; ; \; u_0(\bar{r}) = \frac{e^{ik|\bar{r}-\bar{r}_s|}}{4\pi|\bar{r}-\bar{r}_s|}$$

we may take

$$(1.4) \qquad p = p(\bar{r}, t) = u(\bar{r})e^{-i\omega t}$$

in (1.1) to get the equation

$$(1.5) \qquad \nabla^2 u + \frac{\omega^2}{c^2(\bar{r})} u = 0$$

for the spacial pressure u .

Upon setting

$$(1.6) \qquad f(\overline{r}) = \frac{c_0^2}{c^2(\overline{r})} - 1 \quad , \quad k = \frac{\omega}{c_0} = |\overline{k}| \quad ,$$

where $c(\overline{r})$ denotes the speed of sound in the *interior* of the body B of interest, and c_0 the space of sound in the medium surrounding the body, we may rewrite (1.5) in the form

$$(1.7) \qquad \nabla^2 u + k^2 u = -k^2 f u \quad .$$

Now setting

$$(1.8) \qquad u = u_0 e^W$$

in (1.7) and using Rytov's approximation (see e.g. [1]), we get

$$(1.9) \qquad W(\overline{r},\overline{k},k) = -k^2 \iiint_{\mathbb{R}^3} f(\overline{R}) \frac{\exp[ik|\overline{r}-\overline{R}| + i\overline{k}\cdot(\overline{r}-\overline{R})]}{4\pi|\overline{r}-\overline{R}|} d\overline{R}$$

for the case of (1.2), and

$$(1.10) \qquad W(\overline{r},\overline{r}_s,k) = -k^2 \iiint_{\mathbb{R}^3} f(\overline{R}) \frac{\exp[ik|\overline{r}-\overline{R}| + ik|\overline{r}_s-\overline{R}|]}{(4\pi)^2|\overline{r}-\overline{R}||\overline{r}_s-\overline{R}|} d\overline{R}$$

for the case of (1.3).

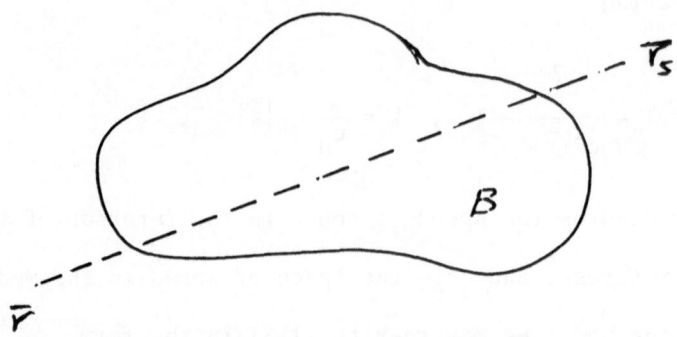

Figure 1.1   Sound Propagation from $\bar{r}_s$ to $\bar{r}$.

It is convenient to use a similar notation for both (1.9) and (1.10). To this end, we assume $\bar{r}$ and $\bar{r}_s$ are distinct (see Fig. 1.1), and take

(1.11) $\bar{k} = k \dfrac{\bar{r}-\bar{r}_s}{|\bar{r}-\bar{r}_s|}$.

In this way, we may also denote the left hand side of (1.9) by $W(\bar{r},\bar{r}_s,k)$.

In Sec. 2 we prove that for sufficiently smooth f

$$W(\bar{r},\bar{r}_s,k) = - \frac{ik|\bar{r}-\bar{r}_s|}{2} \left\{ \int_0^1 f((1-t)\bar{r} + t\bar{r}_s)dt \right.$$

(1.12)

$$\left. + \sum_{m=1}^{M} A_m(\bar{r},\bar{r}_s;f)k^{-m} + 0(k^{-M-1}) \right\}$$

as $k \to \infty$, for the case of (1.9), and

$$W(\bar{r},\bar{r}_s,k) = -\frac{ik}{16\pi} e^{ik|\bar{r}-\bar{r}_s|} \left\{ \int_0^1 f((1-t)\bar{r} + t\bar{r}_s)dt \right.$$

(1.13)

$$\left. + \sum_{m=1}^{M} B_m(\bar{r},\bar{r}_s;f)k^{-m} + 0(k^{-M-1}) \right\}$$

as $k \to \infty$ for the case of (1.10). In (1.12) the A_m and in

(1.13) the B_m are operators which are linear in f, bounded,

and independent of k, and M is a positive integer. The

operators A_m and B_m contain the diffraction terms; they are

generally quite difficult to compute [5] and for this reason, we

propose the use of the following device to circumvent computing

them.

Let us define $Y_M(\bar{r},\bar{r}_s,k)$ by

(1.14) $$Y_M(\bar{r},\bar{r}_s,k) = \frac{2ik^{M-1}}{|\bar{r}-\bar{r}_s|} W(\bar{r},\bar{r}_s,k)$$

for the case of (1.2) and

(1.15) $$Y_M(\bar{r},\bar{r}_s,k) = 16\pi ik^{M-1} e^{-ik|\bar{r}-\bar{r}_s|} W(\bar{r},\bar{r}_s,k)$$

for the case of (1.3). If we apply the frequencies $k_0(=k)$,

$k_1,\ldots,k_M$, where the k_i are distinct, but close to k (e.g.

$.8k \leq k_i \leq k$) and set

(1.16) $$V_M(\bar{r},r_s,k) = \sum_{j=0}^{M} \frac{Y_M(\bar{r},\bar{r}_s,k_j)}{\phi'(k_j)} \;;\; \phi(x) = \prod_{i=0}^{M} (x-k_i) \;,$$

i.e., $V_M(r,r_s,k)$ is the Mth divided difference of the $Y_M(r,r_s,k_j)$ with respect to the k_j , then

$$(1.17) \qquad V_M(\bar{r},\bar{r}_s,k) = \int_0^1 f((1-t)\bar{r} + t\bar{r}_s)dt + O(k^{-M-1})$$

as $k \to \infty$.

For example, Eqs. (1.16) and (1.17) yield

$$\int_0^1 f((1-t)\bar{r} + t\bar{r}_s)dt$$

$$(1.18)$$

$$= \frac{Y_1(\bar{r},\bar{r}_s,k_0) - Y_1(r,r_s,k_1)}{k_0 - k_1} + O(k^{-2})$$

$$= \frac{Y_2(\bar{r},\bar{r}_s,k_0)}{(k_0-k_1)(k_0-k_2)} + \frac{Y_2(\bar{r},\bar{r}_s,k_1)}{(k_1-k_0)(k_1-k_2)} + \frac{Y(\bar{r},\bar{r}_s,k_2)}{(k_2-k_0)(k_2-k_1)}$$

$$(1.19)$$

$$+ O(k^{-3}) \quad .$$

By dropping the $O(k^{-2})$ or $O(k^{-3})$ terms on the right hand sides of these equations, it is possible to ignore diffraction and to use standard "x-ray" algorithms to solve for f[2] .

In applications f is normally not as smooth as we would like and we cannot take M arbitrarily large in (1.12), (1.13) and (1.17). This point is discussed in greater detail at the end of Sec. 2.

2. ASYMPTOTIC EXPANSION OF W

Let W be defined as in (1.9) or as in (1.10). In view of
the definition of f in (1.6), it follows that $f \equiv 0$ outside
of the body B ; we shall assume in addition, that $f \in C^{\infty}(\mathbb{R}^3)$.
This assumption is not valid for real life situations, and we
shall discuss this point later in this section. In the meantime
it suffices to state that approximation to the solutions of (1.9)
or (1.10) are simpler to obtain under this assumption, and the
approximation which one obtains are usually as accurate as those
which one obtains under the assumption of less smoothness on f .

Let us assume the situation in Fig. 1.1. We orient $\mathbb{R}^3$ so
that the x-axis coincides with the line through the points $\overline{r}$
and $\overline{r}_s$, its positive direction being in the direction from $\overline{r}$
to $\overline{r}_s$. Thus we may take $\overline{r} = (-x,0,0)$, $\overline{r}_s = (x,0,0)$, where
$x = |\overline{r} - \overline{r}_s|/2 > 0$. If $-x < \xi < x$, then the point $P_0 = (\xi,0,0)$
is on the line segment joining $\overline{r}_s$ and $\overline{r}$; in a neighborhood of
P_0 ,

$$f(\xi,\eta,\zeta) = \sum_{0 \leq m, 0 \leq n, m + n \leq 2M} \frac{f^{(0,m,n)}(\xi,0,0)}{j!\ell!} \eta^m \zeta^n$$

(2.1)

$$+ R_M(\xi,\eta,\zeta) \quad .$$

In what follows below, we shall be considering non-converging
integrals of the monomials in (2.1), of the form

$$I_{m,n} = \iint_{\mathbb{R}^2} \eta^m \zeta^n F(\eta,\zeta)\, d\eta d\zeta \quad .$$

While these integrals do not converge, in view of the fact that f in (2.1) vanishes in the exterior of B , and $f \in C^\infty(\mathbb{R}^3)$, we may define these integrals $I_{m,n}$ by

$$I_{m,n} = \lim_{\varepsilon \to 0^+} \iint_{\mathbb{R}^2} e^{-\varepsilon\{\eta^2+\zeta^2\}^{1/2}} \eta^m \zeta^n F(\eta,\zeta)\, d\eta d\zeta \quad .$$

For the case of Eq. (1.9), $\bar{k} = k(1,0,0)$, $\bar{R} = (\xi,\eta,\zeta)$, and so $\bar{k}\cdot(\bar{r}-\bar{R}) = k(x-\xi)$. We thus prove

LEMMA 2.1: Under the above assumptions, set

$$(2.2) \qquad I_{m,n} = \frac{-k^2}{4\pi} \iint_{\mathbb{R}^2} \eta^m \zeta^n \frac{\exp[ik|\bar{R}-\bar{r}| - ik(x-\xi)]}{|\bar{R}-\bar{r}|}\, d\eta d\zeta \quad .$$

Then

$$(2.3) \qquad \begin{cases} I_{m,n} = 0 \quad \text{if one of } m \text{ or } n \text{ is odd} \\[2mm] I_{2m,2n} \sim -\dfrac{ik}{2}\, \dfrac{(1/2)_m (1/2)_n}{(m+n)!} \quad . \\[4mm] \left|\dfrac{2i(x-\xi)}{k}\right|^{m+n} \displaystyle\sum_{\ell=0}^{m+n} \binom{m+n}{\ell} \left|\dfrac{i}{2(x-\xi)}\right|^\ell \dfrac{(m+n+\ell)!}{k^\ell} \end{cases}$$

as $k \to \infty$.

PROOF: Using polar coordinates, $\eta = \rho\cos\theta$, $\zeta = \rho\sin\theta$ we have

$$I_{m,n} = -k^2 \int_0^\infty \int_0^{2\pi} \frac{e^{ik\{(x-\xi)^2 + \rho^2\}^{1/2} - ik(x-\xi)}}{4\pi \{(x-\xi)^2 + \rho^2\}^{1/2}}$$

(2.4)

$$\rho^{m+n+1} \cos^m\theta \sin^n\theta \, d\theta d\rho \quad .$$

Hence, clearly $I_{m,n} = 0$ if one of m or n is odd. For the case when m and n are both even, we consider

(2.5) $$I_{2m,2n} = -\frac{k^2}{2} \frac{(1/2)_m (1/2)_n}{(m+n)!} J_{m+n}$$

where

$$J_m = \int_0^\infty \frac{e^{ik\{(x-\xi)^2 + \rho^2\}^{1/2} - ik(x-\xi)}}{\{(x-\xi)^2 + \rho^2\}^{1/2}} \rho^{2m+1} \, d\rho$$

(2.6)

$$= \int_{x-\xi}^\infty e^{ikt - ik(x-\xi)} [t-(x-\xi)]^m [t + (x-\xi)]^m \, dt$$

$$= \int_0^\infty e^{ikt} t^m [t + 2(x-\xi)]^m \, dt$$

$$\sim (\frac{i}{k})^{m+1} \sum_{\ell=0}^m \binom{m}{\ell} (m+\ell)! (\frac{1}{k})^\ell [2(x-\xi)]^{m-\ell}$$

Eqs. (2.3) now follow.

Next, consider Eq. (1.10); we then prove

LEMMA 2.2: Set

(2.7) $$I_{m,n} = -\frac{k^2}{(4\pi)^2} \iint_{\mathbb{R}^2} \eta^m \zeta^n \frac{\exp\{ik|\bar{r}-\bar{R}| + ik|\bar{r}_s-\bar{R}|\}}{|\bar{r}-\bar{R}| \, |\bar{r}_s-\bar{R}|} \, d\eta d\zeta \quad .$$

Then

$$I_{m,n} = 0 \quad \text{if one of } m \text{ or } n \text{ is odd;}$$

(2.8)

$$I_{2m,2n} \sim \frac{-ik}{16\pi x} \frac{(1/2)m(1/2)n}{(m-n)!} \left(\frac{ix}{k}\right)^m e^{2ikx} \cdot$$

$$\cdot \sum_{p=0}^{m} \binom{m}{p} \left(\frac{i}{4kx}\right)^p \sum_{q=0}^{m} \binom{m}{q} (-1)^q \left(\frac{\xi}{x}\right)^{2q}$$

$$\cdot \sum_{s=0}^{\infty} \frac{(2q+1)s}{s!} \frac{(m+p+s)!}{(2ikx)^s} , \quad k \to \infty .$$

PROOF: We again use polar coordinates, as in the evaluation of (2.2), to conclude that $I_{m,n} = 0$ if one of m or n is odd. In the remaining cases we have

(2.9)
$$I_{2m,2n} = - \frac{k^2}{(4\pi)^2} \frac{(2\pi)^{(1/2)}m^{(1/2)}n}{(m+n)!} J_{m+n}$$

where

(2.10)
$$J_m = \int_0^\infty \frac{e^{ik\{(x-\xi)^2 + \rho^2\}^{1/2}} + ik\{(x+\xi)^2 + \rho^2\}^{1/2}}{\{(x-\xi)^2 + \rho^2\}^{1/2} \{(x+\xi)^2 + \rho^2\}^{1/2}} \rho^{2m+1} d\rho .$$

Next, set

$$\rho = |x\xi|^{1/2} \{(t - \tfrac{1}{t})^2 - (|\tfrac{x}{\xi}|^{1/2} - |\tfrac{\xi}{x}|^{1/2})^2\}^{1/2}$$

(2.11)

$$a = |x/\xi|^{1/2} , \quad b = |x\xi|^{1/2} .$$

Then

$$J_m = e^{2ikab}b^{2m}\int_0^\infty e^{2ikbt} t^m(t+2a)^m\left[1-\frac{1}{a^2(t+a)^2}\right]^m \frac{dt}{t+a}$$

$$= ie^{2ikab}b^{2m}\int_0^\infty e^{-2kbt}(it)^m(2a+it)^m\left[1-\frac{1}{a^2(a+it)^2}\right]^m \frac{dt}{a+it}$$

$$\sim i^{m+1}b^{2m}e^{2ikab}\int_0^\infty e^{-2kbt}t^m\left(\sum_{p=0}^m \binom{m}{p}i^p t^p(2a)^{m-p}\right)\cdot\left(\sum_{q=0}^m \binom{m}{p}\frac{(-1)^q}{a^{4q+1}}\right)\cdot$$

(2.12)

$$\cdot\left(\sum_{s=0}^\infty \frac{(2q+1)s}{s!}\frac{(-i)^s t^s}{a^s}\right)dt$$

$$\sim \frac{i}{2kx}\left(\frac{ix}{k}\right)^m e^{2ikx}\left[\sum_{p=0}^m \binom{m}{p}\left(\frac{i}{4kx}\right)^p\right]\cdot\left[\sum_{q=0}^m \binom{m}{q}(-1)^q\left(\frac{\xi}{x}\right)^{2q}\right]\cdot$$

$$\cdot\left[\sum_{s=0}^\infty \frac{(2q+1)_s}{s!}\frac{(m+p+1)!}{(2ikx)^s}\right]$$

from which (2.8) follows.

LEMMA 2.3: Let $k_0,k_1,\ldots,k_m$ be distinct, and set $\phi(x) = (x-k_0)(x-k_1)\cdots(x-k_M)$. Then

(2.13) $$\sum_{j=0}^M \frac{k_n^m}{\phi'(k_j)} = \begin{cases}0, & m=0,1,\ldots,M-1 \\ 1, & m=M\end{cases}$$

PROOF: Let $p_M(x)$ denote the Newton polynomial of degree $\leqq M$ in x which satisfies the equations $p_M(k_j) = y(k_j)$, where $y(x)$ is a given function. The coefficient of x^M in $p_M(x)$ is

(2.14) $$a_M = \sum_{j=0}^M \frac{y(k_j)}{\phi'(k_j)}$$

Hence if $y(x) = x^m$, then $a_M = 0$ for $m = 0,1,\ldots,M-1$, while if $m = M$, then $a_M = 1$. Notice that (2.14) is the Mth divided difference of the $y(k_j)$.

LEMMA 2.4: Let $f \in C^{\infty}(\mathbb{R}^3)$, and let R_M be defined in (2.1).
Then there exists a constant $K = K_M$ such that

$$(2.15) \quad \bar{\varepsilon}_M \equiv \left| -\frac{k^2}{4\pi} \iint_{\mathbb{R}^2} R_M(\xi,\eta,\zeta) \frac{\exp[ik|\bar{R}-r| - ik(X-\xi)]}{|\bar{R}-r|} d\eta d\zeta \right| \leq \frac{K}{k^M} .$$

$$(2.16) \quad \varepsilon_M \equiv \left| -\frac{k^2}{4\pi^2} \iint_{\mathbb{R}^2} R_M(\xi,\eta,\zeta) \frac{\exp\{ik|\bar{r}-\bar{R}| + ik|\bar{r}_s-\bar{R}|\}}{|\bar{r}-\bar{R}| \, |\bar{r}_s-\bar{R}|} d\eta d\zeta \right| \leq \frac{K}{k^M} .$$

PROOF: We consider only the case of (2.15). We omit the case of
(2.16), the analysis of which is similar.

Clearly $\bar{\varepsilon}_M$ is left unchanged if $R_M(\xi,\eta,\zeta)$ in (2.1) is replaced by $\rho_M(\xi,\eta,\zeta) \equiv [R_M(\xi,\eta,\zeta) + R_M(\xi,-\eta,\zeta) + R_M(\xi,\eta,-\zeta) +$
$+ R_M(\xi,-\eta,-\zeta)]/4$. But ρ_M is an even function of both η and
ζ , so that, by (2.1),

$$(2.17) \quad \rho_M(\xi,\eta,\zeta) = \sum_{m=0}^{M+1} \eta^{2m} \zeta^{2M+2-2m} c_m(\xi,\eta,\zeta)$$

where the $c_m(\xi,\eta,\zeta)$ are bounded functions. Substituting
$c_m(\xi,\eta,\zeta)\eta^{2m}\zeta^{2M+2-2m}$ for R_M in (2.15), we get

$$\mu_{M,m} \equiv k^2 \int_0^{\infty} \int_0^{2\pi} \rho^{2M+3} e^{ik\{(x-\xi)^2 + \rho^2\}^{1/2} - ik(x-\xi)} \cdot$$

$$\cdot c_m^*(\xi,\rho,\theta)\cos^{2m}\theta \sin^{2M+2-2m}\theta d\theta d\rho$$

$$(2.18)$$

$$= \frac{k^2}{k^{M+2}} \int_0^{\infty} \int_0^{2\pi} t^{M+1} e^{it}[\frac{t}{k} + 2(x-\xi)]^{M+1} \cdot$$

$$\cdot c_m^{**}(\xi,\frac{t}{k},\theta)\cos^{2m}\theta \sin^{2M+2-2m}\theta d\theta dt$$

where $c_m^{**} = c_m^* = c_m$ under the transformed integrals. But $|c_m(\xi,\eta,\zeta)| \leq c_m$ and $c_m(\xi,\eta,\zeta) = 0$ if $\eta^2+\zeta^2 \geqslant L^2$, and therefore $c^{**}(\xi,t/k,\theta) = 0$ if $|t/k| \geqslant L$. Furthermore

$$\int_0^{2\pi} \cos^{2m}\theta \sin^{2M+2-2m}\theta \, d\theta = 2\pi(1/2)_m (1/2)_{M+1-m}/(M+1)! \text{ and therefore}$$

$$(2.19) \qquad |\mu_{M,m}| \leq \frac{kL \, c_m (1/2)_m (1/2)_{M+1-m}}{k^M (M+1)!} \quad 2\pi \quad .$$

Therefore

$$(2.20) \qquad \bar{\varepsilon}_M \leq \frac{2\pi kL}{k^M (M+1)!} \sum_{m=0}^{M+1} c_m (1/2)_m (1/2)_{M+1-m} \quad .$$

But by taking a Taylor expansion of each $c_m(\xi,\eta,\zeta)$ and integrating the next higher powers $\eta^{2m}\zeta^{2M+2-2m}$ explicitly yields

$$\varepsilon_M \leq |\sum_{m=0}^{M+1} I_{2m,2M+2-2m} \, f^{(0,2m,2M+2-2m)}(\xi,0,0)|2\pi$$

$$(2.21)$$

$$+ \, \varepsilon_{M+1} \, , \, \leq \frac{K_M'}{k^M} + \varepsilon_{M+1} \leq \frac{K_M}{k^M} \quad .$$

This completes the proof.

Completion of the Proofs of (1.12)-(1.17)

The representation (1.12) follows by substituting (2.1) into (1.9) and using Lemmas 2.1 and 2.4, the representation (1.13) follows by substituting (2.1) into (1.10) and using Lemmas 2.2 and 2.4.

The function Y_M defined by (1.14) or (1.15) now takes the form

$$(2.22) \qquad Y_M(\bar{r},\bar{r}_s,k) = a_M k^M + c_{m-1} k^{M-1} + \ldots + c_0 + \delta_M(k)$$

where

$$(2.23) \quad \begin{cases} a_M = \displaystyle\int_0^1 f((1-t)\overline{r} + t\overline{r}_s)dt \\[12pt] \delta_M(k) \sim \dfrac{d_1}{k} + \dfrac{d_2}{k^2} + \cdots \\[12pt] \delta_M(k) = O(1/k) \ , \ k \to \infty \end{cases}.$$

By Lemma 2.3, V_M defined by (1.16) yields

$$(2.24) \qquad V_M(\overline{r},\overline{r}_s,k) = a_M + \left. \in_M(k) \right|_{k=\gamma}$$

where γ is between the largest and smallest of the k_j and k, and where (since (1.16) is the Mth divided difference of the $Y_M(\overline{r},\overline{r}_s,k_j)$)

$$(2.25) \qquad \in_M(\gamma) = \frac{1}{M!} \left(\frac{d}{dk}\right)^M \left. \delta_M(k) \right|_{k=\gamma} .$$

Now, if e.g. $.8k \leqq k_j \leqq k$, $j = 0,1,\ldots,M$, then

$$(2.26) \qquad \in_M(\gamma) = O(k^{-M-1}) \ , \ k \to \infty .$$

Finally, what happens if $f \notin C^\infty(\mathbb{R}^3)$. In this case (2.1) needs to be replaced by

$$(2.1)' \qquad f(\xi,\eta,\zeta) = \sum_{m,n \geqslant 0, m+n \leqq M'} f^{(0,m,n)}(\xi,0,0)\eta^m \zeta^n + R_{M'}(\xi,\eta,\zeta)$$

where $R_{M'}$ might consist of a sum such as

$$(2.27) \qquad R_{M'} = \sum_{\ell,\mu,\nu} c_\ell (\mu,\nu,\xi)|\eta|^\mu |\zeta|^\nu$$

where $0 < \mu,\nu$, $M' < \mu + \nu$ and where μ and ν are not integers.

Let $\sigma = \min\{\mu+\nu: \mu+\nu \neq \text{even integer}\}$. Then (1.17) can no longer hold, instead, for any $M \geq \sigma$, we need to replace (1.17) by

$$(2.28) \qquad V_M(\bar{r},\bar{r}_s,k) = \int_0^1 f((1-t)\bar{r} + t\bar{r}_s)dt + O(k^{-\sigma})$$

as $k \to \infty$ (see [4] Thm. 4.1).

3. AN EXAMPLE

It is not easy to find examples of f which are reasonably realistic and for which the integrals (1.9) or (1.10) can be easily evaluated numerically. One interesting test case would be

$$(3.1) \qquad f(R) = \begin{cases} 1 & \text{if} \quad |\bar{R}-\bar{R}_0| \leq b \\ 0 & \text{if} \quad |\bar{R}-\bar{R}_0| > b \end{cases}$$

however, even this simple case leads to either the evaluation of a highly oscillatory integral, or to a slowly convergent series involving Bessel functions of large argument. A reasonably realistic case is that of

$$(3.2) \qquad f(R) = e^{-b|\bar{R}-\bar{R}_0|^2}$$

and fortunately (1.9) can be readily evaluated for this case to yield numerical results without undue difficulty.

Setting

$$(3.3) \qquad \bar{r} = \bar{R} + \bar{\rho}$$

in (1.9) we thus get

(3.4) $W(\overline{R},\overline{k}) =$

$$\frac{-k^2 e^{-b|\overline{R}-\overline{R}_0|^2}}{4\pi} \iiint\limits_{\mathbb{R}^3} \frac{e^{-b\rho^2+ik\rho-i\overline{\rho}\cdot[\overline{k}-2ib(\overline{R}-\overline{R}_0)]}}{\rho} \, dV(\overline{\rho}) \quad .$$

Let us now take

$$\overline{k} = k(\cos\alpha, \sin\alpha, 0)$$

(3.5)

$$\overline{R}-\overline{R}_0 = |\overline{R}-\overline{R}_0|(\cos\beta, \sin\beta, 0)$$

and let us define the quantities

(3.6)
$$\begin{cases} \theta = \alpha - \beta \\ d = 2b|\overline{R}-\overline{R}_0| \\ \overline{\sigma} = [k^2 - d^2 - 2ikd\cos\theta]^{1/2} \quad . \end{cases}$$

Next, since

(3.7) $dV(\overline{\rho}) = \rho^2 d\Omega(\overline{\rho})$

where $\Omega(\overline{\rho})$ denotes solid angle measure, we use Eqs. (3.6) and the identity

(3.8) $\displaystyle\iint\limits_{\text{Unit sphere}} e^{-i\overline{\rho}\cdot[\overline{k}-2ib(\overline{R}-R_0)]} d\Omega(\overline{\rho}) = 4\pi \frac{\sin(\overline{\sigma}\rho)}{\overline{\sigma}\rho}$

to get

(3.9) $\displaystyle W(\overline{R},\overline{k}) = \frac{ik^2}{2\overline{\sigma}} e^{-d^2/4b}(I_1 - I_2) \quad ,$

where

$$(3.10) \quad \begin{cases} I_1 = \displaystyle\int_0^\infty e^{-b\rho^2 + i(k+\overline{\sigma})\rho} \, d\rho \\[2em] I_2 = \displaystyle\int_0^\infty e^{-b\rho^2 + i(k-\overline{\sigma})\rho} \, d\rho \end{cases} .$$

Now consider the integral

$$(3.11) \quad I(z) = \int_0^\infty e^{-t^2 - tz} \, dt \quad ,$$

in the following cases:

(a) $|z^2/4| \leq 10$.

In this case we can evaluate $I(z)$ from the expression

$$(3.12) \quad I(z) = \frac{1}{2} e^{z^2/4} \left[\sqrt{\pi} - z \sum_{n=0}^\infty \frac{(1/2)_n}{(3/2)_n n!} (-z^2/4)^n \right]$$

which is obtained by expanding e^{-tz} in powers of t , carrying

out termwise integration, and then using Kummer's identity

[3, p. 267].

(b) $|z^2/4| > 10$, $\mathrm{Re}\, z \geq 0$.

In this case expansion of e^{-t^2} in powers of t and termwise

integration yields the asymptotic expansion

$$(3.13) \quad I(z) \sim \frac{1}{z} \sum_{n=0}^\infty (\tfrac{1}{2})_n (-\frac{4}{z^2})^n \quad .$$

(c) $|z^2/4| > 10$, $\mathrm{Re}\, z < 0$.

In this case we note that

$$(3.14) \qquad I(z) = \int_{-\infty}^{\infty} e^{-t^2 - tz} dt - \int_{-\infty}^{0} e^{-t^2 - tz} dt$$

$$\sim \sqrt{\pi} \; e^{z^2/4} + \frac{1}{z} \sum_{n=0}^{\infty} (1/2)_n (-\frac{4}{z^2})^n \quad .$$

Hence, using (3.11) in (3.9), we get

$$(3.15) \qquad W(\bar{R},\bar{k}) = \frac{ik^2}{2\bar{\sigma}b^{1/2}} e^{-d^2/4b} [I(\frac{-i(k+\bar{\sigma})}{b^{1/2}}) - I(\frac{-i(k-\bar{\sigma})}{b^{1/2}})] \quad .$$

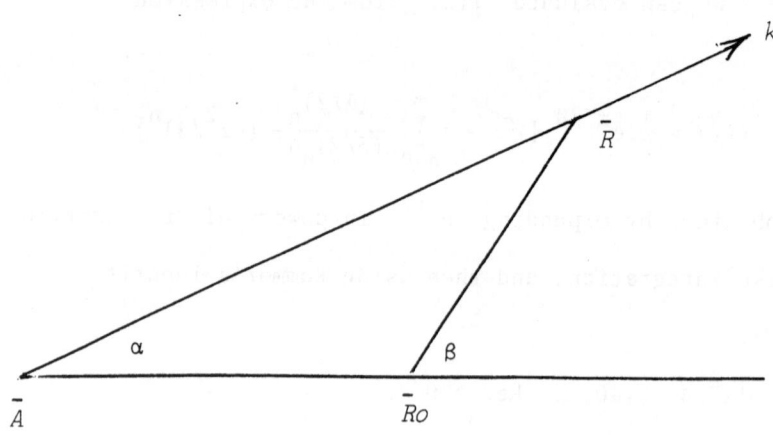

Figure 2.   Ray Through $\bar{R}$.

In view of Fig. 2, if we integrate f along the line $\overline{AR}$
in the direction of $\bar{k}$ we get

$$U(\overline{R},\overline{R}_0,b,\alpha) \equiv$$

(3.16)
$$\int_{-\infty}^{\infty} f((1-t)\overline{A} + t\overline{R})\,|\overline{R}-\overline{A}|\,dt$$

$$= \int_{-\infty}^{\infty} e^{-b|\overline{R}-\overline{R}_0|^2}\,dS(\overline{R}) = \sqrt{\frac{\pi}{b}}\; e^{-b|\overline{R}-\overline{R}_0|^2 \sin^2(\alpha-\beta)} \quad .$$

Let us also set

(3.17)
$$\text{Ryt}(\overline{R},\overline{R}_0,b,\alpha,k) = \frac{2i}{k}\,W(\overline{R},\overline{k}) \equiv \frac{2i}{k}\,W(\overline{R},\alpha,k)$$

where W is given in (3.15), and by taking (1.19) as a model, with $k_0 = k, k_1 = .8k, k_2 = 1.2k$, let us set

(3.18)
$$V_3(\overline{R},\overline{R}_0,b,\alpha,k) = \frac{2i}{.08k}[1.2W(\overline{R},\alpha,1.2k) - 2W(\overline{R},\alpha,k)$$

$$+ .8W(\overline{R},\alpha,.8k)] \quad .$$

Table 3 then gives a comparison of U, Ryt and V_3 for $\overline{R} = (25,0,0)$, $\overline{R}_0 = (5,0,0)$ and $k = 150\pi$, for various values of α .

TABLE 3

Comparison of Rytov and 3 Frequency Approximations

α	Ryt	V_3	U
0	1.218 + .005i	1.256 - .0001i	1.253
$\pi/200$	1.011 + .0043i	1.030 - .00007i	1.029
$2\pi/200$	.573 + .0024i	.568 + .0001i	.569
$3\pi/200$	.218 + .0009i	.212 + .00003i	.212
$4\pi/200$	$(.541 + .002i) \times 10^{-1}$	$(.537 + .0004i) \times 10^{-1}$	$.535 \times 10^{-1}$
$5\pi/200$	$(.824 + .004i) \times 10^{-2}$	$(.909 - .004i) \times 10^{-2}$	$.911 \times 10^{-2}$
$6\pi/200$	$(.684 + .003i) \times 10^{-3}$	$(.1023 - .002i) \times 10^{-3}$	$.105 \times 10^{-2}$
$7\pi/200$	$(.17 + .007i) \times 10^{-4}$	$(.83 - .003i) \times 10^{-4}$	$.821 \times 10^{-4}$
$8\pi/200$	$(.192 - .008i) \times 10^{-5}$	$(.61) - .002i) \times 10^{-5}$	$.437 \times 10^{-5}$

Results for the exact solution u were not known at the time of the writing of this paper.

REFERENCES

[1]   J. S. Ball, S. A. Johnson and F. Stenger, "Explicit Inver-
      sion of the Helmholtz Equation for Ultrasound Insonifi-
      cation and Spherical Detection," Proc. of Houston Con-
      ference on Acoustical Imaging 1(1980).

[2]   A. M. Cormack, "Representation of a Function by its Line
      Integrals, with some Radiological Applications," J. Appl.
      Phys. 34(1963), pp. 2722-2727.

[3]   W. Magnus, F. Oberhettinger and R. P. Soni, "Formulas and
      Theorems for the Special Functions of Mathematical
      Physics," Springer, 1966.

[4]   F. Stenger, "The Asymptotic Approximation of Certain Inte-
      grals," SIAM J. Math. Anal. 1(1970), pp. 392-404.

[5]   F. Stenger, "An Algorithm for Ultrasonic Tomography Based on
      Inversion of the Helmholtz Equation $\nabla^2 u + \frac{\omega^2}{c^2} u = 0$," to
      appear.

THE PROPERTIES AND PERFORMANCE OF A Si-PVF$_2$ OPTICALLY CONTROLLED

ACOUSTIC POINT SOURCE

A. Ayoola and C.W. Turner

Department of Electronic & Electrical Engineering
King's College London
Strand, London, WC2R 2LS

ABSTRACT

An opto-acoustic system for synthesising point sources has been
developed which utilises a monolithic photoconductor-acoustic transducer structure illuminated by a GaAs infra-red source. The characteristics of a 'Si-PVF$_2$' device exhibiting excellent isolation, with
negligible inter-element mutual coupling and intra-element mode
coupling effects, and with emergent beam angles in excess of 60°
(compared with 25° for PZT) are presented. The acoustic power output
per unit area for the optically controlled source has been measured
to be about 10dB below the equivalent value for a conventional plane
wave transducer.

Experimental results on beam profiles, directivity patterns,
'r/λ' ratios, radiation efficiency, and the variation of source
characteristics with optical excitation are presented and compared
with the theoretical values.

Preliminary results from beam-forming experiments with a phased
array of such opto-acoustic point sources are also discussed.

INTRODUCTION

Phased-array imaging and transducer evaluation measurements both
require the synthesis of acoustic point sources. Conventional point
sources fall broadly into the three distinct classes shown in Fig. 1.
These sources suffer from a number of problems depending on the
particular source configuration and operational mode, viz:-

(1) For the discrete sources[1](Fig. 1a), physical fabrication of the
     individual elements is extremely difficult, particularly for
     operation in the extensional mode as this usually entails the
     construction and mounting of very small, slender, structures.
     For this reason, discrete point sources tend to be quite large
     (generally greater than 2 mm in width) with a poor angular
     response and high sidelobe levels.

(2) There are difficulties with the metallization of electrical
     contacts particularly for high-density two-dimensional arrays.

(3) There is a significant degree of coupling to unwanted parasitic
     modes except for the monolithic configurations.

(4) For the trapped-energy mode of operation in monolithic mosaics
     (Fig. 1c) there are electrical bandwidth limitations[2] and a
     significant level of inter-element cross-talk due to untrapped
     eigenmodes propagating laterally across the device.

(5) There are bonding problems and electrical bandwidth limitations
     for monolithic sources employing matching layers[3].

(6) For partially discrete arrays[4,5] dicing is relatively difficult,
     particularly for high density arrays. In addition, there is some
     inter-element mutual coupling due to wave propagation within the
     backing material which is only partially diced.

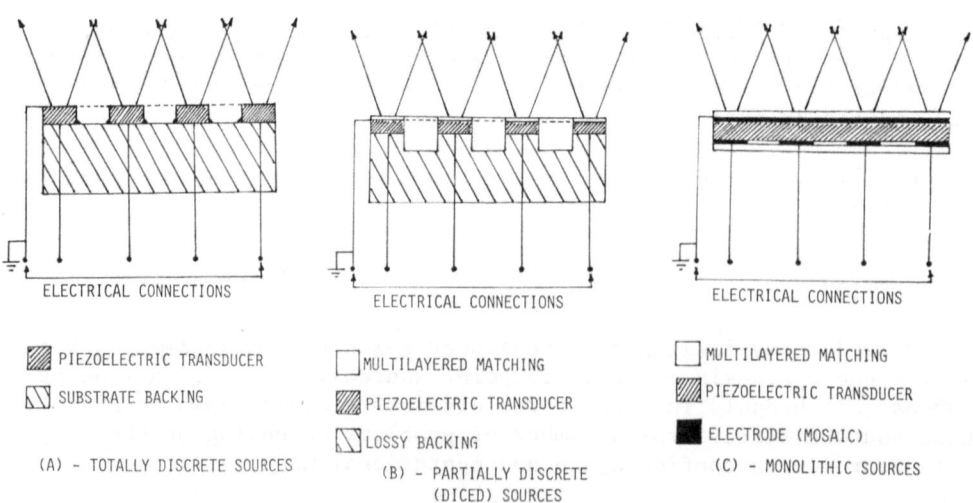

Fig. 1 - Conventional Acoustic Sources

In this paper we report the development of an opto-acoustic point source with beam emergence angles in excess of 60° which also allows optical control of source location. We begin with the theoretical and experimental characterisation of this type of optically-controlled source and then discuss its potential use in phased-array applications.

OPTO-ACOUSTIC POINT SOURCES

The light output from a localised optical source (an optical fibre-coupled infra-red source in this case) is intensity-modulated at a pump frequency f_p and used to switch a single cell of a composite plane piezoelectric transducer-photoconductor structure, electrically driven at a frequency f_c (see Figs. 2a, 2b and 6).

The light modulates the conductivity of the silicon at the pump frequency (typically 50-100kHz) and an acoustic output is generated at the sum and difference sideband frequencies within the illuminated region, which therefore acts as an acoustic source.

The sideband-to-carrier acoustic power ratio for a single cell provides a measure of the point-source radiation efficiency and, in general, is a complex function of the piezoelectric and photoconductor material parameters and the frequency response of the transducer. For a 10:1 photoconductive light-to-dark switching ratio and 50% modulation of the infra-red pump, the theoretical value is typically of the order of -10dB. However, since there is a sizeable level of background carrier insonification from the 'dark' region of the transducer, the total device sideband-to-carrier acoustic power ratio is much lower (typically -30dB).

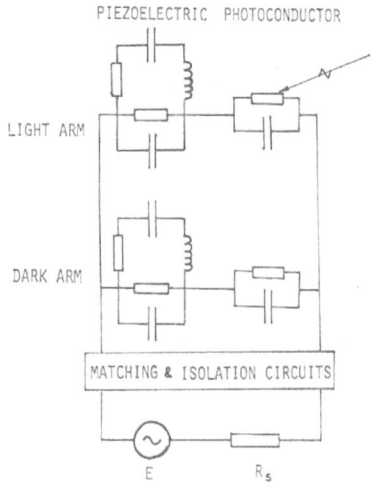

Fig. 2(a) Opto-acoustic synthesis   Fig.2(b) Simplified device
       of point sources   equivalent circuit

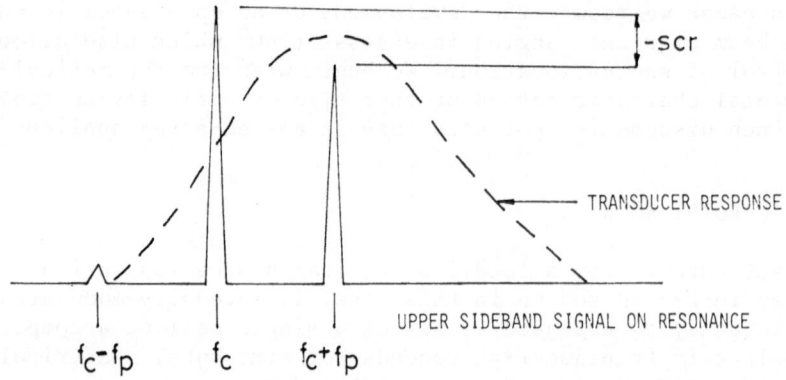

Fig. 3 - Off-resonance technique for suppressing the carrier signal

Although the point-source behaviour at either of the sideband frequencies is unaffected by the presence of the carrier frequency component, from the point of view of reducing overall insonification levels (particularly for medical applications) and easing the signal processing requirements, sideband-to-carrier ratio levels must be as high as possible. One method utilised so far, with some success, has been to operate the device so that the carrier frequency is far from resonance while the sideband frequency is as close as possible to resonance (Fig. 3). The efficacy of such a scheme is presently impaired by the limits on the switching time of the silicon photo-conductive layer, which in turn imposes an upper limit on the pump frequency. It will be shown later that for phased arrays of these point sources a lower sideband-to-carrier ratio can be tolerated.

Before considering in detail the performance of the Si-PVF_2 point source, some background theory will be presented to establish the spatial impulse response limits on this type of point source.

The spatial impulse response of the device, which is directly related to the radiated beam main-lobe angle, can be determined from the individual responses of each layer of the composite represented schematically in Fig. 4.

The output from each stage is obtained by a spatial convolution of the input distribution with the relevant spatial impulse response[6]. For an arbitrary 2-dimensional optical excitation profile $U_i(x,y)$ the resultant 2-dimensional output spatial response profile $U_o(x.y)$ is given by the expression

$$U_o(x,y) = \{U_i(x,y) * \phi_A(x,y) * \phi_B(x,y)\} \qquad \ldots\ldots\ldots(1)$$

where * represents 2-dimensional spatial convolution. Transforming

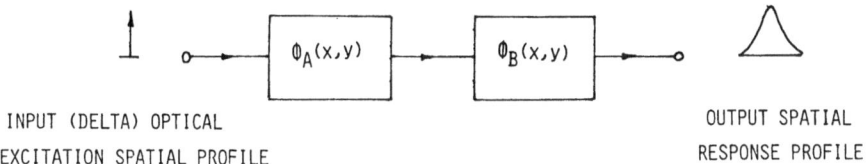

INPUT (DELTA) OPTICAL
EXCITATION SPATIAL PROFILE

OUTPUT SPATIAL
RESPONSE PROFILE

$\Phi_A(x,y)$ = 2-DIMENSIONAL SPATIAL IMPULSE RESPONSE OF PHOTOCONDUCTOR
LAYER

$\Phi_B(x,y)$ = 2-DIMENSIONAL SPATIAL IMPULSE RESPONSE OF TRANSDUCER LAYER

Fig. 4 - Composite spatial impulse response of the opto-acoustic
device

equation (1) into the spatial frequency domain yields -

$$G_o(f_x,f_y) = \{G_i(f_x,f_y) \cdot H_A(f_x,f_y) \cdot H_B(f_x,f_y)\} \qquad \ldots\ldots(2)$$

where $H_A(f_x,f_y)$ and $H_B(f_x,f_y)$ are the respective spatial transfer
functions for each layer.

(1) $\phi_A(x,y)$, the photoconductor layer spatial impulse response is
determined by the photo-carrier lateral diffusion profile and
is given approximately by

$$\phi_A(x,y) = \phi_A(r) = \phi_o n_o \exp(-r/L_n) \quad \text{for circular symmetry within}$$
the photoconductor $\ldots\ldots(3)$

where $L_n = (D_n \tau_n)^{\frac{1}{2}}$ = carrier diffusion length

n_o = carrier concentration at r = 0

D_n = carrier diffusion constant

τ_n = photogenerated carrier lifetime.

(2) $\phi_B(x,y)$, the piezoelectric transducer layer spatial impulse response or transducer point spread function (PSF) has been investigated by a number of authors[7,8]. In the approximate analysis, an
angular spectrum of plane waves approach is used whereby the
spherical waves from the source are decomposed into constituent
plane waves (each plane wave component representing a particular
spatial frequency). The response at the sideband frequency with
water loading and, effectively, air-backing (since the photo-
conductive layer is much thinner than the acoustic wavelength)
is shown in Figs. 5(a) and (b) and $\phi_B(x,y)$, when the effects of
electric field fringing are neglected, is given by the expression

$$\phi_B(x,y) = \phi_B(r) = 2\pi\int_0^\infty \rho T(f_r)J_o(2\pi r\rho)d\rho = H_o(T(f_r)) \qquad \ldots\ldots(4)$$

(for circular symmetry of the transducer layer e.g. 6 mm
crystallographic class materials)
where $\rho = (f_x^2 + f_y^2)^{\frac{1}{2}}$

H_o = Hankel transform of zero order

and $T(f_r)$ = the spatial (angular) transmission response of the
device.

In general, the analytic procedure for determining ϕ_B is very
cumbersome, especially if factors such as mode conversion effects at
the interfaces, electric field fringing, attenuation and multiple
reflections are taken into account. A simpler approach for estimating
the effective spatial response, using the radiated far-field mainlobe
angular width, has given good agreement with the theoretical value
obtained by other authors. The method is ideal for our purposes since
it allows a more rapid characterisation of the point source spatial
response when different piezoelectric materials are used to fabricate
the device.

The minimum point source 'r/λ' ratio attainable under continuous
wave operation, as a result of spatial low pass filtering in the
transducer layer, is given by

$$(r/\lambda)_{min} = \tfrac{1}{2}(\frac{\lambda_2}{\lambda_1}) = \tfrac{1}{2}(\frac{1}{\sin\theta_c}) \qquad \ldots\ldots(5)$$

for $\lambda_2 \geq \lambda_1$

where λ_2, λ_1 are the wavelengths in the piezoelectric and in water
respectively, and θ_c is the compressional wave critical angle for
the two media. The isotropic form of Snell's law has been used (i.e.
low media anisotropicity has been assumed).

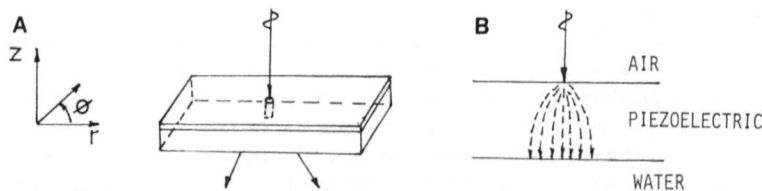

Fig. 5 - (a) Point source configuration

         (b) Electric field fringing within the transducer layer

Therefore, to synthesise an acoustic point source of this type with the smallest possible (r/λ) ratio under c.w. operation, a transducer with the smallest acoustic velocity mismatch with the transmitting medium is required. The calculated results are unaffected by the interposition of planar acoustic matching layers but in practice such layers might be essential to reduce beam spreading effects, due to impedance mismatch, which increase the effective (r/λ) value.

The poled and axially-stretched piezoelectric polymer – Polyvinylidene fluoride (PVF_2) has the smallest velocity mismatch with water and should thus provide the best source 'r/λ' ratios. In addition, the acoustic attenuation within the transducer material itself serves to reduce beam spreading effects and gives good acoustic isolation between neighbouring sources.

In contrast to previous applications of PVF_2 which have used thin films (25-50μm) operated far below the resonant frequency with low transmitter conversion efficiency, thick film samples of PVF_2 operated at half-wave resonance have been used here. With proper electrical impedance matching the conversion efficiency of the $Si-PVF_2$ source can be increased to a level only 9dB below that of the equivalent PZT device, but with a greatly superior (r/λ) ratio.

OPTICALLY CONTROLLED '$Si-PVF_2$' POINT SOURCE EXPERIMENTAL RESULTS

The opto-acoustic device was fabricated from a 2.5 cm square PVF_2 plate, 350μm thick, mechanically clamped to a 10kΩ-cm resistivity n-type silicon wafer, 200μm thick with an evaporated semi-transparent aluminium electrode. The front electrode was silver dag paint. The optical 'point' source consisted of a GaAs infra-red L.E.D. lamp (peak emission at λ_e = 880nm) coupled to a cladded-core, step-index optical fibre (1 mm diameter and NA = 0.54). The experimental set up is shown in Fig. 6.

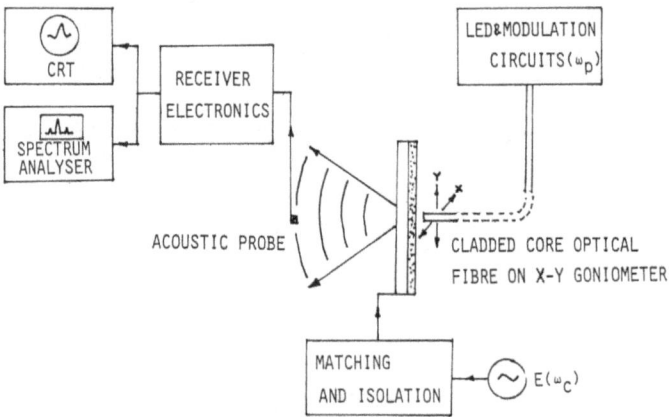

Fig. 6 –The experimental set

The excitation (f_c) and pump (f_p) frequencies used were 2.25MHz and 50kHz respectively. A heterodyne coherent scheme was used to process the upper sideband acoustic signal.

The measured complex impedance of the composite device is displayed in Fig. 7 and is observed to have a large reactive component (at 2.3MHz, R = 48.08Ω and X = 457.5Ω). Maximum drive power at the sideband frequency is delivered to the illuminated cell acoustic radiation resistance if the impedance is conjugate matched. By using a series inductance it is possible to dispense with the use of an r.f. impedance transformer in this case, since the device resistance is already fairly well matched to that of the drive amplifier (50Ω). Fig. 8 shows a graph of the overall device/probe system insertion loss (with no amplification at the probe).

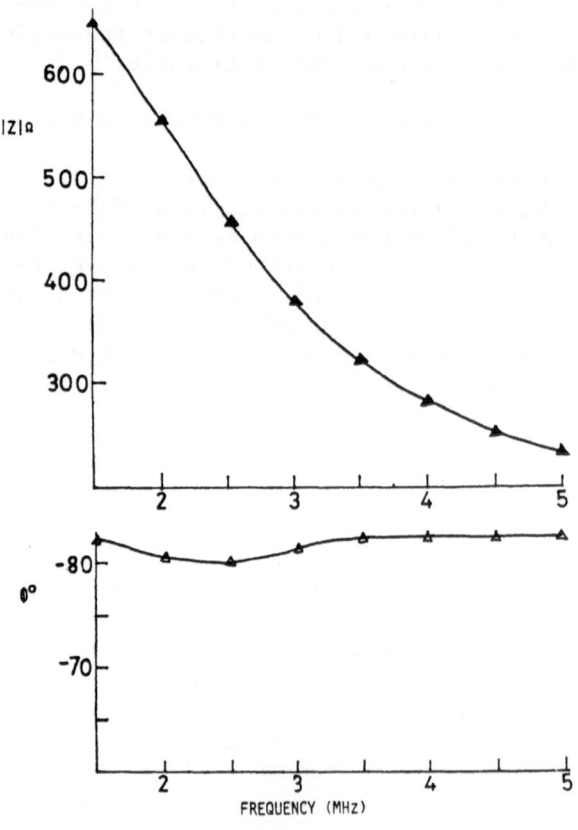

Fig. 7 - 'Si-PVF$_2$' opto-acoustic device complex impedance $|Z|$, ϕ^o

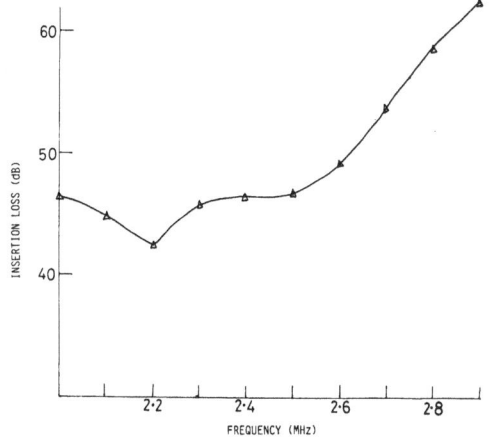

Fig. 8 - System insertion loss
versus frequency

The insertion loss is high as a result of the very small area of the probe receiver ($\simeq$ 2 mm) and the fact that its resonant frequency is approximately 2.0MHz whereas the source is matched for maximum power at 2.3MHz. The measured insertion loss (at the trough) relative to that for the 'Si-PZT' device is approximately 10dB.

With the probe at the central position (range = 20 cm) the sideband-to-carrier ratio was measured to be -18dB for a single cell and equal to -34dB for the whole transducer. The acoustic sideband power radiated from a single cell is estimated to be approximately 1μW for a 30V input electrical excitation voltage. The acoustic field mapping of the point source was carried out with a conventional acoustic probe which could be rotated in the azimuthal plane to correct for any probe directivity effects. Fig. 9 shows the variation of beam profile with axial range for the point transmitter and Fig. 10 is the equivalent polar directivity plot at an axial range of 15 cm. This can be compared with the polar directivity plot of the equivalent 'Si-PZT' device (shown dotted in Fig. 10). The 'Si-PVF₂' device has a much larger emergent beam width, as the theory predicts.

The 'r/λ' ratio, calculated from the experimental results, is equal to 0.98 (corresponding to a mainlobe beam angle of 62°). This is reasonably close to the predicted theoretical value of 0.78, considering that the optical source itself was about 1 mm in diameter, and the effects of electric field fringing, beam spreading and lateral photocarrier diffusion have been neglected. Each of these factors could contribute significantly to the difference between experimental and theoretical 'r/λ' values. It would have been relatively difficult to fabricate a conventional piston source with the same experimental 'r/λ' ratio (the required piston diameter is 1.25 mm). Scanning an acoustic receiver laterally across the transmitter beam at a range in the receiver Fraunhoffer region will give a directivity pattern, which approximates to the receiver polar directivity pattern. Fig. 11 is such an approximate directivity pattern obtained by scanning a PZT receiver, 'r/λ' ratio = 9, laterally across the PVF₂ point source beam

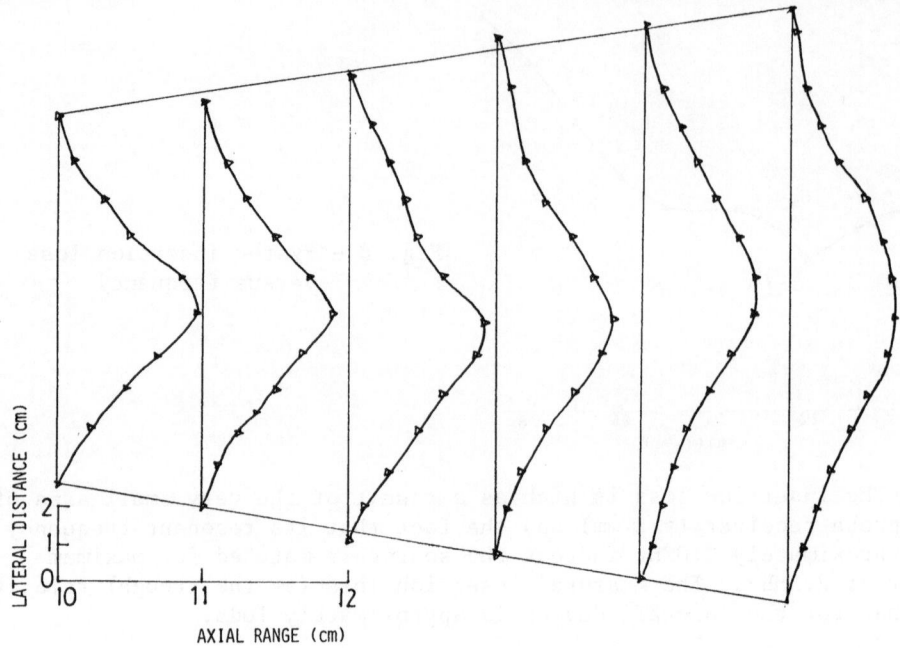

Fig. 9 - Beam Profile for 'Si-PVF$_2$' opto-acoustic transmitter

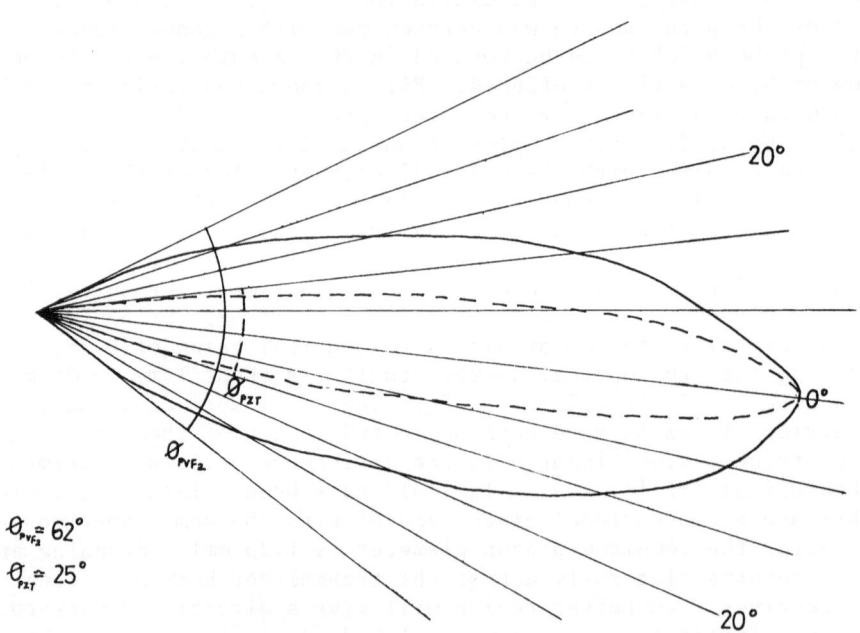

Fig. 10 - Polar directivity pattern for Si-PVF$_2$ device

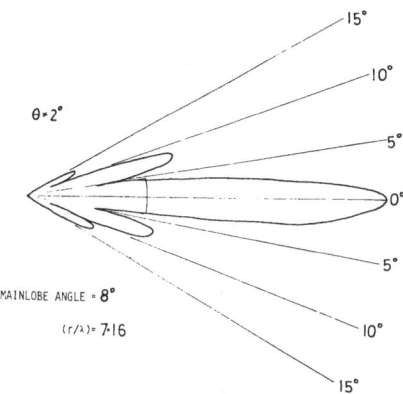

Fig. 11 – Polar directivity plot for conventional PZT-5A receiver

at a range of 15 cm. The sidelobes on either side of the main beam
are apparent and the experimentally obtained 'r/λ' is equal to 7.2
(main lobe half-beam width of 4°) which agrees fairly well with the
theoretical value.

The variation of the point-source radiation characteristics with
the spatial distribution of the optical excitation gives an indication
of the degree of optical control achievable in practice. Using the
GaAs infra-red source with a variable diameter aperture in an opaque
screen, polar directivity plots were again obtained at a range of
15 cm. The resulting polar plots are displayed in Figs. 12 to 15.

Noting that the maximum possible emergent angle cannot exceed
θ_c, the minimum possible 'r/λ' ratio = $(\frac{1}{2})(\lambda_2/\lambda_1) = (\frac{1}{2})(\sin\theta_c)^{-1}$.
From this the theoretical mainlobe angle variation can be deduced to
follow the relation

$$\delta = \lambda_2(\sin\theta)^{-1} \qquad\qquad\qquad\qquad \ldots\ldots(6)$$

As δ increases above λ_2 the mainlobe angle θ decreases and the angular
space between θ and θ_c fills with sidelobes, the number of observable
sidelobes increasing with δ. The radiated acoustic power at the sideband frequency increases with δ, however, and makes the detection of
sidelobes progressively easier. The converse effect might explain
why no sidelobes could be detected for $\delta \leqslant 2$ mm (they could have
fallen below the noise floor of the detection system). Fig. 16 shows
a graph of the variation of experimentally obtained mainlobe beam
angle with excitation diameter over the range $\delta = 1$ mm to 8 mm. There
is fairly close agreement with the theoretical curve.

The conclusion to be drawn is that the spatial response characteristics of the acoustic source are, without doubt, being controlled
optically and that the reasonable agreement between theoretical
predictions and experimental values indicates that extraneous beam
spreading effects are relatively insignificant for the 'Si-PVF₂' device.

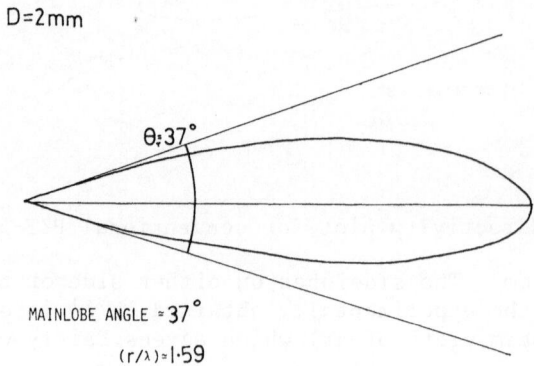

Fig. 13 - Polar plot for D = 4 mm

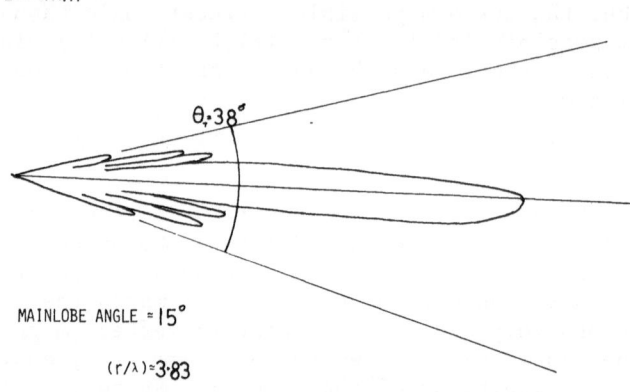

Fig. 12 - Polar plot for D = 2 mm

D=6mm

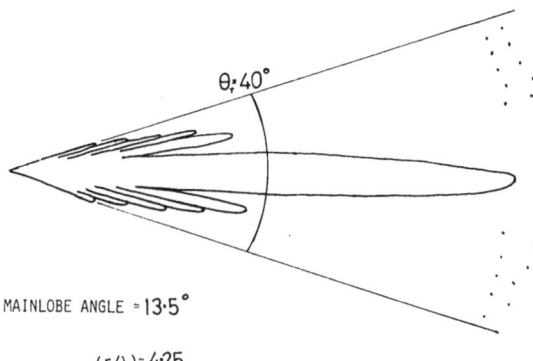

MAINLOBE ANGLE ≈ 13·5°

$(r/\lambda) = 4·25$

Fig. 14 – Polar plot for D = 6 mm

D=8mm

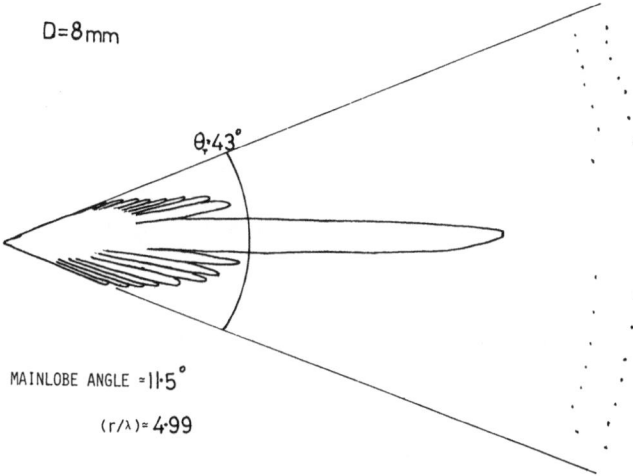

MAINLOBE ANGLE ≈ 11·5°

$(r/\lambda) = 4·99$

Fig. 15 – Polar plot for D = 8 mm

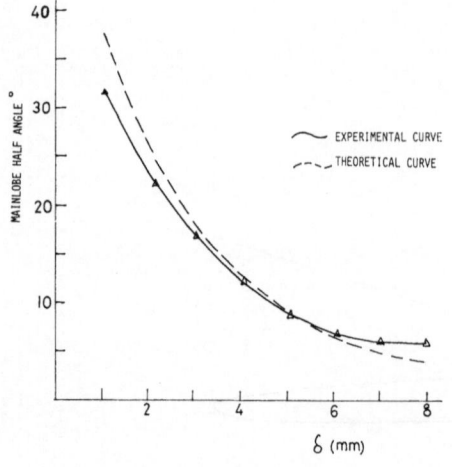

Fig. 16 - Experimental half angle (θ°) versus optical excitation diameter (δ mm) plot

PHASED ARRAY BEAMFORMING WITH THE OPTICALLY-CONTROLLED 'Si-PVF$_2$' TRANSMITTER

Having characterised the individual optically-controlled point sources their utilisation as elements of a phased array can be considered. The phase of the point source output is directly related to the phase of the optical pump and so array phasing can be achieved by controlling the relative phases of the fibre-coupled L.E.D. sources.

The advantages of this type of 'Si-PVF$_2$' opto-acoustic array are

(1) The relative ease of synthesis of the constituent elements of this array. There are no metallization or dicing problems to overcome.

(2) The system gives variability of array geometry, element density and phasing. For example, a high density linear array can be very easily converted to a lower density annular array, to suit the imaging requirements by simply altering the distribution of optical exciters.

(3) Non-uniformities in array response can be corrected for and apodization schemes carried out easily by adjusting the optical pump output from each L.E.D. source.

(4) The excellent electrical and acoustic isolation properties of the Si-PVF$_2$ device suppresses spurious responses.

(5) The spatial impulse response and low intrinsic 'r/λ' ratio reduces the level of grating sidelobes even for marginally under-sampled arrays, and in a 'filled' phased array the total acoustic power sideband-to-carrier ratio tends towards the value for a single cell.

RESULTS OF PRELIMINARY BEAMFORMING TESTS WITH A LINEAR PHASED Si-PVF$_2$
OPTO-ACOUSTIC ARRAY

A 25 element opto-acoustic array with fibres in a '5 x 5' format
has been built (Figs. 17 and 18).

Preliminary tests have been run with a linear array 10 mm wide
containing elements uniformly spaced and individually phased. Pump
signal phasing was carried out using OP-AMP phase shifting networks.
The acoustic output from each fibre-coupled element can be controlled
by trimming the L.E.D. drive circuit, making it possible to compensate
for spatial non-uniformities in both L.E.D. and opto-acoustic device
responses. The phase profiles were calculated to give a focus at a
range of 10 cm using the array phase-focusing formula

$$\Delta\theta = \{\frac{\omega}{V} \cdot \frac{(\Delta x)}{F} \cdot x\} \qquad \qquad \dots \dots (7)$$

where $\Delta\theta$ = phase shift
       ω = angular frequency = $2\pi f$ rads/sec
       Δx = incremental lateral distance
       x = lateral distance from centre line
       F = focal length
       v = acoustic velocity in the medium

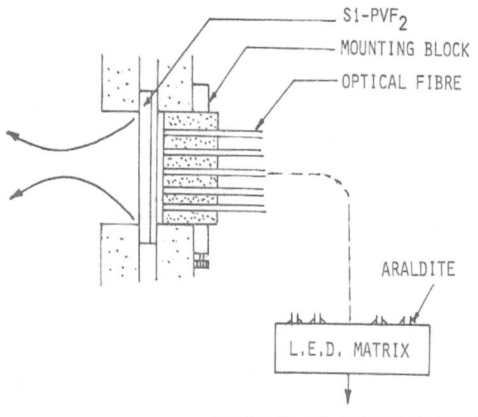

S1-PVF$_2$
MOUNTING BLOCK
OPTICAL FIBRE

ARALDITE

L.E.D. MATRIX

TO DRIVE AND PHASING CIRCUITS

Fig. 17 - Fibre-coupled opto-   Fig. 18 - Opto-acoustic array
          acoustic array on tank   configuration

The beam profiles on the focusing plane have been mapped with an acoustic probe and are displayed in Fig. 19. The focused waist '6dB' width was approximately 8 mm at an axial range of $z \simeq 9$ cm. This can be compared with the diffraction limited value ($\simeq \frac{\lambda F}{D}$) which in this case is 6.5 mm at 2.3MHz. The weak focusing action observed is thus expected and is a direct result of the small array aperture. Right and left-handed steering is apparent in Fig. 19, where the sweep angle has been restricted to prevent oversteering into the tank wall.

A slight asymmetry in sidelobe distribution can be observed in the 'steered' plots. The measured improvement in sideband-to-carrier ratio at the focus over that for a single point source is approximately 11dB.

The theoretical focal plane distribution is drawn in Fig. 20 and is compared with the experimental distribution. For the test array, the theoretical spacing between the grating sidelobes is

$$(\frac{\lambda z_0}{d}) \;=\; 33 \text{ mm}$$

but it is seen that these do not show up clearly because of the generally high sidelobe level away from the central maximum.

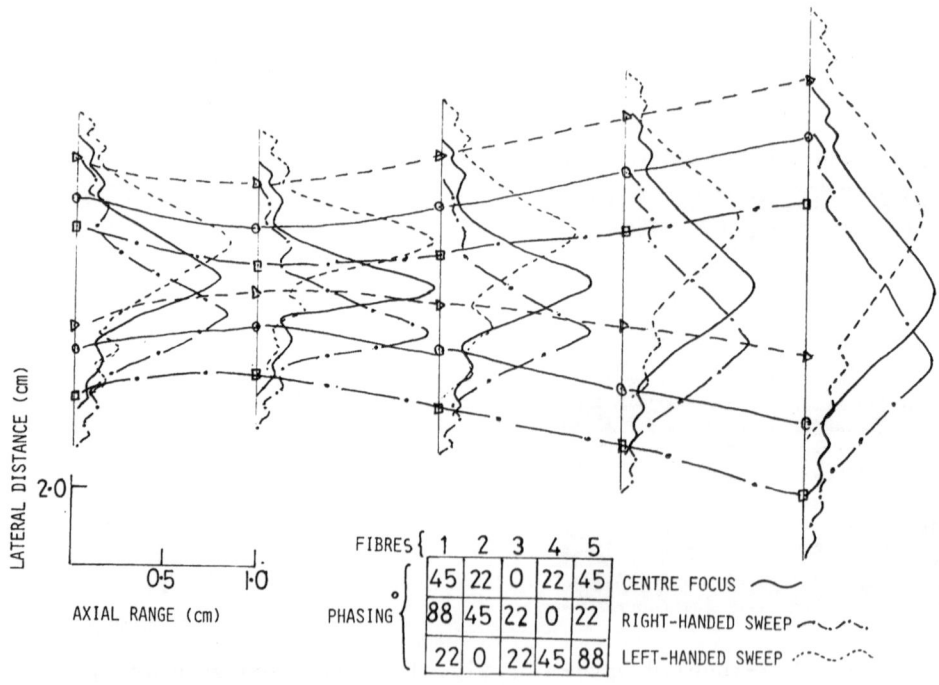

Fig. 19 - Beam profiles for linear opto-acoustic array

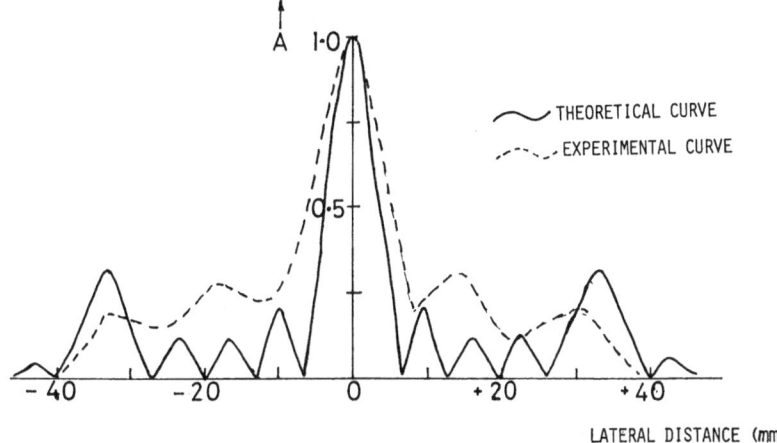

Fig. 20 Focal plane distribution for array

CONCLUSIONS

     An easily implemented opto-acoustical method for synthesising
point sources has been presented. The individual point-source
characteristics for the 'Si-PVF$_2$' device have been investiaged and
found to yield encouraging results, particularly in terms of spatial
response. The preliminary phased array tests have also confirmed the
practical feasibility of an optically controlled, phased acoustic
array. The ease with which the point sources can be addressed, and
their geometrical distribution varied, points to useful applications
in dynamically-scanned and focused two-dimensional arrays, particularly at higher frequencies.

ACKNOWLEDGEMENTS

     The technical assistance of Mr. A. Moore with the mechanical
fabrication of the fibre-coupled L.E.D. matrix is gratefully
acknowledged as is the contribution of Thorn-EMI (Plastics Division)
who provided the plate of thick film PVF$_2$ used.

REFERENCES

1. M. Pappalardo, "Hybrid linear and matrix acoustic arrays",
      Ultrasonics, Vol. 19, No. 2, March 1981.

2. H. F. Tiersten, J. F. McDonald and P. K. Das, "Two-dimensional
      monolithic mosaic transducer array", IEEE Ultrasonics
      Symposium Proceedings, 1977, p408.

3. B. A. Auld, C. DeSilets and G. S. Kino, "A new acoustic array
    for Acoustical Imaging", IEEE Ultrasonics Symposium
    Proceedings, 1974, p24.

4. M. G. Maginness, J. D. Plummer and J. D. Meindl, "An acoustic
    image sensor using a transmit-receive array", Acoustical
    Holography, Vol. 5, p619.

5. N. Takagi, T. Kawaghima, T. Ogura and T. Yamada, "Solid-state
    acoustic image sensor", Acoustical Holography, Vol. 4, p215.

6. J. Goodman, "Introduction to Fourier Optics", McGraw-Hill,
    New York, 1968.

7. B. Noorbehesht, G. Flesher and G. Wade, "Spatial response of
    arbitrarily electroded piezoelectric plates by plane-wave
    decomposition", Ultrasonic Imaging, Vol. 2, No. 2, April 1980,
    pp102-121.

8. M. Ahmed, "The response of piezoelectric face plates in ultrasonic
    imaging systems", IEEE Trans. Sonics and Ultrasonics SU-25,
    1978, pp330-339.

EFFECT OF MECHANICAL ABERRATIONS ON THE

RESOLUTION OF FRESNEL ZONE PLATE TRANSDUCERS

B. Noorbehesht

Department of Electrical Engineering
University of Houston
Houston, Texas 77004

M. Mortezaie and G. Wade

Department of Electrical & Computer Engineering
University of California
Santa Barbara, California 93106

C. Schueler

Advanced Applications Department
Santa Barbara Research Center
Goleta, California 93117

ABSTRACT

Ultrasonic imaging transducers have been proposed in which a piezoelectric plate is used in conjunction with a Fresnel Zone Plate (FZP) to produce a focused beam of ultrasound. Focusing is achieved with these devices by activating the piezoelectric plate with an electrode pattern in the shape of an FZP. The focal plane distribution of these transducers cannot, in general, be predicted by diffraction theory alone. This is because the acoustic field pattern in front of the transducer is not an exact replica of the FZP electrode pattern. In this paper we show that elastic wave generation and propagation inside the piezoelectric plate cause the acoustic field pattern directly in front of the plate to differ from the electrode pattern. These effects are called mechanical aberrations. We show that mechanical aberrations cause the focal plane distribution to be broader than that predicted by diffraction theory. The effects on resolution of material properties and other device parameters are discussed. Experimental results are also presented.

INTRODUCTION

Ultrasonic imaging devices have been proposed in which a piezoelectric plate is used in conjunction with a Fresnel Zone Plate (FZP) to produce a focused beam of ultrasound [1,2]. This focused beam must be scanned over the object plane to produce an image. One way to do this is to scan the FZP pattern over the surface of the transducer. A scheme for Zone Plate scanning has been proposed by Wang and Wade [3], in which a photoconductive layer is used in conjunction with the piezoelectric plate. A scanning light beam carrying the FZP pattern can be used to control the photoconductor. Here we concentrate on problems associated with producing the focused acoustic beam.

One measure of effectiveness of FZP transducers is their resolving power. This may be determined by the size of the generated focal spot. The focal plane distribution cannot, in general, be calculated by considering diffraction theory alone, as is usually done in the optical case [4]. In the acoustical case we should also take into account the spatial response of the piezoelectric plate, i.e. its ability to generate an acoustical nearfield pattern which replicates the electrode (FZP) pattern.

In this paper, we use a linear spatial piezoelectric transducer model developed previously [5] to construct a spatial transfer function for an FZP transducer system. The transfer function can be represented as the product of two factors. The first is denoted H_A, and relates the acoustical field distribution in front of the transducer to the electrode pattern. The second is denoted H_D, and represents the effect of wave propagation from the transducer surface (z=0) to the FZP focal plane (z=f). H_A is called the aberration-limited transfer function because it represents the effect of mechanical wave-spreading in the transducer [5]. This is analogous to aberrations in optical imaging systems. H_D is the diffraction-limited transfer function.

We show that the effect of mechanical aberrations is to alter the shape of the diffraction-limited system spatial transfer function in such a way as to lower the cut-off spatial frequency. Because of this, the focal plane distribution is broadened, resulting in poorer resolution. The new cut-off frequency is related to the material properties of the piezoelectric plate. From this information, we theoretically determine the best resolution obtainable from any piezoelectric material. Experimental measurements of the acoustical focal plane distribution for a one-dimensional FZP transducer are presented and compared with the theoretical predictions.

SPATIAL TRANSFER FUNCTION OF THE FZP TRANSDUCER

To study the resolution characteristics of an FZP transducer, we treat the transducer as a linear spatial system. The
electrode pattern on the piezoelectric plate is the input and the
focal plane distribution is the output. This is a valid model
because both the piezoelectric effect and wave propagation are
linear phenomena over a wide range of amplitudes. If the near-
field acoustic distribution were an exact replica of the electrode pattern, then the focal plane distribution could be obtained by standard techniques, as in optics. The focal plane
distribution under these conditions could be predicted accurately
by applying the Fresnel diffraction formula to the electrode
pattern. This process may be characterized by H_D, a linear
spatial transfer function, which represents the effect of wave

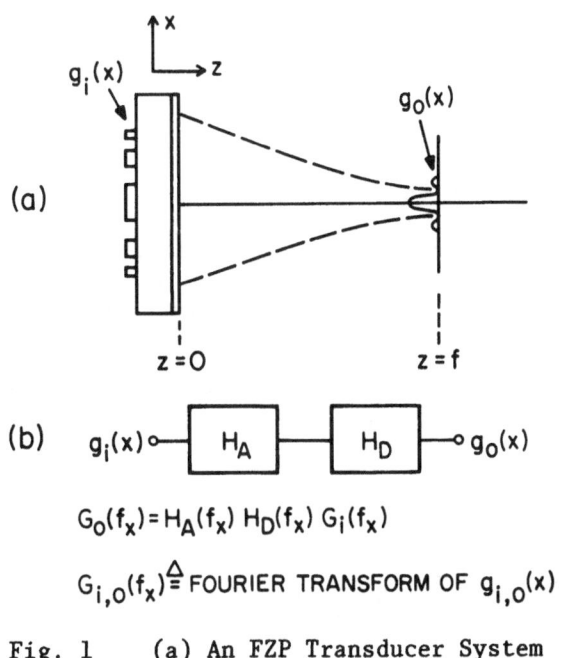

$$G_0(f_x) = H_A(f_x) \, H_D(f_x) \, G_i(f_x)$$

$$G_{i,0}(f_x) \overset{\Delta}{=} \text{FOURIER TRANSFORM OF } g_{i,0}(x)$$

Fig. 1   (a) An FZP Transducer System

              (b) Linear System model of (a)

propagation from the z=0 to the z=f plane. In reality, however, the nearfield acoustic pattern is not exactly the same as the electrode pattern, as discussed earlier. This effect may be characterized by H_A, another linear spatial transfer function. Consequently, an FZP transducer system such as the one shown in Fig. 1(a) may be represented by the linear model depicted in Fig. 1(b). In this analysis, we assume variation in the x-direction only. Generalization to two dimensions may be obtained by simple mathematical manipulations [6].

DIFFRACTION LIMITED TRANSFER FUNCTION

H_D characterizes the effect of wave propagation from the z=0 plane to the z=f plane. Using the planar spectrum approach [7], H_D may be represented by:

$$H_D(f_x) = \text{rect}(\frac{\lambda f_x}{2})\exp[j(2\pi f/\lambda)(1-f_x^2 \lambda^2)^{1/2}] \tag{1}$$

rect ($\cdot$) represents the rectangle function, which is equal to unity when the absolute value of its argument is less than or equal to 1/2, and is zero otherwise.

ABERRATION LIMITED TRANSFER FUNCTION

$H_A(f_x)$ can be written as:

$$H_A(f_x) = |H_A| \exp(j\angle H_A) \tag{2}$$

Explicit magnitude and phase of H_A can be found in reference [5]. Although both of these quantities are important to determine the focal plane distribution, we only use the magnitude of H_A because it provides adequate qualitative information about the resolution of the device. $|H_A|$ is plotted in Fig. 2 as a function of normalized spatial frequency, $v_x = f_x\lambda$. f_x is the spatial frequency and λ is the wavelength in water. (To obtain the result in Fig. 2, only compressional waves have been taken into account. The contribution due to shear waves has been neglected, because they are weakly coupled in this case.)

The cut-off spatial frequency v_{xc} is given by:

$$v_{xc} = f_{xc} \lambda = \frac{v}{(c_{11}^E/\rho)^{1/2}} \tag{3}$$

v is the sound velocity in water, ρ is the density of the piezoelectric material, and c_{11}^E is one of its elastic constants.   The
plot in Fig. 2 has been obtained for a PZT-5A transducer resonant
at 3.4 MHz and in contact with water.

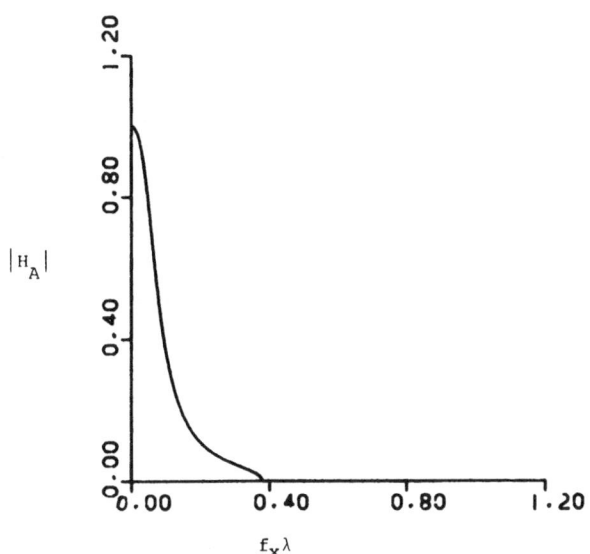

Fig. 2   Magnitude of the aberration limited transfer function
           versus normalized spatial frequency.

EFFECT OF ABERRATIONS ON RESOLUTION

     The transmission function $g_i(x)$ of a one-dimensional, posi-
tive FZP pattern is shown in Figure 3. The parameters given in
that figure are related to λ and the FZP focal length f by the
following equations:

$$x_n = [nf\lambda + (n\lambda/2)^2]^{1/2} \tag{a}$$

$$x(n) = (x_{n-1} + x_n)/2 \tag{b}$$

$$\Delta x(n) = x_n - x_{n-1} \tag{c} \quad (4)$$

$$D = 2x_N \tag{d}$$

$$N = -2(f/\lambda) + [4(f/\lambda)^2 + (D/\lambda)^2]^{1/2} \tag{e}$$

N is the number of zones. Each zone consists of a pair of electrodes situated symmetrically about the x = 0 line. D is the width of the zone plate. If D, λ, and f are known, equation 4e can be used to obtain the required number of zones. The value for N, however, must be rounded to the nearest integer.

Using equations (4), the Fourier transform of $g_i(x)$ can be written as:

$$G_i(f_x) = \sum_{n=1}^{N} [2\Delta x(n)\cos(2\pi x(n)f_x)] \left[\frac{\sin(\pi\Delta x(n)f_x}{(\pi\Delta x(n)f_x)}\right] . \tag{5}$$

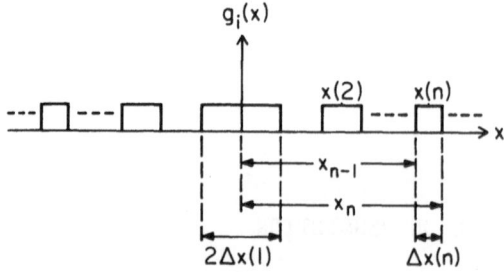

$g_i(x)$

Fig. 3   Transmission function $g_i(x)$ of a one-dimensional FZP Pattern.

The focal plane distribution $g_o(x)$ can be found from the following:

$$G_o(f_x) = H_A(f_x)H_D(f_x)G_i(f_x) \qquad (6)$$

$$g_o(x) = \int_{-\infty}^{+\infty} G_o(f_x)\exp(j2\pi f_x x)df_x \qquad (7)$$

Figure 4 shows the magnitude of $G_i(f_x)$ as a function of normalized spatial frequency for three values of N with $f = 80\lambda$. Figure 5 shows the magnitude of $G_o(f_x)$. Here, due to aberrations, most of the high spatial frequency content of the FZP pattern is lost. Figure 6 shows variations of focal plane intensities (for the same three values of N) as a function of normalized lateral distance x/λ. Dashed curves represent the aberration-free case, i.e. where $H_A(f_x) = 1$. These are simply the Fresnel diffraction patterns of the FZP. It can be seen that the resolu-

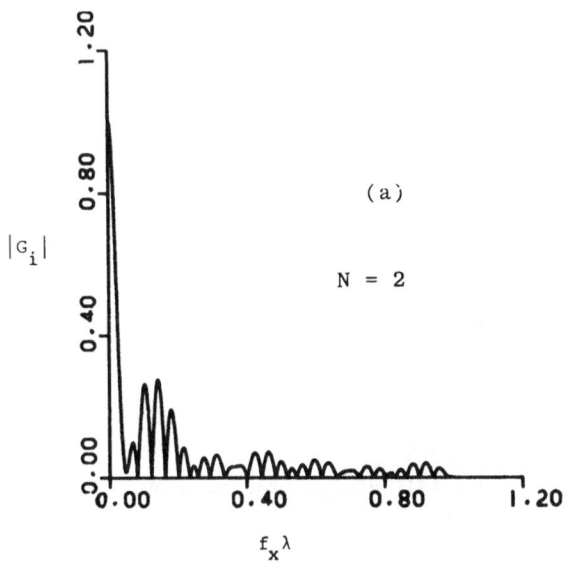

Fig. 4   Spectral content of an ideal FZP Pattern for:

(a) N=2   (b) N=4   (c) N=6

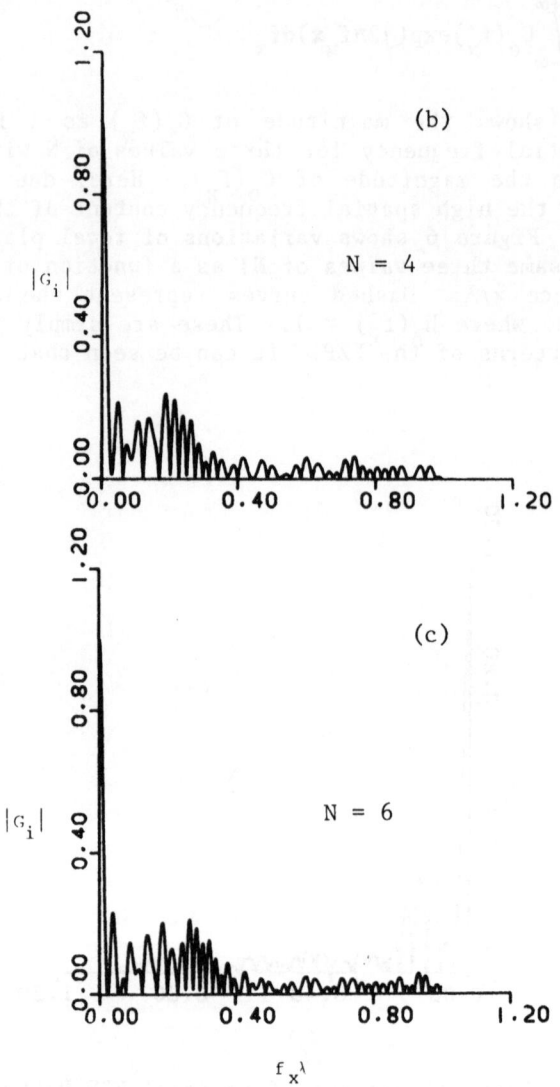

Fig. 4 - Continued

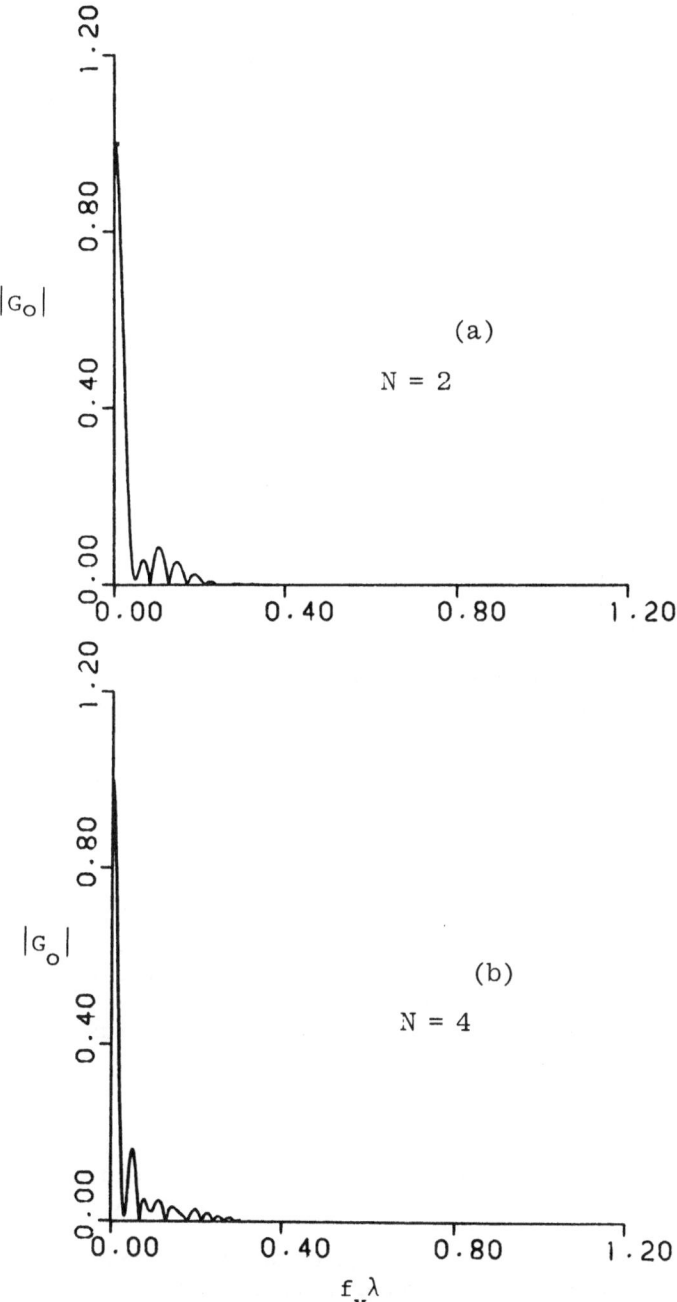

Fig. 5   Spectral Content of an FZP Pattern reproduced in acoustic form by a piezoelectric transducer for:

(a) N=2   (b) N=4   (c) N=6

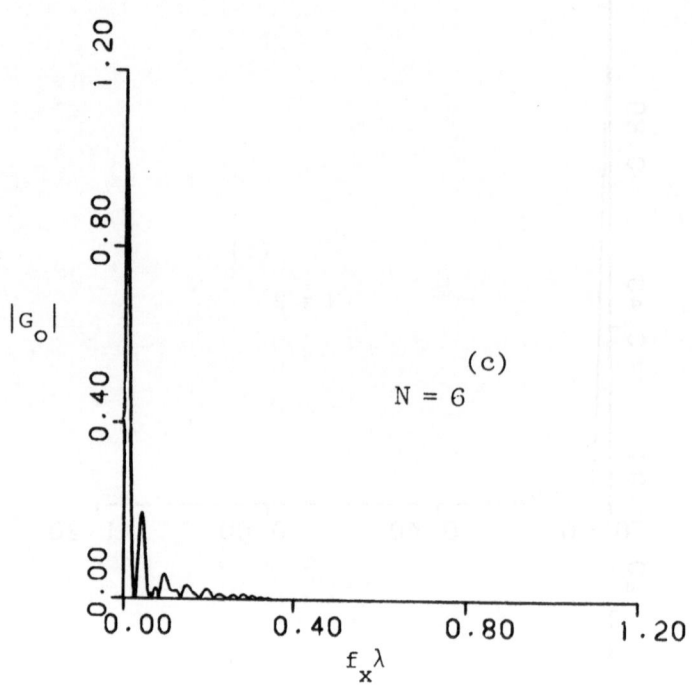

Fig. 5 - Continued

tion, as defined by the width of the 40% intensity points (Rayleigh criterion), increases as N increases. The solid curves represent focal plane intensities when aberrations are taken into account. Due to attenuation of high spatial frequencies inside the transducer, as indicated in Figure 5, the widths of the focal spots are increased. This results in worse resolution than in the previous case. Also, since spatial frequencies above ν_{xc} are absent, increasing N beyond a certain limit does not improve the resolution. This optimum value of N may be found from properties of the transducer material and the propagation medium as well as the specifications of the zone plate (i.e. f and λ).

Finally, the crosses in Figures 6(a)-(c) represent the results of experimental measurements taken with a one-dimensional FZP electroded transducer, for the N=2, 4, and 6 zone cases. In the next two sections we discuss the experimental technique, and we compare the theoretical and experimental results shown in fig. 6.

EXPERIMENTAL TECHNIQUE

The experiment employed a 3×3 cm^2 PZT-5A transducer (Model Channel 5500 from Channel Industries, Goleta, California). Aluminum was evaporated on both sides of the transducer. One side was photolithographically etched to give a positive FZP pattern. The other side remained fully electroded. The pattern was one-dimensional as depicted in Fig. 7, and the transducer was operated at its resonant frequency of 3.4 MHz. Wires were connected to both sides of the transducer from an RF generator. The transducer was mounted on the acoustic tank with the fully electroded side facing the water, as indicated in Figure 7.

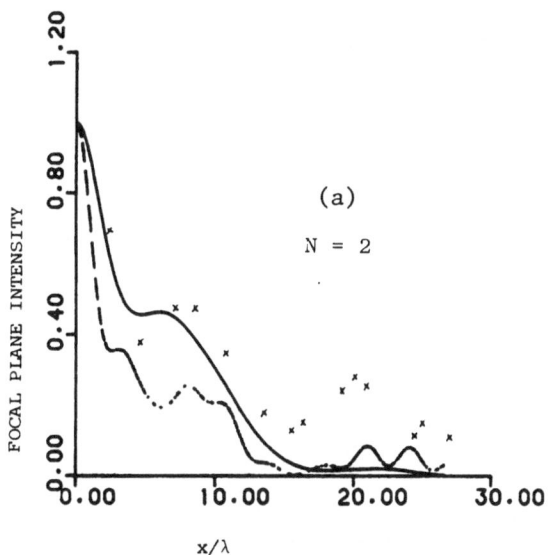

Fig. 6   Focal plane intensities of an FZP transducer for: (a) N=2   (b) N=4   (c) N=6

Dashed curves represent the ideal diffraction limited case, while solid curves correspond to the case when aberrations are included. Finally, crosses are the results of the experimental measurements.

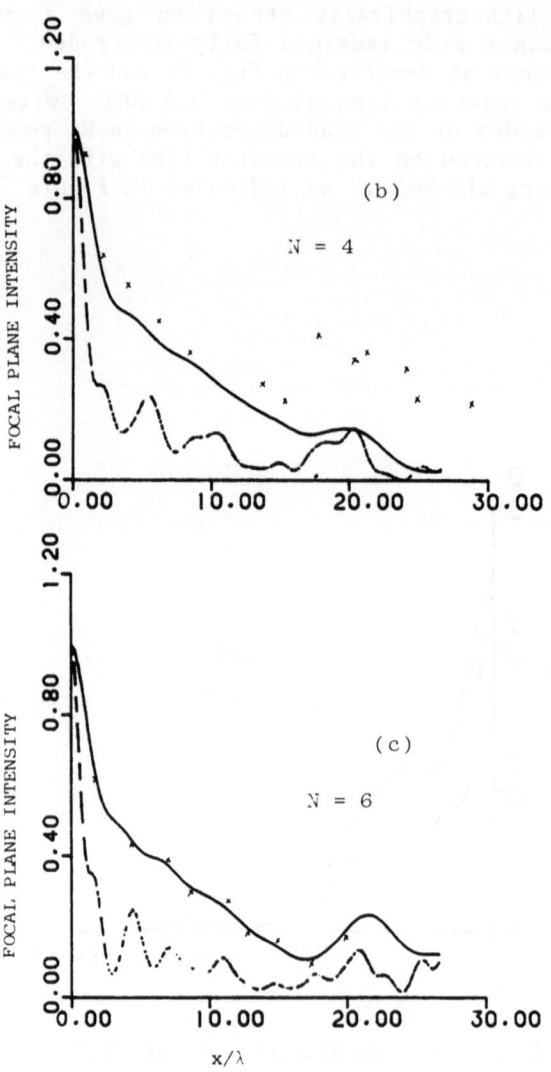

Fig. 6 - Continued

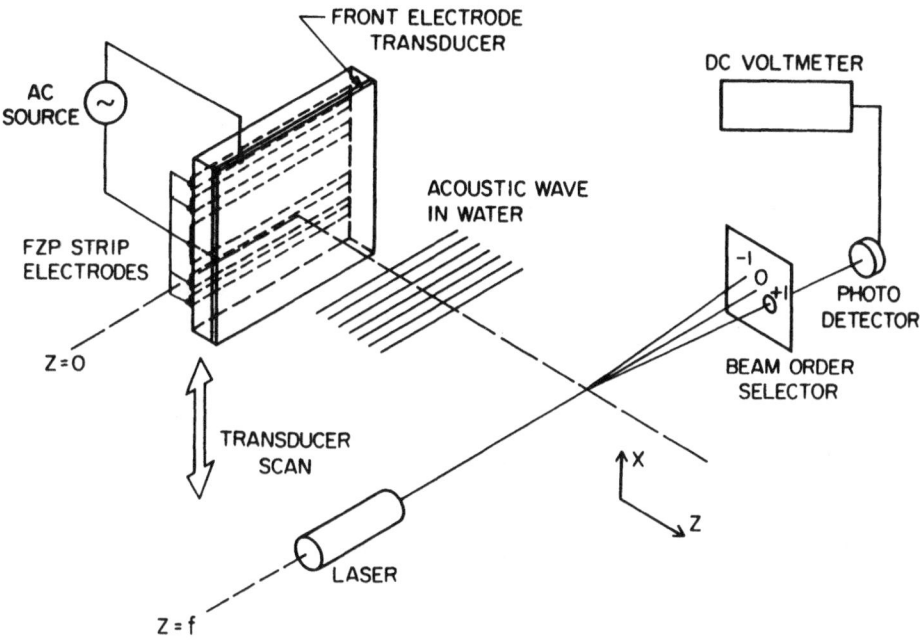

Fig. 7   Experimental Setup. Not explicitly shown is a water-
         filled plexiglas tank. The tank is moved vertically
         (x-axis) to obtain the transducer scan indicated. The
         laser, beam order selector, and detector are outside
         the tank. The laser beam enters the plexiglas tank
         along the y axis, as indicated, interacts with the
         acoustic waves in the water, and then exits the tank to
         be measured at the photo-detector.

        Debye-Sears diffraction of light by sound was used to mea-
sure acoustic intensities [8]. For low acoustic field intensities, the first diffracted order of light intensity is proportional to the acoustic field intensity. A laser beam was passed
through the focal plane of the FZP transducer, parallel to the
transducer plane and along the long axis (y-axis in Fig. 7) of
the zone stripes. The light was diffracted at the acoustic
focal-plane of the transducer and a pin hole was used to select
the first diffracted order. This light was detected by a photo-
detector, and its intensity measured by a dc-voltmeter.

The scan along the x-axis was achieved by manual vertical (x-axis) translation of the tank. The zone plate stripes (y-axis) were assumed to be long enough for end effects to be negligible. The experimental results for N=2, 4, and 6 positive zone plates are shown in Fig. 6(a)-(c), for comparison to theory. The crosses represent the experimental results.

Although the experimental results are close to the theory, at least three error sources should be mentioned. First, the diffracted beam intensity is not exactly linearly related to the acoustic field intensity. Second, the PZT-5A transducer material may not be perfectly uniform in piezoelectric characteristics [2]. Third, refraction at the tank walls may not be identical at every x-axis scan position of the tank. Therefore, the position of the light beam relative to the transducer may have changed slightly as the tank was moved. Also, the infinite plate assumption made to simplify the theory is not exactly valid, either along the x or y-axis. This is especially true as we increase the number of electrodes on a transducer plate to obtain an FZP of higher zone number. In this case, the outer zones are quite near the x-axis edge of the transducer plate, and the infinite plate assumption becomes less accurate.

DISCUSSION

We note first that the theoretically predicted trend of increasing resolution with increasing FZP zone number (a well-known diffraction effect) is verified by the experimental results shown. But most satisfying is that the experimental resolution of the focal spot for N=2, 4, and 6 is best predicted by the theory when aberration is taken into account.

The correlation between experiment and theory would be better if the theory were to account for the effect of a finite transducer. In addition, the experimental error sources previously mentioned certainly account for some of the discrepancies between theory and experiment. Clearly, however, mechanical aberrations worsen the diffraction-limited resolution, as predicted by including the aberration-limited transfer function in the computation of piezoelectric transducer response to the FZP zone-plate pattern. The experimental results appear to corroborate the theory quite well in this respect.

CONCLUSION

We have presented both theoretical calculations and experimental measurements of the focal-plane resolution for an FZP piezoelectric transducer. The results show that the focal-plane resolution is closely predicted by taking into account both the diffraction of the electrode pattern and mechanical aberrations caused by the piezoelectric material.

Therefore, aside from the secondary effects of shear waves and mode conversion, and the effect of the finite transducer size, we have shown how the resolving power of an FZP transducer may be properly characterized. This is an important step in the design and implementation of any ultrasonic imaging system that uses FZPs and piezoelectric transducers.

ACKNOWLEDGEMENTS

The authors wish to thank the National Science Foundation, which supported parts of this work under grant NSF ECS-79-13259 at the University of Houston, and grant NSF ECS-79-18779 at the University of California, Santa Barbara. We also want to thank Tracy Hamilton, of the Electrical and Computer Engineering Department at the University of California, Santa Barbara, for carefully typing the manuscript.

REFERENCES

1. S.A. Farnow and B.A. Auld, "Acoustic Fresnel Zone Plate Transducers," App. Phys. Lett., Vol. 25, pp. 681682, 1974.

2. K. Wang, V. Burns, G. Wade, and S. Elliott, "Opto-Acoustic Transducers for Potentially Sensitive Ultrasonic Imaging," Opt. Eng., Vol. 16, No. 5, pp. 432-439, Sept/Oct 1977.

3. K. Wang and G. Wade, "A Scanning Focused-Beam System for Real-Time Diagnostic Imaging," Acoustical Holography, Vol. 6, N. Booth Ed., Plenum Press, New York, pp. 213-228, 1975.

4. D.J. Stigliani, Jr., R. Mittra, and R.G. Semonin, "Resolving Power of a Zone Plate," J. Opt. Soc. Am., Vol. 57, pp. 610-613, 1967.

5. B. Noorbehesht, G. Flesher, and G. Wade, "Spatial Response of Arbitrarily Electroded Piezoelectric Plates by Plane Wave Decomposition," Ultrasonic Imaging, Vol. 2, pp. 102-121, 1980.

6. Generalization to two dimensions may be achieved by taking
    advantage of the crystallographic symmetry of the piezoelec-
    tric ceramics. The two dimensional transfer function
    $H(f_x, f_y)$ or $H(f_r f_\theta)$ (in polar coordinates) may be obtained
    from $H(f_x)$ by a rotation of $H(f_x)$ around the $H(f_x)$ axis.
    This is mathematically equivalent to replacing f_x in $H(f_x)$
    by f_r. Due to the symmetry, there is no dependence on f_θ.
    The corresponding point spread function $h(r,\theta)$ is obtained
    by an Inverse Fourier Bessel transform of $H(f_r, f_\theta)$. It
    should be noted that $h(r,\theta)$ cannot be obtained from $h(x)$ by
    a simple rotation of $h(x)$ about the $h(x)$ axis.

7. J.W. Goodman, Introduction to Fourier Optics, McGraw-Hill,
    New York, 1976, pp. 54.

8. R. Adler, "Interaction Between Light and Sound," IEEE Spec-
    trum, Vol. 4, No. 5, pp. 42-54, 1967.

SIDELOBE REDUCTION OF THE RING ARRAY FOR

USE IN CIRCULARLY SYMMETRIC IMAGING SYSTEMS*

Bernard D. Steinberg and Ajay K. Luthra

Valley Forge Research Center
University of Pennsylvania
Philadelphia, PA   19104

INTRODUCTION

The ring or annulus array offers certain attractive properties for use in circularly symmetric imaging systems. Its excitation function is defined by

$$i(r,\phi) = ae^{j\phi} \sum_{n=0}^{n=N-1} \delta(\phi - \frac{2\pi n}{N}) \qquad (1)$$

$ae^{j\phi}$ is a continuous annulus of radius a while the second term is the sum of Dirac delta-functions that are nonzero at discrete angles $2\pi n/N$ (N is the number of elements in the array and n = 0,1,...N-1). Its attractive properties are:

1. The underlying circular symmetry of the ring results in nearly circular symmetry on the radiation sphere, i.e., except for the discreteness of the array the radiation pattern is only a function of the polar angle θ and not the azimuthal angle ϕ. This property is generally advantageous in a scanning or imaging system in which there is no preferred azimuth.

2. Although the angular spacing of elements on the ring is periodic, as given in (1), the apparent spacing from any direction of view other than the zenith is not uniform. This nonuniformity of apparent element spacing permits thinning the

*This work was principally supported by Interspec, Inc. and by the Office of Naval Research, the latter under Contract No. N00014-79-C-0505.

array, i.e., mean interelement spacing $> \lambda/2$, where λ is the wavelength, without introducing grating lobes (à la the interferometer) [*]. This property is advantageous because it permits a larger aperture and therefore a finer resolving power from a fixed system cost. The reasoning is as follows: given a limited number N of elements and accompanying circuits associated with a given system cost, the size L of the square array having $\lambda/2$ element spacing is $\lambda\sqrt{N}/2$. The available beamwidth $\Delta\theta$ always is about λ/L, which equals $2/\sqrt{N}$, and is limited, therefore, by N. Furthermore, the resolving power improves only as $N^{1/2}$. A bilinear crossed array of length L has a smaller beam cross-section: $2L = N\lambda/2$ and $\Delta\theta \simeq 4/N$. The periodic ring array is still better because the $\lambda/2$ spacing limitation is removed.

3. Because of the circular symmetry, beamforming and scanning phase shifts and/or time delays need only be calculated for a single angular interval $2\pi/N$. The same set of values can be made to pertain to every sector by rotating the angular frame of reference (in the computer or signal processor) in steps of $2\pi/N$. This property can result in a significant reduction in the data handling requirement in a digital processing system.

4. The beamwidth of a ring array is theoretically smallest for a given size aperture for which the requirement of circular symmetry is imposed. This is because the elements are dispersed most widely from the center subject to the given condition. This property leads to the finest available resolving power from a circularly symmetric aperture of fixed size.

One reason for the relative lack of interest in the ring array is the loss of simplicity in the calculations or settings of the phase shifts or time delays as compared with the periodic linear or two-dimensional array. This disadvantage largely vanishes with the advent of digital control and is further obviated by the third feature discussed above. The major disadvantage is the very poor sidelobe properties of the annulus. Given its attractive properties the question of whether its side radiation pattern can be improved to the point of acceptability becomes relevant. The object of this paper is to demonstrate that the answer often is yes.

The paper opens with an approximate derivation of the monochromatic properties of the periodic ring array, showing that its

---

*Grating lobes result from linear periodicities in the aperture excitation function when the interelement spacing exceeds $\lambda/2$. This is because the far-field radiation pattern is the Fourier transform of the excitation function, and the Fourier transform of a periodically sampled function also is periodic.

theoretical beamwidth is $0.7\lambda/L$ (as compared to $0.88\lambda/L$ for the uniformly illuminated aperture) and that its largest sidelobe is -8 dB (as compared to -13 dB). By using wide signal bandwidth, multiple rings, shaped waveforms, aperture tapering and matched filter detection it is shown that (1) all sidelobes can be reduced to about -30 dB or less, or (2) sidelobes more than 5 or 6 beamwidths from the main lobe can be suppressed below -40 dB if some liberties are permitted in the nearby sidelobes. The numbers of elements in all the arrays described below are 100 or less.

THE CONTINUOUS RING

     The relevant properties of the array are most easily studied by examining the continuous ring antenna having constant excitation. It is represented by the first term in (1). Its radiation properties closely approximate that of the discrete array when N is large. The far-field, monochromatic radiation pattern is easily calculated in any polar cut of the radiation field, i.e., in a plane intersecting the radiation sphere and containing the polar axis (see Figure 1): It is the Fourier transform of the equivalent line excitation on the diameter of the array in the polar plane, when the beam is pointed along the polar axis, and it is a similar integral with the integrand modified by a linear phase term for other beam directions [1]. It is evident that, for the annulus, i(x) is proportional to the differential arc length.

$$ds = [(dx)^2 + (dy)^2]^{\frac{1}{2}} \tag{2}$$

Since $x^2 + y^2 = a^2$ is the locus of the annulus it is evident that

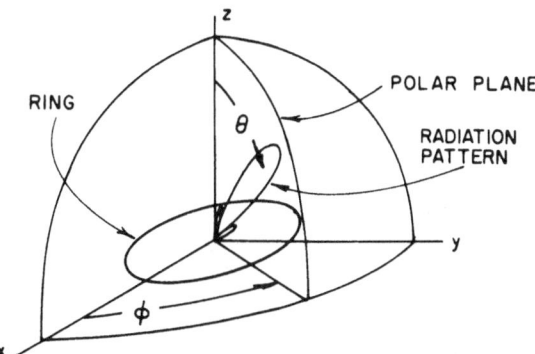

Figure 1.   Ring Array and its Radiation Pattern in A Polar Plane.

$$i(x) \alpha \frac{1}{\sqrt{a^2 - x^2}} \qquad (3)$$

which is sketched in Figure 2.

It is well known that sidelobe reduction is achieved by tapering an aperture excitation so that the energy density is clustered near the central region and reduces gradually towards the edges of the aperture. Quite the contrary situation is evident in Figure 2. The sidelobe level would be expected to be between the -13 dB level of the uniformly excited aperture and the 0 dB level of the two-element interferometer, in which the entire excitation is accumulated at the edges. That this is true is found from the Fourier transform of (3) which is

$$f(u) = J_0 \left(\frac{2\pi a u}{\lambda} \right) \qquad (4)$$

where $u = \sin \theta$, θ is the angle from the axis of the main lobe, and $J_0(\cdot)$ is the zeroth order Bessel function, the first lobe of which is 8 dB below the value at the origin. The beamwidth Δu is related to the second central moment σ_x^2 of the aperture excitation defined by

$$\sigma_x^2 = \frac{\int x^2 i(x) dx}{\int i(x) dx} \qquad (5)$$

through the relation

$$\Delta u = \frac{\lambda}{4\sigma_x} \qquad (6)$$

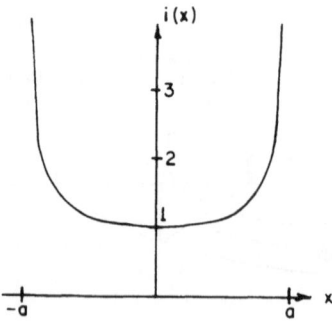

Figure 2.   Equivalent Line Excitation for The Ring Array.

Equation (6) evaluates to 0.7 $\lambda/2a$ for the annular ring of diameter 2a.

While the beamwidth is attractive compared to the uniformly excited aperture ($0.88\lambda/L$, where L is the length of the aperture) or to the triangularly tapered aperture ($1.25\lambda/L$), the high side-lobe level makes the ring relatively unattractive for array design for an imaging system. Sidelobe control techniques are available, however, under certain circumstances and these are discussed in the following sections.

Figure 3 shows the calculated monochromatic radiation pattern in a polar cut of a periodic ring array of 100 equally spaced elements. Its first sidelobe is −8 dB. This figure and all subsequent radiation patterns correspond to a 10 cm diameter array focused 3.5 mm from its normal axis at a distance of 10 cm. The mean wavelength is 0.6 mm, corresponding to an ultrasonic center frequency of 2.5 MHz. The assumed propagation medium is water.

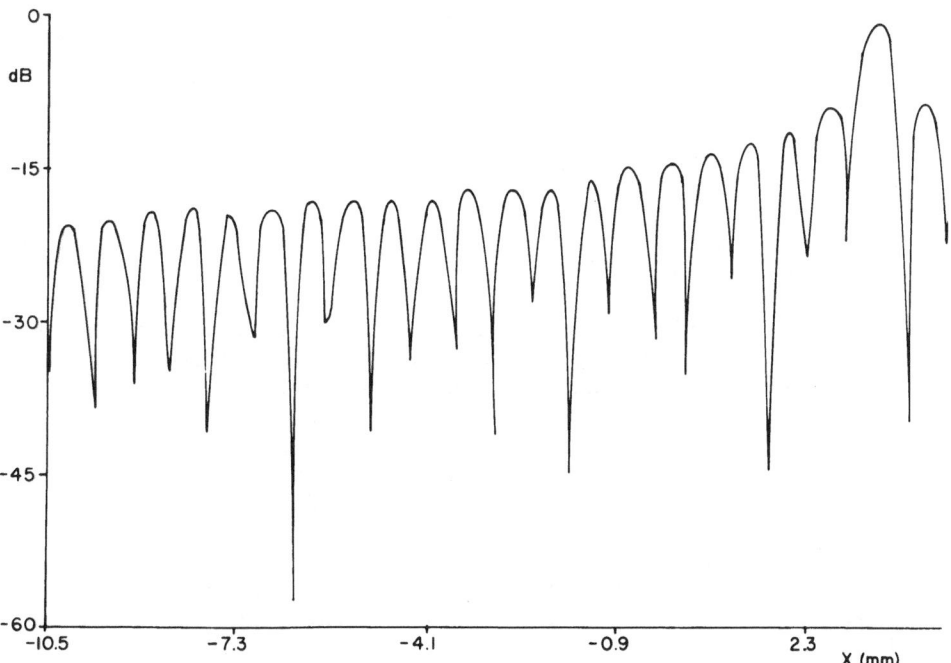

Figure 3. Monochromatic Radiation Pattern of 100-Element Ring
          Array, 10 cm Diameter, Focused at 10 cm Distance.
          WAVELENGTH IS 0.6 mm.

484   B. D. STEINBERG AND A. K. LUTHRA

Typical ultrasonic transducers are wide band. Wideband transmission or reception requires time-delay/steering or focusing
instead of the simpler phase shift steering or focusing. While
delay steering increases system complexity, wideband signalling
materially decorrelates signals arriving off the beam axis and
thereby reduces the side radiation level of the pattern. The
computed patterns of Figure 4 show the effect for bandwidths of
approximately 30, 50 and 70%. The signal waveform is

$$[1+\cos(2\pi f_m t)]\cos(2\pi f_c t)\text{rect}(1/f_m) \tag{7}$$

It is one cycle of a raised-cosine amplitude modulation of a carrier
sinusoid. The nearest sidelobe dropped to -12 dB and the sidelobes
fall off thereafter toward the -36 to -42 dB region. The detector
is an envelope detector plus integrator.

MATCHED FILTER DETECTION

In certain problems it is possible to use matched filter
detection [2], in which cases the pattern properties can be further
improved. In Figure 4 the output signal was calculated as the
integral of the amplitude of the received wave for a duration equal
to one pulse. Applying a matched filter detector further reduces
the sidelobes as shown in the computed pattern of Figure 5. This
figure represents the 50 percent bandwidth case. Shown also is
the corresponding curve of Figure 4. It is seen from Figure 5
that the first sidelobe has been reduced by a few dB and the nearby
sidelobes by as much as 13 dB.

Matched filtering is generally useful in sidelobe suppression,
provided that the problem permits its use. One limitation to its
use is system complexity and cost, which are evidently greater than
a simple envelope detector and integrator. The second is a more
fundamental limitation pertaining to the frequency-selective
attenuative properties of such media as biological tissue. In such
media, the radiated waveform can be markedly different from the
received waveform. This is not serious provided that the designer
knows a priori the nature of the received waveform so that the
proper matched filter detection waveform can be employed. In a
large aperture system, however, focused in the very near field, as
in the illustrations in this paper, the differential distances
from target to array elements may be sufficient to cause the
waveforms received by the different elements to differ considerably.
The value of matched filter detection under such circumstances is
questionable.

ANGULAR APERTURE TAPER

If the clutter or scattering centers are confined to a narrow slice of the volume passing through the array, instead of being spread throughout the volume, a further improvement can be obtained through the use of an amplitude taper. The effective excitation along the diameter in the array lying in the plane containing the scatterers was given by (3) for the continuous ring. The same function expresses the effective number density of elements, $n(x)$, along the same diameter. Now weight the elements according to their x-coordinates by the weighting factor $w(x)$. The equivalent current excitation along the diameter becomes

$$i(x) = n(x)w(x) \qquad (8)$$

and the radiation pattern in the polar plane containing the clutter is the Fourier transform of (8). The key factor responsible for the high sidelobes of Figure 5 are the edge properties of the distribu-

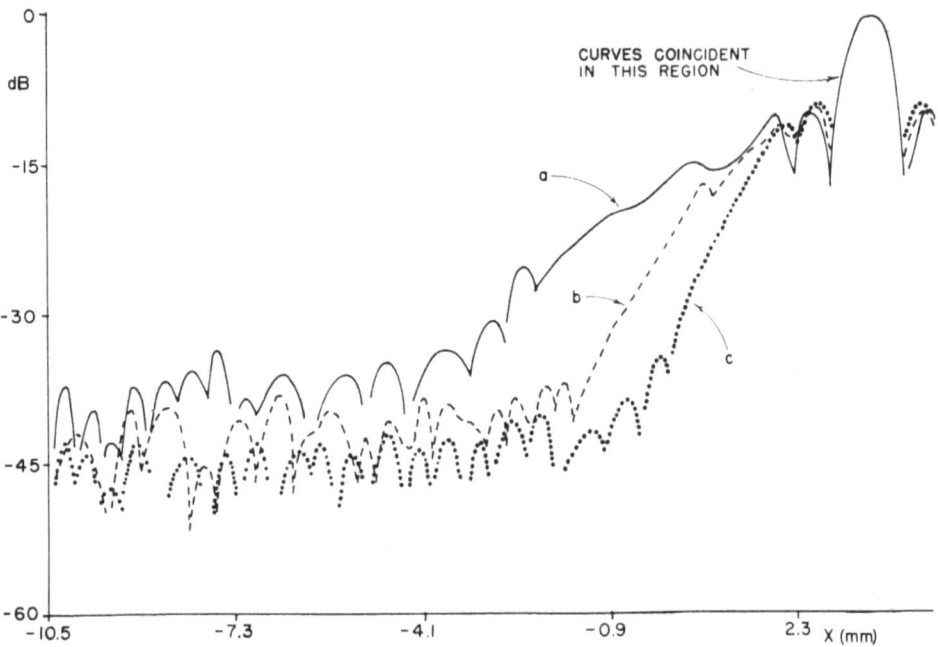

Figure 4.   Radiation Patterns of Array of Figure 3 using Wide Bandwidth Waveform and Envelope Detection plus Integration: (a) 30% Bandwidth, (b) 50%, (c) 70%.

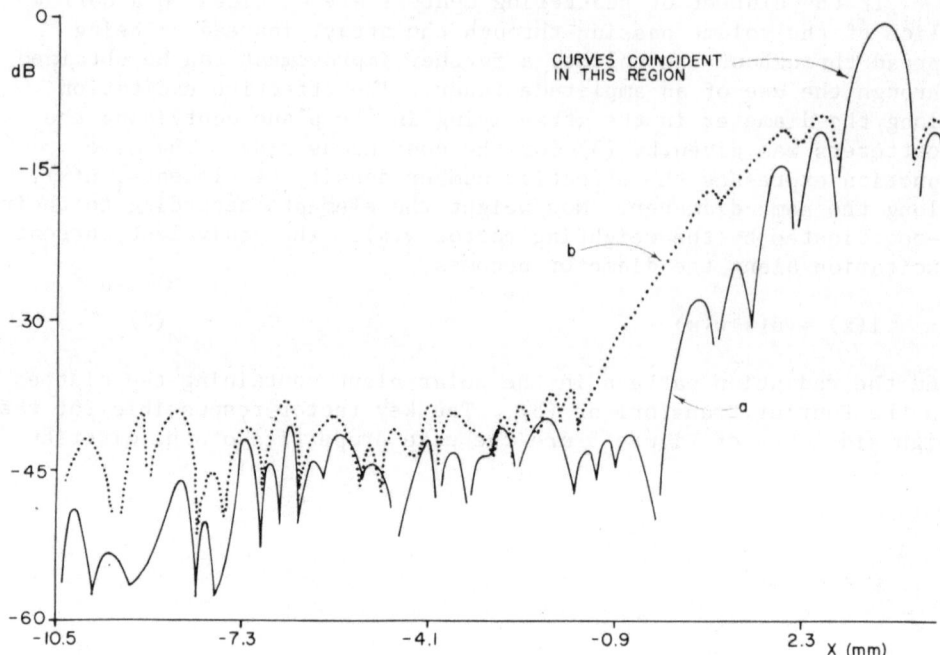

Figure 5. Radiation Pattern of Array of Figure 3 using 50% Band-
          width Waveform: (a) Matched Filter Detection, and
          (b) Envelope Detection plus Integration.

tion given by (1). As the distance from the center increases, the
excitation grows without limit and then abruptly terminates. This
undesirable characteristic is easily alleviated in (8) through a
suitable choice of w(x). Figure 6 is a calculated radiation pattern
showing the salutory effect for

$$w(x) = (a^2 - x^2)^{1/2} \tag{9}$$

which results in a uniform effective excitation function. Matched
filtering again is used in this case. The bandwidth is 50%. The
peak sidelobe has been reduced by 9 dB.

RADIAL APERTURE TAPER

    Angular tapering is not generally useful because scatterers
are not generally confined to a thin slice in a volume. Radial
tapering, however, is generally useful and can be applied provided

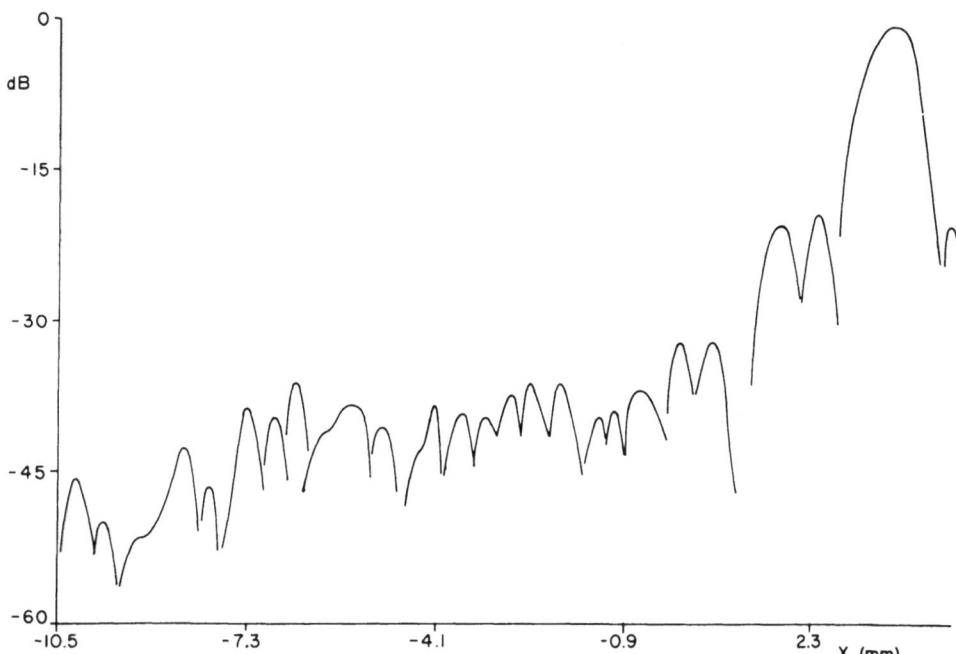

Figure 6.   Radiation Pattern of Array of Figure 3, using 50% Bandwidth Waveform, Matched Filter Detection, and the
Weighting Function Given in (9).

that additional concentric rings are added. The additional rings
allow a trade-off between the distant sidelobe levels and the nearby
sidelobes. Figure 7 shows the calculated radiation pattern for a
five-ring array of 95 elements. The numbers of elements are 5, 10,
15, 25 and 40 in rings of diameter 2, 4, 6, 8 and 10 cm, respectively. All sidelobes have been reduced to about -30 dB or below.

DISTANT SIDELOBES

All the preceding figures show 15 mm of the radiation pattern
from the axis of the main lobe which, at a distance of 10 cm, is an
angle scan of 0.15 rad or about 10°. Figure 8 shows the sidelobe-
envelope properties for the next 30 mm. In all but one of the
arrays the sidelobes rise materially and develop grating plateaux.
The sidelobe plateaux levels vary between -18 and -26 dB. The
phenomenon occurs approximately 25 mm away from the target, which
is an angle of ¼ rad or about 15°. Only the multiple-ring array

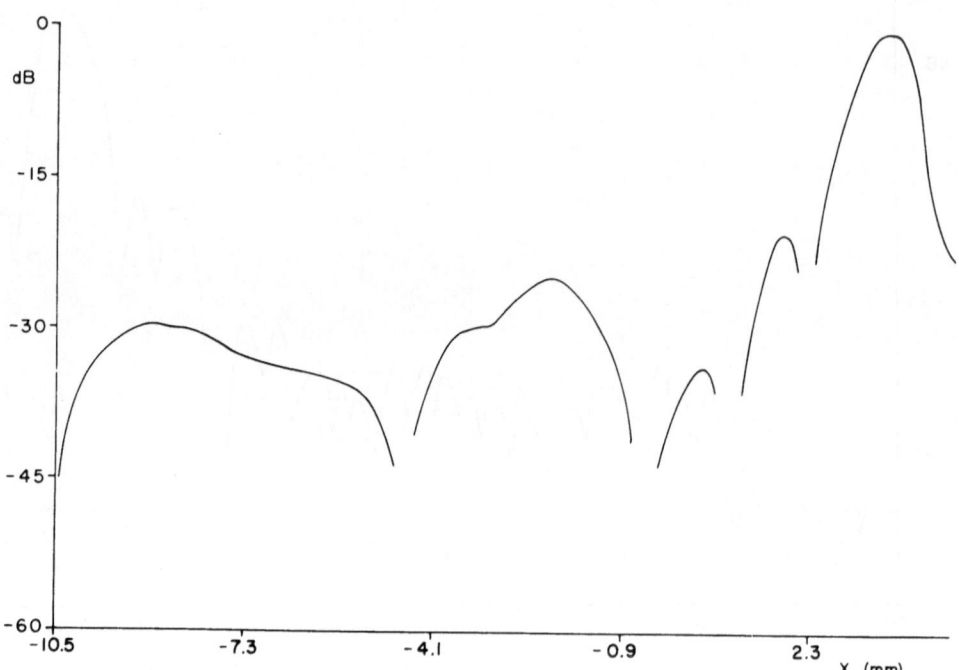

Figure 7. Multiple-Ring Array, 50% Bandwidth Waveform, Matched
Filter Detection, 95 Elements. Ring Diameters are
2, 4, 6, 8 and 10 cm.

maintains a uniform level in the distant sidelobe pattern.

This broad, high sidelobe level region is a weak, grating-lobe
phenomenon related to an apparent quasi-periodicity in element
spacing. In a narrowband, periodic line array with element spacing
$d > \lambda$, grating lobes are spaced by $\Delta u = \lambda/d$. The element spacing
along the ring for the 100-element, 10 cm diameter array is π mm.
However, the apparent spacings betweeen elements varies with position in the array relative to the observing point. Since the
apparent spacings are not identical, the energy that would normally
pile up into a grating lobe is spread into a low-level plateau.
The grating plateau can be moved further from the main lobe by
decreasing the interelement spacing, which means increasing the
number of elements and, therefore, the system cost, or by distrib-
uting the available elements into multiple rings of varying
diameters, as in Figure 7. There the desirable angular symmetry
properties of the system remain, while the apparent spacing

distribution of pairs of adjacent elements becomes broadened. It is this characteristic which accounts for the uniform distant-sidelobe pattern of the multi-ring array shown in Figure 8.

CONCLUSIONS

The desirable radial-symmetric properties of the ring array can be preserved and the normally undesirable sidelobe properties can be mitigated through the use of the following techniques:

1. Use the largest bandwidth waveform that the system constraints and the components permit.

2. Use matched filter detection, provided that system constraints and frequency-sensitive attenuation properties of the medium permit its use.

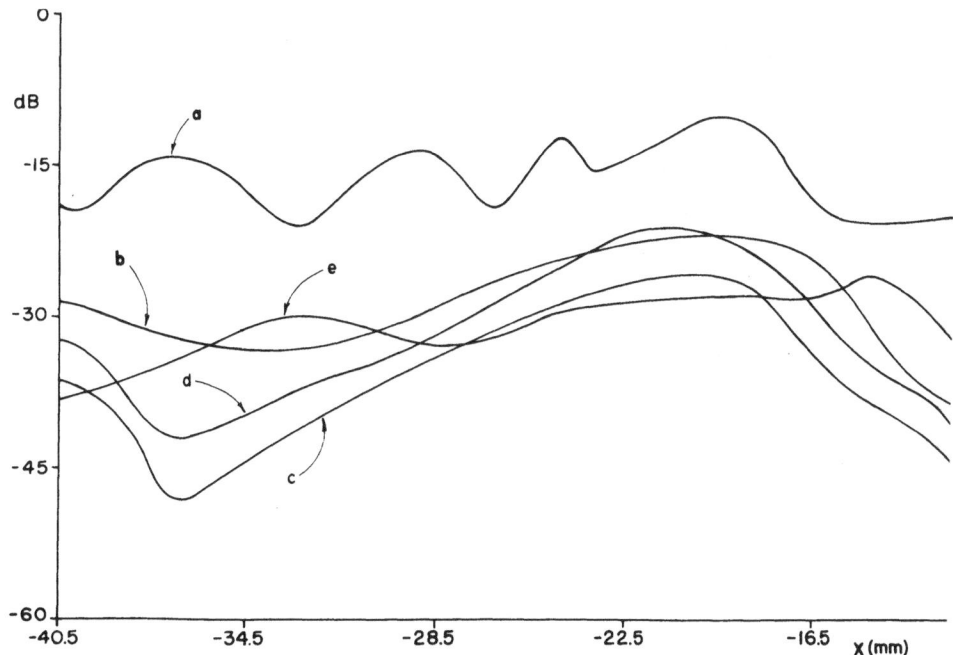

Figure 8.   The Envelope of the Distant Sidelobes of the Array of Fig. 3 (a-d) using (a) Monochromatic Radiation, (b) 50% Bandwidth Waveform plus (c) Matched Filter Detection plus (d) Angular Aperture Taper, and (e) the Array of Fig. 7.

3. Distribute the available transducer elements in multiple concentric rings of different diameters.

4. Experiment with radial aperture taper to achieve the desired shape of the side radiation pattern.

SUMMARY

Although the ring array offers many attractive features (circular symmetry, ease of data handling, absence of grating lobes in large, high resolution, thinned designs) its high sidelobe level limits its general utility in imaging systems. It is shown that wide bandwidth (30-70%) waveforms and matched filter detection can reduce the peak sidelobe from -8 dB to -15 dB, and that aperture taper can further reduce all lobes to about -30 dB or less. These computed results are for 10 cm diameter arrays consisting of 95-100 elements focused at a distance of 10 cm and operated at 2.5 MHz in water.

REFERENCE

[1]   Bernard D. Steinberg, <u>Principles of Aperture and Array System Design</u>, John Wiley and Sons, New York (1976).
[2]   John B. Thomas, <u>An Introduction to Statistical Communication Theory</u>, John Wiley and Sons, New York (1969).

ANOMALOUS QUANTIZATION ERROR LOBES IN PHASED ARRAY IMAGES

P.A. Magnin, F.L. Thurstone, and O.T. von Ramm
Duke University
Durham, North Carolina 27706

ABSTRACT

Time quantization errors result from the discrete time increment delay lines used to steer and focus phased array ultrasonic imaging systems. These quantized time delays, required for the phasing of the individual elements within the transducer aperture, create a phase grating which can produce anomalous quantization error lobes within the field of view of the imaging system [1,2]. The manifestation of these lobes can be seen in computer simulations of the azimuthal impulse response and in scans made of point reflectors as an increased sidelobe amplitude which can limit the dynamic range of the imaging system.

A simulation program predicting the position and amplitude of these anomalous lobes has been developed to study the effects of transmitted pulse length and shape on the amplitude of these lobes. Three temporal integration processing techniques for decreasing the anomalous lobe amplitudes are discussed. These involve changing the phase error grating from frame to frame thereby varying the amplitude and position of these lobes. The processors also have the ability to reduce the speckle contrast and therefore the potential to further improve the quality and usefulness of medical images.

INTRODUCTION

Improvements in the quality of biological images made with phased array sector scanners bring the resolution in soft tissue images increasingly close to the diffraction limit. A number of very significant difficulties remain, particularly when imaging biological tissues, which must be overcome if such a resolution limit is to be obtained. One form of degradation in phased array images is caused by the presence of phasing errors caused by the

quantized time delays used to steer and focus the ultrasound.
These quantized time delays can cause a phase error grating over
the elements in the transducer aperture which produces anomalous
lobes within the field of view of the imaging system. Another
difficulty, although seemingly unrelated to phasing errors, which
may limit the resolution of phased array imaging systems, is the
presence of speckle noise. Speckle noise, which results from
coherent interference in the image, is primarily responsible for
the mottled appearance of phased array images. It has been shown,
in the case of laser speckle, that significant decreases in the
resolution, as perceived by human observers, results from the
presence of speckle noise [3]. Such phasing errors, like the
speckle problem, can cause not only a reduction in the perceived
azimuthal and range resolution but also a degradation of the gray
scale resolution which can be particularly important in biological
images.

        Two recent studies [1,2] examining the effects of the time
quantization errors used continuous wave approximations to the
pulsed phased array imaging systems which leads to a significant
overestimation of the severity of these lobes. However, even in
the pulsed case the anomalous quantization error lobes can result
in ambiguous responses which limit the dynamic range of the imaging
system. The dynamic range of an ultrasonic imaging system, at any
given angle, can be defined as the ratio of sidelobe or the grating
lobe amplitude, at that angle, to the mainlobe amplitude [4].
Marked reductions in the dynamic range, resulting from time
quantization errors can be seen in the B-mode images and are
dependent on the length of the ultrasonic pulse. The present study
examines the effect of the transmitted pulse bandwidth on the
severity of the anomalous lobes. In addition, three processing
schemes for reducing the anomalous response and at the same time
reducing the contrast of the speckle noise are presented.

        A phased array simulation program which calculates the
azimuthal impulse response of a 32 channel imaging system for any
given pulse shape, pulse length, and transducer aperture has been
developed. It simulates the receiver's response to a point source
fixed, on-axis, at a focal point of the imaging system, when the
scan is steered off-axis using discrete time increment delays over
a range of 50°. The point source is assumed to produce a perfectly
spherical wave at the surface of the phased array transducer. As
the simulated waves arrive at the transducer elements, they are
compressed from 60 dB to 20 dB and summed in time. The peaks of
these sums, in a specified neighborhood about the focal point, are
calculated and displayed, along with the pulse waveform and its
spectrum, using a Calcomp Plotter. The logarithmic compression is
included to model the compression used in the Duke University
scanner [5]. The compression permits the relatively large dynamic
range echos to be displayed on television monitors. Simulations

without logarithmic compression show reduced sidelobe and anomalous
error lobe amplitude; however, the presence and position of these
lobes remains unchanged.

    Figure 1a shows a typical impulse response from the Duke
phased array system with time delay quanta of 125 nsec. The figure
is the image received from a 2.25 mHz center frequency point source
held in a fixed, on-axis position, 10 cm from the phased array
transducer. The point source is highly damped and its transmitted
pulses are synchronized with each receive line of the phsed array
system. Figure 1b shows the measured luminance response of the
image in Figure 1a as a function of azimuthal angle. At +17° and
at -15.5.° one can see the anomalous error lobes which result from
the time quantization errors. As a comparison, Figure 1c shows the
simulated azimuthal impulse response for the same situation at the

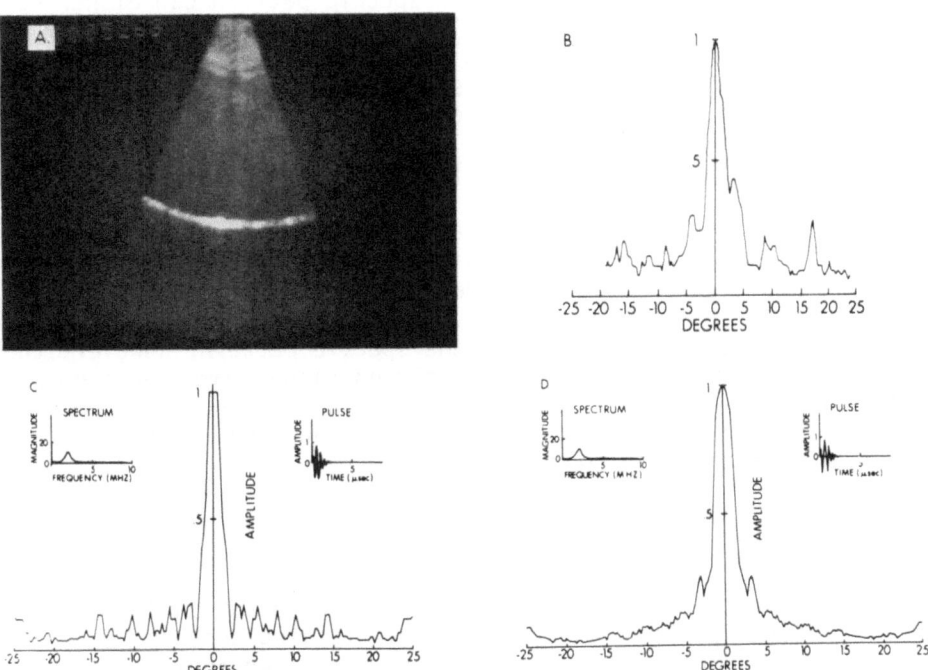

Fig. 1: Panel A shows a typical impulse response for the phased
          array system; this response is peak detected and graphed
          in panel B to give the azimuthal impulse response.
          Panel C shows the simulated azimuthal impulse response
          for the situation in panel A, and panel D shows the
          simulated azimuthal impulse response when the delay
          quantum is infinitesimally small.

range of the point source. Note the anomalous lobes at ±14.5° and
at ± 10.5. Figure 1d shows the "zero delay error" simulation in
which the delay quantization error is zero. Note the comparatively
smooth, monotonically decreasing, envelope on the sidelobe
amplitude

in the absence of delay error. The "zero delay error" simulation
demonstrates the very low amplitude sidelobes one would expect to
see around ±15° for a 19 mm, 32 element aperture receiving the
2.25 mHz pulse shown. At ± 25° one begins to see the grating
structure caused by the sampled aperture in both the "zero delay
error" simulation (1d) and the 125 nsec delay error simulation (1c).

The schemes for reducing the anomalous response all involve
altering the time quantization errors, and therefore the complex
sum over the array aperture in a series of consecutive video
frames. As a result of the alteration of the complex sum, these
schemes cause changes in the speckle patterns of the tissue being
imaged. The resultant motion of the speckle pattern can either be
integrated visually or electronically prior to display. This
integration results in a reduction in the speckle contrast or a
smoothing of the overall graininess in the images [6]. Although
the processing schemes presented here are not actually done simultaneously, the rapidity with which the processed frames are displayed
effectively makes the system a real time processing technique.

The Duke University phased array imaging system has the ability
to vary the entire set of delays used on each channel during a scan
at video frame rates. This provides the opportunity to vary the
quantization errors among successive frames. Observation of the
display indicates some flicker of the speckle pattern; however, the
observer can quickly differentiate between target movement and
speckle pattern movement. At these rates, a great deal of speckle
and anomalous error grating lobe smoothing can occur in the
persistence of the scan conversion system.

METHODS

All of the algorithms used to break up the quantization error
grating effect involve averaging four overlapping sectors [Table 1].
In each control scan, and in each of the four overlapping sectors of
the processed scans, 100 lines are written in a 45° arc. All scans
had a 15 cm maximum depth. In the first processor, the desired
delays are calculated as they would be in a simple scan; however,
instead of rounding the calculated delay to the nearest delay
quantum, the values are rounded to the nearest 1/4 delay quantum.
Four separate overlapping sectors are then calculated such that at
any given focal point in the scan, the average of the four sets of
delays is equal to the calculated delays rounded to the nearest 1/4

delay quanta. For example, if the desired delay in a given focal zone for a particular channel is 12.31 time delay quanta, the simple scan would have rounded this number to the nearest quantum (12.0). The processor algorithm, however, would use 12 time delay quanta in three of the four overlapping sectors and 13 time delay quanta in the fourth. The average then, in this focal zone, would be the calculated delay value rounded to the nearest 1/4 delay quantum. The position of the values which are rounded upward are randomized among the four scans to avoid the possibility of a net movement of the desired focal point. In the analysis which follows, the scans produced using this first processing algorithm are compared to control scans which simply round the calculated delay to the nearest delay quanta for the same transmit and receive foci used in the processed scans.

Table 1:

		No. of Sectors Averaged	$\frac{Signal}{Noise}$
Processor 1	1/4 delay quanta rounding	4	1.490
Control 1	simple delay quanta rounding	1	1.471
Processor 2	1/4 delay quanta rounding, interlaced transmit	4	1.492
Control 2	simple delay quanta rounding, simple transmit	1	1.362
Processor 3	simple rounding, interlaced transmit and receive	4	1.572
Control 3	simple rounding, simple transmit and receive	1	1.459

The second algorithm, in addition to the rounding scheme described above, involves the interlacing of the transmit foci so that the focus of the transmitted sound will vary from frame to frame over the four overlapping scans. Ten receive focal points were used but they remained fixed from one frame to the next. In each sector, 4 different transmit foci occur on any 4 adjacent lines; however, the position of the focus for a particular line was different for each of the 4 overlapping sectors. The receive delay values for each element in the phased array are calculated in the same manner as in processor 1. This second processing algorithm produces a scan which is compared to a control scan made using the same 4 transmit foci on each set of 4 adjacent lines and a simple rounding scheme to calculate the nearest delay quantum. Both processed and control scans used the same 10 receive foci.

The third processing routine varies the position of the 10 receive foci over the four overlapping sectors. The "control" scan used the same 4 transmit foci as the 4 processed scans but used the 10 receive delays from only the first sector of the 4 overlapping sectors. In addition, the 4 transmit foci used in each sector of the processed scan (on every 4 adjacent lines) were interlaced so that the position of the transmit foci on any given line varied for each of the 4 overlapping sectors in the processed scan. For this processor, the nearest 1/4 delay quantum round off scheme, described for processor 1, was not used.

Six scans using the three processors and corresponding control scans of a Radiation Measurements Inc. tissue equivalent phantom consisting of graphite particles in an augar gel were made. This tissue phantom provided a very speckled image with few, if any, resolvable targets. The scans were recorded on video tape from which suitable sequences were transferred to an Eigen Video Disc Model 16. The desired frame was selected and transferred to a Vidco Graphic Memory where it was digitized. Digital image data was then read into a Digital VAX 11/780 for analysis. The signal-to-noise ratios as previously defined [7] and expressed as:

$$SNR = \mu/\sigma \qquad (1)$$

where: μ = the mean luminance of the speckle pattern,

σ = the standard deviation of the speckle pattern

and the gray level histograms of the two-dimensional tissue phantom images were calculated to test the ability of the processors to smooth speckle. Cross-correlation coefficients between the four frames in each processor were calculated for "targetless" tissue phantom images to test the ability of the processors to vary the speckle pattern.

Four impulse responses were obtained using the first and third processors and their respective controls by scanning (in the receive mode) a 2.25 mHz, 1/4" diameter active source transducer (Panametrics) located 10 cm from the phased array. The active source was pulsed on each sector line. The phased array transmit circuitry remained inactive throughout the scan. Since the impulse response, as described above, employs only the receive circuitry, no differences between the impulse response produced using the second processor and that produced in the control scan for the second proccesor are to be anticipated.

The dependence of the error lobe amplitude on the number of transmitted cycles was investigated by studying a series of simulations in which the transmitted pulse length was increased from 3 cycles duration to the continuous wave case. The delay quantization steps were also varied in this experiment to investigate the dependence of anomalous grating lobe amplitude on this parameter.

RESULTS

Degradation of the impulse response from the simulated ideal "zero delay error" case is clearly seen in Figure 2 at 19.5° when the delay quantization time is 105 nsec and the transmitted pulse is only a 5 cycle, 2.0 mHz burst. The anomalous error grating lobes at ±19.5° are 15.1 dB below the main lobe amplitude. This is typically only 2 dB below the first sidelobe. In the "zero delay error" case, the amplitude at ±19.5° is 34.2 dB below the mainlobe.

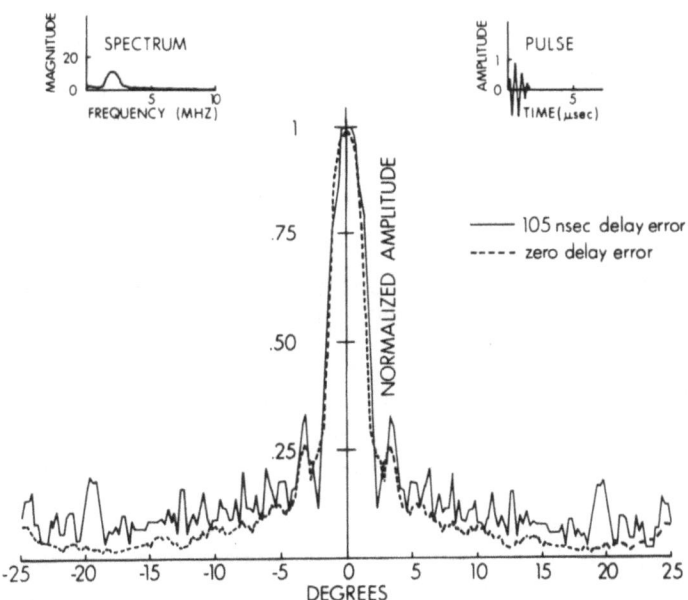

Fig. 2: Solid lines show the simulated azimuthal impulse response for a 19 mm aperture emitting a 5 cycles burst of 2.0 mHz ultrasound. Drawn in dashed lines, is the "zero delay error" response when an exponentially weighted 3 cycle burst of 2.0 mHz ultrasound is transmitted. Note the large error grating lobes at ±19.5°.

     The azimuthal impulse responses for the 1st and 3rd processors have been graphed and displayed next to their controls in Figure 3. Both processors demonstrate the ability to reduce the amplitude of the anomalous lobes which appears near -15° and +16°. In addition they also consistently reduce the amplitude of the third sidelobe which appears near 7 degrees. In the first and third control scans the error grating lobe is most clearly seen at -15.5°. These lobes are only 10.7 dB down from the main lobe and comparable to the amplitude of the first sidelobe. Processor 1 decreases the -15.5° error grating lobe by 6.9 dB where as the third processor decreases this lobe by 3.0 dB.

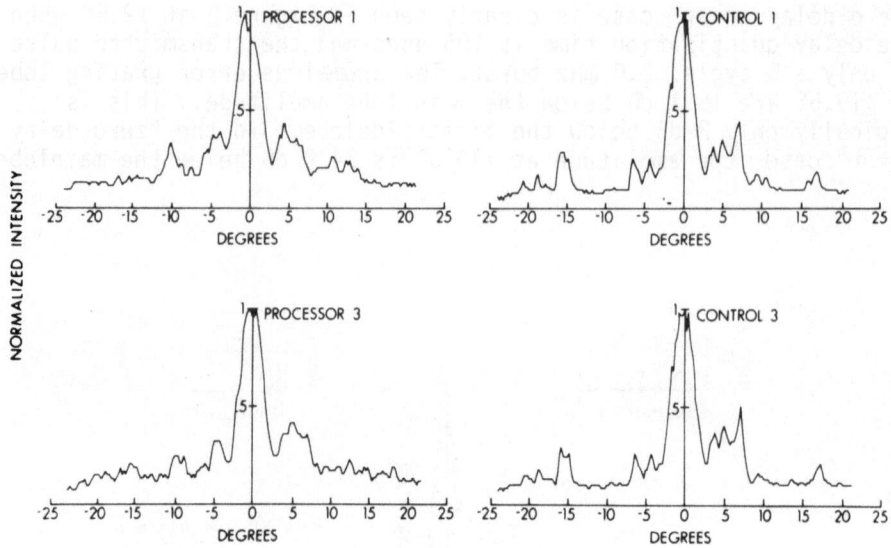

Fig. 3: Azimuthal impulse responses of processors 1 and 3 next
         to their respective controls.

     Figure 4 shows the gray level histograms of tissue phantom
images for the three processors and their corresponding controls.
The processors decrease the standard deviation of the histogram,
concentrating the gray levels more closely about a normalized gray
level of 0.2. This is evidenced in the decreased contrast of the
processed scans. For the first processor, the variance of the gray
level histogram was decreased by 53% when compared with its control
histogram. The second processor's histogram shows a 57% decrease
in variance and the third processor, a 9% decrease in variance when
compared with their respective controls. The signal-to-noise ratio
(as defined in Eq. 1) is the inverse of the speckle contrast and is
increased by 1.3% using processor 1, 9.5% using processor 2 and by
7.7% using processor 3 [Table 1]. The cross-correlation coefficients for the six possible combinations of the four frames in each
processor were averaged to determine a measure of the ability of
each processor to vary the speckle pattern of a "targetless" image.
The first processor had an average cross-correlation coefficient
between frames of 0.78, the second processor 0.59, and the third
processor 0.69.

     Figure 5 demonstrates the dependence of anomalous grating lobe
amplitude on the transmitted pulse length. These simulations show
that continuous wave approximation to the pulsed imaging system
tends to overestimate the severity of the lobes. The degree of
overestimation depends upon the length of the delay quantum

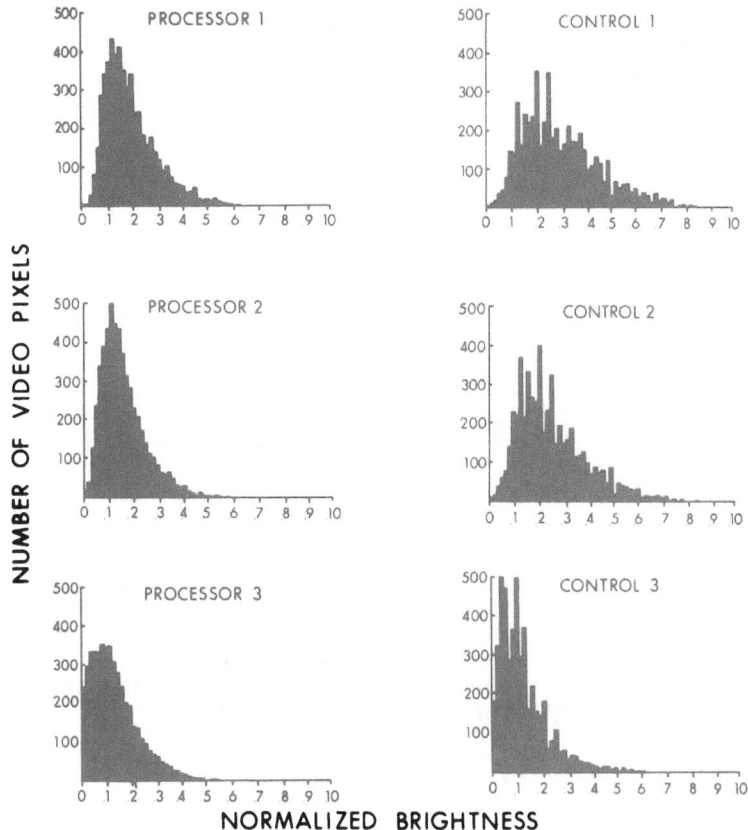

Fig. 4: Gray level histograms for the tissue phantom for each
         of the three processors and their respective controls.

relative to the element spacing in wavelengths. For the particular
aperture-delay quantization combination used in the simulations
displayed in Figure 5, the error grating lobe amplitude, for the
continuous wave case is only 3.9 dB below the mainlobe when the
delay quantum is 100 nsec. If three cycles are used, the error
grating lobe is 10.9 dB below the mainlobe amplitude. When the
typical exponential rise and decay of the transmitted pulse from
damped piezoelectric transducers are included in the simulation,
the error grating lobe is seen to be 12.4 dB below the main lobe.

DISCUSSION

    The receive impulse response of an acoustic imaging system is
the electrical amplitude output of the system as a function of the
azimuthal angle when an on-axis point source is imaged. Although

ultrasonic phased array imaging systems are in general neither
linear nor shift invariant, the azimuthal impulse response can
provide a comparative figure of merit which, within reasonable

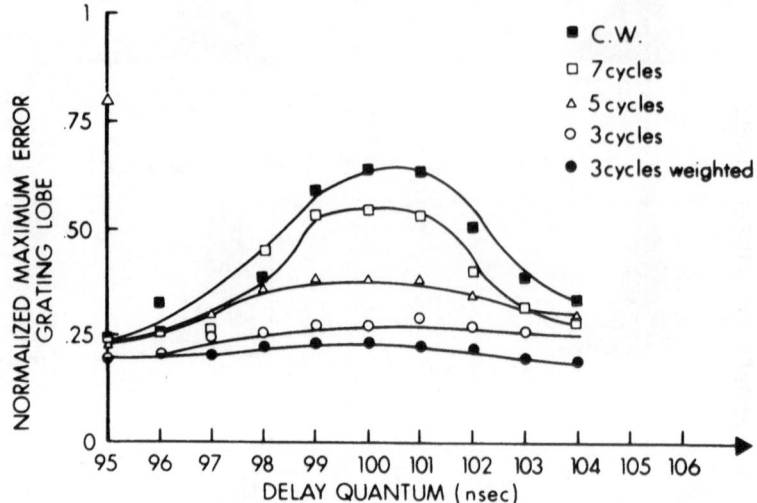

Fig. 5:   Error grating lob amplitude plotted against the delay
          quantum length for five transmit signals.

boundaries, characterizes the azimuthal response of the system in
the receive mode. Since the range resolution is usually significantly higher than the azimuthal resolution, the performance limits
in both resolution and dynamic range are generally determined by
the azimuthal impulse response function.

     In narrow band systems, these error grating lobes can cause
anomalous signals as much as 19.1 dB above what could be obtained
in the absence of delay error [Fig. 2]. Figure 2 shows how poor
aperture-delay quantization matching can exacerbate the problem.
For this simulation 5 cycles of 2 mHz sound was received by a
phased array system with delay quantization steps of 105 nsec. Not
all aperture-delay quantum combinations produce such severe anomalous grating lobes. This particular combination was chosen to
demonstrate a worst case situation.

     One of the difficulties in using the impulse response to model
the error grating lobe effect is that even though a severe grating
lens may be present for a given angle, if that grating lens
produces a null at the angle of the point source, a large error
grating lobe will not appear. This, however, does not mean that
this particular grating is benign. If, instead of imaging a point
source, one imaged a diffuse object the ambiguities caused by this
grating lens would become immediately noticeable.

In the case of pulsed ultrasound, the amplitude of these quantization error lobes is decreased due to the limited number of overlapping cycles. Figure 5 is a plot of the simulated maximum amplitude of the anomalous lobes above the "zero delay error" case over a small range of quantization times for 5 different pulse lengths. It is clear from these simulations that as the pulse length decreases, the amplitude of the anomalous lobes decreases. The lowest anomalous lobes are produced, in this simulation, when the burst is limited to 3 cycles and the onset and decay of the pulse are exponentially weighted to approximate the actual pulse emitted from a phased array system. Simulations using continuous wave approximations are likely to greatly overestimate the severity of the quantization sidelobes.

The focusing delays added to the steering delays also help to reduce the amplitude of these sidelobes. One can view the focusing delay function as adding an essentially random delay error to the grating lens. This random delay error contribution will serve to break up the uniformity of the gratings and smooth the diffraction pattern. Unfortunately as the range of the focal point increases, the number of delay quanta subtended by the focusing function becomes small and the randomness of its contribution to the delay error grating becomes less significant.

The 9.5% increase in the signal-to-noise ratio and the relatively small average cross-correlation coefficient of 0.59 found for processor 2 indicates that interlacing the transmitted foci on each line and using the 1/4 delay quantum rounding scheme has the greatest has potential of the three processors for improving the acquisition of gray level information from tissue scans. The 1.3% increase in the signal-to-noise ratio and the fairly high average cross-correlation coefficient found in processor 1 indicates that the 1/4 delay quantum rounding scheme does change the complex sum over the transducer elements enough to provide a modest reduction in the contrast of the speckle noise, however, the change is relatively small when compared with an interlaced transmit focus scheme. Finally, the third processor which used a conventional 1/2 delay quantum rounding scheme and interlaced both transmit and receive foci, showed an increase of 7.7% in the signal-to-noise ratio of the tissue phantom and an average cross-correlation coefficient of 0.69. When comparing the relative proformances of the second and third processors, it appears that the quarter delay rounding scheme is a more effective technique for decorrelating speckle patterns and increasing the signal-to-noise ratio than a receive foci interlace scheme. Certainly a combination of all three processors would produce better results than any single processor.

CONCLUSIONS

     Anomalous lobes resulting from delay quantization errors in
pulsed phased array systems are greatly reduced in amplitude from
those predicted for continuous wave systems and tend to decrease in
amplitude with increasing bandwidth. This is due to the reduced
number of overlapping pulses seen at the transducer elements when
steering off-axis. Both the first and third processing routines
reduce the amplitude of the anomalous lobes. The comparatively
modest reduction in the anomalous error lobe amplitude found with
processor 3 can be attributed to very small changes, among the
overlapping sectors, in the delays making up the focal zone in
which the active source is imaged. Although the impulse response
analysis is not applicable to the second (interlaced transmit)
processor, one would expect it to reduce the quantization lobes in
a manner very similar to that seen in the third processor.

     Gray level histograms, signal-to-noise ratio calculations and
cross-correlation coefficients for the three processors indicate
that in addition to decreasing the amplitude of the quantization
error lobes, the processors change the complex sum over the
elements and therefore reduce the speckle contrast. Such a
reduction in the speckle contast improves the observer's ability to
perceive subtle differences in tissue grey levels and thereby may
enhance the diagnostic utility of two-dimensional ultrasonic
imaging.

REFERENCES

1. Beaver, W.L., "Phase Error Effects in Phased Array Beam
    Steering," 1977 Ultrasonics Symposium Proceedings, p. 264-267,
    IEEE Cat. #77CH1264-15U.

2. Bates, K.N., "Tolerance Analysis for Phased Arrays."
    Acoustical Imaging: Visualization and Characterization,
    9:239-262, 1979, Proceedings of the Ninth International
    Symposium, Plenum Press, New York, N.Y.

3. Kozma, A. and Christensen, C.R., "Effects of Speckle on
    Resolution," J. Opt. Soc. Am., Vol. 66, No. 11, 1257, 1976.

4. von Ramm, O.T. and Smith, S.W., "A Multiple Frequency Array
    for Improved Diagnostic Imaging," IEEE Transactions on Sonics
    and Ultrasonics, SU-25, No. 6, 340-345, 1978.

5. von Ramm, O.T. and Thurstone, F.L., "Cardiac Imaging Using a
    Phased Array Ultrasound System I: System Design," Circulation
    Vol. 53, February, 1976.

6. Goodman, J.W., "Some Fundamental Properties of Speckle,"
    J. Opt. Soc. Am., Vol. 66, No. 11, 1145-1150, 1976.

7. Burckhardt, C.B., "Speckle in Ultrasound B-Mode Scans,"
    IEEE Transactions on Sonics and Ultrasonics, SU-25, No. 1,
    1-6, 1978.

MULTI-ELEMENT ARRAYS FOR NDE APPLICATIONS

Robert C. Addison, Jr.

Rockwell International Science Center
1049 Camino dos Rios
Thousand Oaks, CA   91360

ABSTRACT

Ultrasonic phased or time delay steered arrays have been widely used for diagnostic medical applications for the past several years. There has been no similar widespread use of these arrays for Non Destructive Evaluation (NDE) applications. This paper discusses the different requirements of the two applications and how this affects the design of an array.

Several 32 element arrays operating at a center frequency of 2.25 MHz have been fabricated and tested. These arrays have a transient response that is somewhat longer than that produced by commercially available single element transducers. However, the element to element variation in the response is quite low; preliminary measurements indicate that the shape of the responses are virtually identical and the peak to peak amplitude varies less than 10%. The design and fabrication of these arrays are described. Measurements of the array performance are compared to theoretical predictions.

INTRODUCTION

Beginning over a decade ago systems and techniques for obtaining ultrasonic images from phased or time delay steered arrays of transducers at frequencies in the 1 to 5 MHz range became available in research laboratories.[1-3] As this technology progressed, increasing numbers of applications were discovered in the diagnostic medical fields of echocardiography and the monitoring of fetal development. Now in these medical fields, the use of linear arrays is well established and several commercial sources of such systems are

available. In view of this widespread acceptance of arrays in the
medical field it is at first somewhat puzzling that there is so
little use of the arrays in the field of non-destructive evaluation
(NDE). The frequencies of operation are comparable and the technology for making transducers is applicable in both areas. Thus the
question to be answered is what are the differences in the requirements of the two application areas that have prevented the widespread use of phased arrays for NDE applications.

COMPARISON OF MEDICAL IMAGING AND NDE IMAGING

In order to clarify this question it is useful to compare the
requirements for NDE imaging and for medical imaging. There are
some very fundamental differences between the two application areas.
First of course is that tissue can only support scalar waves and
thus confines the wave mode used in medical imaging to longitudinal
waves. In NDE, the materials are elastic and support longitudinal,
shear, and surface wave modes. The presence of these modes can complicate the interpretation of an image since mode conversions and
re-radiation from creeping waves can produce degradations in the
imaging process. In practice the effects of these modes on image
quality can generally be ignored or eliminated with time gating. It
is only for the quantitative interpretation of scattering data from
defects that these effects must be included.

The next item may be more of a marketing consideration than a
technical one but does have a significant impact on the acceptability of arrays for NDE imaging. The diversity of body shapes that
are encountered in medical imaging encompasses a small range when
compared to the range of part shapes encountered in NDE imaging.
Even in medical imaging one frequently has separate specialized
systems for abdominal scanning, for echochardiography, and for
breast scanning. In NDE, the situation is more complicated with a
number of different industries such as areospace, nuclear power,
railroads, pipelines, etc. that have totally different requirements
regarding the material types, size of flaws, and size of structures
to be inspected. Even within a single industry such as areospace,
several specialized systems are required for aircraft structural
assemblies and several more for gas turbine engine components.
These requirements have made it a difficult and challenging task to
devise a generic inspection system that can address a wide range of
inspection tasks. Thus in NDE instead of a single array system, one
has to have a multitude of different systems, each of which addresses the specific requirements of a particular inspection task.

The techniques and requirements for coupling acoustic energy
into the medium of interest are also significantly different. The
surface of the human body is flexible and has an acoustic impedance
that can be matched reasonably well with a gel couplant. This permits a phased array transducer to be used on irregular surfaces and

also to be twisted and tilted relative to the surface to obtain different views of an object. Thus contact scanning is the preferred mode of operation. In NDE the materials are predominantly rigid and no couplant is available to match the impedance of the material. It is possible to scan an array transducer over a flat surface that has been smoothed through a grinding operation, but if the surface has been sand blasted, has machining marks, or has some curvature, contact is lost and the beam quality is seriously degraded. In general, immersion scanning is the preferred mode to use with phased arrays in NDE imaging. Immersion scanning permits one to vary the water path so that there is no problem with the ring down from the transducer, although a similar problem does occur at the front face of the metal because of the severe impedance mismatch between the metal and the water.

Next it is useful to consider the character and utility of the images that are obtained in the two application areas. In the medical field the image consists of a continuous flow of patterns with significant internal structure containing many different shades of gray or intensity variations. In many situations the image is dynamic and the motion of objects in the image is of diagnostic value. Typically the normal internal structures are delineated in the image and abnormalities such as tumors, cysts, or other disease states are detected and diagnosed by recognizing changes in the normal pattern. For moving objects such as valve leaflets in the heart, disease states can be detected by noting changes in the normal pattern of motion. The interpretation of images of this sort relies heavily on the pattern recognition capability of the eye-brain complex. It is helpful in studying these images to have a frame rate comparable to television frame rates so that the eye can track the features of the image as the viewpoint is moved about. This permits the determination of the spatial relationships of the imaged structures in the presence of clutter and artifacts that are introduced by scattering within the medium and manifested as speckle in the image. Further the objects are generally three dimensional and sometimes difficult to interpret using a single frame of a B-scan image. By changing the viewpoint through a rotation and/or translation of the array transducer, the region covered by the displayed B-scan is moved about. The viewer's eye-hand coordination comes into play and with some experience it is possible to obtain a reasonably accurate estimate of the size and shape of a three dimensional object. Thus in the case of medical imaging the use of phased arrays with a flicker-free frame rate considerably enhances the utility of the image by coupling it to the pattern recognition capability of the human eye-brain complex as well as decreasing the time required to obtain an image.

In contrast NDE images of an object do not contain a continuium of patterns nor are the objects dynamic but the images are more likely to contain a few strongly scattering stationary objects which

are separated by distances that are large compared to their dimensions such as the image shown in Fig. 1. There are frequently no
patterns to be recognized and the gray scale variations take place
over small distances. In this case the viewer is being required to
determine size, shape, orientation and location of the flaw in a
more or less featureless background. Pattern recognition ability or
eye-hand coordination do not come into play for these types of
images. The image provides a mapping function for locating defects
and can also provide orientation information if a defect has at
least one dimension that can be resolved. An analogy can be drawn
between NDE imaging and the night sky, whereas medical imaging is
more like a daylight scene that contains many overlapping objects
and a continium of intensities.

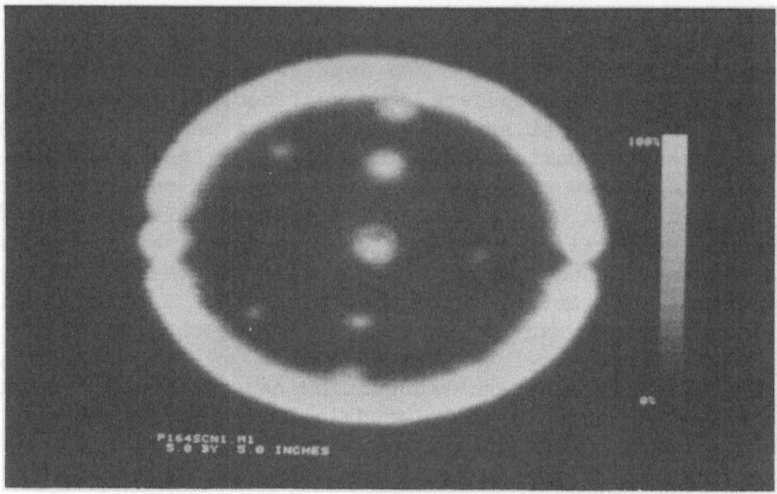

Fig. 1.   Acoustic C-scan image of a titanium disk containing nine
          defects. As is typical of NDE images, the defects are all
          separated by distances that are large compared to their
          dimensions.

     The impedance and velocity differences of the tissue and metal
media are the sources of significant changes in the wave propagation
characteristics of two media. Although there is some variability,
it is reasonable to assume that the average acoustic velocity in
tissue is about the same as that of water, 1.5 mm/μsec, with deviations measured in percentages. In contrast the velocity variation
in metals and other structural materials that are examined using NDE
techniques is very large with velocity values ranging from a low of
3 mm/μsec for a graphite fiber reinforced epoxy material to a high
of 10 mm/μsec for a silicon nitride structural ceramic. For metals
used in the areospace industry such as aluminum and titanium, velocities of approximately 6 mm/μsec are typical. Choosing these

latter materials as examples for discussion, we have an immersion
media, water, that has a velocity only one fourth as large as that
of the metal. Thus there will be significant refraction of the
acoustic beam as it enters the metal and there is the possibility
that it will not enter at all because it is incident at an angle
larger than the critical angle of approximately 14 degrees. Inside
the metal the wavelength becomes four times the wavelength in the
water and the waves will be diffracting proportionately faster. The
faster velocity complicates the problem of ring-down from the front
surface because the signal from an object at a specific distance
below the surface will occur four times faster in the metal medium
than in a tissue medium while the ring-down will require the same
time in either case. Thus the diffraction phenomena behaves as if
the distance to the object, measured in water, is four times the
actual distance in the metal, whereas the arrival times of the
signals from the object are only one fourth what they would be in
the water.

The objects studied via medical imaging techniques are distinguished by small changes in impedance of the order of 10 or 15
percent whereas the flaws visualized via NDE imaging have very large
impedance variations that can differ by factors of ten from that of
the host material. The impedance difference between the host
material and the water creates a very strong front face reflection
with the associated loss in signal strength.

The complications that result from the refraction caused by the
4 to 1 velocity difference is illustrated in Fig. 2 for a sample
that has a flat front face and a fan beam scanning from -10 degrees
to +10 degrees. Note that the internal angles are -44 degrees to
+44 degrees. A considerable distortion of the image will result if
these internal angles are not properly taken into account when the
data is displayed. In principle this sort of correction presents no
difficulty, although in practice one must accurately know the position of the transducer relative to the surface and if the surface
has some curvature this also has to be known. For an inspection
task on many identical parts appropriate jigs can be fabricated to
orient the part and the curvature of the suface can be programmed
into the display. One of the advantages of a phased array is the
ability to adjust the phases of the elements to compensate for the
focusing or defocusing at a curved metal interface. This is particularly important when scattering data is being acquired to obtain
quantitative information about a flaw where the technique requires
that plane wavefronts be incident on the flaw.

In Fig. 2b the same pseudo-polar scale is used to display the
radiation pattern that results from a 1/4 in. wide, 2.5 MHz transducer after it enters the metal. A decibel scale is superimposed.
It can be seen that at the -10 dB points the pattern is 2 in. wide
thus the lateral resolution is very poor for these transducer

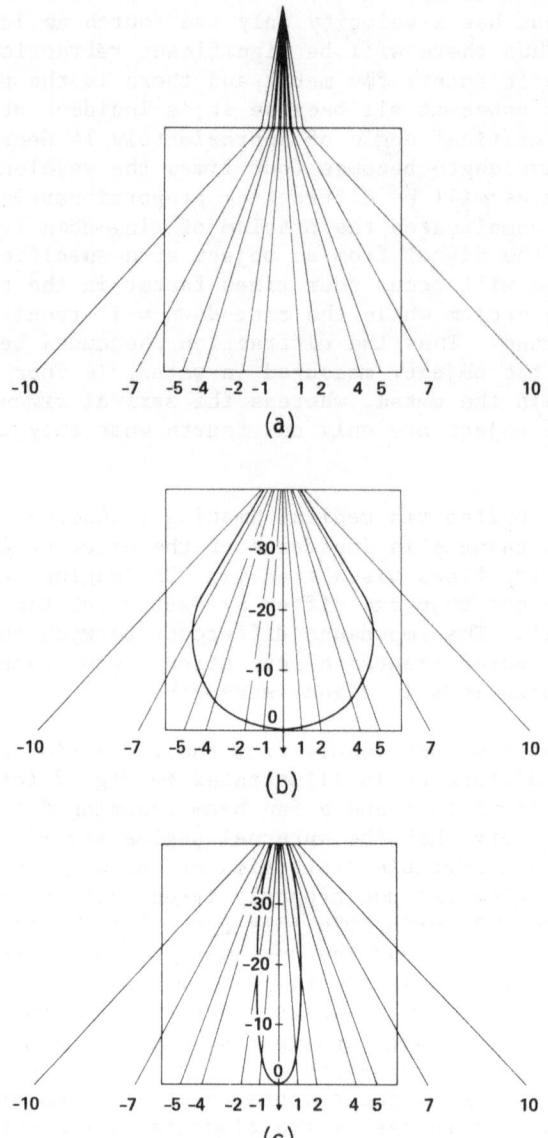

Fig. 2.   (a) Illustration of refraction of a fan beam emanating from
a phased array when it enters a metal having a velocity
four times that of the immersion medium. (b) Angular
radiation pattern produced in the metal by a 16 element,
2.5 MHz array with an aperture of 0.25 in. and a 2 in.
water path.   (c) Angular radiation pattern produced in the
metal by a 32 element, 2.5 MHz array with an aperture of
0.5 in. and a 2 in. water path.

parameters. Figure 2c shows the same sort of plot for a 1/2 in.
wide, 5 MHz transducer where the beam is only 3/4 in. wide at the
-10 dB points. This leads to the conclusion that for NDE applications it is desirable to use higher frequencies and incorporate more
elements into the array than is necessary in medical imaging
applications.

In addition to the loss of lateral resolution, the increase in
velocity also results in a loss of longitudinal resolution. An
object that is one inch away in water will return an echo in about
34 μsec, whereas in a metal like titanium this echo will return in a
little over 8 μsec. Thus one is forced to go to higher frequencies
to achieve adequate longitudinal resolution of closely spaced objects. Particularly troublesome is the low level of the flaw signal
relative to the front face echo. In the case of a 1200 micrometer
diameter spherical void, the front face signal must decay to a level
that is 40 dB below its peak value before the flaw signal is detectable without some sort of post-processing. Single element transducers for NDE applications have been manufactured that can resolve
a flaw of this sort when it is within 0.050 in. or 0.4 μsec from the
front surface. This is far beyond the reported performance of array
transducers which generally require times measured in tens of microseconds to decay in amplitude by 40 dB. This makes it almost impossible to use real time arrays for NDE except in cases where the flaw
signals are very strong or the flaws are located several inches below the surface. In non-real time applications where digital post-
processing can be used, flaws within 1/4 to 1/2 in. of the front
surface can be detected using a 2.5 MHz array.

In view of the difficulties that are encountered when using
phased arrays for NDE applications, the question of their utility
vis a vis single element conventional transducers must be addressed. A phased array offers two advantages over conventional
transducers: increased scanning speed and increased beam agility.
The increased speed does not have to result in a flicker free image
as in medical applications but only offer frame times of 10 to 60
sec. to make it useful. The term beam agility is used to describe
the ability of the array to shape the wavefronts to compensate for
the focusing caused by a curved metal surface. It also refers to
the ability to steer the beam in different directions and, with a
sufficiently large array, scan a beam over a curved part so that it
always enters the surface parallel to the local surface normal,
commonly referred to as a contour following mode.

FABRICATION OF MULTI-ELEMENT ARRAYS

General Approach

A modest program has been established to investigate the performance of multi-element arrays both theoretically and

experimentally. The initial goals have been to obtain an accurate
model of the transducer element and of the array, to fabricate elements with the predicted performance, and to obtain arrays containing elements having nearly identical performance. Particular
emphasis has been placed on those features that are of paramount
importance in NDE applications. The current literature on piezoelectric phased arrays has been used as a guide in developing
fabrication techniques suitable for our application. Specifically
the work of Desilets, Fraser and Kino,[4] of Larson,[5] of Hanafy,[6] and
of Defranould and Souquet[7] has been helpful.

Modeling of Array Performance

A set of models of array performance has been developed that
permits prediction of array performance. These models encompass the
lowest modes of vibration of the array element, the one dimensional
electrical and mechanical behavior of the array element, the angular
radiation pattern of the element, and the angular radiation pattern
of the complete array of elements.

Modes of Vibration

The array element is considered to be sufficiently long that
this dimension can be neglected in an analysis of the modes of
vibration. The two lateral dimensions of the element are comparable
in magnitude and play a strong role in determining the frequency and
character of the many modes of vibration that are allowed. Since
most of these modes are undesirable for generating an ultrasonic
beam in an adjacent fluid, the desired mode is selected by controlling these lateral dimensions to isolate the desired mode to as
large an extent as possible consistent with practical constraints.
A simplified model of the two lowest order modes was derived by Onoe
and Tiersten[8] and results in the modal curves shown in Fig. 3. The
lower mode accurately predicts the frequency of the element vs its
aspect ratio whereas the upper curve can ony be used as a guide
because the effects of nearby higher order modes, which have not
been taken into account, produce significant changes in the curve.[9]
Superimposed on the curve are the experimental points obtained from
a series of transducers having different aspect ratios. The agreement between the lowest order mode and the experimental points is
excellent. The experimental points for the next highest order mode
do not fall on the theoretical curve but are in agreement with the
experimental results of Larson[5] and Desilets.[10] The experimental
points provide a convenient technique for selecting a sufficiently
small aspect ratio to obtain adequate frequency separation between
adjacent modes. This allows the transducer to have adequate bandwidth for producing a short tansient response without the excitation
of spurious modes. An aspect ratio of 0.5 was chosen since this
provides adequate bandwidth, is convenient and is close to the value

of 0.57 that is reported to maximize the electromechanical coupling
coefficient.[11]

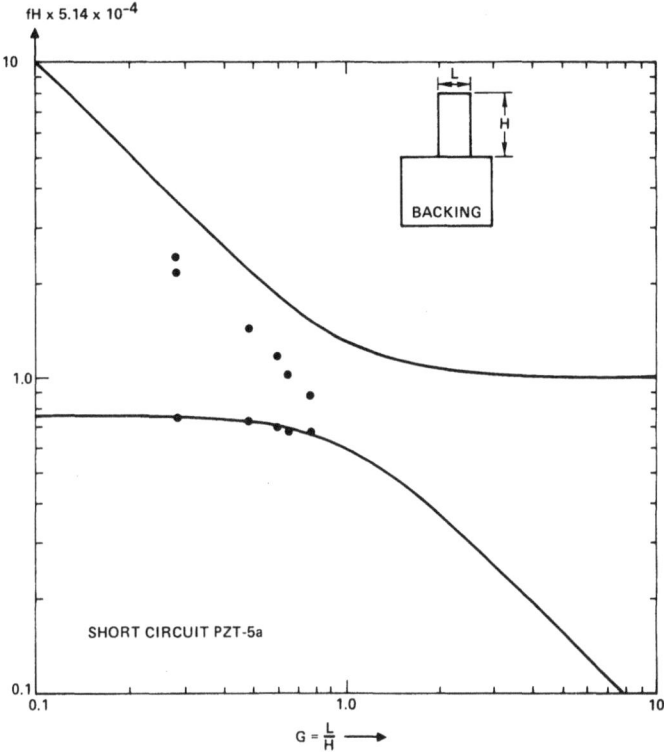

Fig. 3. Comparison of theory to experimental measurements of
         resonant frequencies of piezoelectric array elements having
         different aspect ratios.

One Dimensional Electrical and Mechanical Behavior

The electromechanical characteristics of a transducer element
must be obtained to predict its complex electrical impedance, insertion loss and transient response. A suitable model for this analysis has been developed by Krimholtz, Leedom and Matthaei (KLM)[12] and
is shown in Fig. 4. The transducer is modeled as a three port network, with one port representing the electrical input to the element, and the other two representing ultrasonic energy leaving the
piezoelectric element and entering the backing and fluid respectively. The only assumptions inherent in the model are that there is a
single mode of vibration and that the one dimensional analysis used
to obtain the propagation velocity is valid. The model is basically
a center tapped transmission line coupled to the electrical port of
the transducer via an ideal transformer with a frequency dependent
turns ratio and a frequency dependent series reactance. The model

is easy to work with because the addition of impedance matching
layers and backings to the transmission line can be handled by the
well developed scattering matrix techniques used in network
analysis. In this paper we are only concerned with the complex
electrical impedance and the transient response. The use of the
model to minimize the insertion loss has been treated in Ref. 4.
Here the array is backed with an absorbing backing and has a thin
layer of polyurethane over its face.

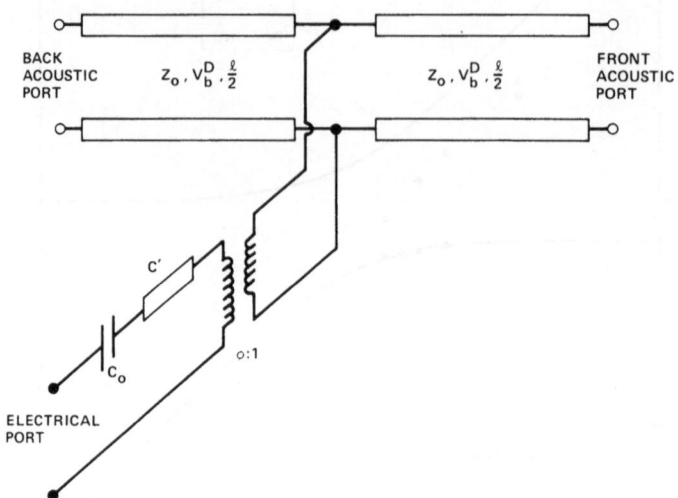

Fig. 4.   Schematic of KLM transducer model.

The complex electrical admittance of a PZT-5a transducer element
with width = 0.015 in., length = 0.500 in. and thickness = 0.029
in., a backing impedance of 26×10^6 kg/sec-m^2, and operating into
water at a frequency of 2.2 MHz is shown in Fig. 5. The solid curves are obtained from the KLM model and the dashed lines are experimentally measured values. Note that the peak value of the real part
of the admittance, G, occurs at the same frequency and conductance
for both theory and experiment. The additional resonances that are
seen at higher frequencies in the experimental data correspond to
higher modes that are not included in the theoretical model. The
chief discrepancy between the two curves is a narrowing of the
predicted bandwidth.

In Fig. 6 the theoretical and experimental transient responses
of the element are compared. The experimental transient response is
longer and implies a narrower bandwidth for the element. This discrepancy is of some concern since the transient response must be
short for an array to be useful in NDE applications. It is more
revealing to compare the two transient responses after they have
been rectified and plotted on a log scale as shown in Fig. 7. Here

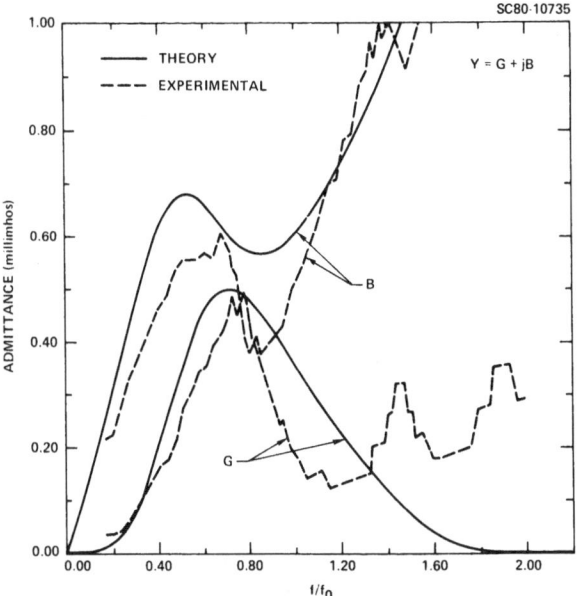

Fig. 5. Comparison of theory and experimental measurements of the
complex electrical admittance of a piezoelectric array as a
function of frequency.

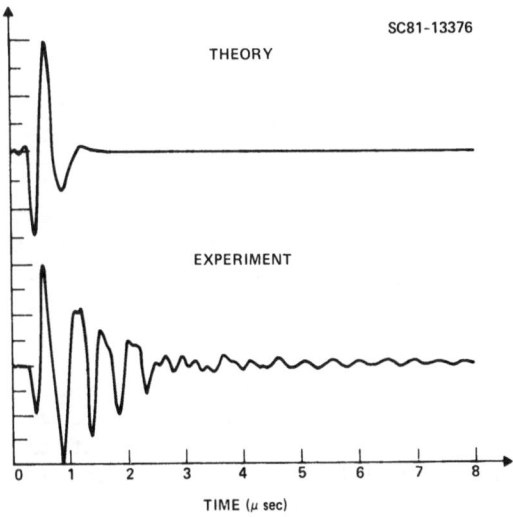

Fig. 6. Comparison of theory and experimental measurements for the
transient response of a single array element. Responses
are plotted using a linear vertical scale.

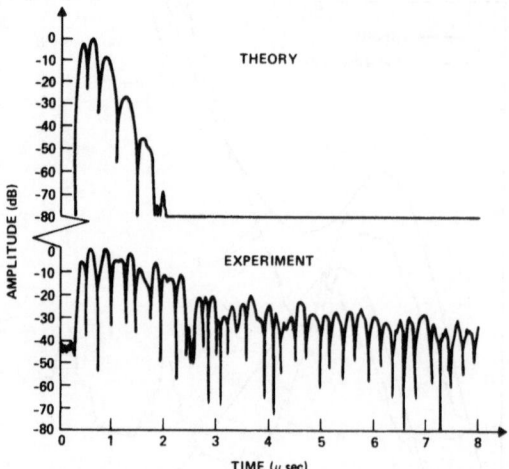

Fig. 7.   Comparison of theory and experimental measurements for the
transient response of same transducer element as shown in
Fig. 6. The absolute value of the response is plotted
using a logarithmic vertical scale.

we can see that the time required for the transient response to
decay to a level 30 dB, 40 dB and 50 dB below its peak response does
not agree with the theory well at all. Several possible sources of
the longer transient response can be identified. These include
energy that is not fully absorbed in the backing material and leaks
out over a relatively long period of time, interfacial waves that
propagate back and forth along the boundary between the backing and
the transducer elements, surface waves that propagate back and forth
along the covering material on the front of the transducer elements,
energy in other spatial modes of the transducer element that is converted to the piston mode and leaks out, and nearest neighbor inter-
actions of the transducer elements via coupling in the fluid medium. Efforts have been made, of course, to minimize these effects
but they seem to persist to a certain extent in all array transducers.

The exact shape of the transient response varies slightly among
transducer elements of he same array. Therefore we have measured
the times required for the transient response of an element to decay
to a specified, set of amplitudes that are measured relative to the
peak transient response. This permits us to generate the curve
shown in Fig. 8 that shows the average decay time of an element and
also the spread in the decay times for the set of transducer elements. Such a curve is useful for specifying the transient response
of an array and also providing information regarding the uniformity
of the response from element to element. For the array shown two

standard deviations of the time required to decay to 30 dB below the peak value correspond to a variation of 10.7% from the mean value.

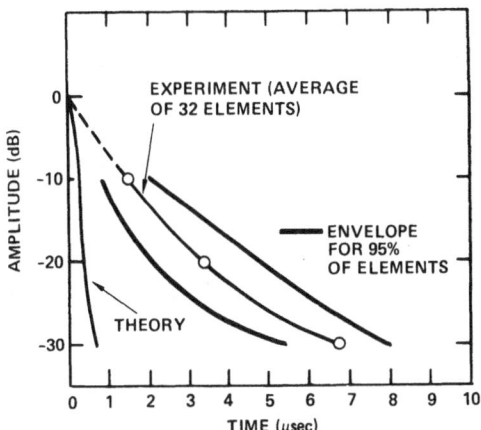

Fig. 8. Comparison of the theoretical decay time of the transient response of an array element with the average decay time of 32 elements measured experimentally. The envelope is two times the standard deviation of the experimentally measured values.

The entire ensemble of transient responses for a different but similar array are shown in Fig. 9. It can be seen that all responses are nearly identical. Measurements of the peak to peak ampltudes were made and it was found that two standard deviations correspond to a variation of 9.8% from the mean value. This performance is excellent in terms of the uniformity of the element response of the array.

CONCLUSIONS

There are a number of significant differences in the performance requirements of a phased array suitable for medical applications. The chief differences are related to the large velocity differences of the materials that are encountered in NDE applications and to the strength of the targets relative to the front face echo. These two conditions require that phased arrays for NDE have large apertures to minimize diffraction effects within the materials of interest and short transient responses to enable the detection of flaws close to the surface. We have found that the experimentally measured transient responses of arrays that have been purchased and also those that have been fabricated in our lab are significantly longer than

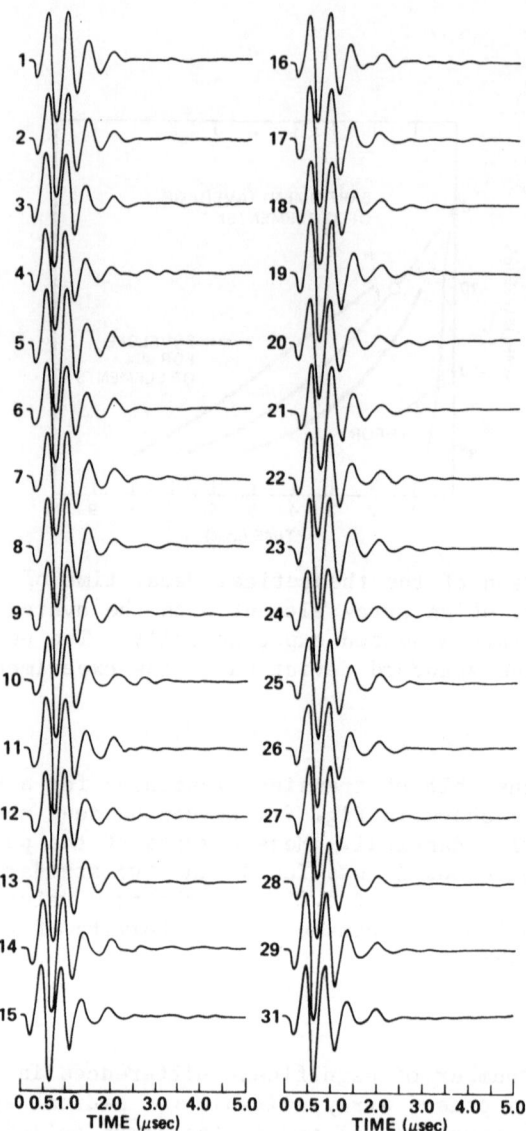

Fig. 9.   Ensemble of transient responses of a piezoelectric array
          illustrating reproducibility of element response.

the theoretical transient response. Several possible reasons for
this have been mentioned. It is clear that additional experimental

work is required to fabricate array elements with transient responses that are in agreement with theoretical predictions.

References

1. J.C. Somer, "Electronic Sector Scanning with Ultrasonic Beams," Proceedings of the First World Congress on Ultrasonic Diagnostics in Medicine, June 2-7, 1969, Vienna, Austria.
2. J.C. Somer, "New Processing Techniques for Instantaneous Cross-Sectional Echo-Pictures and for Improving Angular Resolution by Smaller Beams," Proceedings of the Fourth Congress of the International Society for Ultrasonic Diagnosis in Opthalmology, May 6-9, 1971, Paris, France.
3. F.L. Thurstone and O.T. von Ramm, "A New Ultrasound Imaging Technique Employing Two-Dimensional Electronic Beam Steering," Acoustical Holography, 5, P.S. Green, Ed., Plenum Press, New York, pp. 249-259 (1974).
4. C.S. Desilets, J.D. Fraser and G.S. Kino, "The Design of Efficient Broad-Band Piezoelectric Transducers," IEEE Transactions on Sonics and Ultrasonics, SU25, pp. 115-125 (1978).
5. J.D. Larson, "A New Vibration Mode in Tall, Narrow Piezoelectric Elements," 1979 Ultrasonics Symposium Proceedings, IEEE Cat. No. 79CH1482-9SU, pp. 108-113 (1979).
6. A. Hanfy, "Dead Zone Elimination in Acoustic Arrays," Ultrasonic Imaging, 2, pp. 302-312 (1980).
7. Ph. Defranould and J. Souquet, "Design of a Two Dimensional Array for B and C Ultrasonic Imaging System," 1977 Ultrasonic Symposium Proceedings, IEEE Cat. No. 77CH1264-1SU pp. 259-263 (1977).
8. M. Onoe and H.F. Tiersten, "Resonant Frequencies of Finite Piezoelectric Ceramic Vibrators with High Electromechanical Coupling," IEEE Trans. of the Professional Technical Group on Ultrasonic Engineering, UE-10, July 1963.
9. H.F. Tiersten, Private Communication.
10. C.S. Desilets, "Transducer Arrays Suitable for Acoustic Imaging," Ginzton Laboratory Report No. 2833, Stanford University, June 1978.
11. J. Sato,, M. Kawabuchi and A. Fukumoto, "Dependence of the Electromechanical Coupling Coefficient on the Width-to-Thickness Ratio of Plank-Shaped Piezoelectric Transducers Used for Electronically Scanned Ultrasound Diagnostic Systems," J. Acoust. Soc. Am, 66, pp. 1609-1611, Dec. 1979.
12. D.A. Leedom, R. Krimholtz and G.L. Matthaei, "Equivalent Circuits for Transducers Having Arbitrary Even- or Odd-Symmetry Piezoelectric Excitation," IEEE Transactions on Sonics and Ultrasonics, SU-18, pp. 128-141, July 1971.

STUDY OF THE NORMAL MODES OF VIBRATION OF THE PIEZOELECTRIC

TRANSDUCERS USED IN ACOUSTIC IMAGING APPLICATIONS

C. Bruneel, B. Delannoy, H. Lasota* and J.M. Rouvaen

Laboratoire O.A.E. - ERA CNRS N°593
Universite De Valenciennes - 59326 Valenciennes
Cedex - France

I - INTRODUCTION

The performances of acoustical imaging devices are now much more limited by the insufficient understanding of the ultrasonic transducers behaviour rather than by imperfections in the associated electronics, as shown by the growing interest among transducer operation studies. For example, M. PAPPALARDO (1) measured the frequency spectrum of a freely vibrating transducer, in terms of the width (W) to thickness (t) ratio, but was unable to interpret all the observed vibration modes. For this purpose, the finite element mathematical model has been used by J. SATO (2) in a narrow range of W/t values, and the transducer behaviour at a single frequency was deduced therefrom. To our opinion, a simpler approach may be taken, by measuring and studying the eigen resonance modes, to give a more physical picture of the phenomenon. The eigen modes arise from the resonance of Lamb waves propagating along the three major orthogonal directions of the transducer. For acoustical imaging applications, like biomedical ones, the width and thickness are of the same order of magnitude, so that resonances along these two directions occur in the same frequency band.

Moreover, the present study shows that when the width becomes very narrow compared to the wavelength, the optimum thickness of the transducer departs from the classical half-resonance- wavelength value, computed from the bulk longitudinal wave velocity. The phase velocity of the S_o Lamb wave may rather be used.

---

*Permanent Address: Institute of Telecommunication
Technical Univ. of Gdansk, 80952 Gdansk, Poland.

II - THEORETICAL STUDY OF THE RESONANCE MODES

When the transducer dimensions become smaller or nearly equal to the acoustic wavelength in the ceramic transducer material, neither bulk waves nor surface waves may propagate. Only guided waves of the Lamb type (3,4) may occur with symmetrical (S_n) or antisymmetrical (A_n) mechanical displacement fields, where n is the (positive or null) integer standing for the order of the mode.

Inside the transducer arrays used in acoustical imaging, the length L of the elementary transducers is generally much higher than the wavelength, but the thickness and width are of the wavelength order of magnitude. The Lamb waves may then propagate parallel to the transducer surfaces along y and z directions (see fig.1) with phase and energy velocities varying with the thickness e pertaining to the case (fig. 1), either w or t. The resonances along x direction may be discarded here, since they occur at frequencies much lower than the working one.

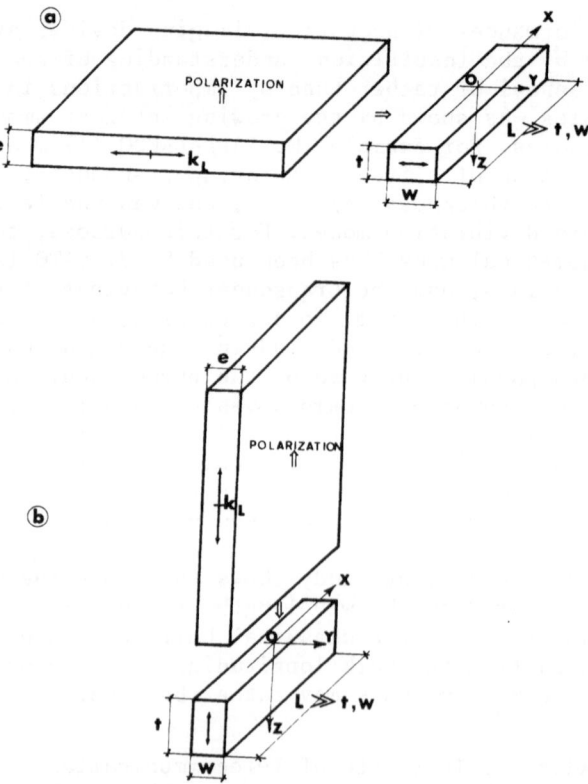

Fig. 1 - Lamb waves propagation medium and transducer geometry
a) width resonance   b) thickness resonance

   The eigen modes are characterized here as arising from Lamb
wave résonances along y and z, that is width and thickness (fig. 1).
In acoustical imaging, the width W is imposed by the special sampling period of the transducer array, but the thickness t is a parameter to be adjusted to its optimal value. So normalized variables
like f.W and W/t may be used in the study, the resonance diagram
becoming a plot of W/t against f.W. Such theoretical curves are
shown in fig. 2, as obtained from the dispersion curves published
for a PZT-4 ceramic by FURGASON and NEWHOUSE (5). Assuming isotropy
of the material along y and z directions, the same curves have been
used for resonances along thickness t (e = w) and along width w
(e = t). The notations S_n^j and A_n^j are used for the different symmetrical and antisymmetrical modes, where n is the the order of the
modes (0, 1, 2, ...) and j the rank of the resonance (the pertaining
width being so equal to $(2j + 1).\Lambda/2$, with j = 0 for the fundamental).

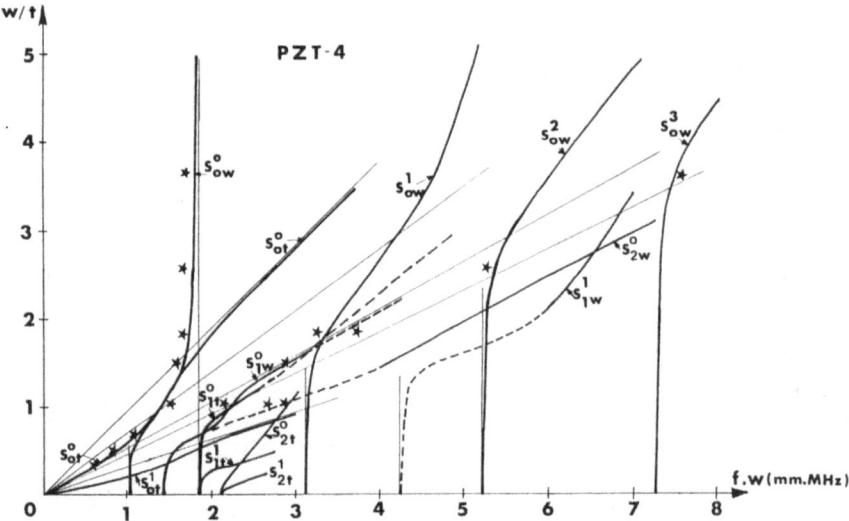

Fig. 2 : Theoretical (—— , ---) and experimental (*) results
         for symetrical modes of the PZT 4 and P1-60 transducers.

   However, only symetrical modes are shown in fig. 2, since they
appear physically more probable, the stresses and strains being
piezoelectrically generated in a symetric fashion for a freely vibrating transducer. This assumption has been verified experimentally. If Λ_n stands for the wavelength of the nth order symetrical
Lamb wave with phase velocity C_n, the resonance along thickness t
are given by :

$$t_n^j = (2j + 1)\frac{\Lambda_n}{2} = (2j + 1)\frac{C_n}{2f} \qquad (1)$$

so that :

$$\frac{w}{t_n^j} = \frac{2}{(2j + 1) C_n} \cdot (f.W) \qquad (2)$$

and the resonances along width W by :

$$W_n^j = (2j + 1) \frac{\Lambda_n}{2} = (2j + 1) \frac{C_n}{2f} \qquad (3)$$

so that :

$$\frac{W_n^j}{t} = \frac{(2j + 1)}{2} \cdot C_n \cdot \frac{1}{ft} \qquad (4)$$

III - EXPERIMENTAL RESULTS

The experiments have been performed using piezoelectric ceramic transducers of P1-60 type from Quartz et Silice (France), a
material which is very similar to the PZT-4, whose dispersion curves are known (5).

The dimensions and working frequency have been chosen so as to
cover a large spectrum of f.W and W/t values. The length L = 20 mm
is sufficient to reject the resonances along this dimension to the
very low end of the frequency spectrum and to restrict so our analysis to the bidimensionnal case (directions W and t only).

The series f_s and parallel f_a frequencies have been measured
for each freely vibrating transducer, where the transducer electrical impedance goes to, respectively, very low and very high values.
For obtaining the correct determination of the f.W value, the anti-
resonance frequency f_a is needed, since the electrical impedance of
a very large transducer becomes also "infinite" when its thickness
is an odd multiple of the half-wavelength.

The experimental values of f.W and W/t for the significant
resonances have been also shown on the theoretical curves of fig.2.
It is seen there that a single resonance occurs along the
transducer thickness, given by the S_0^0 Lamb mode, so that it may be
designated as $S_0^0(t)$, when the condition W/t < 1 holds.
In the opposite case, it may be distinguished between resonances along thickness t and width W by using the radiation pattern
of the transducer. The resonance modes along thickness t are the
source of the useful radiation (often termed as the "longitudinal"
wave), but those along width W contribute to a parasitic radiation,
from which parasitic sidelobes are produced in the radiation pattern of the transducer, more and more slanted from the transducer

axis when the phase velocity of the Lamb wave becomes more and more closer to that of the propagation medium (6) (water for example). At particular frequencies, both modes exist with different weighting factors. The radiation pattern allows then for the distinction between the modes. In fig. 3, for W/t nearly unity, are may distinguish at a frequency f = 609 kHz the modes shown on fig. 2. (f.W = 1.5 mm MHz and W/t = 1) namely $S_0^0(t)$ radiating along transducer axis and $S_0^0(W)$ radiating with a slant angle greater than 40°.

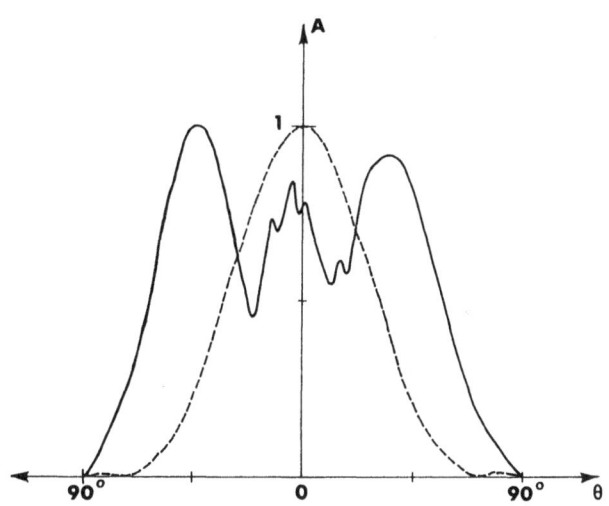

Fig. 3 - Calculated (---) and measured (——) directivity pattern
        for a transducer : W/t ≅ 1   f = 609 kHz
        f.W ≅ 1.5 mm MHz

At a frequency f = 819 kHz (f.W = 2.1 mm MHz), the fig. 2 predicts a possible simultaneous resonance of $S_1^0(t)$ and $S_1^0(W)$, as shown by fig. 4.

A more significative example is given in fig. 5. The free resonance frequency at 687 kHz for W = 4.55 mm (f.W = 3.1 mm MHz and W/t = 1.8) may be attributed from fig. 2 to several resonances, namely $S_0^1(W)$, $S_1^0(W)$, $S_1^0(t)$. The diagram of fig. 5 shows that the mode $S_0^1(W)$ radiating at a 46° angle dominates over the near axis radiating $S_1^0(W)$, and $S_1^0(t)$.

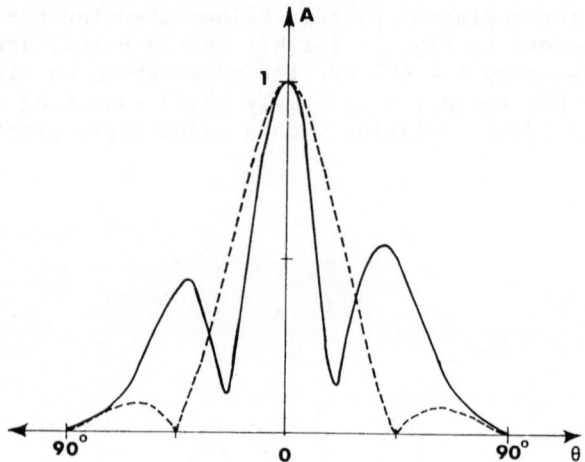

Fig. 4 - Calculated (---) and measured (——) directivity pattern for a transducer : W/t ≅ 1   f = 819 kHz
f.W ≅ 2.1 mm MHz

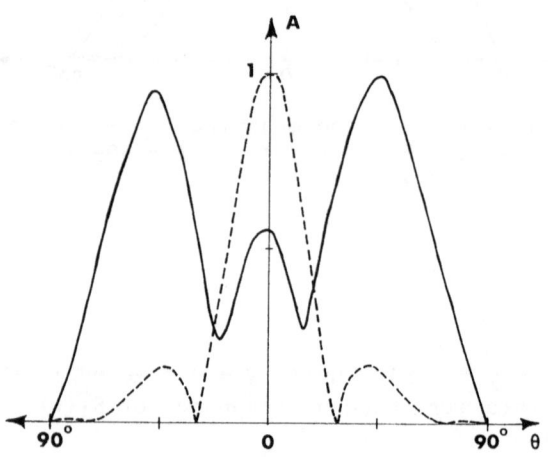

Fig. 5 - Calculated (---) and measured (——) directivity pattern for a transducer : W/t = 1.8   f = 687 kHz
f.W = 3.1 mm MHz

It may also be verified that, for a larger value of W/t ratio, namely 3.7, at a 818 kHz frequency (f.W = 7.5), the higher modes re-
sonating along thickness tend to dominate (fig. 6), leading to the
so called longitudinal" vibration mode of the transducer. However,
the $S_o^3(W)$ mode still exist, but with a smaller relative weight.

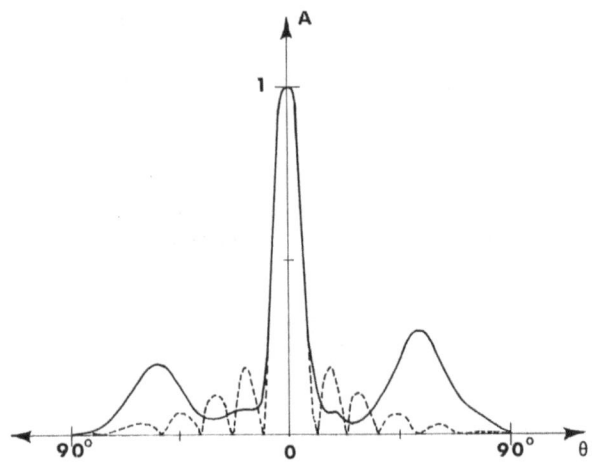

Fig. 6 - Calculated (---) and measured (——) directivity pat-
         tern for a wide transducer : W/t ≅ 3.7   f = 818 kHz
         f.W ≅ 7.5 mm MHz

     The theory and the experiments therefore fit very well, and
a global physical insight to the transducer behavior is obtained.

IV - CONCLUSION

     The theoretical model, well supported by the experimental re-
sults, shows that the resonances along the width W must be suppressed, since they introduce severe perturbations in the acoustic radiation patterns of the transducers. In a preceding work, it has
been shown that such a task may be accomplished by cutting the
transducer into a number of elements, parallel to the x axis (6).
Moreover, it is possible to distinguish between three working zones
for the transducers and deduce practical rules therefrom. In fact :
     - for large f.W values, say greater than 7 mm MHz in our case,
the transducer works practically in its fundamental "longitudinal"
thickness mode, and no special problem arises, the thickness t being given as the half-wavelength for the bulk longitudinal wave
in the piezoelectric ceramic.

- for small f.W values, say lower than 1 mm MHz, the transducer also in a thickness mode, which is no longer related to the bulk longitudinal wave velocity, but rather to that of the zeroth order symetrical Lamb wave. This later velocity being always smaller than the former one, this lead to thinner transducers than in the first case. Such a situation applies particularly for acoustical imaging where the f.W values are currently in a range smaller than 1 mm MHz.

- lastly, for intermediary values, the parasitic modes along the width W play a significant role and must be suppressed, a reasonable cutting criterion giving the number N of elements being:

$$\frac{f.W}{N} = f.W' < 1 \text{ mm MHz}$$

This allows each individual transducer to work in a single well defined vibration mode along thickness t if $W'/t < 1$.

The results are given here in normalized form (f.W and W/t variables), so that they may be applied to transducers working at different frequencies, with different widths, made of PZT-4 like ceramic materials.

REFERENCES

1) M. PAPPALARDO
   Journal of sound and vibration 52 (4) 579 - 586 (1977).
2) J. SATO, M. KAWABUCHI and A. FUKUMOTO
   J. Acoust. Soc. Am. 66 (6) Dec. (1979).
3) H. LAMB
   "On waves in an elastic plate"
   Proc. Roy. Soc. London Serie A, 93 - 114, 1917.
4) I.A. VIKTOROV
   " Rayleigh and Lamb Waves"
   Plenum Press New-York 1967.
5) E.S. FURGASON and V.L. NEWHOUSE
   I.E.E.E. Trans. Sonics Ultrason., SU-20, 360 - 364, (1973).
6) B. DELANNOY, C. BRUNEEL, F. HAINE and R. TORGUET
   J. Appl. Phys. 51 (7) 3942 - 3948, July (1980).

LINEAR IMPULSE HOLOGRAPHY*

B. P. Hildebrand, A. J. Boland, M. L. Cochran

Spectron Development Laboratories, Inc.
3303 Harbor Blvd, Suite G-3
Costa Mesa, California 92646

INTRODUCTION

A continuing problem in the gas utility industry is the uncertainty of the location of gas distribution pipe. As a result, unnecessary digging is often undertaken when a pipe is to be exposed for repair or inspection. A more serious consideration is accidental breakage when digging occurs for other utilities such as telephone or electricity, or at construction sites. It would, therefore, be useful to have a device capable of locating these pipes and displaying the location in direct relationship to the surface. This paper describes the results of research performed in the development of such a system.

TECHNICAL DISCUSSION

The basis for the proposed underground pipe location system is linear impulse holography using an acoustic source. The configuration of the system is shown in Figure 1.

The source emits an impulse of energy which travels through the medium to the object from which it reflects to the receiver array. The time-of-flight (TOF), of the impulse is measured at each receiver. An expression for the TOF may be obtained as shown in Equation 1.

---

* This research is sponsored by the Gas Research Institute under Contract 5080-353-0335.

529

$$t = \frac{r_o + r}{c}$$

where   (1)

$$r_o = (z_o^2 + x_o^2)^{\frac{1}{2}},$$

and

$$r = (z_o^2 + [x - x_o]^2)^{\frac{1}{2}}$$

$$c = \text{velocity of propagation}$$

When this expression is plotted, a hyperbolic curve is generated as shown in Figure 2.

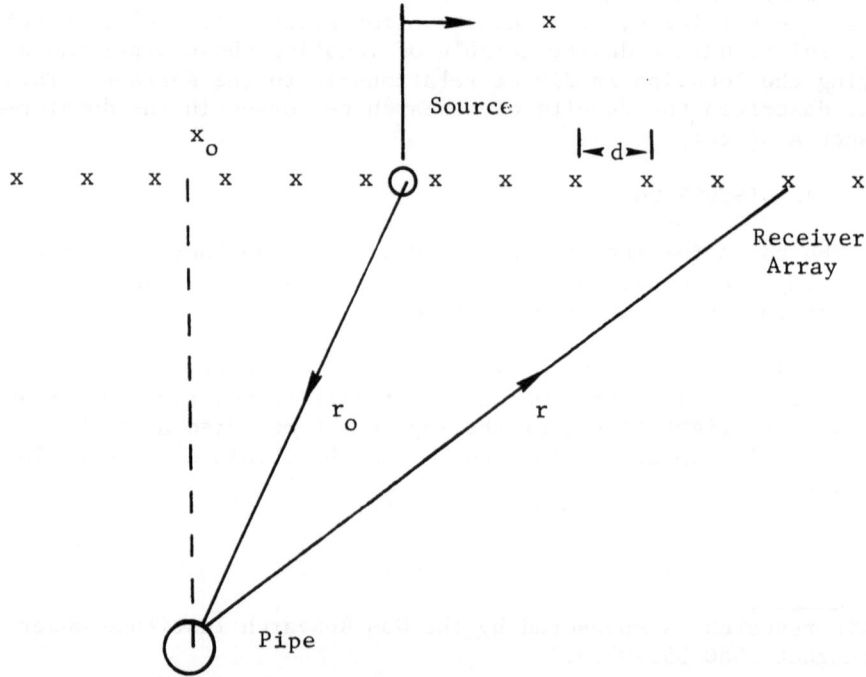

Figure 1.   Cross-Section of the Linear Impulse Holography System.

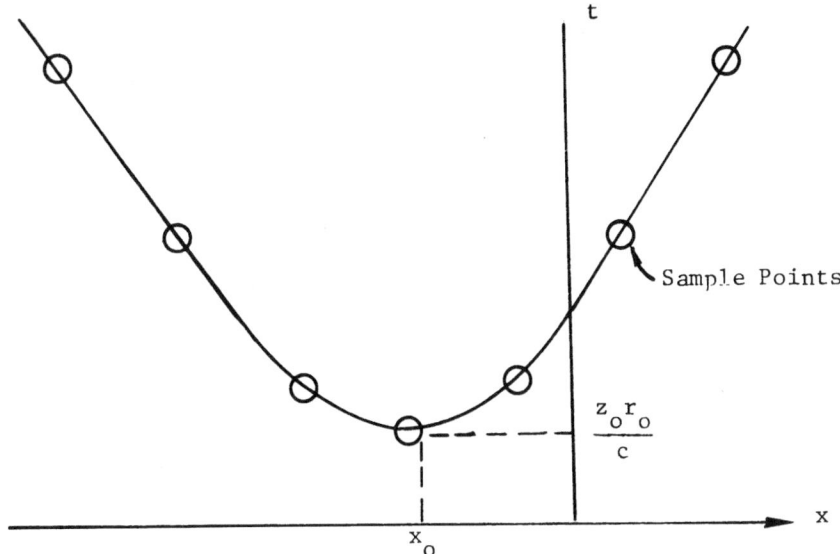

Figure 2.   Time-of-Flight Profile Across a Line Object.

Holography is based upon the measurement of phase rather than
time delay[1].   Therefore, the TOF profile is converted to phase by
the simple conversion

$$\phi = \omega t, \tag{2}$$

where ω = radian frequency.

A plot of this equation vs x would yield a curve similar to Figure
2.   However, if the sine or cosine are taken, the result is

$$f(x) = \cos \left\{ \left[\frac{r_o}{c} + \frac{1}{c} \; (z_o^2 + [x - x_o]^2)^{\frac{1}{2}} \right] \omega \right\} \tag{3}$$

A plot of this function, shown in Figure 3, yields curves equivalent to a Fresnel zone pattern[2].   As is well known, such a pattern
forms the foundation of holography since, upon being photographed
and presented to a collimated coherent light wave, it will focus
it to a point.   Thus, it appears that by proper presentation of
TOF data, it can be converted to a holographic format.

The choice of ω is arbitrary within the bounds of the sampling
theorem.   For example, the pattern shown in Figure 3 could be generated only if there are sufficient number of receivers.   Obviously,

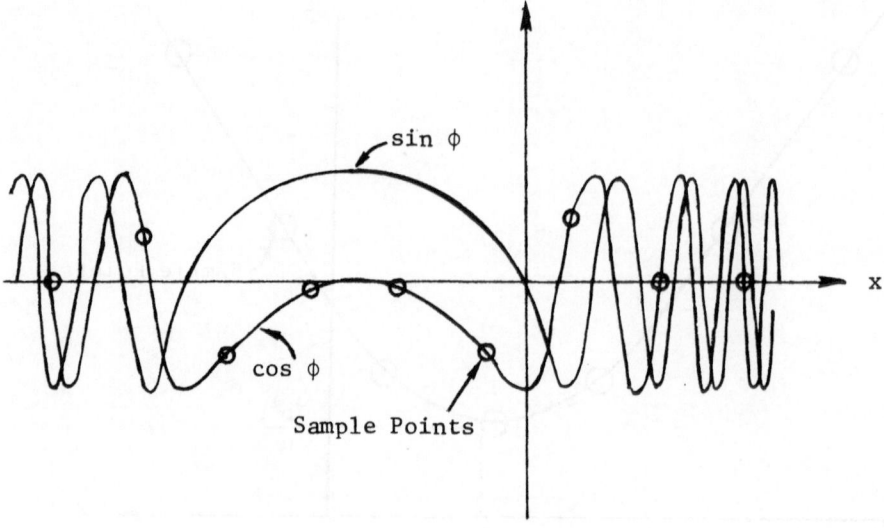

Figure 3. Time-of-Flight Profile Converted to a Phase Profile.

this example does not have enough, since the last cycles on the
right are not adequately sampled. Thus, they would not be reproduced. For a smaller ω, the pattern would spread out to accomodate
the receiver spacing. The relationship between the receiver spacing
and ω can be derived by the sampling theory, resulting in the expression

$$\omega \leq \frac{\pi c}{d} = \omega_o \qquad\qquad (4)$$

where d = receiver spacing.

As an example, if the receiver spacing is 10cm (4 in.), and
the velocity of propagation is 300m/sec (1000 ft/sec),

$\omega \leq 9425$ rad or 1500 Hz.

Thus, this receiver spacing allows impulse holography equivalent
to 1500 Hz. continuous wave. The lateral resolution can now be
found by the equation

$$\delta = \frac{\lambda z_o}{L} , \qquad\qquad (5)$$

where L = Nd = length of the array, and N = number of receiver
elements.

Using Equation 4 to define λ, Equation 5 reduces to

$$\delta = \frac{2z_o}{N} .$$ (6)

Thus, 32 receiver elements spaced 3cm for an array length of 1m will provide a resolution of 6.25cm at a depth of 1m.

The depth of the pipe will be computed by using the minimum TOF and the expression

$$(TOF)_{min} = \frac{z_o + r_o}{c} = \frac{z_o + \sqrt{z_o^2 + x_o^2}}{c} .$$ (7)

We can estimate x_o, by noting the receiver, x_i, with the shortest TOF, substitute this value into Equation 7 and compute z_o. The subsequent reconstruction of the image will therefore be carried out at this depth. Since the estimate of x_o is uncertain within the receiver spacing, d, the depth estimate will be uncertain within

$$\Delta = \frac{x_i}{c(TOF)_{min}} \cdot d$$ (8)

where x_i = receiver position yielding $(TOF)_{min}$.

Thus, the depth estimate depends upon the position of the object, being minimum when it is centered under the array.

There may be instances where the velocity of sound in the medium is not known, even approximately. It is possible to derive this information by estimating the slope of the asymptote to the hyperbolic TOF curve. The expression for the asymptote to the hyperbola is

$$t = \frac{2}{c} (x - x_o - r_o).$$ (9)

Therefore, once the computer has estimated the aymptote from the TOF curve, the velocity of sound can be obtained from its slope

$$m = \frac{2}{c}$$ (10)

The value of c, obtained in this way can then be used in equation 7 and in the reconstruction algorithm.

The image is formed in a computer by means of the backward wave propagation (BWP) algorithm, rather than by the diffraction

of light[3]. This circumvents the necessity for producing a transparency and using a laser and optics. The line of data can be reconstructed in less than one second and displayed in the time it
takes to move the system to the next position. The new line of data
is then taken, reconstructed and displayed next to the first. In
this way a complete image is built up in real time. The display can
be in isometric format or in plan and cross-section views.

THEORY

It is not obvious that a 1-D reconstruction actually takes a
cut across the object. The following mathematical analysis suggests
that it does so within certain limits. Consider Figure 4.

A transducer at (x,y) sends and receives a pulse reflected by
the object $0(\varepsilon,\eta)$. The complex amplitude of the echo is

$$f(x,y) = \iint 0(\varepsilon,\eta)\exp\ (-2jk_1r_1)\ d\varepsilon d\eta \tag{11}$$

where

$$r_1 = \left[(x-\varepsilon)^2 + (y-\eta)^2 + z_o^2\right]^{\frac{1}{2}}$$

This information can be back propagated as shown in Figure 4(b) to
retrieve an image of the object, by the Equation

$$I(u,v) = \iint f(x,y)\ \exp\ \left\{-jk_2r_2\right\}\ dxdy \tag{12}$$

where

$$r_2 = \left[(x-u)^2 + (y-v)^2 + z_1^2\right]^{\frac{1}{2}}$$

Substituting Equation 11 into Equation 12 yields

$$I(u,v) = \iiiint 0(\varepsilon,\eta)\ \exp\ \left\{-j\ 2k_1\left[(x-\varepsilon)^2 + (y-\eta)^2 + z_o^2\right]^{\frac{1}{2}}\right.$$
$$\left. + k_2\left[(x-u)^2 + (y-v)^2 + z_1^2\right]^{\frac{1}{2}}\right\}\ d\varepsilon d\eta\ dxdy. \tag{13}$$

If we let $k_2 = 2k_1$, $z_1 = z_o$ and use the Fresnel approximation to
the square roots, we obtain

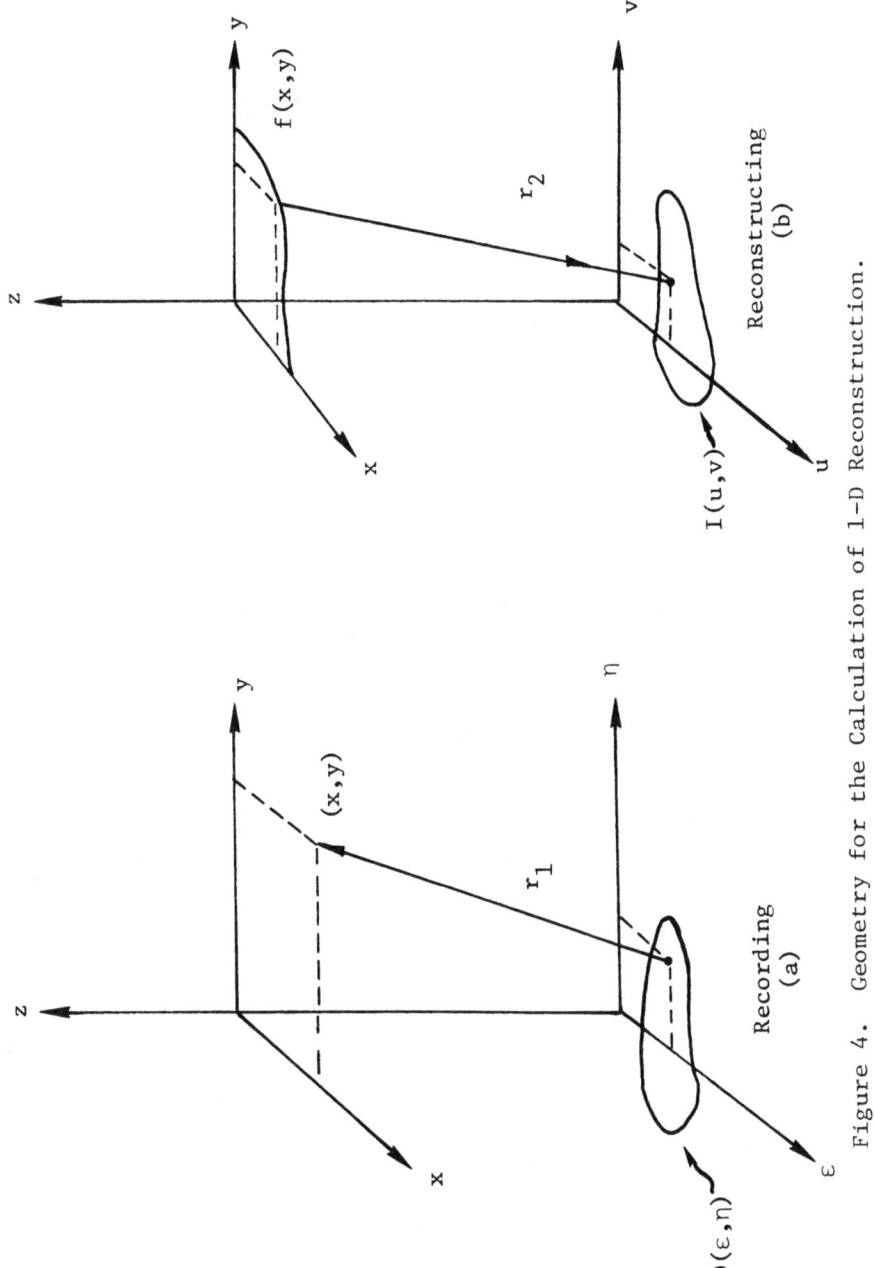

Figure 4. Geometry for the Calculation of 1-D Reconstruction.

$$I(u,v) \quad \iiiint O(\varepsilon,\eta) \, \exp\left[-j \, \frac{k_1}{z_o} \, (\varepsilon^2 + \eta^2 - u^2 - v^2)\right]$$

$$\exp \, j\left[\frac{2k_1}{z_o} \, x \, (\varepsilon-u)\right] \quad \exp \, j\left[\frac{2k_1}{z_o} \, y(\eta-v)\right] \, d\varepsilon d\eta \quad dxdy. \tag{14}$$

Since $\int \exp \, [ja \, (p-q)] dp \, \propto \, \delta(p-q)$, we have

$$I(u,v) \, \propto \, \iint O(\varepsilon,\eta) \, \exp\left[-j \, \frac{k_1}{z_o} \, (\varepsilon^2 + \eta^2 - u^2 - v^2)\right]$$

$$\delta(\varepsilon-u) \, \delta(\eta-v) \, d\varepsilon d\eta,$$

which becomes

$$I(u,v) \, \propto \, O(u,v). \tag{15}$$

That is, the image is a perfect replica of the object. Now, suppose we reconstruct along the line $y = y_1$ only. Then, we get the result

$$I(u,v) \, = \, \exp\left[j \, \frac{k_1}{z_o} \, (v-y_1)^2\right] \int O(u,\eta)$$

$$\exp\left[-j \, \frac{k_1}{z_o} \, (\eta-y_1)^2\right] \, d\eta. \tag{16}$$

The kernal of this integral oscillates rapidly for $\eta \gg y_1$. Therefore, only a narrow strip about y_1 contributes to the integral[4]. As a rule of thumb, the integral will have finite value for

$$|\eta-y_1| \, \leq \, \sqrt{2\lambda_1 z_o}, \tag{17}$$

where

$$\lambda_1 \, = \, \frac{2\pi}{k_1}$$

Thus, the cross section below the line $y = y_1$ is selected and imaged. Further insight may be obtained by making a change of variable in Equation 16.

Let

$$\sqrt{\frac{k_1}{z_o}} \, (\eta-y_1) \, = \, \sqrt{\frac{\pi}{2}} \, \phi.$$

Then we have

$$I(u,v) = \exp\left[j\,\frac{k_1}{z_o}\,(v-y_1)^2\right]\sqrt{\frac{\pi z_o}{2k_1}}\int O(u,\phi)\exp(-j\tfrac{\pi}{2}\phi^2)d\phi. \quad (18)$$

Suppose the object is bounded by $B(\varepsilon,\eta)$ as shown in Figure 5. Then at any particular ε, the integral of Equation 16 has the lower and upper limits $1(\varepsilon)$ and $m(\varepsilon)$ respectively. Then Equation 18 becomes

$$I(u,v) = \exp\left[j\,\frac{k_1}{z_o}\,(v-y_1)^2\right]\sqrt{\frac{\pi z_o}{2k_1}}\int_a^b O(u,\phi)$$

$$\exp\,(-j\,\tfrac{\pi}{2}\,\phi^2)\,d\phi$$

$$(19)$$

where

$$a = \sqrt{\frac{2k_1}{\pi z_1}}\,[1(\varepsilon)-y_1]$$

$$b = \sqrt{\frac{2k_1}{\pi z_o}}\,[m(\varepsilon)-y_1]$$

If we assume that

$$O(\varepsilon,\eta) = 1 \text{ for } (\varepsilon,\eta) \subset B(\varepsilon,\eta)$$

$$= 0 \text{ elsewhere}$$

we have the integral

$$G(\varepsilon) = \sqrt{\frac{\pi z_o}{2k_1}}\int_a^b \exp\,(-j\,\tfrac{\pi}{2}\,\phi^2)\,d\phi \qquad (20)$$

This integral can be plotted as a function of ε as shown in Figure 6. Thus, if y_1 falls between $1(\varepsilon)$ and $m(\varepsilon)$ as shown in Figure 5, the integral has a value, although it oscillates about a mean of $\sqrt{\pi z_o/k_1}$. If y_1 falls outside this window, it value falls very quickly to insignificance. Since $1^{-1}(y_1)$ and $m^{-1}(y_1)$ define the boundary of the flaw beneath the line $y = y_1$, the dimension of the object is well defined. The reflectivity or strength of the echo is not well defined because of the fluctuation of the integral.

The conclusion is that linear holography can provide a very good measure of the cross section of the object, but not of its

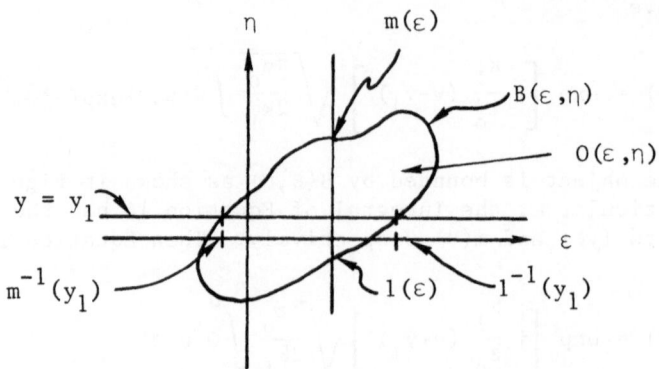

Figure 5. Symbolism for Equation 19.

reflectivity. As a final example consider the triangular object
drawn in Figure 7. In this case, we see that $l^{-1}(\varepsilon) = y_1/m_2$ and
$m^{-1}(\varepsilon) = y_1/m_1$. Referring to Figure 6, we see that $G(\varepsilon)$ has value
only for $y_1/m_2 \leq \varepsilon \leq y_1/m_1$ or $\varepsilon_3 \leq \varepsilon \leq \varepsilon_5$. Within this interval
$G(\varepsilon)$ oscillates but remains large, but outside this range $G(\varepsilon)$ falls
to a small value. Since ε_3 and ε_5 coincide with the edges of the
object, a precise measurement of its cross-section can be made.
If $0(\varepsilon,\eta) \neq 1$, the value of $G(\varepsilon)$ within the window will be even
more variable, but the boundaries will remain precise. Thus, the
display should probably be binary, since grey scale could be mis-
interpreted. All of this obtains only if the object is several
wavelengths across. If it is not, no inference about the size
of the object can be drawn. This will usually be the case when
looking for gas pipe less than about 15 cm in diameter. The location and orientation of the pipe will, however, be accurately
determined.

EXPERIMENTAL RESULTS

    An experiment simulating the proposed system has been set up,
and preliminary results obtained. An SDL 2-D mechanical scanner
and PDP 11/23 computer were mated to a microprocessor controlled
ultrasonics package developed for EPRI under contract RP606-7.

    A single broadband transducer is mounted on an x-y scanner
mechanism controlled by the computer. A short focus lens attached
to the transducer diverges the sound into a wide cone so that echos
will be received over a wide angle. The computer moves the transducer to selected points in the x-direction and instructs the
pulser to energize it by applying a large voltage pulse. The sound
spreads in all directions, and reflects from the pipe. The drive
pulse turns on the time-interval counter, and the reflected pulse,
after passing through the gate, turns it off. The resulting TOF

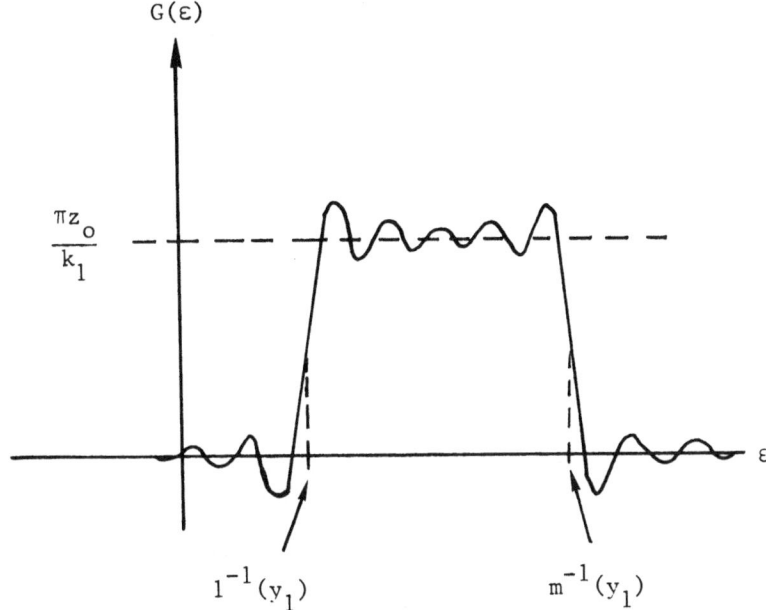

Figure 6.   Value of the Integral of Equation 20.

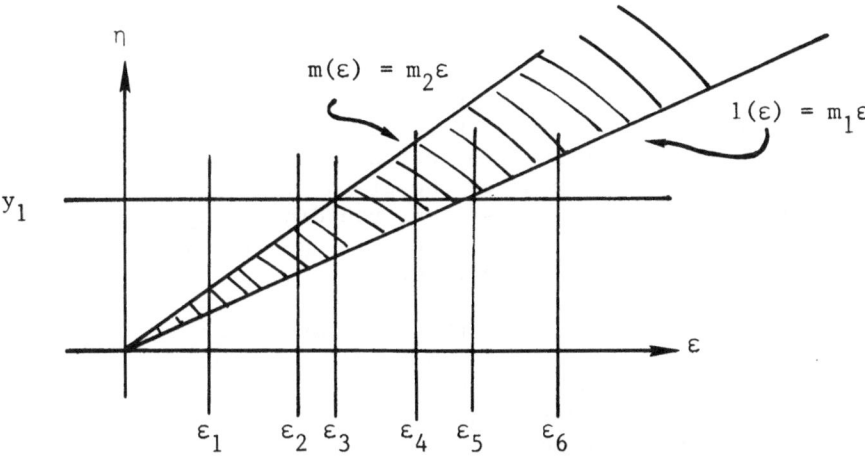

Figure 7.   Example of a Triangular Object.

is recorded in the computer. Each line of data represents sufficient information to reconstruct a cross-section of the reflector.

The following figures show the various steps along the way to the image. Figure 8 is a computer plot of the TOF to a 10 mm diameter steel rod; note the parabolic shape of the curve. The sharp drop at the edges occurs when signal is lost at the limits of the cone of sound. Figure 9 is a photograph of the display of cos φ for the TOF data shown in Figure 8. **Figure 10 shows a** plan view of the image of the 10 mm rod after the BWP algorithm has operated on the data. The steel rod ended at the bottom of the photograph. Figure 11 shows the image from data obtained from a bent 15 mm copper tube. Note that the crimp at the bend resulted in inadequate data making it look as though the tube was actually severed.

The system can be commanded to take as many samples along a line as desired (in powers of 2). The reconstruction program automatically selects the correct frequency according to Equation 4, so that aliasing does not occur. The computer can also be instructed to perform automatic focussing by searching for the minimum TOF, computing the depth z_0, and reconstructing at that plane. In this way it is possible to image slanted pipes without searching different depths. An example of this is shown in Figure 12, where the pipe slanted upward by about 5 cm from one end to the other.

In the field, such a system may have to work with interfering objects, such as rocks or other pipes. To simulate such a condition, a number of ball bearings were interspersed about the pipe as shown in Figure 13. The resulting image is shown in the auto-focus mode in Figure 14(a). Note, that in some places, the program obviously chosen to focus on a ball, resulting in a defocussed pipe. Figure 14(b) shows the same data reconstructed at fixed focus at the depth of the pipe. The noisiness is practically dissappeared. Finally, in Figure 14(c) a fixed focus on one of the balls, reveals the one-dimensional nature of the system, yielding elongated features. This series of images reveals the surprising tolerance of this system to interfering point scatterers.

Intuitively, it would seem that this system would not work well for complicated objects. As a test, we imaged a pair of pliers and a monkey wrench and obtained the results shown in Figures 15(a) and (b). We consider these to be a remarkable demonstration of the robustness of the method.

CONCLUSIONS

We have demonstrated that linear impulse holography is a very powerful technique for imaging structures that tend also to be

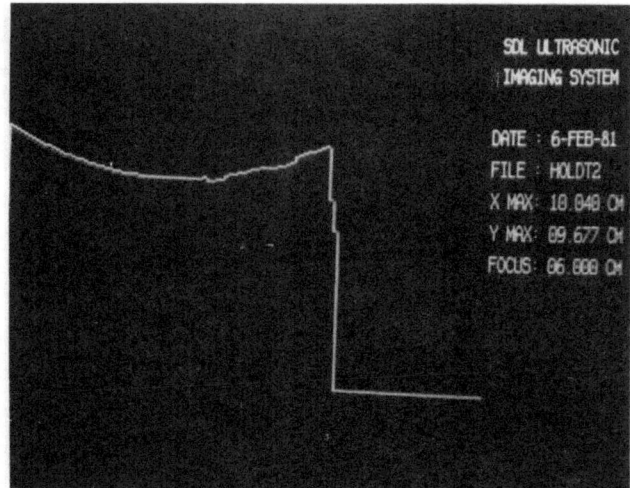

Figure 8.   Experimental Time-of-Flight Profile.

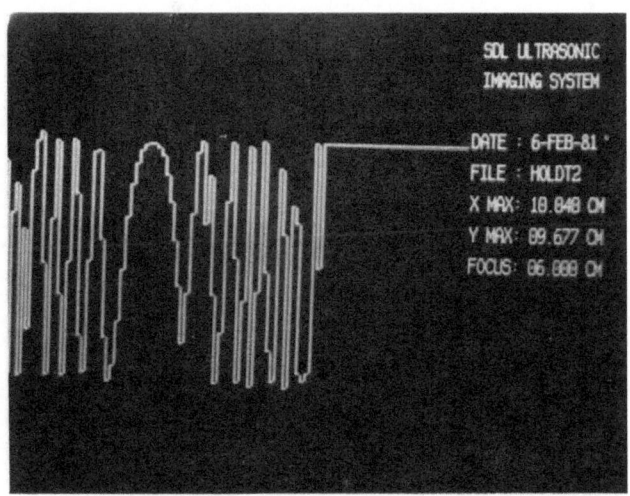

Figure 9.   Experimental Phase Profile.

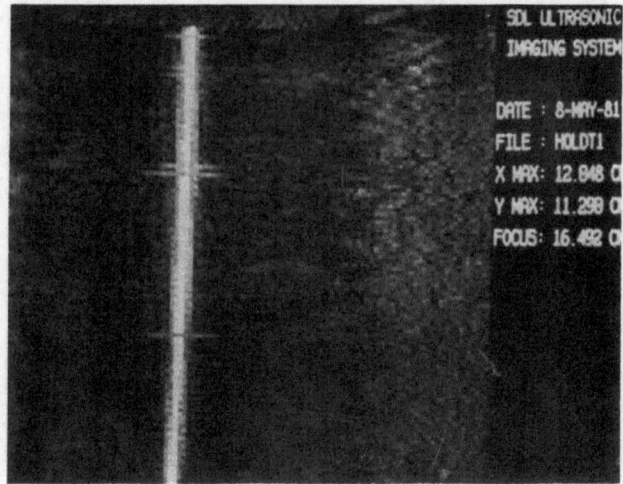

Figure 10.   Image of A 10 mm Rod.

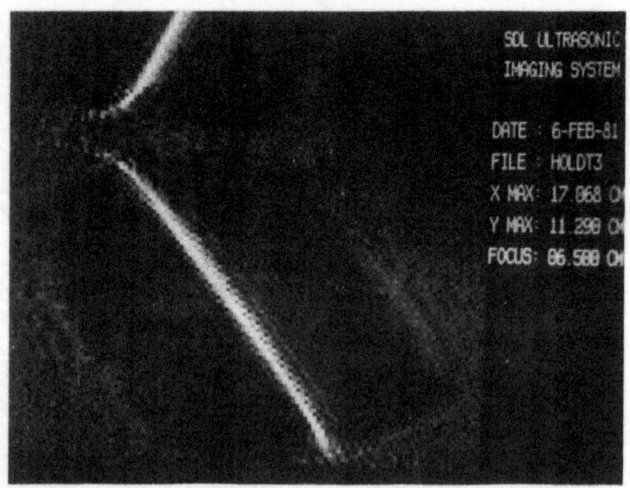

Figure 11.   Image of a Bent Copper Tube.

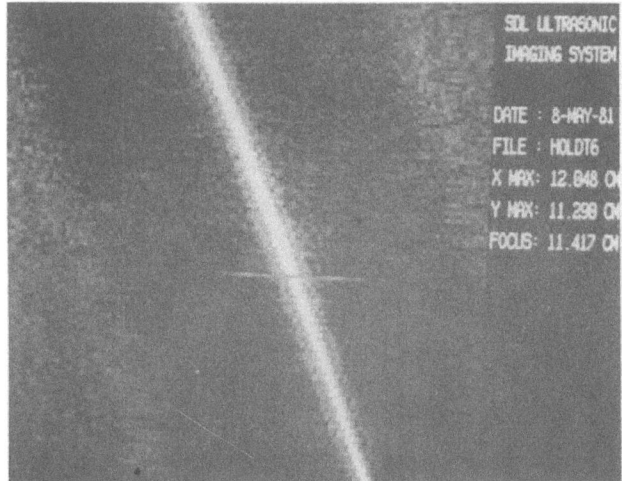

Figure 12.   Image of an Upward Slanting Rod.

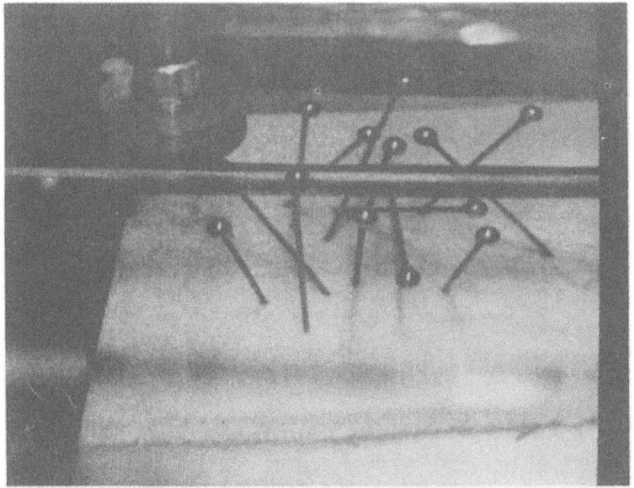

Figure 13.   Photograph of a Rod Surrounded by Interfering Balls.

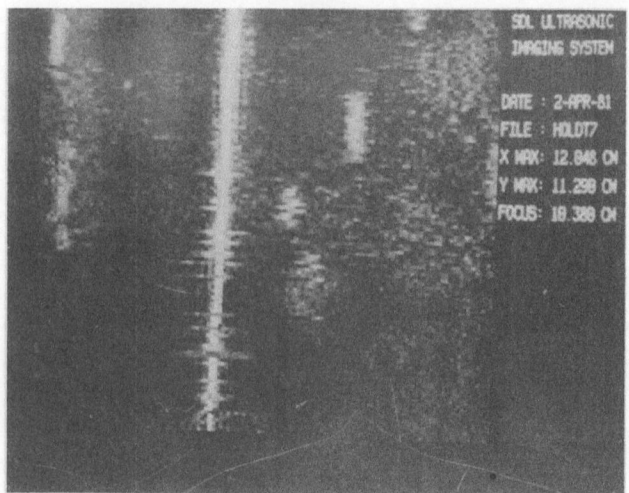

Figure 14a. Image of the Rod Surrounded by Interfering Balls taken
             with Autofocus.

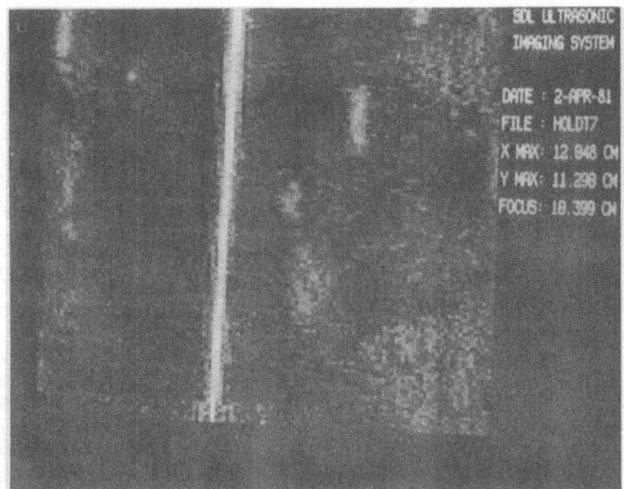

Figure 14b. Image of the Rod Surrounded by Interfering Balls taken
             with Fixed Focus on the Rod.

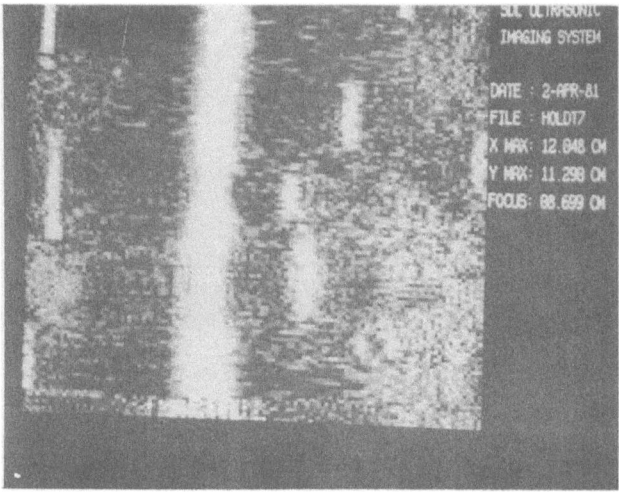

Figure 14c.   Image of the Rod Surrounded by Interfering Balls taken
with Fixed Focus on One of the Balls.

linear. A great advantage of this method is that reconstruction
can be done in essentially real time. Using a standard mini-computer,
a line of 128 points reconstructs in 2 seconds. Hardwired systems
could significantly shorten this time. The full two-dimensional
reconstruction can be performed later, if desired, and if time is
available. Thus, even where a two-dimensional reconstruction is
required, a one-dimensional stack could be built up while the scan
proceeds. This would allow a much quicker evaluation of the object
to be carried out, without waiting for the complete reconstruction.

The next phase of this program will evaluate the concept in
soils before a full scale engineering prototype is built.

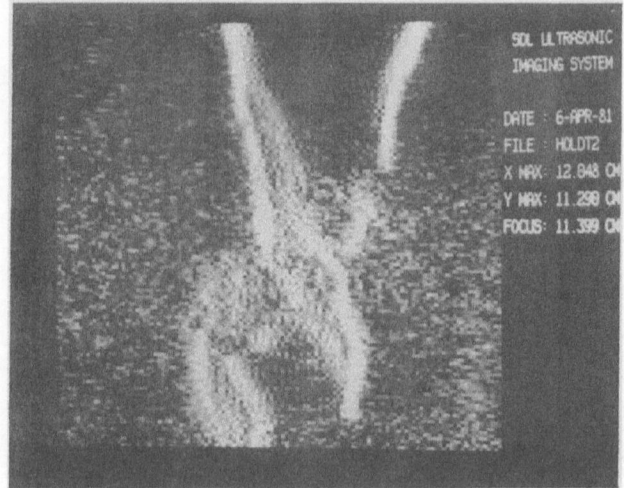

Figure 15a.   Image of a Pair of Pliers.

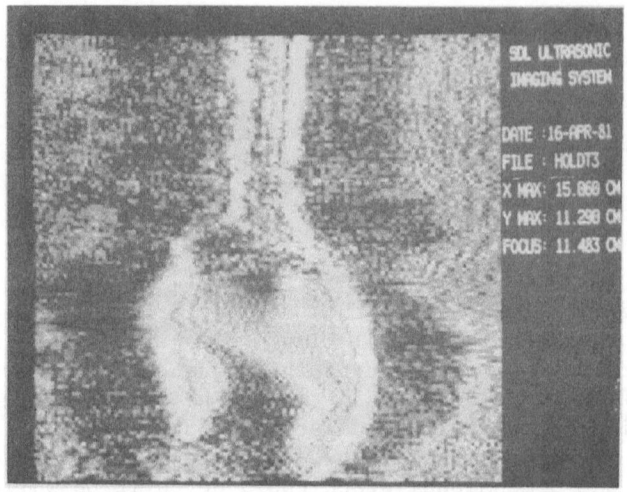

Figure 15b.   Image of a Monkey Wrench.

REFERENCES

1. Gabor, D., "A New Microscope Principle", <u>Nature</u>, 161:777, (1948).
2. Leith, E. N., and Upatnicks, J., "Reconstructed Wavefronts and
   Communication Theory", <u>J. Opt. Soc. Am.</u>, 52:1123, (1963).
3. Boyer, A. L., Hirsch, P. M., Jordan, J. A., Lesem, L. B., and
   Van Rooy, D. L., "Reconstruction of Ultrasonic Images by
   Backward Wave Propagation", <u>Acoustical Holography</u>, Plenum
   Press, 6:333, New York, N.Y., (1971).
4. Papoulis, A., "Systems & Transforms with Application in Optics",
   McGraw-Hill, New York, N.Y., (1968).

ACOUSTIC PASSIVE REMOTE TEMPERATURE SENSING

Theodore Bowen

1202 Calle Gardenias
Tucson, Arizona 85705

ABSTRACT

Passive non-invasive temperature measurement of the
interior of a body may be accomplished by using the
acoustic thermal noise spectra of the body. The noise
power spectrum received by an acoustic transducer coupled
with the surface of the body is measured and mathematically deconvoluted to provide the temperature-depth distribution in an interior region determined by the directivity
of the transducer. While this technique may be utilized
for any reasonably homogeneous material or animal body,
typical parameters are discussed for a system designed
to sense temperature profiles in soft tissue.

INTRODUCTION

There are many situations in medical diagnosis and
treatment, industrial processing, geophysical exploration,
and other fields where the temperature inside a material
body is desirable to measure, but it is not practical to
insert a probe beyond the surface of the body. In medical diagnosis, the usefulness of temperature measurement
at the few places available for probe insertion is well
established. In recent years, thermograms produced by
infrared camera equipment and other surface temperature
measurement have shown promise as a means of detecting
breast cancer lesions.[1] A technique which extends temperature measurement to all soft-tissue parts of the body
offers promise as a powerful new diagnostic tool.

In medical therapy, a non-invasive temperature monitoring technique would be useful in almost any procedure involving heating or cooling of the soft tissues of the body. For example, hyperthermia has been found to be a promising technique, either alone or in combination with other modalities, for the treatment of cancer.[2] However, its effectiveness is very sensitive to the temperature which is reached, becoming more effective as one approaches 45º C, but tissue necrosis becomes a serious problem if the temperature goes above 45º C. Therefore, a non-invasive method of monitoring temperature profiles is important if hyperthermia is to have wider potential.

In many manufacturing processes involving the curing, heating or cooling of massive solid or semi-solid bodies, non-invasive temperature measurement would permit quality control monitoring of internal temperatures which has not previously been possible. Non-invasive temperature measurements in homogeneous earth formations such as rock or salt might find application in geothermal exploration or in conducting mining explorations in such a manner as to avoid costly complications now encountered where drilling inadvertently gets into water-bearing strata.

The acoustic passive remote temperature sensing system[3] is based upon the mathematical analogies between acoustic and electromagnetic radiations. In the case of electromagnetic radiation, it is well known that any surface at any absolute temperature $T > 0$ emits "black-body" radiation. A broad band of frequencies is emitted from 0 to an upper limit determined by the temperature T. For example, for objects at temperatures in the neighborhood of room temperature, the frequencies extend through radio and microwave frequencies into infrared frequencies. Many temperature measuring systems determine temperature of the surface of a "black-body" by measuring the intensity of all or some portion of the "black-body" radiation, comparing the apparent color of the radiation from the surface of unknown temperature to that of a surface of known temperature. Infrared cameras measure surface temperature by the intensity of infrared emissions. Less well known are microwave radiometers, which determine temperature by measuring the intensity of "black-body" radiation in the microwave frequency region. When only the microwave frequencies are measured, the total intensity is directly proportional to the absolute temperature T of the "black-body" surface. It is possible to measure temperatures down to within a few degrees from absolute zero with a microwave radiometer.

If the body which is radiating the thermal microwave radiation has a thickness comparable to the attenuation length for the radiation, and if the attenuation length varies with frequency, then a measurement of the apparent radiation temperature as a function of frequency gives information on the temperature-depth profile. This technique for microwave passive remote temperature sensing of upper atmospheric temperatures has been elaborated, since the pioneering work by R. H. Dicke and collaborators[4] in 1946, by many authors.[5,6] Barrett and Myers[7] have applied the microwave technique to subcutaneous temperatures in human and animal tissues. However, microwaves having wavelengths on the order of a few centimeters must be employed, so the lateral spatial resolution is poor for this application.

Acoustic remote temperature sensing appears feasible in many applications where the microwave method is unsuitable, primarily because (a) much shorter wavelengths may be employed when acoustic radiation is utilized and (b) in many cases, the acoustic attenuation coefficient is more uniform and predictable than the microwave attenuation coefficient. In the following sections, acoustic passive remote temperature sensing will be discussed with reference to non-invasively measuring temperature-depth profiles in human or animal soft tissues. The same general principles are applicable to measurements within any material body which is reasonably homogeneous; i.e., (a) variations of density and speed of sound are less than about ±5% and (b) variations of acoustic attenuation coefficient (at a given frequency) are less than about ±15%. Most bodies suitable for acoustic-echo imaging would also be suitable for acoustic passive remote temperature sensing.

REMOTE SENSING WITH THERMAL RADIATION

The general mathematical principles of microwave and acoustic remote temperature sensing are similar. Although there is extensive scientific literature on[4-6] microwave remote sensing of atmospheric temperature, the basic principles are probably unfamiliar to many. This section presents a qualitative discussion of the detection of thermal radiation, whether acoustic or electromagnetic "black-body" radiation, and of the effect upon the power spectrum if the temperature of successive layers is not uniform.

If a surface is "black," e.g., perfectly absorbing,

the intensity of the emitted radiation depends only upon the temperature T. If the receiving device has a highly directional response in a "beam" region, and if the "black-body" surface extends across the entire beam, the received power is unaffected by the relative distances between detector and "black-body" surface or by the orientation of the "black-body" surface (provided only that it intercept the entire beam); hence, the received power can be directly interpreted as a "black-body" temperature of the radiating surface. In the case of a microwave radiometer, the directional beam response is achieved by a suitable antenna, often a parabola followed by a conical horn section. In a pyrometer, the optical system provides the directionality characteristic.

Suppose the body which is at temperature T does not appear "black," that is, it does not completely absorb incident radiation. If the fraction absorbed is A, then the intensity of the emitted "black-body" radiation must be the same fraction A of the amount expected for a perfect "black-body." This can be deduced from very general arguments based upon the conditions which must prevail when thermodynamic equilibrium is established. The factor A is less than unity if some incident radiation is reflected, but corrections for reflection can be made small by suitable impedance matching. Of prime interest in this discussion is the situation where the factor A is less than unity because some of the radiation passes through to reach subsequent layers in the same body. Consider a body with a large number of identical layers, each of which absorbs one-half the radiation intensity incident at its depth. Then the amount absorbed by layers 1, 2, ..., n, . . . would be

$$\frac{1}{2}, \frac{1}{4}, \frac{1}{8}, \ldots, \frac{1}{2^n}, \ldots$$

because each layer receives one-half as much radiation as the preceding layer. Now, when emission is considered, each layer emits equally at one-half the "black-body" rate corresponding to temperature T. However, as viewed from the outside, the first layer contributes fully, but one-half of the radiation from layer 2 is absorbed in passing through layer 1 for a net contribution of $\frac{1}{4}$. The relative amount contributed to emission to the outside by layers 1, 2, 3, . . ., n, . . . would be

$$\frac{1}{2}, \frac{1}{4}, \frac{1}{8}, \ldots, \frac{1}{2^n}, \ldots$$

When all these fractions, which represent emission relative to a perfect "black-body", are summed, the total is
the same as for a normal "black-body." Thus, for a surface in which radiation is gradually absorbed with increasing depth, the emission is still that of an ideal
"black-body" at temperature T provided that all absorbing layers are at the same temperature T. When temperature is measured with pyrometers or infrared cameras, the
depth in which the radiation is absorbed is generally
very small, so that the assumption of uniform temperature
is very good. The microwave radiometer, however, has
found application to situations where the temperature
cannot be assumed uniform throughout the range of depth
where the radiation is absorbed.

Suppose, in the example of the preceding paragraph,
that the absolute temperatures of layers 1, 2, . . .
were T_1, T_2, . . . As was already mentioned, the microwave radiation intensity is proportional to the absolute
temperature, so the contribution from the nth layer must
be proportional to T_n, and the relative emission from
layers 1, 2, 3, . . ., n, . . . is

$$\frac{T_1}{2}, \frac{T_2}{4}, \frac{T_3}{8}, \qquad \frac{T_n}{2^n}, \ldots$$

The sum of the above series would predict a total
emission corresponding to an apparent "black-body"
temperature T_a, where

$$T_a = \frac{1}{2}T_1 + \frac{1}{4}T_2 + \frac{1}{8}T_3 + \ldots + \frac{1}{2^n}T_n + \ldots$$

The temperature T_a is a weighted average of the temperature of the various layers, the first layer having a
weight of one-half, the second one-fourth, and so forth.
Suppose, by utilizing a different band of frequencies,
the relative absorption in (and emission from) the layers
is altered, say one-fourth is absorbed in each layer.
Then the total emitted radiation corresponds to an
apparent temperature given by

$$T_b = \frac{1}{4}T_1 + \frac{3}{4}(\frac{1}{4})T_2 + \frac{3}{4}(\frac{3}{4} \cdot \frac{1}{4})T_3 + \ldots + (\frac{3}{4})^{n-1} \cdot \frac{1}{4}T_n + \ldots$$

$$= \frac{1}{4}T_1 + \frac{3}{16}T_2 + \frac{9}{64}T_3 + \ldots$$

Notice that the relative weights of successive layers
are different; deep layers contribute more heavily in

T_b than T_a, and, in general, T_a and T_b are not equal.

The above example illustrates why the apparent temperature is a function of the attenuation coefficient, which in turn is a function of the frequency of the radiation. It also indicates why successive depths contribute with differing weights as one varies the frequency. Generalizing, if one obtains a sufficient number of apparent temperatures T_a, T_b, - . . . in different frequency (and attenuation coefficient) intervals, one can solve the simultaneous equations for T_1, T_2, ... The calculation of the temperature of a deep layer, such as T_3, involves a suitable linear combination of T_a, T_b, . . . , so the error in T_3 would be a similar combination in quadrature of the errors in T_a, T_b, ... Mathematical techniques for going from the directly measured T_a, T_b, . . . to the inferred temperature distribution T_1, T_2, . . . as a function of depth have been extensively developed.[6]

The error ΔT in measuring a noise temperature T is given by[8]

$$\Delta T = T/(Bt)^{1/2} \tag{1}$$

where B is the frequency bandwidth and t is the measuring time. If the amplifier and input termination introduce additional noise their equivalent noise temperatures must be added to the apparent radiation temperature in computing the total noise temperature, T for insertion in Eq. (1).

ACOUSTIC REMOTE TEMPERATURE SENSING IN SOFT TISSUE

Since the temperature-depth profile along a reasonably well-defined line is usually desired, an acoustic transducer with an approximately parallel or weakly-focused beam pattern, as shown in Fig. 1, would be employed. Note that the radiation temperatures of successive layers are properly measured independent of the particular transducer beam pattern, provided that each layer extends completely across the beam. The amplifier shown in Fig. 1 should contribute the least possible noise, since its noise temperature must be included in the value of T used in Eq.(1), and also must be subtracted from the measured noise temperature to determine the actual temperature. For highly stable operation, it may be necessary to repeatedly switch the amplifier input to a reference noise source. The

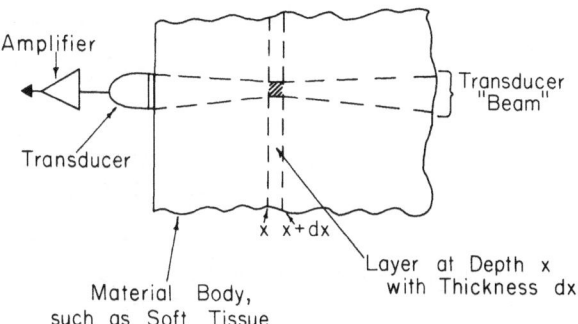

Fig. 1.   Sketch of a material body, such
          as soft tissue, and the acoustic
          transducer and amplifier to sense
          the acoustic thermal noise rad-
          iated toward the transducer by
          successive layers within the body,
          such as the layer illustrated.

amplifier output would be measured by a noise power
spectrum analyzer with the frequency spectrum divided
into a finite number of bands such that the precision
given by Eq.(1) is tolerable. For example, if the bandwidth B = 0.1 MHz, the measurement time t = 100 seconds,
and body temperature T = 310°K (neglecting the amplifier
noise temperature), then ΔT = 0.1^{O}K.

The range of acoustic frequencies whose noise power
must be measured is determined by considering the relation of attenuation coefficient to frequency. It is
well established that attenuation of ultrasound in body
tissue is approximately proportional to frequency, so
that one can write the contribution of thermal noise
power by the layer between x and dx as $kbf \cdot T(x)$
exp(-bfx)dx, where k is the Boltzmann constant, b is a
constant characteristic of tissue (a typical value for
soft tissue is b = $.2(cm-MHz)^{-1}$), f is the frequency,
and T(x) is the temperature at depth x. From this expression, one can show that this layer makes its maximum
contribution to the received noise power at a frequency
f(max) = 1/(bx). At a lesser depth, the maximum is
located at a higher frequency. This is the basis of the
connection between the temperature distribution with
depth x and the power spectrum with frequency f. The
important range of frequencies is given by the expression
for f(max).

Suppose the temperature distribution is desired to
a depth of 16 cm. in soft tissue. Then the lowest frequency f_1 approximately equals 1/(.2)(16) = 0.3 MHz.
The smallest distance is determined by the fact that
large thermal gradients are not expected in tissues due
to limits on the magnitudes of possible heat sources
and sinks: 0.5 cm. might be reasonable corresponding
to the highest frequency f_2 = 1/(.2)(0.5) = 10 MHz.

Note that the wavelengths corresponding to the
frequencies f_1 and f_2 in soft tissue are approximately
0.5 cm. and 0.015 cm. This is very convenient since any
device to create a directional beam sensitivity pattern
must have dimensions large compared to a wavelength.
This being the case, a diameter d = 5 cm. of the acoustic
transducer might be conveniently selected. The frequencies involved are also conveniently high so that it is
reasonable to measure the power spectrum with bandwidths
of 0.1 MHz or more, and temperature measurement errors
given by Eq.(1) for 100 second measurements would be
on the order of 0.1^{O}K or less, as was already suggested
earlier in this section.

SENSITIVITY AND RESOLUTION

If acoustic intensity is attenuated as exp(-bfx), where f is the frequency and x is the depth, then b is approximately a constant for soft tissue. If we wish to be exact, b = b(x). A new variable u can then be defined by

$$u \equiv \int_0^x b(x) \, dx \tag{2}$$

It can be shown[5] that the acoustic noise power spectrum actually measures temperature as a function of u; i.e., T(u). If b(x) = b = constant, then Eq.(2) gives

$$u = b \, x, \tag{3a}$$

so T(u) can be directly interpreted as a function of the actual depth x.

Suppose b(x) varies somewhat with depth (perhaps $\pm15\%$ would be typical for many soft tissues). We can still state that

$$u \simeq \bar{b} \, x \tag{3b}$$

where $\bar{b}$ is an appropriate average attenuation coefficient. Assuming that the main error in $\bar{b}$ is due to inaccurate measurement and individual variability, the fractional error in determining x from u is independent of the depth x. The depth resolution of temperature variations measured by the noise power spectrum can be shown to be proportional to depth, so the uncertainty in b(x) contributes, relatively, the same at all depths.

If we adopt the approximate Eq.(3b), then it can be shown that the apparent temperature $\bar{T}(f)$ at frequency f due to a temperature distribution T(x) is given by

$$\bar{T}(f) = \bar{b}f \int_0^\infty T(x) \, \exp(-\bar{b}fx) \, dx, \tag{4}$$

If T(x) = T = const., then Eq.(4) gives $\bar{T}(f)$ = T = const. at all frequencies. Suppose θ(x) and $\bar{\theta}(f)$ are the deviations from a constant reference level, such as normal body temperature; then Eq.(4) applies equally well to θ(x) and $\bar{\theta}(f)$ instead of T(x) and $\bar{T}(f)$.

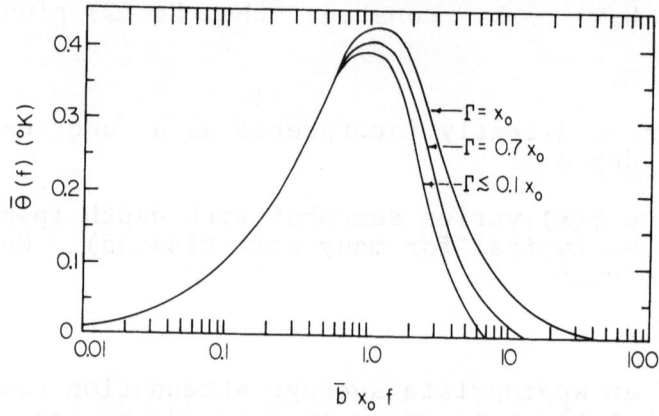

Fig. 2. Apparent temperature increment $\bar{\theta}$ (f) versus
the dimensionless frequency $\bar{b}x_0f$ defined in
the text for temperature distributions $\theta(x)$
of width Γ, the area under the $\theta(x)$ curve is
the same and the peak temperature $\theta_\Gamma = x_0/\Gamma$.

Figure 2 shows the function $\bar{\theta}(f)$ as a function of the dimensionless variable $\overline{bx_0}f$ for bell-shaped temperature distributions centered at depth x_0 of the form:

$$\theta(x) = \theta_\Gamma [1-(x-x_0)^2/\, \Gamma^2] \quad \text{if } |x-x_0| \leq \Gamma, \qquad (5a)$$

$$= 0 \qquad\qquad\qquad \text{if } |x-x_0| > \Gamma. \qquad (5b)$$

In Fig.(2), the peak temperature deviation θ was adjusted for each value of the width Γ so that the area under the $\theta(x)$ curve was constant. For $\Gamma \leqslant 0.1\ x_0$, the radiation noise temperature $\bar{\theta}(f)$ measures only the area under the $\theta(x)$ curve; i.e., the response is proportional to the product $\theta_\Gamma \cdot \Gamma$. For $\Gamma \geqslant 0.7x_0$, the $\bar{\theta}(f)$ distribution changes rapidly as the width Γ changes. So the depth resolution at depth x_0 is on the order of $\pm 70\%$ of x_0. Since the curves in Fig.2 are universal for any depth x_0 provided that the width Γ be scaled in proportion to x_0 and the frequency be scaled in inverse proportion to x_0, the resolution widens in direct proportion to the depth. This characteristic can be significantly improved by simultaneously measuring acoustic noise temperature along other directions, such as from the opposite side of the body.

CONCLUSIONS

An acoustic passive remote temperature sensing system is capable of non-invasively measuring temperature-depth profiles in soft tissue to accuracies on the order of $0.1\ K$ in 100 second measurements. Roughly, the lateral resolution at any depth x_0 would be $\gtrsim 10\%$ of x_0 and the longitudinal resolution would be 70% of x_0; these figures could be improved by simultaneously acquiring data along other beam paths.

When a body is at uniform temperature, a measurement of temperature employing acoustic thermal radiation converges upon the correct temperature at all acoustic frequencies. The method therefore is very sensitive to temperature anomalies in a body which normally has uniform temperature; any variation of apparent temperature with frequency must be due to temperature non-uniformity.

Since acoustic attenuation is frequency dependent in all materials, this passive remote temperature sensing system is applicable to any reasonably homogeneous material or animal body.

REFERENCES

1. Part VIII. Thermography and Cancer Detection: A
    Critical Evaluation in "Thermal Characteristics
    of Tumors: Applications in Detection and Treat-
    ment," R. K. Jain and P. M. Gullino, eds., Annals
    N.Y. Acad. Sci. 335:383-474 (1980).
2. Part V. Hyperthermia: Characterization of Tumor
    Temperature and Environmental Parameters in Which
    Heat is Selectively Destructive to Neoplasms,
    ibid., pp. 180-297.
3. T. Bowen, Passive Remote Temperature Sensor System,
    U.S. Patent 4,246,784, Jan. 27, 1981.
4. R. H. Dicke, R. Beringer, R. L. Hyhl, and A. B. Vane,
    Atmospheric Absorption Measurements with a Micro-
    wave Radiometer, Phys.Rev. 70:340-348 (1946).
5. L. D. Kaplan, Inference of Atmospheric Structure
    from Remote Radiation Measurements, J.Opt.Soc.
    Amer. 49:1004-1007 (1959);
    A. H. Barrett and V. K. Chung, A Method for the
    Determination of High Altitude Water Vapor
    Abundance from Ground-Based Microwave Observations,
    J.Geophys.Res. 67:4259-4266 (1962);
    M. L. Meeks and A. E. Lilley, The Microwave Spectrum
    of Oxygen in the Earth's Atmosphere, J.Geophys.
    Res. 68: 1683-1703 (1963);
    W. J. Welch, A Selective Review of Ground Based
    Passive Microwave Radiometric Probing of the
    Atmosphere, in "Atmospheric Exploration by Remote
    Probes," vol. 2 (Committee on Atmospheric Sciences,
    National Academy of Sciences, Washington, D. C.
    (Jan. 1969) pp. 369-396.
    J. B. Snider and E. R. Westwater, Radiometry in
    "Remote Sensing of the Troposphere," V. E. Derr,
    ed., Supt. of Documents, U. S. Government Printing
    Office, Washington, D. C. (Aug. 15, 1972) pp.15-1 -
    15-33.
6. E. R. Westwater and O. N. Strand, Inversion Tech-
    niques, ibid., pp. 16-1 - 16-13;
    S. Twomey, Introduction to the Mathematics of Inver-
    sion in Remote Sensing and Indirect Measurements,
    Elsevier, Amsterdam (1977).
7. A. H. Barrett and P. C. Myers, Subcutaneous Tempera-
    tures: A Method of Noninvasive Sensing, Science
    190:669-671 (1975);
    P. C. Myers, A. H. Barrett, and N. L. Sadowsky,
    "Microwave Thermography of Normal and Cancerous
    Breast Tissue," reference 1, pp. 443-455;
    J. Edrich et al., Imaging Thermograms at Centimeter
    and Millimeter Wavelengths, reference 1, pp. 456-474.

8.   S. O. Rice, Filtered Thermal Noise-Fluctuation of
     Energy as a Function of Interval Length, J.Acoust.
     Soc.Amer. 14:416 (1943).

THE SPEED OF SOUND AS A THERMAL

IMAGE CT SCAN PARAMETER

Richard L. Nasoni, Theodore Bowen, Mark W. Dewhirst,
Howard B. Roth and Robert Premovich

Division of Radiation Oncology
University of Arizona Health Sciences Center
Tucson, Arizona   85724

ABSTRACT

The speed of sound in mammalian (canine) tissue was measured
as a function of temperature to determine whether these measurements
have the stability, repeatability and freedom from artifact to
qualify speed as a scan parameter in a CT ultrasounc thermal mapping
system. This system would provide thermal dosimetry for the treatment of cancer by hyperthermia; a modality which is receiving acceptance by the medical community, but has been hampered by inadequate dosimetry. Measurements were made of the speed of sound
as a function of temperature using a 5 MHz pulse transmission velocimeter on a number of canine tissues both in vitro and in vivo.
Measurements were also made in vivo on extracorporeal blood and on
a number of malignant tumors. Biochemical fractions were taken
from the insonified volumes, and speed fit to a multiple regression
model involving these fractions and temperature. Conclusions are:
(1) the speed of sound as a function of temperature can be measured
in vivo with stability and freedom from artifact, (2) relatively
large variation from tissue to tissue makes speed an improbable
thermal scan parameter; however, (3) the temperature coefficient
of ultrasound may have possible use in thermal dosimetry.

INTRODUCTION

There is mounting evidence that hyperthermia (the use of elevated temperatures) may provide an effective mode of therapy for
malignant tumors especially when combined adjuvantly with conventional modes of therapy such as radiation or chemotherapy.
Dethlefson and Dewey (1981), Jain and Gulleno (1980), and Streffer

563

et al. (1977), among others, have recently reviewed the techniques
and progress in this field. From these reviews it is clear that
the successful application of this mode of therapy will be largely
conditional on the ability to surmount the two main physical problems associated with it: heat delivery and thermal dosimetry.
Heating methods are varied and largely depend on tumor involvement
and site as well as other factors. They include: ultrasound,
microwave, radiofrequency (using either Joule or induction heating),
inhalation of heated gases, immersion in heated liquids, and extracorporeal heating of the patient's blood. Although none of these
methods has proven to be completely effective, ultrasound, because
of its ability to be focused in tissue, has shown promise in many
cases, Lele (1980).

       With respect to thermal dosimetry, its adequacy is necessary
not only for reasons of safety (hyperthermic temperatures can be
destructive to normal tissue and often border on lethality), but
also for careful documentation and control. Methods of thermal
dosimetry have been extensively reviewed by Cetas and Connor (1980).
At present, they mainly depend on conventional means such as thermistor and thermocouple probes and IR cameras; but, these suffer
from a number of limitations: they may be interactive with the
heating modality; they are, in the case of probes, invasive; and,
they provide surface area or point temperature information only.

       An inconventional method which would obviate these short-
comings, for certain geometries to be presently described, has been
proposed by Bowen et al. (1979) and Johnson et al. (1979) and involves a novel use of ultrasound. This method which capitalizes
on thermal variations in the propagation properties of ultrasound
in tissue, would provide, when used in conjunction with computerized
tomography, noninteractive and noninvasive two dimensional on-line
information during therapy. However, before extended applications
can be attempted, the acousto-thermal properties of tissues must
be investigated, especially in vivo, to determine whether they can
be measured with sufficient accuracy, repeatability, and freedom
from artifact. The thermo-acoustic property investigated and reported on in this article is speed and its derivative with respect
to temperature, i.e., the ultrasound temperature coefficient.

TEMPERATURE MEASUREMENTS BY ULTRASONIC COMPUTED TOMOGRAPHY

       The idea of using the ultrasonic properties of a tissue to
measure its temperature was apparently first suggested by Sachs
and Janney (1977). In their system (TAST), the change in the time
of flight in a pulsed transmission sensing beam through a small
tissue volume irradiated and heated by a focused ultrasonic beam
is used to determine tissue parameters as a diagnostic aid. The
change in the time of flight is proportional to a factor which
contains the essential information concerning the tissue. These

authors concluded that if the temperature dependence of this factor could be determined, TAST could, in principle, be used as a thermometer. However, as conceived, this system would provide temperature information only at a point or along a line.

The use of computed tomography would enable temperature, or change in temperature, information to be reconstructed in either two or possibly three dimensions. The role of ultrasound in computerized tomography has been recently reviewed by Kak (1979). The purpose of the ultrasound CT scan is to reconstruct cross-sectional images using acoustic parameters. The two main parameters investigated are the attenuation coefficient and the index of refraction. The last is equivalent to velocity or time of flight, and it appears to be the parameter which is most easily reconstructed because of the difficulty in taking divergence effects into account in attenuation CT scanning. Greenleaf et al. (1974, 1975) constructed the first tomograms; these were followed by Carson et al. (1976), Jakowitz and Kak (1976) and Glover and Sharp (1977). In non-destructive testing, ultrasonic CT time of flight reconstruction has been suggested as a means of mapping stress fields in metals by Hildebrand and Hufferd (1977).

Because of the problems encountered in refraction by bone and in transmission through gas filled volumes such as lungs the reconstructions are best confined to cross-sectional areas which are mainly composed of soft tissue. These could be the abdomen or the breast. (It is also possible to reconstruct cross-sectional areas of limbs. For a discussion of the problems arising in ultrasonic CT scan of hard tissues such as bone see Carson (1977).)

A system in which the principles of ultrasonic CT reconstruction might be used to map the change in the temperature field as hyperthermia is applied is shown schematically in Figure 1. (This system is shown for its heuristic value only.) Let us suppose that the patient's body cross-section with interior tumor heated by some modality (not shown in the figure) is centered with respect to a circular array of transducers coupled to the body surface by some suitable fluid coupling medium. In the figure one of the transducers is shown to be in a transmit mode while the remainder are in a receive mode. In this way a fan beam is generated which is partially subtended by the patient's body and in particular the tumor volume. If the transducers are electronically switched such that each in turn is put in a transmit mode while the others are either switched (i.e., the one previously in a transmit mode) or remain in a receive mode, an interlaced pattern is generated. These interlaced fan patterns will serve as the basis for the reconstruction. Now, if the speed of sound as a function of temperature is known over the range of temperature of interest (body temperature to about $45^{\circ}C$) and if the speed of sound as a function of temperature is the same for all soft tissues, the temperature

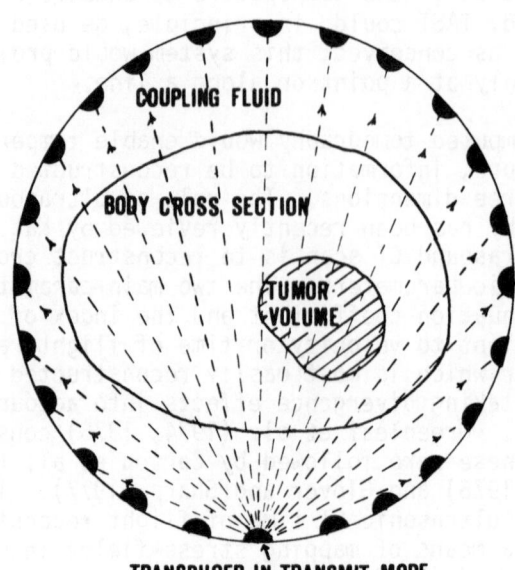

Fig. 1.   A schematic of a possible CT ultrasonic scan system for
           reconstructing a changing temperature field during a
           hyperthermic treatment. The body cross-section is com-
           posed of soft tissue.

variation within a reconstruction resolution element (pixel) can
be determined. (It is assumed that the base temperature prior to
heating are determinable from fixed point temperatures.) Thus,
a map of the changing temperature field over the body cross-section
including the tumor will be generated. In addition, corrective
action can be taken if at any time the temperature in any region
of the normal tissue exceeds a prescribed safe limit. However,
as was previously indicated, before this can be implemented, the
assumption concerning the equality of the speed vs. temperature
curves for all soft tissues must be investigated as well as the
determination of whether speed as a function of temperature can be
measured with sufficient accuracy, repeatability and freedom from
artifact. The remainder of this article will be concerned with
the answers to these questions.

EXPERIMENTAL TECHNIQUE AND METHODOLOGY

Velocimeter

     The basic methods for in vitro and in vivo measurements were
the same; they involved the use of an acoustic velocimeter using
pulse transmission. The tissue under examination was placed be-

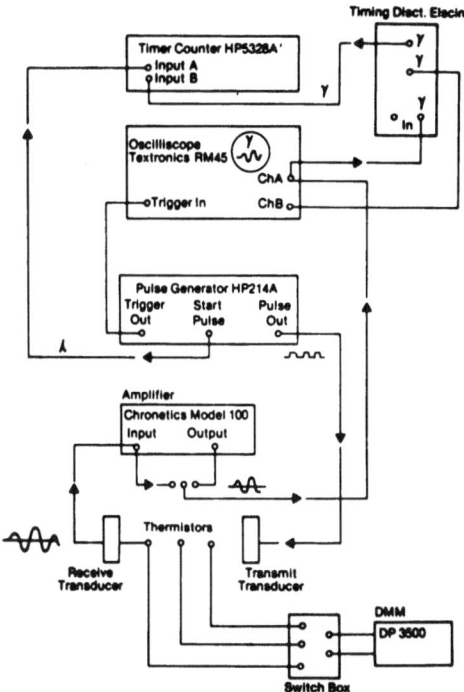

Fig. 2.   Diagram of the electronic arrangement.

tween and in contact with two matched 5 MHz, one half inch diameter
disc PZT5A transducers which were axially aligned and attached to
the jaws of a large (20 inch mainframe, 7 inch jaws) stainless
steel vernier caliper. Figure 2 shows a block diagram of the
electronic arrangement. A pulse generator (HP214A) put out a rectangular pulse train, each pulse of which both excited the transmit
transducer and concomitantly started an electronic counter timer
(HP5328A). The acoustic pulse generated by the transmit transducer
travelled through the tissue and excited the receive transducer
which produced an electrical signal in the form of a decaying
sinusoid. This signal was amplified, if necessary, (Chronetics
Nanologic System 100) and fed into a constant fraction timing discriminator (Elscint STDN-1). The leading edge of the signal activated the timing discriminator which then sent a sharp stop pulse
to the electronic counter timer. The time registered by the counter
thus included the time of flight of the acoustic pulse plus electronic delay time which was calibrated out using the precise speed
temperature data of Del Grosso and Mader (1972). This action
repeats itself with the pulse repetition frequency of the generator
and thus the counter timer's interval averaging feature was used
to increase timing resolution to a fraction of a nanosecond. The
speed of sound is easily calculated by dividing the time of flight

Fig. 3.   In vitro arrangement. Tissue shown is a kidney prior
           to immersion for the beginning of a run.

into the transducer's separation read directly from the vernier
caliper to 0.001 inch. A dual channel oscilloscope was used to
monitor the position of the time pick-off signal from the discriminator relative to the received electrical signal. This should,
of course, remain constant through the duration of a run. Temperatures in the insonified tissue volume were monitored using implanted
thermistor probes (YSI 513) held in place by a holder attached to
a vernier jaw.

In Vitro Technique

In vitro runs were made in an arrangement of the equipment
which allowed accommodation of whole organs which were freshly
excised (about 5 to 30 minutes from termination of the animal to
the commencement of data taking), ligated to prevent fluid loss,
and chilled to impede tissue degradation. The organ was suspended
in a plastic bag from the mainframe and between the jaws of the
caliper and the transducers brought in contact by closing the jaws.
Figure 3 shows the assembly before immersion into a water bath
which was heated using a water circulator (Haake E52). The tissue

was then continuously heated from chilled state (about 20°C) to a
45°C bath temperature at a rate of 0.25°C per minute while the time
of flight and temperature data were recorded at about one minute
intervals. In some of the later runs, when a heater/circulator
(Lauda RC-3B) became available, the tissue after being taken to
45°C was then cooled to about to 37°C by using an exchange coil in
the tank. In this way both heating and cooling curves were generated on the tissue.

To correct for thermal lag within samples, a linear, homogeneous thermal model utilizing suitable boundary conditions (temperature varying linearly with time) was used to calculate the
average temperature along the insonified path. This model yields
the temperature distribution at a time t as

$$\Theta (r,t) \simeq Kt + \eta - \frac{Kl^2}{a^2} r(1-r) \tag{1}$$

where

     r = normalized position along the path
     Θ = temperature
     a^2 = thermal diffusivity
     l = path length
     K = rate of change of temperature
     η = initial temperature

This model was independently tested and shown to hold with
reasonable accuracy for in vitro kidney tissue. Measurement of
temperature at two positions along the path allows the average to
be calculated. Since cooling was non-linear in time, this in vitro
temperature model was inapplicable and simple averaging had to be
applied.

In Vivo Techniques

In vivo techniques can be conveniently divided into those
dealing with tumors, normal soft tissue and blood.

Malignant Animal Tumors

In vivo measurements of the speed of sound versus temperature
were conducted, with one exception, at body temperature by placing
the transducers in contact with surface or subcutaneous tumors.
The speed of sound in these tumors could not be measured over an
extended temperature range because their treatment was subject to
a rigid clinical protocol. However, repeated measurements were
made on tumors which had responded to treatment and had, therefore,
changed in size. In the case of one non-protocol tumor, measurements
were made on cooling mammary tumor which had been heated using rf
current.

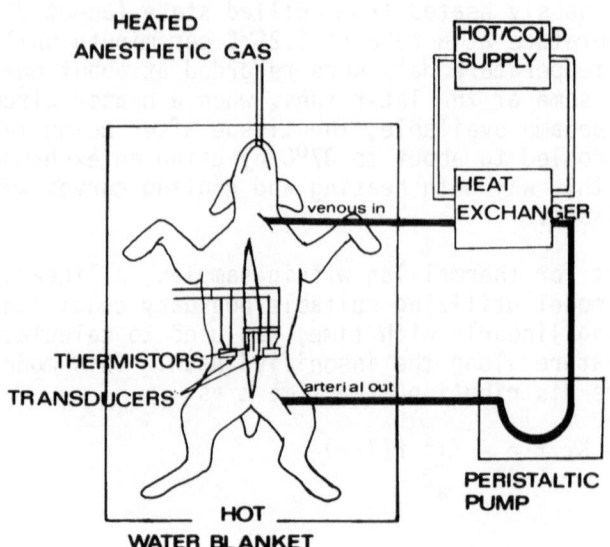

Fig. 4.   Diagram of the in vivo extracorporeal heating arrange-
          ment. The transducers are inserted into the body cavity
          and directly onto the organ being studied.

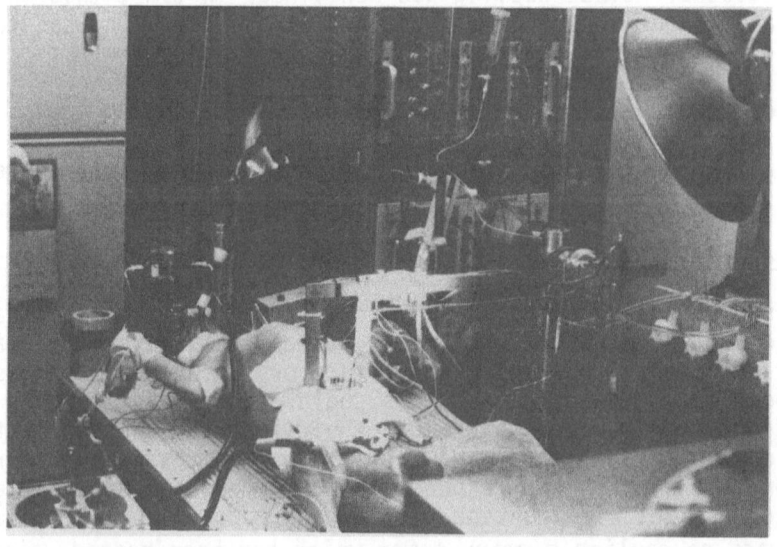

Fig. 5.   An in vivo experiment in progress.

Normal Soft Tissue

The measurement of the speed of sound as a function of temperature involved considerably more complicated techniques. Figure
4 shows the configuration used for these measurements. Blood was
pumped from a cannulated artery through a peristaltic pump and
heat exchanger (Travenol Miniprime) and returned via a cannulation
into a jugular vein. Extracorporeal heating or cooling of the
blood was effected through exchange with a hot/cold fluid supply
(Lauda RC-3B). The transducers were introduced into a mid-line
incision and placed on the tissues under study. By successive
heating and cooling a number of organs could (in principle) be
measured on a given animal. Additional heating was provided by
the breathing of heated anesthetic gases and contact with a heated
blanket. Thermistors were placed in the arterial feed as well
as in the tissue under study. Rectal temperature was also monitored.
Figure 5 shows an in vivo experiment in progress.

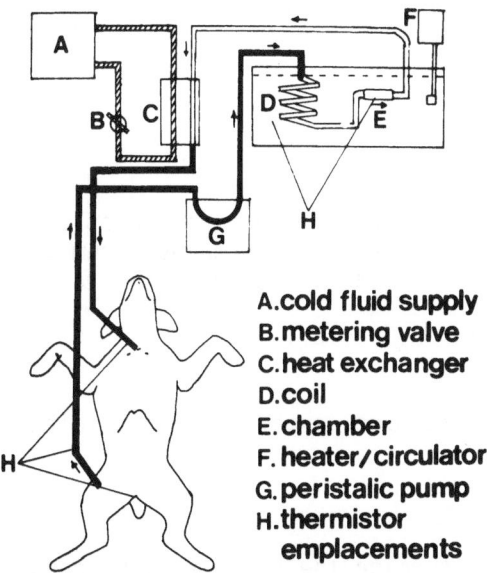

A. cold fluid supply
B. metering valve
C. heat exchanger
D. coil
E. chamber
F. heater/circulator
G. peristalic pump
H. thermistor
   emplacements

Fig. 6.   Diagram of the experimental arrangement for the measurement of the speed of sound in extracorporeal blood.
Blood is heated to measurement temperature and then
cooled to body temperature for the return flow.

Blood

Figure 6 shows the method used for measuring the speed of sound as a function of temperature in extracorporeal blood. The blood taken from the femoral artery was heated from body temperature to the required temperature by exchange in the coil D in the tank. The water in the tank was heated by the heater/circulator F (Haake E52). The measurement of the time of flight was made by placing the transducers in contact with the chamber E. This chamber is cylindrical (about 2 cm diameter) and made of then (0.01 inch thick) silastic sheating. Because the water temperature in the bath was near that of blood, there was little temperature gradient across the acoustic path. The blood was cooled back to body temperature in the heat exchanger C and entered a jugular vein. In this way very little physiological trauma was experienced by the animal. Temperatures were recorded by thermistors at the position indicated.

ANALYSIS

Data Analysis

The data from the in vitro and in vivo results were fit after data smoothing to cubic polynomials using least fit techniques. Thus, speed for each run was fit to an equation of the form

$$V = A_0 + A_1 T + A_2 T^2 + A_3 T^3 \qquad\qquad (2)$$

where

V = Speed
T = temperature

By further data smoothing and the use of cubic splines, Gerald (1978), the first derivative was obtained from the original data. This led for each run to an equation of the form

$$\frac{dV}{dT} = B_0 + B_1 T + B_2 T^2 + B_3 T^3 \qquad\qquad (3)$$

where $\frac{dV}{dT}$ is the ultrasound temperature coefficient.

Biochemical Analysis

In order to determine the relative importance of biochemical components in determining the speed, a cylindrical coring of the insonified path in the tissue was made after the completion of all in vitro runs. From this coring, the wet weight fractions of the following components were assayed: water, lipid, collagen and non-collagenous protein. These are considered to be the most sig-

nificant in determing the speed of sound. The water fraction was
determined by lyophyzing the tissue to dryness. Total lipid was
determined by a modification of a test devised by Folch et al.
(1951), collagen by a test introduced by Stegeman (1958), and the
non-collagenous protein by a test due to Lowry et al. (1951).

The data was then fit to a multiple regression model using
the equation

$$V = \alpha_0 + (\alpha_1 W + \alpha_2 L + \alpha_3 C + \alpha_4 P) + (\beta_0 + \beta_1 W + \beta_2 L + \beta_3 C + \beta_4 P)T \qquad (4)$$

where

    V = speed
    T = temperature
    W = water fraction
    L = lipid fraction
    C = collagen fraction
    P = non-collagenous protein fraction

(these are wet weight fractions)

This regression was run using the SPSS program, Nie et al.
(1975), on a DEC10 computer. Altogether, twenty runs were included
which represented about 1500 data points.

RESULTS AND DISCUSSIONS OF ERROR

Results

A representative result for the speed of sound vs. time and
the temperature vs. time for an in vivo kidney run is shown in
Figure 7. Figures 8,9,10 and 11 present the speed of sound vs.
temperature data for muscle, liver, kidney and spleen respectively.
The paucity of in vivo liver data is attributable to the difficulty
in using this technique on that organ. Because of its attachment
to the diaphragm, the liver exhibits motion which is strongly coupled
to breathing which is very heavy at elevated temperatures. In vivo
muscle data is also difficult to obtain using this technique because
of the problems encountered in elevating muscle temperature by
extracorporeal heating techniques. Because of its tendency to
deflate upon excision or shock, spleen is also a difficult organ
to measure. Figure 11 shows the spleen data as composed of two
groups with different slopes and magnitudes. Two of the curves in
the lower group indicate negative slope. This may be attributable
to a high lipid content as revealed by the biochemical test (3.5
to 4.5 times that of the spleens with a positive slope). Figure
12 shows the combined results for in vitro and in vivo runs for the

kidney, liver and muscle data. Since the spleen data was ambiguous, it was not included in the figure.

These curves demonstrate a similarity in slope and relative position suggesting agreement between in vitro and in vivo measurements. All of the curves discussed in the figures mentioned above were drawn from cubic fits to the data. In all cases, the goodness of fit as measured by χ^2, was less than 0.05 so that the error between fit and actual data at each point over the range is of the order of 0.1 meter/second.

The result of the measurement of the speed of sound in extracorporeal blood is shown in Figure 13. The result compares favorably with that of others as reported in two recent compilations in the literature by Goss, Johnston and Dunn (1978, 1980).

Table 1 shows the results of measurements on various tumor types. Stability during measurement was good (the same reading holding as long as ten minutes) as was repeatability even in view of the shrinkage of some tumors due to the protocol therapy treatments.

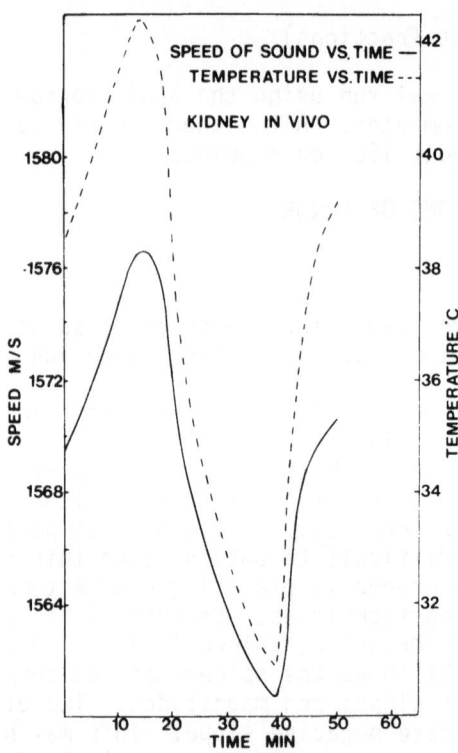

Fig. 7.   Speed of sound versus time and temperature versus time for a representative in vivo kidney run.

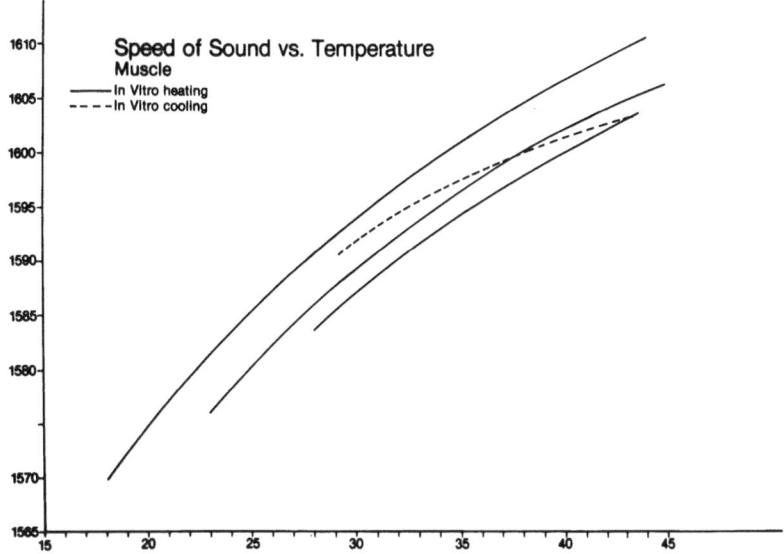

Fig. 8. Speed of sound versus temperature for muscle. The muscle
is canine quadriceps femorus with measurements made
perpendicular to the fiber.

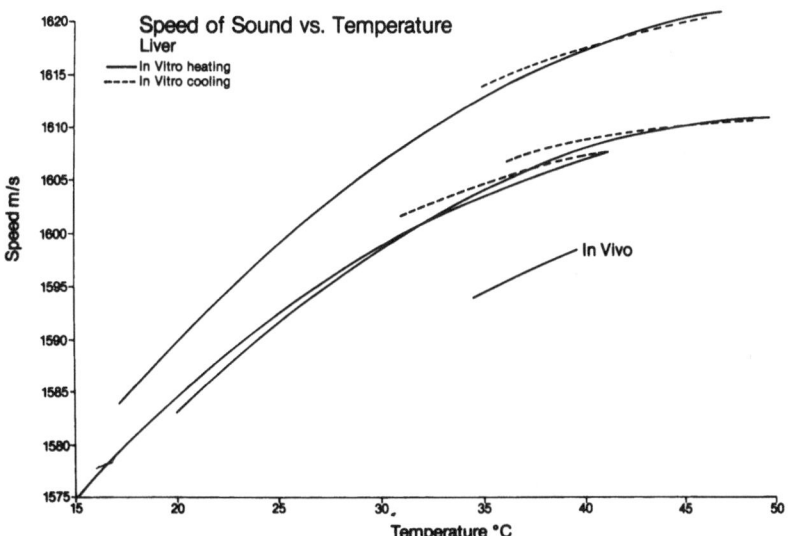

Fig. 9. Speed of sound versus temperature for liver. In vivo
run as indicated.

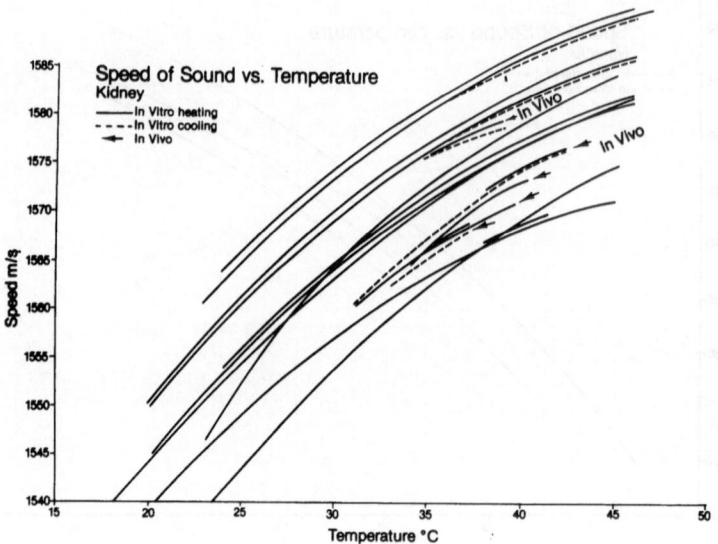

Fig. 10.   Speed of sound versus temperature for kidney. In vivo
           runs as indicated; all other runs are in vitro.

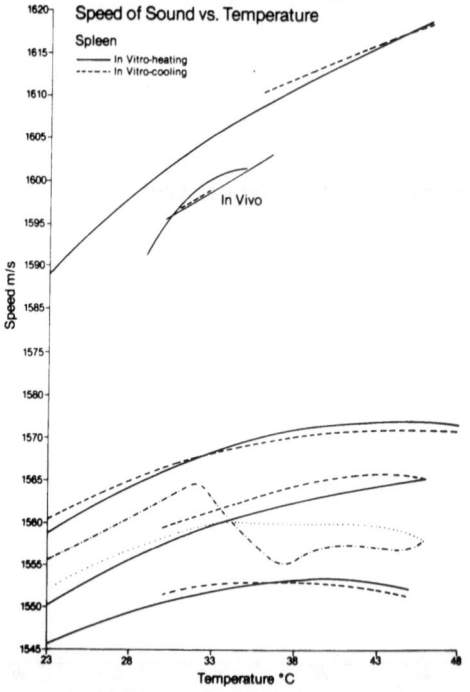

Fig. 11.   Speed of sound versus temperature for spleen. In vivo
           runs as indicated. Two runs in the bottom group are
           singled out by separate designation (by dots and dashes
           followed by dots) since they are atypical.

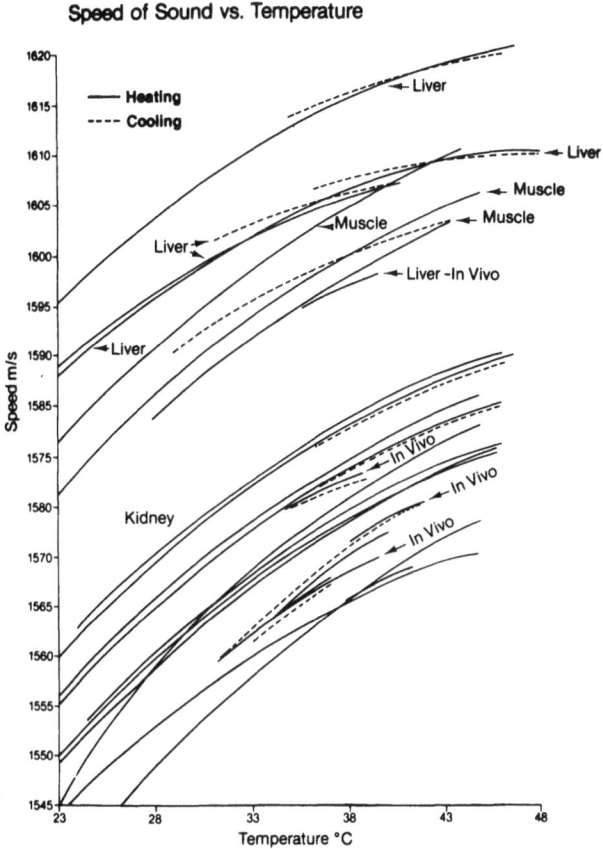

Fig. 12.   The combined results for in vivo and in vitro tissue
            runs. In vivo runs are as indicated in the figure.

        Figure 14 shows a composite of in vivo speed temperature mea-
surements including one run on tumor which was outside of protocol.
Note the wide variation in spleen results which, as mentioned
earlier, is a difficult organ on which to make measurements.

        Figure 15 shows the averaged ultrasound temperature coefficient
(first derivative of the speed with respect to temperature) for
all in vitro kidney runs (11) and for all in vitro kidney, liver,
muscle (and one spleen) runs (19). The error bars are drawn from
the standard deviation calculated over the runs at each temperature.
These averages and standard deviations were derived from the cubic
fits to the first derivatives obtained by cubic splines. Goodness
of fit, as measured by χ^2, was of the order of 0.15. Also shown
on the figure is the temperature coefficient of ultrasound for

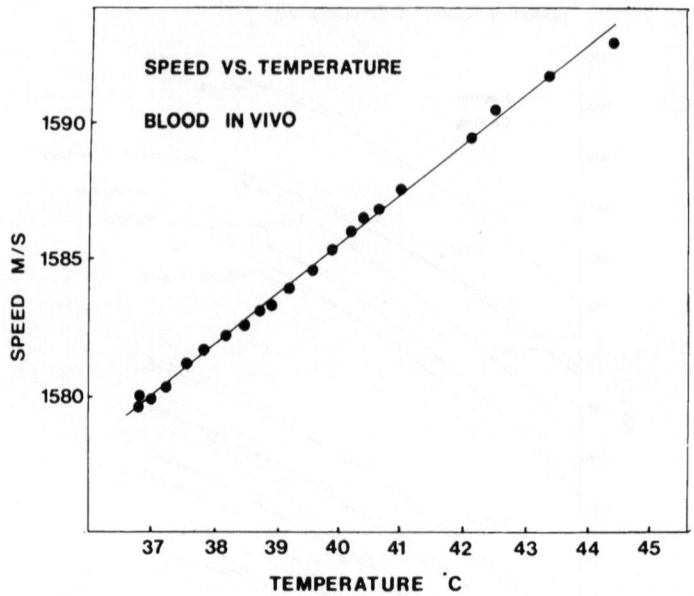

Fig. 13.   The result of the extracorporeal speed of sound in blood
          measurement.

Table 1.   Speed of Sound in Various Tumors
          (Approximately Body Temperature)

Tumor Type	Site	Speed m/s
Malignant mast cell	Abdominal surface	1556
Squamous cell carcinoma	Leg	1525
Malignant melanoma	Leg	1557
Mixed mammary	Chest surface (nodular)	1576
Thyroid	Thyroid region	1559
Mammary	Surface	1589
Giant cell	Foot	1575

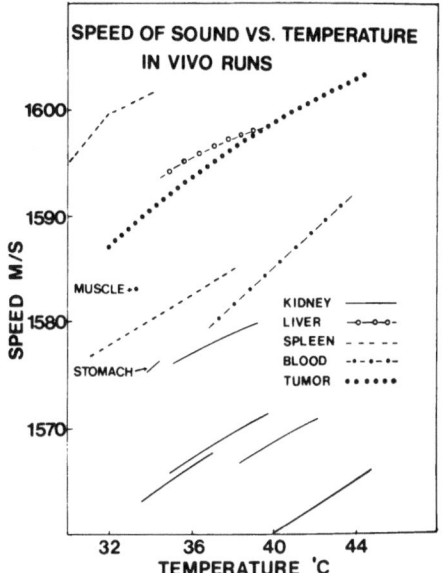

Fig. 14. Combined in vivo results.

pure water calculated from the fifth order polynomial fit of Del Gross and Mader (1972).

Table 2 shows the results of the multiple regression fit. Almost 81% of the variance is explained by the independent variables, the remaining 19% is presumably due to errors, tissue architecture and other unassayed constituents.

Of the explained variance, 90% is explained about equally by the combinations of lipid and temperature. The remaining variables play decreasingly lesser roles. It is noteworthy that the model predicts a negative contribution by lipid to the ultrasound temperature coefficient.

Discussion of Error

Two kinds of errors arise in these experiments: systematic and statistical. Systematic errors due to the use of plastic bags, etc., can easily be accounted for; errors arising due to the use of temperature averaging can be shown to be small (less than 1 m/s). Statistical or random error is principally introduced through the

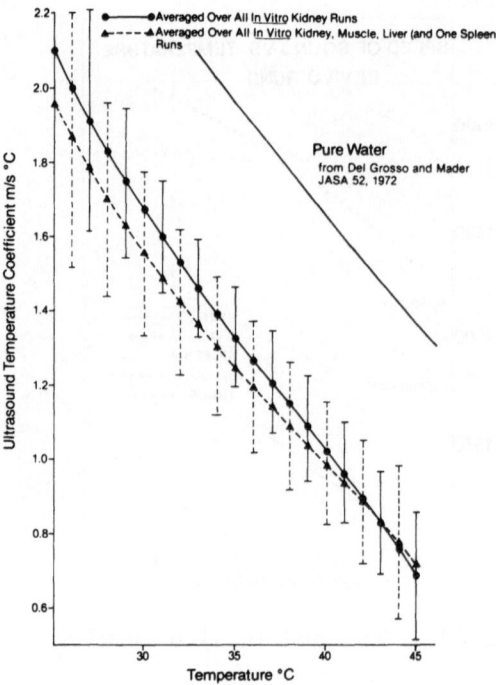

Fig. 15.   Ultrasound temperature coefficient as a function of
           temperature.   Averaged values over runs as indicated.
           Error bars are the standard deviations over the runs.
           The temperature coefficient of ultrasound velocity for
           pure water taken from the data of Del Grosso and Mader
           (1972) is shown for comparison purposes.

counter timer and length measurements and is of the order of 3 m/s.
However, the high resolution of the timer counter (of the order of
less than 1 rs) renders the point-to-point uncertainty within a
run much less (of the order of 0.1 m/s). This is important in the
determination of the slope, a possible CT parameter. The variation
from run to run as seen in figures 7 to 12 is due mainly to differences in the biochemical content of the insonified tissue volume
and to tissue architecture.

CONCLUSIONS

    We can conclude from the data presented that the speed of
sound as a function of temperature can be measured in vivo with
stability and freedom from artifact and that when good transducer
contact and stable conditions are possible, accuracy can be obtained.
For soft tissues such as kidney and liver, in vivo and in vitro

Table 2.   Multiple Regression

Equation $V = \alpha_0 + (\alpha_1 W + \alpha_2 L + \alpha_3 C + \alpha_4 P) + (\beta_0 + \beta_1 W + \beta_2 C + \beta_4)T$

where

V = speed
W,L,C,P = water, lipid, collagen, non-collagenous protein wet
weight fractions

Proportion of Variance Explained by Variables 80.9%

R = due to LT + due PT + due to W + due to CT + due to T +
due to L + due to C

100% = 45.1% + 44.9% + 3.5% + 2.60% + 1.36% + 1.11% +
0.99% +...

Prediction Equation

$V = 1747.80 - (251.06W + 193.86L + 421.06C) +$
$(1.07 - 3.44L + 13.55C + 0.754)T$

Standard Error 8.89 m/s
Number of cases 1501
Number of runs 20

data have approximately the same magnitude and slope indicating
that it may be desirable to build up the necessary tissue data
using in vitro tissue. The relatively wide variation in the speed
versus temperature curves for tissue types make it improbable that
speed can be used as a CT scan parameter for thermal dosimetry.
However, because of the apparent repeatability in the slope, the
temperature coefficient of ultrasound may have possible use in
thermal dosimetry. Nevertheless, the negative role played by lipid
in the slope, as evidenced in the regression equation, and in
actual measurements on highly lipidous tissues, Nasoni et al.
(1979), must be taken into account.

REFERENCES

Bowen, T., Connor, W. G., Nasoni, R. L., Pifer, A. E., and Sholes,
R. R., 1979, "Ultrasound Tissue Characterization II," M.
Linzer, ed., NBS Spec. Publ. 525, U.S. Government Printing
Office, Washington, D.C., pp. 57-61.
Carson, P. L., Oughton, T. V., and Hendee, W. R., 1976, "Ultrasound
in Medicine," D. N. White and R. W. Bames, ed., Plenum Press,
New York, 2:391-400.
Cetas, T. C., and Connor, W. G., 1978, Med. Phys., 5:79-83.

Dethlefson, L. A., and Dewey, W. C., 1981, JNCI, Monograph, Vol. 60.
Del Grosso, V. A., and Mader, C. W., 1972, J. Acoust. Soc. Am.,
     52:1442-1446.
Folch, J., Ascoli, I., Lees, M., Meath, J. A., and Le Baron, F. N.,
     1951, J. Biol. Chem., 191:833-841.
Gerald, C. F., 1978, "Applied Numerical Analysis, 2nd Ed.," Addison
     Wesley, Reading, Mass.
Gover, G. H., and Sharp, 1977, IEEE Trans. Sonics and Ultrasonics,
     SU-24:229-334.
Goss, S. A., Johnston, R. L., and Dunn, F., 1978, J. Acoust. Soc.
     Am., 64:423-457.
Goss, S. A., Johnston, R. L., and Dunn, F., 1980, J. Acoust. Soc.
     Am., 68:93-108.
Greenleaf, J. F., Johnson, S. A., Lee, S. L., Herman, G. T., and
     Wood, 1974, "Acoustical Holography," Plenum Press, New York,
     6:591-603.
Greenleaf, J. F., Johnson, S. A., Samoya, W. F., and Duck, F. A.,
     1975, "Acoustical Holography," Plenum Press, New York,
     8:71-90.
Hildebrand, B. P., and Hufferd, D. E., 1977, "Acoustical Holography,"
     Plenum Press, New York, 7:245-262.
Jain, R. K., and Gulleno, P. M., 1980, Thermal characteristics of
     tumors: applications in detection and treatment, Annals
     NY Acad. of Sciences, Vol. 335.
Jakowitz, C. V., and Kak, A. C., 1976, School Elec. Eng., Purdue
     Univ., West Lafayette, Indiana, Res. Rep. TR-EE 76-26.
Johnson, S. A., Greenleaf, J. F., Rajagopalan, B., Bahn, R. C.,
     Baxter, B., and Christensen, D., 1979, "Ultrasonic Tissue
     Characterization II," M. Linzer, ed., NBS Spec. Publ. 525,
     U.S. Government Printing Office, Washington, D.C., pp. 235-
     246.
Kak, A. C., 1979, Proceedings of the IEEE, 67:1245-1272.
Lele, P. P., 1980, Radiat. Environ. Biophys., 17:205-217.
Lowry, O. H., Roseborough, N. J., Fan, A. L., and Randall, R. J.,
     J. Biol. Chem., 193:265-275.
Nasoni, R. L., Bowen, T., Connor, W. G., and Sholes, R. R., 1979,
     Ultrasonic Imaging, 1:34-43.
Nie, N. H., Hull, C. H., Jenkins, J. G., Steinbrenner, K., and
     Bent, D. H., 1975, "Statistical Package for the Social
     Sciences, 2nd Ed.," McGraw Hill, New York.
Sachs, T. D., and Janney, C. D., 1977, Phys. Med. Biol., 22:327-340.
Stegeman, H., 1958, Z. Phiol. Chem., 311:41-45.
Streffer, C., van Beumengen, D., Dietzel, F., Rottinger, E., and
     Robinson, J. E., 1977, Cancer therapy by hyperthermia and
     radiation, Proceedings of the Second International Symposium,
     Essen, Germany.

FOCUSED ACOUSTIC BEAMS FOR ACCURATE

PHASE MEASUREMENTS

S. D. Bennett, D. Husson, and G. S. Kino

Ginzton Laboratory
Stanford University
Stanford, California 94305

INTRODUCTION

There are many nondestructive evaluation tasks in which the ability to determine the state of stress in a component would be invaluable. Of the techniques potentially at our disposal, only the use of ultrasonic probes offers any real possibility of measuring stress below the immediate surface region. Acoustic measurements based on the acousto-elastic effect have been successfully demonstrated for determining the cross-sectional variation of stress for specimens in plane strain.[1] Until now there has been little success in measuring stress distribution through the thickness of a specimen. In this paper we outline a theory and an initial experiment with a new technique which is capable of determining the distribution of stress in a solid body in three dimensions.

The method we have adopted is a differential phase contrast technique based on the use of a focused acoustic beam to probe the stressed region of interest. An unfocused essentially parallel reference beam propagates along the same axis as illustrated in Fig. 1. The difference in phase shift that the two beams experience as they propagate through the specimen is measured. Suppose the diameter of the reference beam is D and the diameter of the focused beam at its focal plane is d_s ($d_s \ll D$). If both beams are moved perpendicular to their common axis a distance $x > d_s$, then the change in phase shift of both beams in the region outside the focal plane is essentially the same. In the region near the focus, however, a small movement, $x \ll D$, will not change the phase shift of the unfocused beam; if there is a

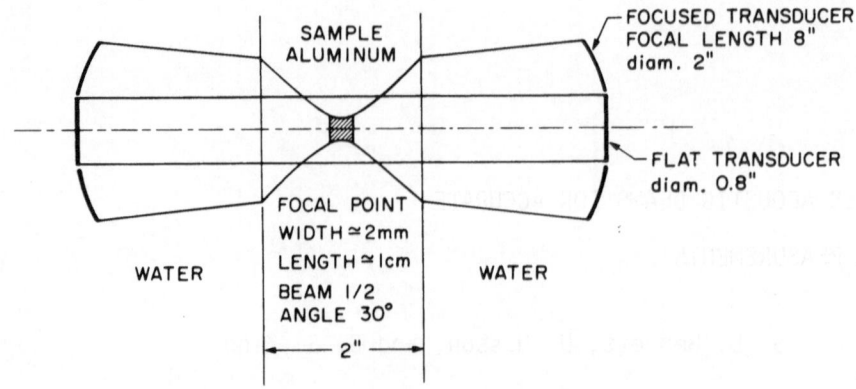

Fig. 1.   Basic system configuration.

change in stress in this region, the phase shift of the focused beam will change. A similar argument holds for movement in the axial direction. Consequently, by moving the two beams together, we can measure stress field variations both along and perpendicular to the axis. For longitudinal waves, the strongest phase shifts occur for axially-directed stress components, while for shear waves we might expect the largest phase shifts to occur for stress components in the direction of particle motion.

In Section 2 we will outline a theoretical analysis[2] which confirms that the difference in phase is largely associated with the region of the focus. This conclusion is confirmed by experiments carried out to measure the stress around a crack tip in a test specimen. Our theory enables us to predict the sensitivity of the system to particular components of stress. We find that for a longitudinal wave acoustic beam, the stress component in the direction of propagation has the greatest influence on the phase shift.

THEORETICAL ANALYSIS

The analysis of acoustoelastic effects generally depends upon the third order elastic constants of the material (the Murnaghan constants). In the past, such analyses have been extremely involved. We have developed a new technique based on a Lagrangian[3] description of the motion of an acoustic wave propagating in a stressed medium, coupled with an energy perturbation method[4,5] which we believe is easier to use and is more flexible in its application than previous analyses. In this paper we will outline the essential details of the calculation and indicate its relevance to the experiments; a full analysis will be published elsewhere.

An acoustic wave propagating in a stressed medium experiences a phase shift due to the presence of the static stress. The aim of the following calculation is to determine an expression for this phase shift. We consider two different situations:

(a)   an acoustic wave propagating through the medium in the absence of static stress; in the expressions which follow quantities associated with this case are given the superscript (0);

(b)   an acoustic wave propagating in the same medium but now in the presence of static stress;

In a Lagrangian system of coordinates, the equations of motion for these two cases are

$$\nabla \cdot \overset{\leftrightarrow}{T} = \rho_0 \frac{\partial \vec{v}}{\partial t} \qquad \text{for the stressed medium} \qquad (1)$$

$$\nabla \cdot \overset{\leftrightarrow}{T}^0 = \rho_0 \frac{\partial \vec{v}^0}{\partial t} \qquad \text{for the unstressed medium} \qquad (2)$$

$\overset{\leftrightarrow}{T}$ and $\overset{\leftrightarrow}{T}^0$ are the two stress tensors, and we are using the dyadic notation of Auld.[5] $\vec{v}$ and $\vec{v}^0$ are the particle velocities, and ρ_0 is the density of the unstressed medium. After we multiply Eq. (1) by $\vec{v}^{0*}$, and the conjugate of Eq. (2) by $\vec{v}$, integrate over the volume of interest and use Gauss' theorem, it follows that

$$\cdot \int_s (\vec{v}^{0*} \cdot \overset{\leftrightarrow}{T} + \vec{v} \cdot \overset{\leftrightarrow}{T}^{0*}) \cdot d\vec{s} = - \int_V (\nabla_s \vec{v}^{0*} \cdot \overset{\leftrightarrow}{T} + \nabla_s \vec{v} \cdot \overset{\leftrightarrow}{T}^{0*}) \, dV \qquad (3)$$

where $*$ denotes the complex conjugate, V is an arbitrary volume, and s is the corresponding enclosing surface.

We choose s and V such that s is an equi-phase surface (the transducer) for $\vec{v}$, $\vec{v}^0$, $\overset{\leftrightarrow}{T}$, $\overset{\leftrightarrow}{T}^0$. In this case, by writing $\nabla_s \vec{v} = j\omega \overset{\leftrightarrow}{S} \approx j\omega \overset{\leftrightarrow}{S}^0$ and $\nabla_s \vec{v}^{0*} = -j\omega \overset{\leftrightarrow}{S}^{0*}$, the phase shift due to the static stress may be expressed in the form

$$\Delta\Phi = - \frac{\omega}{4P} \int_V \overset{\leftrightarrow}{S}^{0*} \cdot (\overset{\leftrightarrow}{T} - \overset{\leftrightarrow}{T}^0) \, dV \qquad (4)$$

where P is the total power in the acoustic beam. Here we have made the additional assumption that the displacement field velocity $\vec{v}$ or strain $\vec{S}$ of the acoustic wave is not modified by the presence of the static stress. The elastic energy $\mathscr{E}$ as a function of Green's strain tensor E_{ij} (now using tensor notation) is

$$\mathscr{E} = \frac{1}{2} C_{ijk\ell} E_{ij} E_{k\ell} + \frac{1}{6} C_{ijk\ell mn} E_{ij} E_{k\ell} E_{mn} \tag{5}$$

where the Green's strain tensor is defined by the relation

$$E_{ij} = \frac{1}{2} \left(\frac{\partial u_i}{\partial a_j} + \frac{\partial u_j}{\partial a_i} + \frac{\partial u_m}{\partial a_i} \frac{\partial u_m}{\partial a_j} \right) \tag{6}$$

where u_i is the displacement and a_i and a_j are the coordinates in the unperturbed system. $C_{ijk\ell}$ and $C_{ijk\ell mn}$ are the appropriate elastic constants.

In the Lagrangian formulation, the stress can be written in terms of the energy in the form

$$T_{ij} = \frac{\partial \mathscr{E}}{\partial \dfrac{\partial u_i}{\partial a_j}} \tag{7}$$

We can use this relationship to express $\Delta\Phi$ as a function of the static stresses σ_{ij}. We substitute the first order acoustic strain as $S_{ij} = E_{ij}$. Finally, for a focused acoustic beam with its focal point at (X,Y,Z), the phase shift due to a static stress field $\sigma_{ij}(x,y,z)$ is

$$\Delta\Phi(X,Y,Z) = \omega \int_V \sigma_{ij}(x,y,z) \, F_{ij}(X - x, Y - y, Z - z) \, dx \, dy \, dz \tag{8}$$

where F_{ij} is a function of the beam geometry only.

We have developed a computer model to calculate the phase shift that would be expected for a focused Gaussian beam propagating in a variety of stress fields. The form of the Gaussian beam was chosen to match closely the conditions of the experiments we will describe in the next section. The beam has a half angle of 30^0 in aluminum with a wavelength corresponding to an excitation frequency of 5 MHz . In Fig. 2 we plot the calculated values of phase shift for two different stress functions. The results for other stress fields, not shown here, are all smaller in amplitude than these by almost an order of magnitude. For a beam propagating along the z axis, it is clear from these results that the system is more sensitive to a σ_{zz} stress than to any other component of stress. Knowing the geometry of the beam, we are also able to predict that our experimental system should be sensitive to stress over a region about 7 mm long and .8 mm in diameter, this estimate being based on the 3 dB level of the peaked sensitivity function.

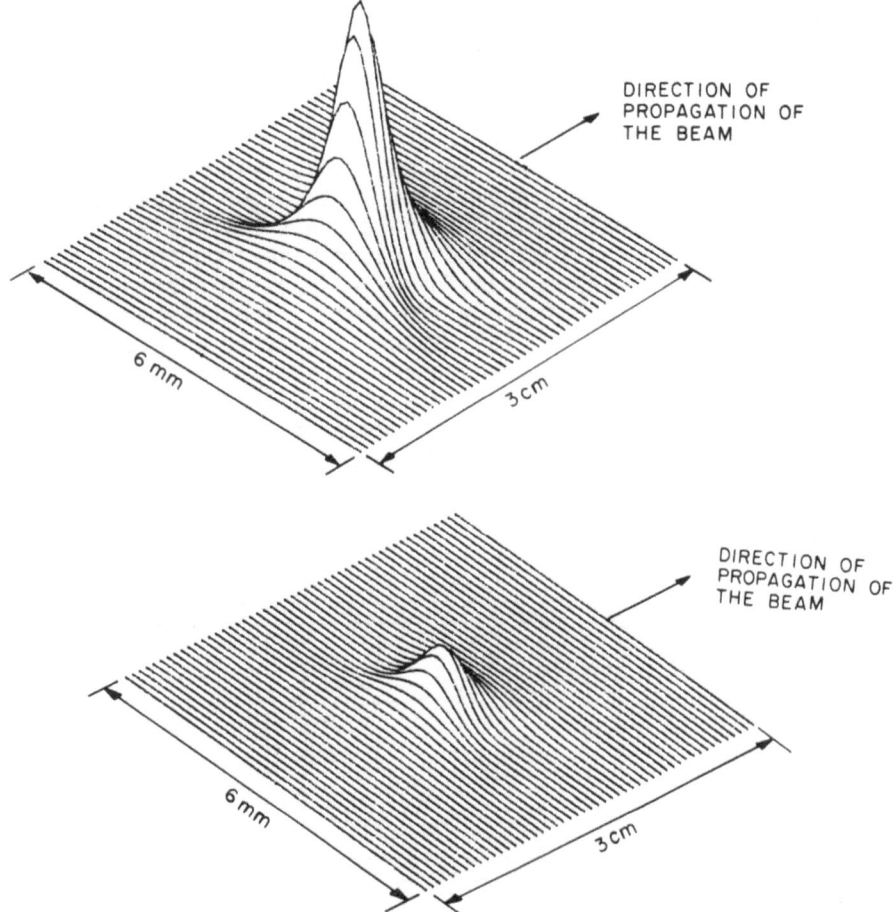

Fig. 2.   (a) Theoretical response of focused beam to stress fields; and (b) Response to a stress perpendicular to the direction of propagation of the beam.

EXPERIMENTAL SYSTEM

Transducer Design

     In order to demonstrate the feasibility of measuring stress
in the manner we have suggested, an experimental system operating
at 4 MHz was constructed. In Fig. 3, the transducer geometry is
indicated; the focused annular transducer has a radius of
curvature of 8" and an aperture of 2" . These dimensions were
chosen so that, allowing for the foreshortening of the focal
distance by the velocity difference between water and aluminum, a
region two inches thick in the aluminum could be probed in the
axial direction. As the focusing beam passes from the water to
the aluminum, there is significant spherical aberration, and for
this reason the aperture of the beam was restricted to give a
convergence angle of about 7^0 in water, or 30^0 in the
aluminum. In fact, we believe that substantially larger angles
would be acceptable and would thus offer the possibility of a
shorter depth of field than the current arrangement. The
transducer material was PZT-5A and was backed with silicon carbide
loaded epoxy to give the system a somewhat larger bandwidth than
would otherwise be possible. The transducers have relatively
large areas and hence a small shunt capacitive impedance. Some
electrical matching with a transformer was possible bringing the
input impedance to approximately 20 ohms . The reference beam
was provided by means of a flat piston transducer mounted
coaxially with the focused transducer. Its diameter was chosen so
that it would spread only slightly due to diffraction over the
propagation path from the input transducer to the output.
Consideration was given to the problem of cross-talk between the

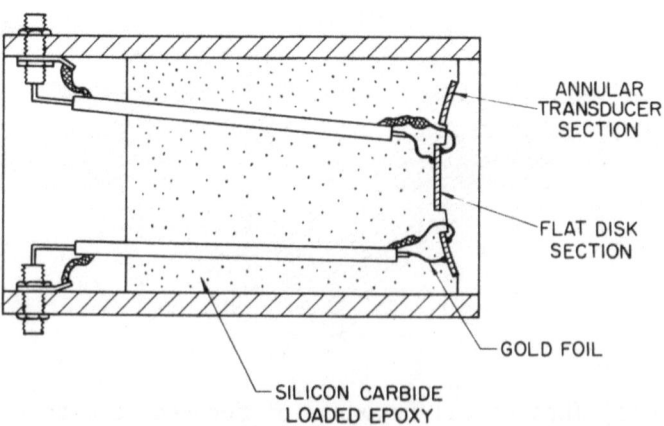

Fig. 3.   Transducer construction.

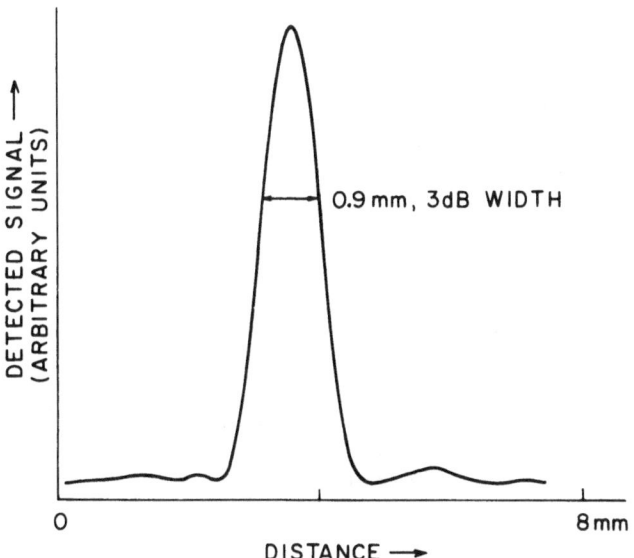

Fig. 4. Measured transverse beam profile at focal plane in aluminum block.

two types of transducers; it is clearly undesirable for part of the reference beam to appear on the receiving focused transducer and vice versa.

The two pairs of transducers were mounted in a simple mechanical jig, capable of being scanned manually whilst retaining the confocal alignment of the transducers. Each of the transducers was tested before any stress measurements were made; the beams were probed using a small scatterer in the water bath and were found to have profiles close to theoretical expectation. The process was repeated for the beams focused inside an aluminum medium by scanning a block with a small hole drilled in it through the beam. A typical result is shown in Fig. 4. The cross-section of the block was much larger than that of the beams and so the edges of the block had little or no effect on the measured spot size. The measured spot width in the aluminum was 0.9 mm and the axial distance between the 3 dB

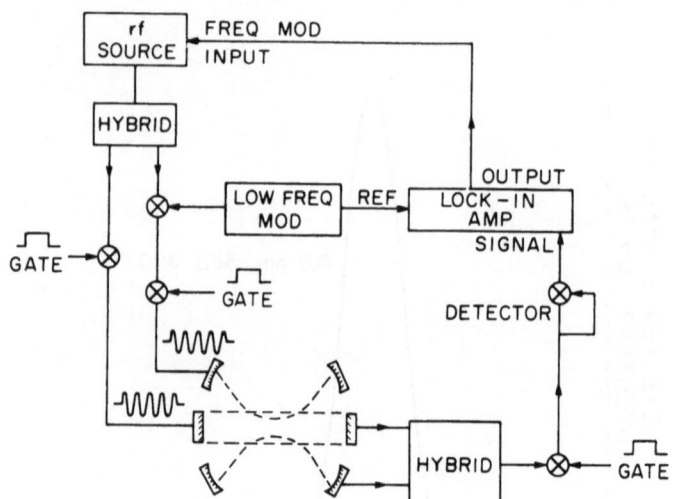

Fig. 5. Phase measurement system schematic.

points was 9 mm . Based on paraxial approximations, these are
just slightly larger than one would expect theoretically.

MEASUREMENT SYSTEM

     It has been demonstrated that very accurate relative phase
measurements can be made using a closed-loop system in which the
phase difference between two rf tone bursts is translated to a
low frequency (typically 1 kHz), where the difference is detected
using a lock-in amplifier.[6] The output from the lock-in amplifier
is used to drive the frequency modulation input of the rf signal
source in such a way that the frequency is shifted until the two
tone bursts are in phase quadrature. A simple measurement of this
lock frequency is then all that is needed to deduce the phase
difference between the pulses. The system used in our
measurements is shown schematically in Fig. 5. It is an essential
feature of this technique that the tone bursts propagate largely
over a common path; in this way small disturbances of the system,
such as vibration, cause equal disturbances in both beams and are
thus cancelled out. In the case of our focused beam system, one
may describe the two paths as being approximately equivalent,
except in the region of the focus, and it is precisely the
differences in phase shift for the two beams in this region that

we seek to measure. It follows from Eq. 8 that the phase shift experienced by a propagating beam is given by the relation

$$\phi = \omega(L_1 - L_2)/V \tag{9}$$

where L_1 and L_2 are constants associated with each beam. Therefore, by differentiation, it follows that

$$\Delta\phi = \Delta\omega(L_1 - L_2)/V + \omega(\Delta L_1 - \Delta L_2)/V \tag{10}$$

It is interesting to note that if the phase difference is kept constant by means of a feedback loop, the frequency change required is large when $L_1 \approx L_2$.

It is easy to arrange for this to be the case in our experiment and hence we are able to increase the sensitivity of the system significantly over previously reported results[1,6] in which the two paths are necessarily of different lengths.

The stability of the closed-loop system and the drift of frequency due to unwanted disturbances and electronic noise was determined by simply measuring the lock frequency over a period of time with the system left as it would be in a stress measurement, but without scanning the specimen. We have found from tests of this type that the maximum drift of the frequency over the period of a typical stress-scanning experiment was less than 0.2% of the center frequency. The frequency shifts we are concerned with in the stress experiments are of the order of 5% of the center frequency; this system drift is therefore well below the level where it would be troublesome.

EXPERIMENTAL STRESS MEASUREMENTS

For the purpose of establishing the feasibility of the technique, we have scanned the stress field associated with the tip of a crack in a test specimen. The specimen, which is shown in Fig. 6, consists of an aluminum block with a slot cut in it. The stress at the crack tip was controlled by means of a bolt which was used to force the crack open.

Our experimental procedure consisted of scanning the beam transversely, as indicated in Fig. 7, for several different axial planes. It was necessary to carry out a series of scans in the absence of any applied stress as a control experiment since there is a significant phase shift measured when the beams are in the region of the crack, due to diffraction of the reference beam. In future experiments we plan to use two focused beams which may be scanned either together or independently. The phase differences

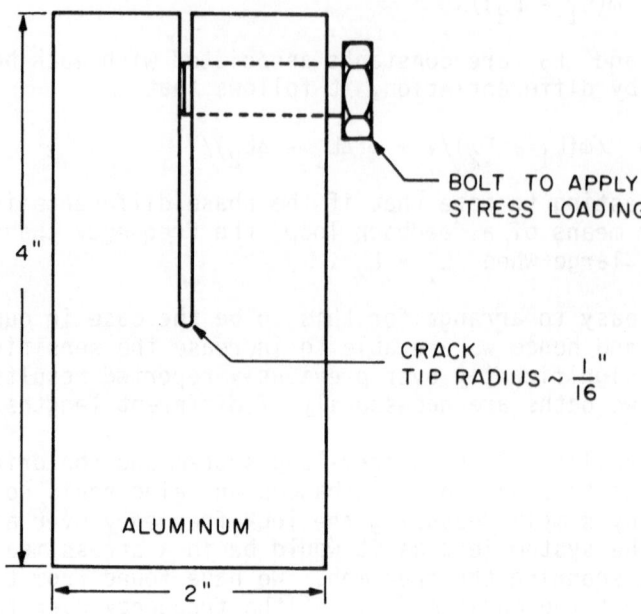

Fig. 6.   Sample used in stress experiment.

would depend upon the different propagation conditions at the two foci, but now it would be possible to bring the reference focused beam close to such obstructions as a crack or a weld, without serious diffraction effects interfering with the measurement.

Having obtained a set of control scans, stress was applied to the crack by tightening the bolt. The scanning procedure was repeated in exactly the same spatial positions as before. The data resulting from these two sets of measurements was reduced by finding the fractional change in frequency in each case and then subtracting corresponding scan lines. In this way, we are able to plot a quantity for each scan line which depends <u>only</u> on the phase shift induced by the application of stress to the crack. This experiment was repeated on several different occasions and gave extremely consistent results.

Our experimental results have been compared qualitatively with a numerical solution of the problem. The numerical results were obtained by Barnett and Tolf[7] for the case of a sharp crack

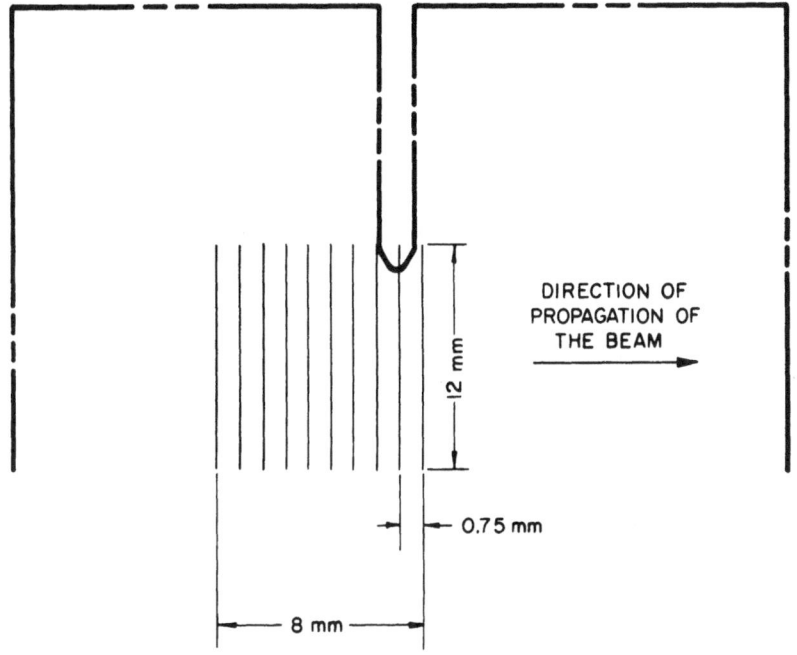

Fig. 8.   Specimen configuration and scan geometry for stress
          experiment.

of the same relative geometry as our test specimen, using a
collocation technique. These numerical results were convolved
with a function representing the experimental beam profile. We
would expect only partial agreement since the tip of the crack in
our test specimen had a radius of 1.6 mm , and was thus rather
far from the case of an ideal sharp tip. Figure 8 shows a
comparison of the results and remarkably good qualitative
agreement is seen.

CONCLUSIONS

     We have established that stable and accurate local
measurements of differential phase shift are possible using a
focused acoustic beam to define the region of sensitivity. A
reference acoustic beam is used which, when combined with the
focused beam, forms a phase bridge which is extremely insensitive
to spurious disturbances.

     A theoretical analysis has been carried out which indicates
that the phase shift experienced by a focused longitudinal

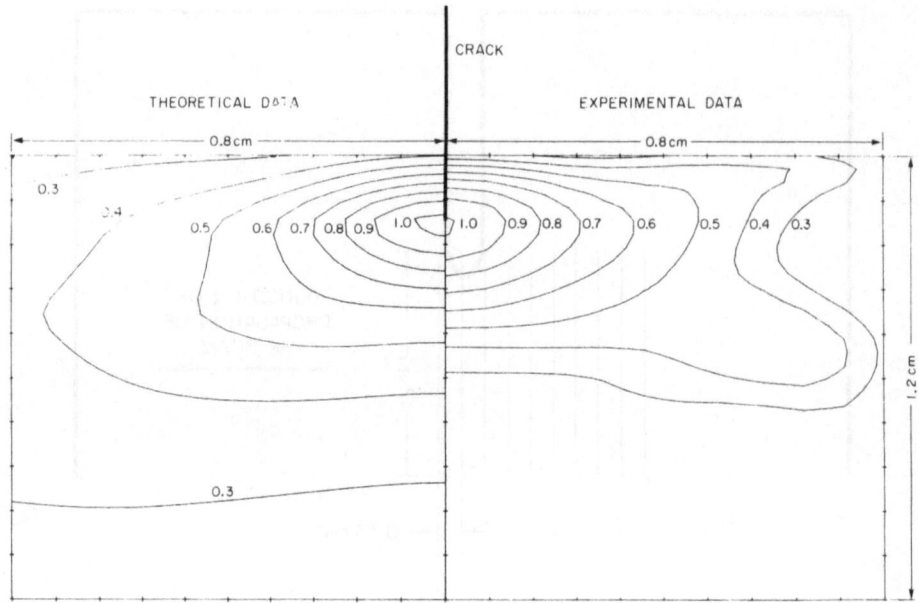

Fig. 9. Comparison of experimental data and theoretically
        determined stress distribution. On the right hand side
        of the figure, contours of constant measured phase
        difference are plotted (arbitrary magnitude scale). On
        the left hand side, contours of theoretically determined
        constant stress are plotted (magnitude scale
        arbitrary).

acoustic beam is principally sensitive to axial components of
static stress in the region of the focus. It is likely that other
stress components may be measured by using appropriately polarized
focused shear wave beams; we are currently investigating this
possibility.

    The feasibility of making practical measurements based on
this idea has been clearly demonstrated using a test specimen
consisting of an artificial crack, loaded by means of a bolt. The
sensitivity of the measurement is sufficiently high to encourage
us to extend the measurements to situations of more practical
interest. We plan to carry out experiments in the near future on
steel samples with real fatigue cracks in them. An important
extension of the technique which we are currently working on is
the development of calibration methods to give accurate
quantitative data from our experiments.

ACKNOWLEDGMENTS

We are grateful to D. M. Barnett and E. G. Tolf for providing us with the numerical results for stress in the region of a crack tip.

This research was supported by EPRI Contract No. EPRI-609-1 and AFOSR Contract No. F49620-79-C-0217.

REFERENCES

1. G. S. Kino et al, "Acoustic Measurements of Stress Fields and Microstructure," Journal of Nondestructive Evaluation, Vol. 1, 1980.
2. D. Husson and G. S. Kino, to be published.
3. Y. C. Fung, "Foundations of Solid Mechanics," Prentice Hall, 1965.
4. B. A. Auld and G. S. Kino, "Normal Mode Theory for Acoustic Waves and its Application to the Interdigital Transducer," IEEE Trans. ED-18, 898-908, 1971.
5. B. A. Auld, "Acoustic Fields and Waves in Solids," Vol. II, Wiley, 1973.
6. D. B. Ilic, G. S. Kino and A. R. Selfridge, "Computer-Controlled System for Measuring Two-Dimensional Acoustic Velocity Fields," Rev. Sci, Instrum, 50(12), 1527-1531, December, 1979.
7. E. G. Tolf and D. M. Barnett, "A Proper Collocation Method for Analyzing Edge Cracks in Stressed Solids," in preparation for submission to Journal of Applied Mechanics.

HIGH RESOLUTION ULTRASONIC TESTING SYSTEM USING DYNAMIC FOCUSING AND SIGNAL CORRELATION

Jun Kubota, Junichi Ishii, and Soji Sasaki

Hitachi Research Laboratory
Hitachi, Ltd.
Hitachi-shi, Ibaraki-ken 319-12   Japan

INTRODUCTION

Progress in fracture mechanics has made it necessary to evaluate flaw sizes exactly (within 1 or 2 wave lengths) in order to predict lifetimes of structural materials. To meet this requirement, the azimuthal resolution of ultrasonic testing must be improved, without increasing handling difficulty. Phased array(i), holography, and synthetic aperture are presently the best techniques available for improving the resolution(2). In addition, the signal processing array(3) has been studied as a method for acquiring ultrasonic beams of sharp directivity without enlarging the transducer size.

This study deals with dynamic focusing using an annular array, and with beam compression using correlation processing, which produces a narrower directivity than that achieved with linear processing. The paper describes: beam compression in the Fraunhofer Zone by means of multiplication (correlation) processing; some characteristics and examples of ultrasonic testing applications using beam focusing with an annular array; and combinations of the methods.

THEORY OF CORRELATION

Fig.1 shows a setting of the multiplicative array transducers, placed at the origin 0 (r=0). The transmitting transducer (size:2a) produces the Fraunhofer Zone (far field) at the point $P(r,\theta)$ where a point reflector is located. Since the interval between the receiving transducers is D, distances between then and the point P are $r-\Delta r$ and $r+\Delta r$ respectively, where $\Delta r \simeq (D/2)\sin\theta$. The transmitting transducer is driven by a pulsed signal, and the waveforms of the

597

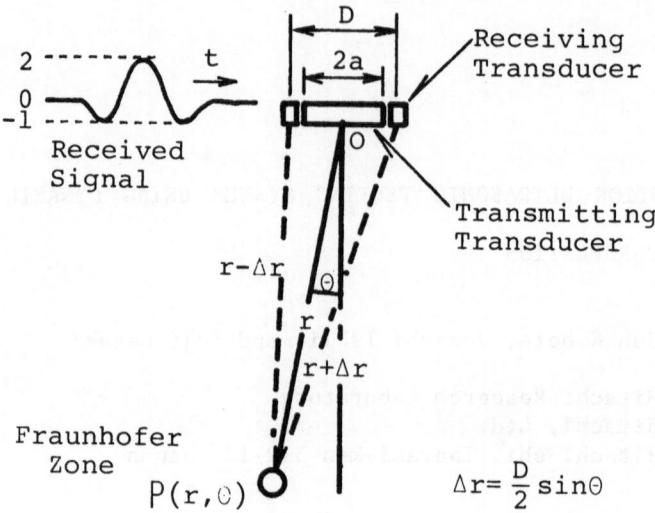

Fig. 1. Simplified geometry of transducers.

echo signals at the receiving transducers are as shown in Fig.2. Assuming the receiving transducers are non-directional, a received signal can be expressed as

$$\psi_i(t)= H(t-\frac{2r\pm\Delta r}{c})\ Re(S(t-\frac{2r\pm\Delta r}{c}))\quad (i=1,2) \text{ ------ (1)}$$

where H(t): Modulation (Window) Function; Re( ) denotes a real part of S(t); c: sound velocity; and

$$S(t) = \exp(j\omega t) \text{ ----------- (2)}$$

where $\omega(=2\pi f)$: angular frequency.

When the reflector is on the central axis of the transducers, the echo signals are received at the same time or in phase. Multiplying the received signals as shown in Fig.2, the product signal $(\psi_1\times\psi_2)$ becomes

$$\psi_1\times\psi_2=\frac{1}{2}H_1(t-\frac{2r+\Delta r}{c})\times H_2(t-\frac{2r-\Delta r}{c})\times(\cos 2\omega(t-\frac{2r}{c})+\cos\frac{2\omega\Delta t}{c}) \text{ --- (3)}$$

Eliminating the 2ω carrier component from the product signal by a low pass filter, the expected value of the product signal can be obtained.

$$\overline{\psi_1\times\psi_2} = \frac{1}{2}H_1(t-\frac{2r+\Delta r}{c})\times H_2(t-\frac{2r-\Delta r}{c})\times\cos\frac{2\omega\Delta r}{c} \text{ ----------- (4)}$$

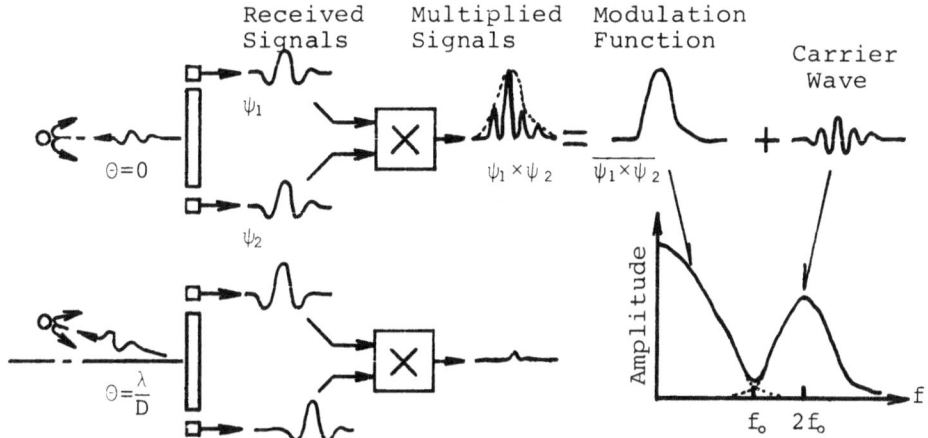

Fig. 2.   Process of multiplication.

Since Eq.(4) is a function of only the time difference $\Delta r/c$, it is considered to be the correlation function of ψ_1 and ψ_2, and is called the correlated signal in the following.

If the reflector moves off-axis or the transducers are scanned, the correlated signal vanishes four times as rapidly as that of the focused beam of the same aperture does. When the pulse echo system is working, the directivity pattern of the correlated signal becomes the product of the directivity of the transmitting transducer and that of Eq.(4).

$$R = \left| \frac{2J_1(X)}{X} \right| \cos(kD \cdot \sin\theta) \ \text{---------------------} \ (5)$$

The beam width calculated from Eq.(5) is about one fourth that of a far field of the same aperture circular piston, as shown in Fig.3. Since the modulation function H(t) has a time width as short as two periods of the carrier component of the received signal and has a Hanning's Window-like shape, the side-lobe level is about 20dB less than the peak of the main beam.

However, in actual applications, we often must use the near field (Fresnel Zone) in ultrasonic testing under the conditions of higher frequency and large aperture for improving resolution. Supposing the Fresnel Zone is as shown in Fig.3, the directivity pattern of the system does not follow Eq.(5), but conforms to Eq.(4).

Thus the sidelobe becomes larger in this case. Beam focusing is essential to reduce the sidelobe. Fixed focusing means such as concave transducers or acoustic lenses, however, limit the range

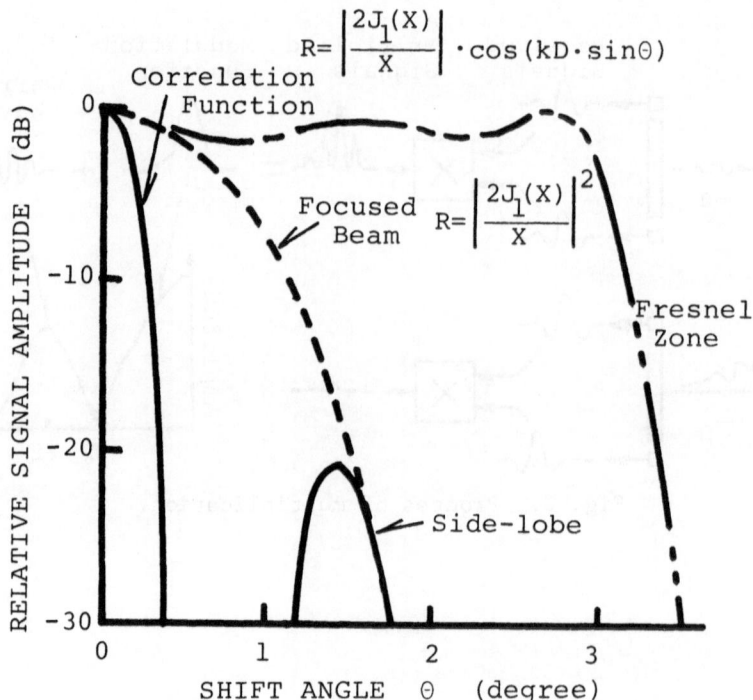

Fig. 3.   Directivity patterns of circular transducer and
          correlation.   (Pulsed wave, 3.5MHz, calculated).

capabilities of high resolution testing. Therefore, the variable
focusing technique was developed using a Fresnel Zone patterned
annular array concave transducer with four segments. Fig.4 shows
some Schrielen photographs of pulsed ultrasound beams in water,
focused at various ranges according to the delay time differences$\Delta\tau$
between the driving pulses applied to neighboring segments. Thus,
the focused region can be controled electrically by varying $\Delta\tau$.

CONSTRUCTION OF THE SYSTEM

     Fig.5 shows a block diagram of the high resolution ultrasonic
testing system, while Fig.6 shows the search unit of the system,
prepared for the angle beam technique (45° angled shear wave). The
system can be used as either the dynamically focused B-scan system,
or the correlation processed B-scan, which displays correlated
signals of the echoes from the focused region.
     Microprocessor controls the delay time differences ($\Delta\tau$) which
are related to the operation of the phased array transmitters
connected to the annular array transducer. The received signals at

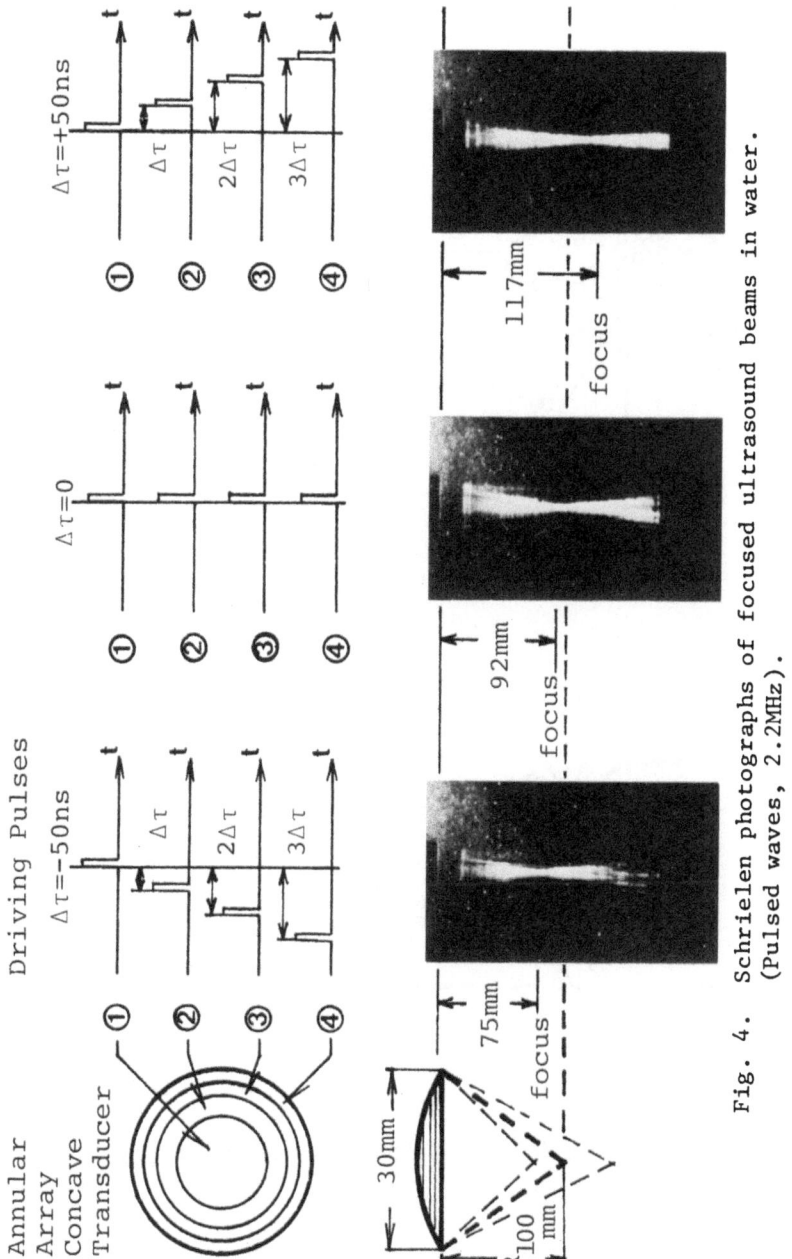

Fig. 4.   Schrielen photographs of focused ultrasound beams in water.
(Pulsed waves, 2.2MHz).

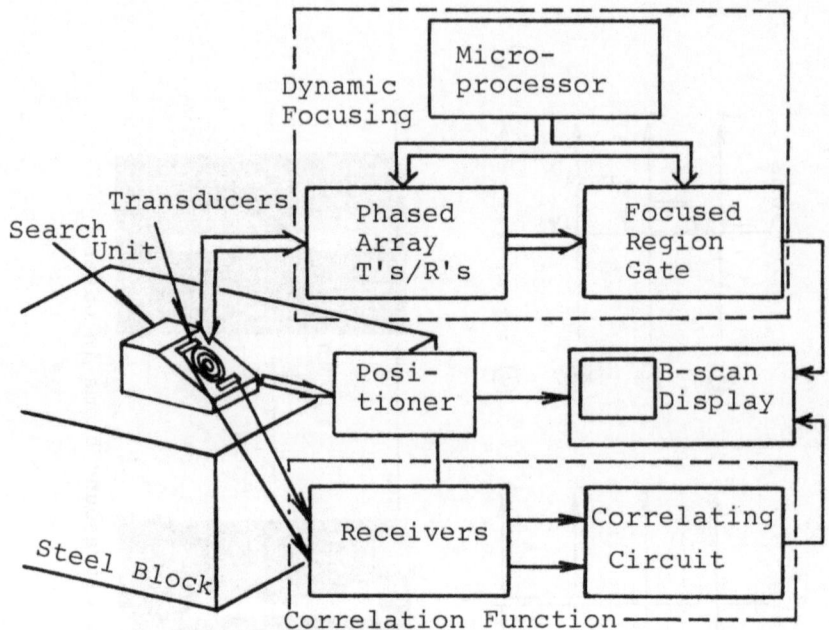

Fig. 5.   Block diagram of the system.

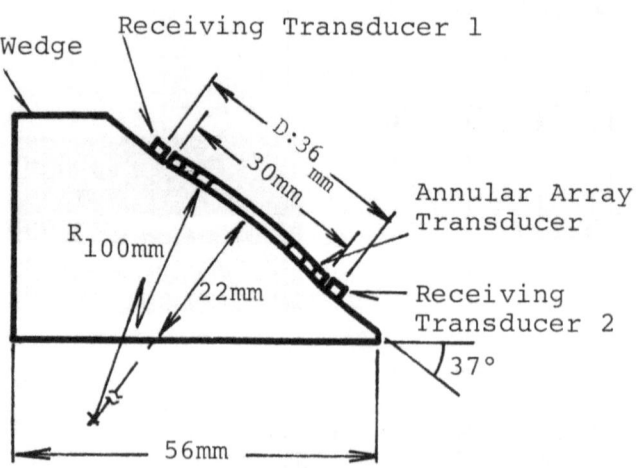

Fig. 6.   Forty-five degree angled shear wave search unit.

the array are amplified first by the receivers and next fed to the focused region gate, which lets only the echoes from the focused range pass through.

On the other hand, when the correlation technique is used for

processing the received signals according to Eq.(4), receivers and
the correlating circuit are operating and sending a correlated
signal of received echoes to a display unit.

The B-scan display can show either dynamically focused images
of phased array or high resolution images obtained by the transmitted focused beam and the correlation.

DYNAMIC FOCUSING IN THE NDE

Fig.7 shows the variation of the focal length and the corresponding focused region at 5MHz, in responce to the delay time
difference $\Delta\tau$.

Avoiding a signal cancellation due to phase interference, $\Delta\tau$
is limited as follows.

$$|\Delta\tau| \lessapprox T/4 \quad ------------------- \quad (6)$$

where T (=1/f): period.

Front and back ends of the focused region, calculated by the
empirical equation for a distance amplitude characteristic curve of
a fixed focusing transducer(4), agreed fairly well with the experimental results. The usable range of the focused region of the

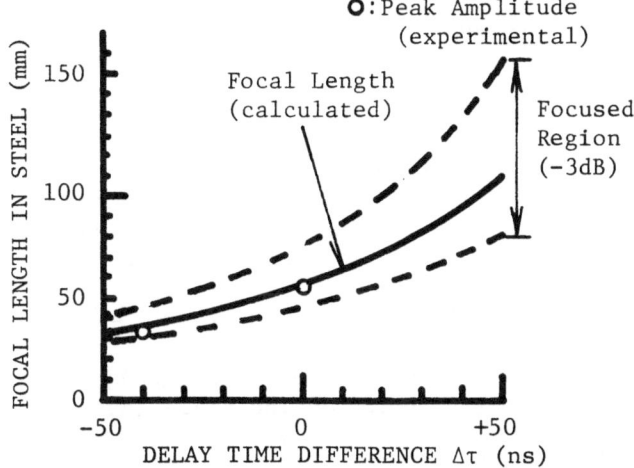

Fig. 7. Focal length versus delay time difference.
         (Forty-five degree angled shear wave annular
         array transducer, 5MHz).

transducer proves to be from 27 to about 150mm at 5MHz, though
distance amplitude compensation is necessary for practical use.

The beam width d of the focused beam varies according to $\Delta\tau$,
as shown in Fig.8. The minimum beam width (-6dB) in the focused
region (diffraction limit) is approximated with the following

experimentally derived expression(5),

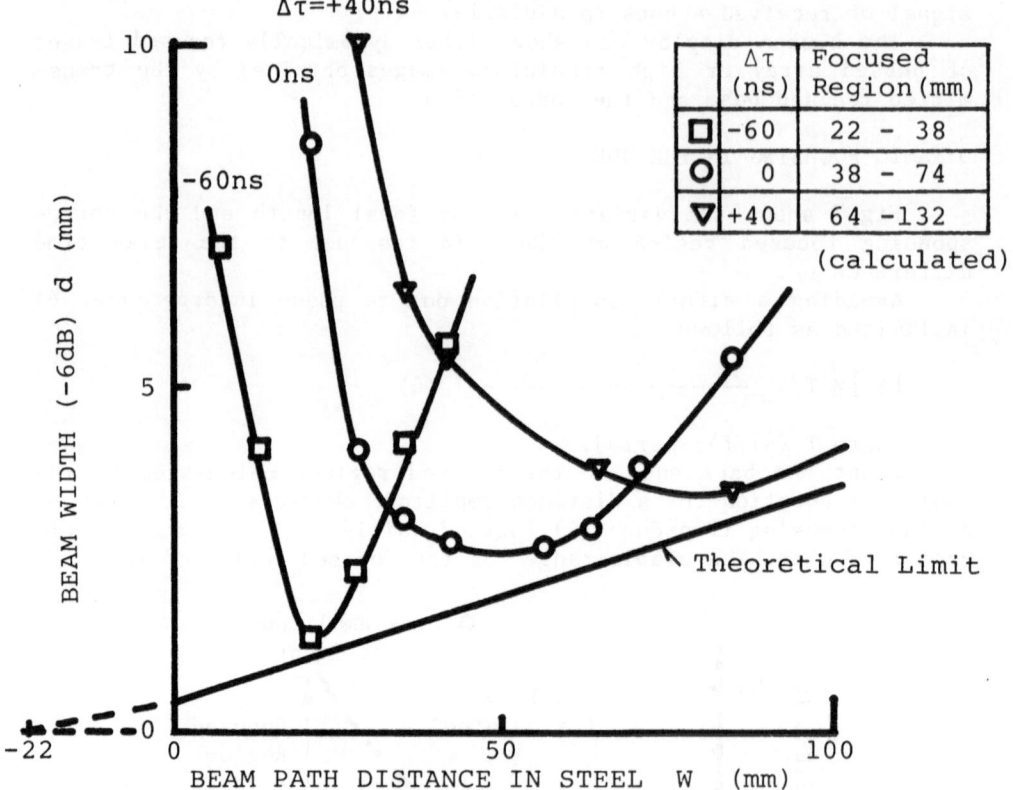

Fig. 8.   Beam width versus beam path distance in steel.
           (Forty-five degree angled shear wave annular
           array transducer, 3.5MHz, ϕ30mm).

$$d=1.41(w\times(\cos\theta_r/\cos\theta_i)+W\times(\tan\theta_r/\tan\theta_i))\times\lambda/D \quad ---------- \quad (7)$$

where w: beam path length in the wedge (mm); θ_r : angle of
refraction (degree); θ_i : angle of incidence (degree); W: beam
path distance in steel (mm); λ: wave length in wedge (mm);
and D: transducer diameter (mm).

Since the actual received signal contains a considerably wide
spectrum, especial for the 50% lower frequency component, the
experimental values are larger than expected.

Fig.9 shows the B-scan images of 1mm diameter side-drilled
holes in steel, where the beam is focused on various range and the
displayed echo level is from the peak to -3dB. The experiment was
made with the search unit scanned once along the Test-Piece. Fixed
focusing makes the resolution significantly poorer in both the
further and nearer regions from the focused zone. On the other
hand, dynamic focusing brings about a high resolution nearly that

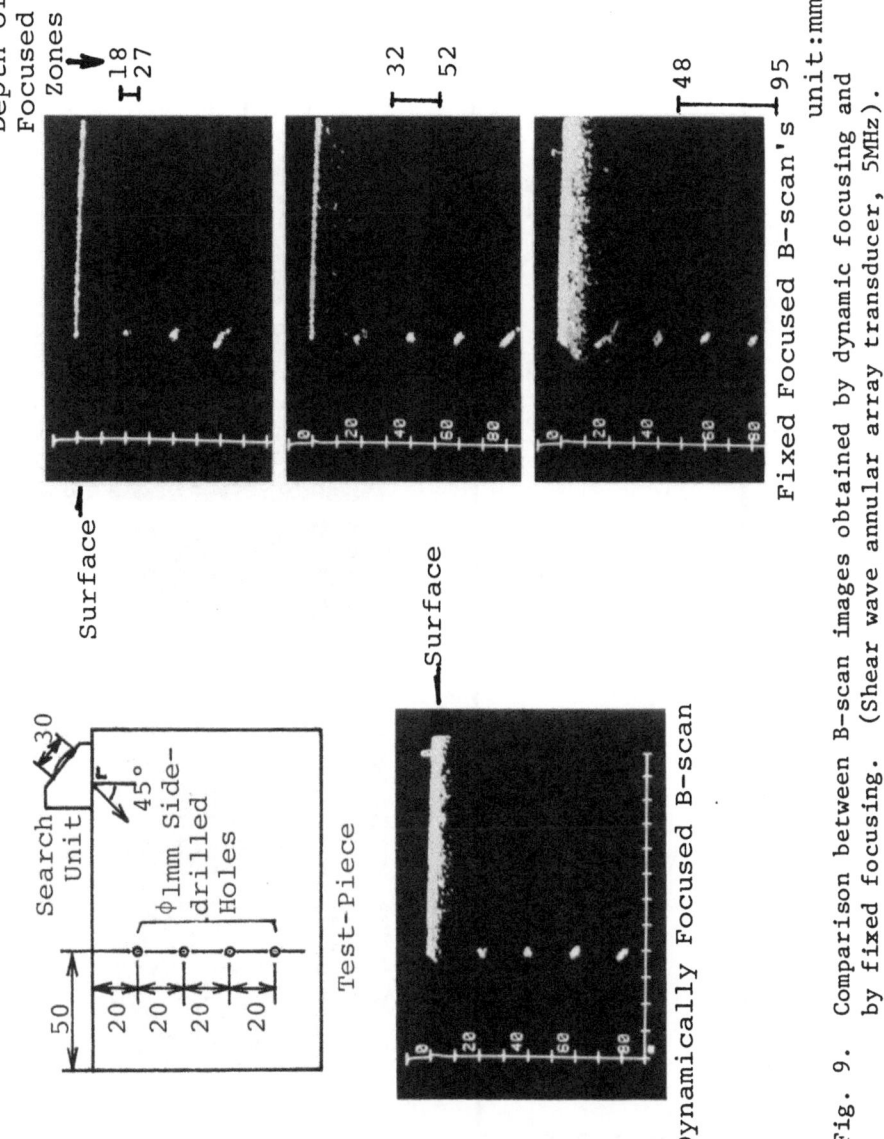

Fig. 9. Comparison between B-scan images obtained by dynamic focusing and by fixed focusing. (Shear wave annular array transducer, 5MHz).

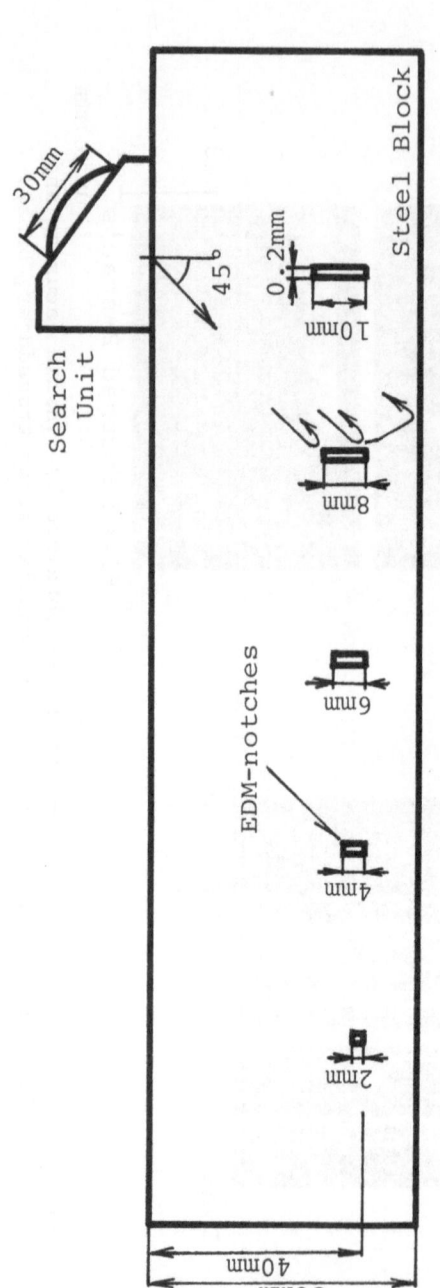

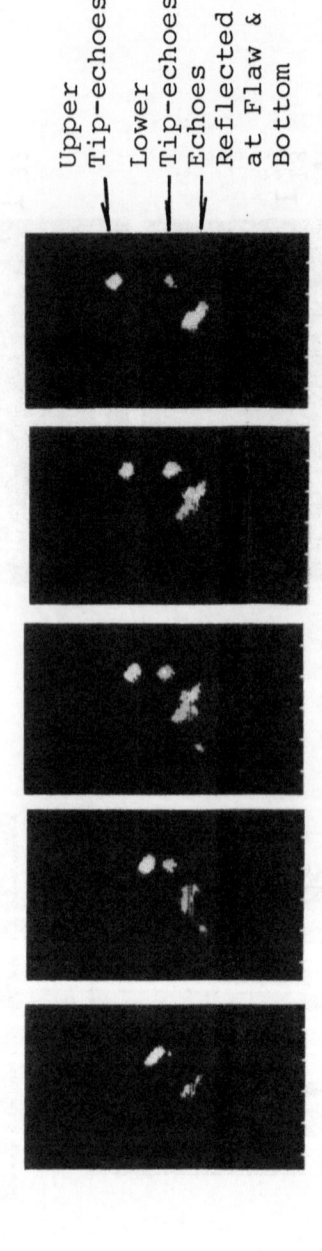

Fig. 10.   Example B-scan images of tip-echoes reflected from EDM-notches.
(Shear wave annular array transducer, 5MHz).

of the diffraction limit over the displayed range.

Improved resolution makes it possible to separate the echoes from very close targets. Fig.10 shows examples of B-scan images formed by echoes from tips of notches, having heights ranging from 2 to 10mm.

As it is clear from the above mentioned results, dynamic focusing improves the ability of the ultrasonic testing on both resolution and extension of focused region.

CORRELATION OF ECHOES FROM FOCUSED FIELD

Fig.11 shows the comparison between the estimated flaw sizes obtained by techniques of beam focusing and beam focusing plus signal correlation. The flaws are square-shaped and flat-bottomed.

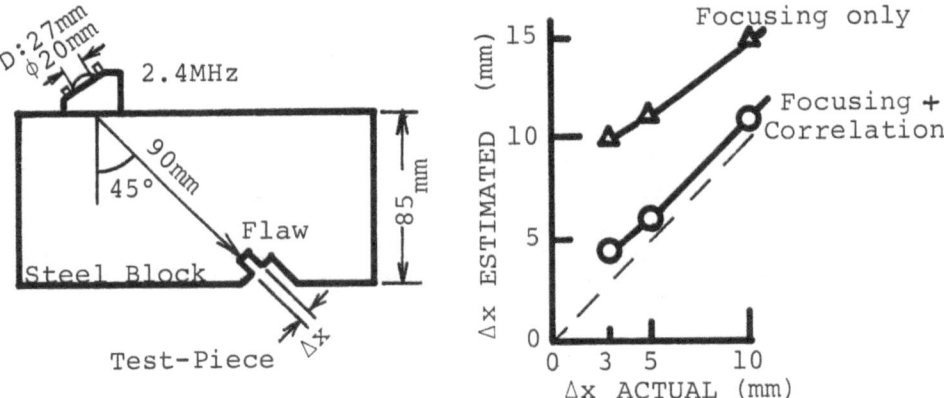

Fig. 11.   Comparison between estimated flaw sizes obtained by beam focusing plus signal correlation and by the conventional method.   (Shear wave, 2.4MHz).

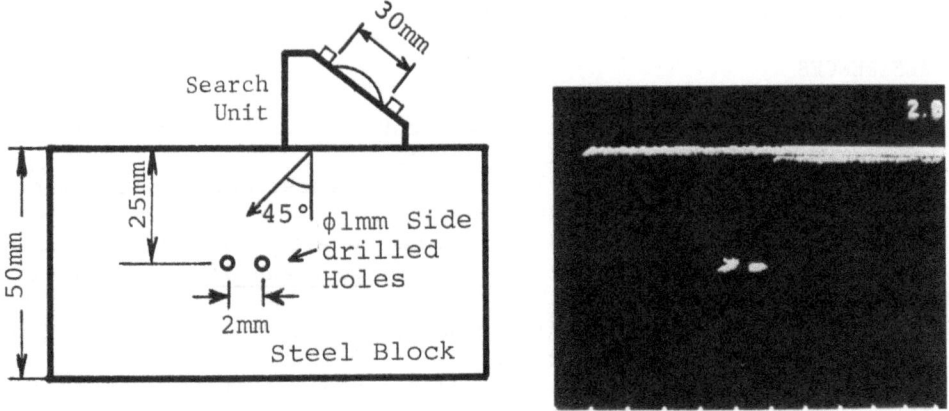

Fig. 12.   The B-scan display of side-drilled holes obtained by beam focusing plus signal correlation.   (Shear wave, 3.5MHz).

Though the aperture of the transducer is smaller, the frequency
lower, and the beam path distance longer than those of the previous
experiments, it is obvious that the flaw sizing error due to the
beam spread, arising in the former technique, could be markedly
reduced in the latter one.

Fig.12 shows B-scan images of two side-drilled holes (each
1mm) whose centers are 2mm apart (1mm-gap). These images were
obtained by the correlated signals (from the peak to -6dB). Since
the -6dB beam spread of the transducer of the same condition (36mm
diameter circular piston) for 45° angled shear wave is 1.5mm at the
depth of 25mm from the surface, the echoes of the holes of 1mm-gap
cannot be separated. On the other hand, it is clear in Fig.12 that
the images of those echoes were separated by means of the correlation method.

CONCLUSION

A high resolution B-scan ultrasonic testing system has been
developed using an annular array for dynamic focusing and a correlation processor for beam compression. Preliminary results of the
study on the focused region and beam width showed that the prediction closely approximated the focused sound field of the phased
array.

The B-scan images obtained by dynamic focusing demonstrated
the suitability of the focused beam technique in a long depth range
in steel.

The correlation processing could compress the beam width of
the focused beam. The side-lobe level of the correlated signal can
be reduced by focusing the ultrasonic beam in the near field of the
same aperture transducer. This ultrasonic testing system covers a
long range with sharp directivity.

It is expected that increased resolution will contribute to
improving accuracy of defect sizing, and securing the reliability
of welded components, nuclear equipments, and others.

REFERENCES

(1). J. C. Somer, Electronic Sector Scanning for Ultrasonic Diagno-
       sis, Ultrasonics 21: 153 (1968).
(2). G. S. Kino, Acoustic Imaging for Nondestructive Evaluation,
       Proc. of IEEE 67: 510 (1979).
(3). D. D. Lobdell, A Nonlinear Processed array for Enhanced Azi-
       muthal Resolution, IEEE Trans. Son. and Ultrason. su-15:
       202 (1968).
(4). J. Krautkramer and H. Krautkramer, "Ultrasonic Testing of Mate-
       rials", 2nd Ed., Springer, NY (1977).
(5). K. Kimura, Design Method of Angled Shear Wave Focusing Trans-
       ducer (in Japanese), Research Materials of NDI No.2644,
       Japan Soc. of NDI (1977).

EDDY CURRENT PHASOGRAPHY

H. Dale Collins, T.J. Davis L.J. Busse,
R.P. Gribble
Battelle Northwest Laboratories
P.O. Box 999
Richland, Washington 99352

ABSTRACT

A new concept, phasography, has been developed showing
excellent potential for high resolution imaging of flaws using
eddy current data. The process may be thought of as an eddy current
equivalent to acoustic holography. Phase multiplication of the
detected coherent signal is used to simulate a test wavelength one
to two orders of magnitude smaller than the actual wavelength,
permitting the use of image reconstruction techniques. Since this
process is performed on the detected data, the eddy current penetration depth remains unchanged. It is also possible to apply this
phase multiplication technique to low frequency acoustic holography,
resulting in a test which combines excellent penetration of difficult
materials with high resolution images. Preliminary experimental
results graphically illustrate this unique imaging technique. Eddy
current phasograms and their reconstructed images of small internal
simulated defects in 1 mm thick stainless steel plates are presented
in this paper.

CONCLUSIONS

The experimental results graphically illustrate the excellent
potential of phasography for the quantification, classification and
high resolution imaging of flaws using eddy current and acoustic
data.

Frequency and phase multiplication of the received flaw signal
appears to synthetically reduce the construction or test wavelength
by the multiplication factor. The apparent reduction in wavelength

performs a synthetic frequency translation hologram (i.e., phasogram). The phasogram constructed with the high synthetic frequency is similar to a hologram of the same frequency. This unique feature is illustrated in Figure 8 of the experimental results.

This concept (i.e., phasography) allows inspection at the lower test frequencies with excellent depth penetration and the resulting reconstructed image at the higher synthetic frequency with increased magnification, etc.

INTRODUCTION

This paper discusses the theory and application of scanned phasography using both electromagnetic (i.e., eddy current) and acoustical radiation. The acoustical experiments were conducted to verify the basic parameters associated with phasography that are extremely difficult to determine using eddy current type probes, etc.

The theory parallels the work of Hildebrand and Haines in scanned acoustic holography.[1,2] The image location equations and magnifications are derived with the phase multiplication factor which appears to synthetically reduce the construction wavelength.

The principles of electromagnetic nondestructive testing are well defined in the literature.[3,4] Specifically, eddy currents are generated within an object to be inspected by induction from an adjacent coil by an alternating current. Eddy currents then generate magnetic fields which couple to the coil at the same frequency as that of the excitation current, but which may be of different phase. The phase and amplitude of the induced voltages depend upon the characteristics of the object under test. The phase relationships may be measured by the appropriate signal processing circuits.

The flow of eddy currents in the test object is governed by the skin effect phenomenon. The currents decrease exponentially with depth, depending on the test object shape, thickness and electromagnetic properties. In addition to the decrease of current amplitude as depth below the surface increases, the phase angle of the current increasingly lags the excitation signal. The normal linear and phase multiplied relationships between depth and phase angle in the idealized plane wave case is shown in Figure 1. The phase angle increases with object dpeth and this one parameter is utilized to construct the eddy current phasograms.

Figure 2 illustrates the sequence of events that occur in constructing a typical phasogram before and after phase multi-

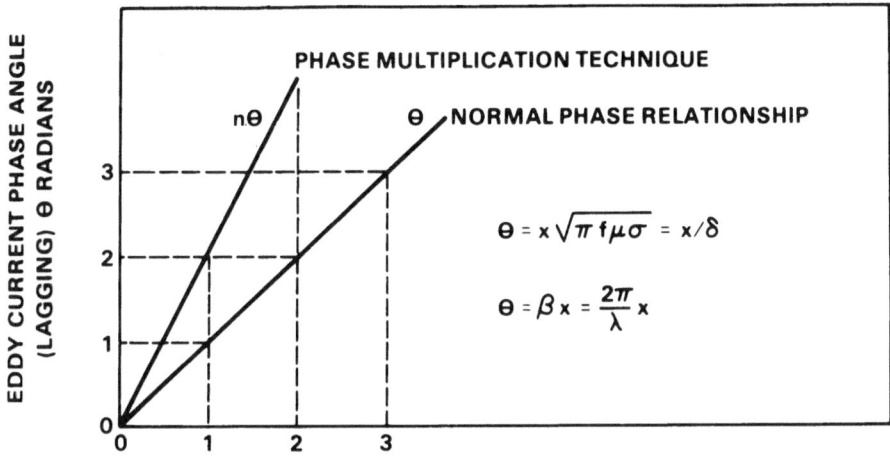

Figure 1. Variation of Eddy Current Phase Angle as a Function of
           Depth

plication. The upper sequence shows the eddy current coil scanning
across the point defect and recording a two dimensional one fringe
circle. The single fringe indicates the scanned aperture dimensions
are insufficient to produce a multi-fringe phasogram that can be
reconstructed as shown in Figure 2. This restricted aperture occurs
in many applications where the defect is either near the surface or
confined by geometry. The solution is shown in the lower sequence
of Figure 2. All phase values obtained in the general approach are
synthetically multiplied by some expansion factor (i.e., 9 in this
example). Multiplying the phase values produces a new expanded
aperture phasogram with the typical multi-fringe (Zone lens)
pattern that is so familiar in acoustic holography.

     Thus, through synthetic wavelength reduction (i.e., phas-
ography), we have been able to construct unique two dimensional
images of eddy current defects.

DESCRIPTION OF THE PHASOGRAPHIC IMAGING SYSTEM

     The simplified block diagram of Figure 3 shows one way of
implementing the method for eddy current testing. The search coil
is excited at a test frequency of ω_1. Any off-null or flaw signal
from the eddy current bridge is applied to a phase multiplier.
Both the frequency and phase of the signal are multiplied by a
preselectable number n in the phase multiplier. The phase of the
resulting signal is then detected with respect to a reference
signal of the same frequency ($n\omega_1$). The detected phase information

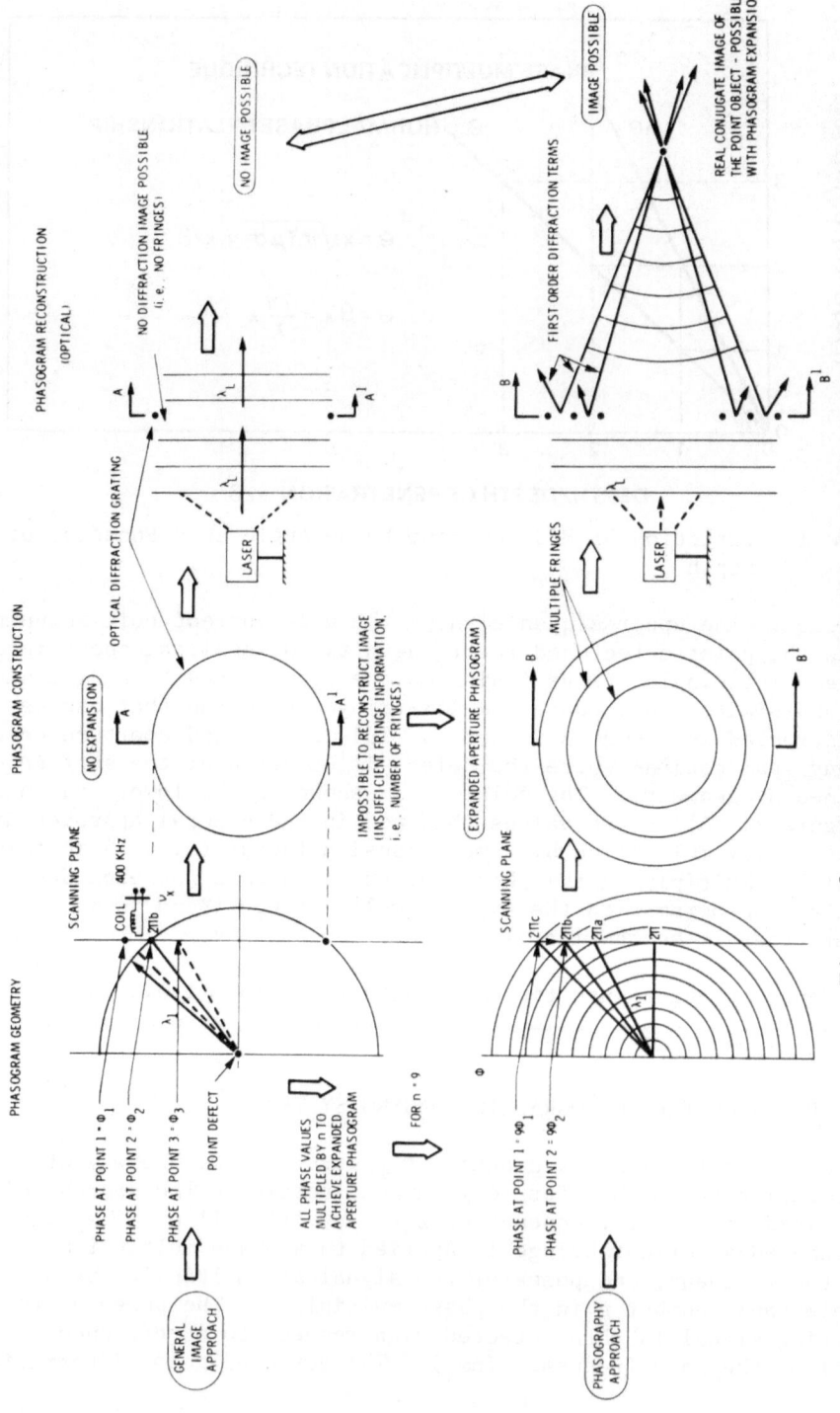

Figure 2.   Simplified Diagram of Phasography

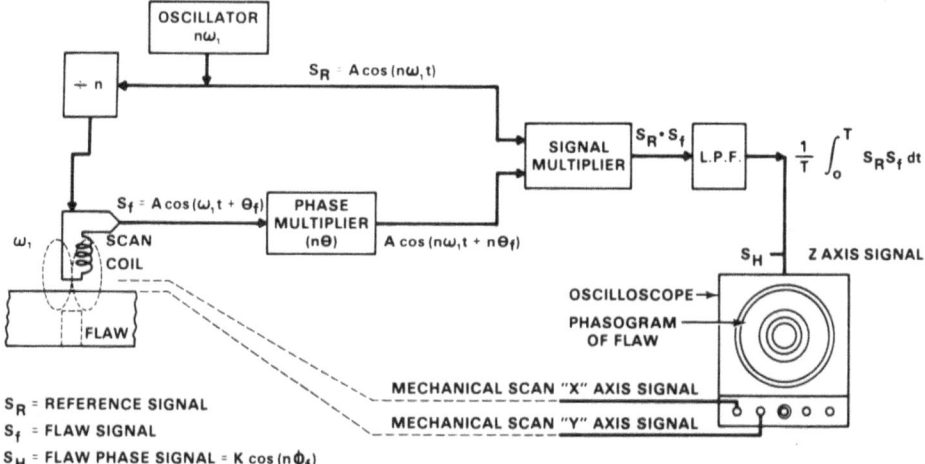

S_R = REFERENCE SIGNAL
S_f = FLAW SIGNAL
S_H = FLAW PHASE SIGNAL = $K \cos(n\phi_f)$

Figure 3. Simplified Block Diagram of Eddy Current Phasography
           System

is used as a write command on a storage oscilloscope. A phasogram
consisting of a unique set of fringe patterns will be written on
the display as a flaw is scanned. The image of the flaw may be
reconstructed from the phasogram by techniques similar to those
used in holography. A transparency of the phasogram is illuminated
with coherent light and treated as a diffraction grating to produce
an image. Alternately, the image could be generated by computer
methods.

ANALYSIS OF SCANNED PHASOGRAPHY

     This section presents an analysis of phasography employing
simultaneous focused (or point) source-receiver scanning. The
analysis used is similar to that of Hildebrand and Haines.[1] The
image location equations are derived for the various scanning
techniques that are used in phasography.

     The phasogram construction and reconstruction geometry used
in the analysis is illustrated in Figure 4. The phase at the
receiver point (x,y,z) during the phasogram construction is

$$\phi(x,y,z) = \phi_0(x,y,z) - \phi_r(x,y,z) \tag{1}$$

and

$$\phi(x,y,z) = \frac{2\pi n}{\lambda_s} \left[r_0 + r_1 - r_2 \right] \tag{2}$$

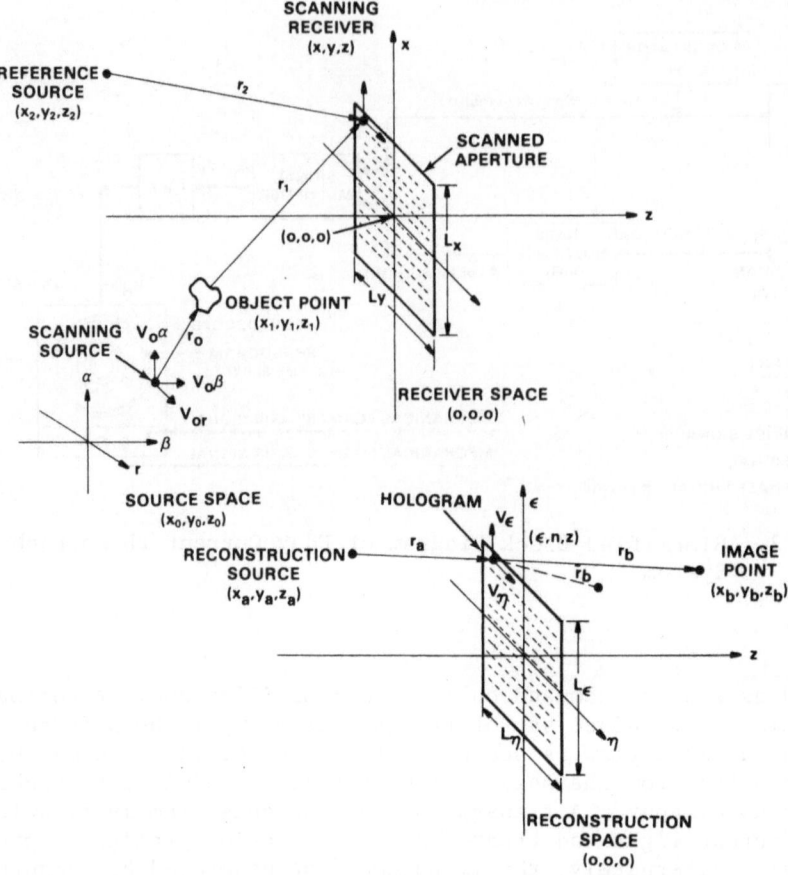

Figure 4.   Geometry for Scanned Phasography

The phase at the receiver point (x,y,z) after illumination of of the phasogram by the reconstruction source is

$$\phi_1(x,y,z) = \pm \frac{2\pi n}{\lambda_S} \left[r_0 + r_1 - r_2 \right] - \frac{2\pi}{\lambda_L} r_a \qquad (3)$$

where

        n = phase multiplication factor

        λ_S = construction wavelength

        λ_L = reconstruction wavelength

        +   refers to the conjugate image

        −   refers to the true image.

If the phase from Eq. (2) is to focus at the image point (x_b, y_b, z_b), then

$$\phi_1(x,y,z) = \frac{2\pi}{\lambda_L} r_b \tag{4}$$

which is termed the Gaussian-image sphere. The usual procedure is to expand the distance terms (r_a, r_b, r_1, r_2 and r_0) in a binomial series and equate coefficients of x, y and z. We expand the distance terms about the origin of the (x,y,z) system and the distance r_0 is expanded about the α, β, γ system. The area in which the receiver scans is assumed small with respect to the distances and is centered at the (x,y,z) origin. A similar restriction holds for the source motion. The first order terms yield the Gaussian image location equations:

$$\frac{1}{r_b} = \pm \frac{n}{m^2} \frac{\lambda_L}{\lambda_S} \frac{1}{r_1} + \frac{1}{r_0} - \frac{1}{r_a} \tag{5}$$

$$\frac{x_b}{r_b} = \pm \frac{n}{m} \frac{\lambda_L}{\lambda_S} \left\{ \frac{x_1}{r_1} + \frac{(x_1 - x_0)}{r_0} \right\} - \frac{x_a}{r_a} \tag{6}$$

$$\frac{y_b}{r_b} = \pm \frac{n}{m} \frac{\lambda_L}{\lambda_S} \left\{ \frac{y_1}{r_1} + \frac{(y_1 - y_0)}{r_0} \right\} - \frac{y_a}{r_a} \tag{7}$$

where $r_2 = \infty$ (plane wave reference beam) and the phasogram magnification $m = m_x = m_y$.

If we assume simultaneous source-receiver scanning configuration, the approximate image-to-phasogram distance is given by Eq. (8). The phasogram appears to have been constructed at the

$$r_b = \frac{\lambda_S}{n} \frac{m^2}{\lambda_L} \frac{r_1}{2} \tag{8}$$

smaller synthetic wavelength (i.e., λ_S/n). This reduces the effective distance by $1/n$ as compared with a hologram constructed with λ_S. The phasogram will have the same image distance as a hologram constructed with the equivalent λ_S/n wavelength. The phasogram constructed at the lower frequency simulates a higher frequency hologram.

The lateral image magnification for the same scanning configuration is easily derived from the image location equations and is given by Eq. (9). The image magnification is effectively

$$M_L = \frac{2\lambda_L}{\lambda_{S/n}} \left(\frac{1}{m}\right) \frac{r_b}{r_1} \tag{9}$$

increased by the synthetic wavelength ($\lambda_{S/n}$) and will be identical with a hologram constructed at this wavelength.

The analysis implies we have constructed a synthetic frequency translation hologram (i.e., phasogram). The concept is still in the preliminary research phase and additional work is necessary to answer the sticky question about resolution.

PHASOGRAM RECONSTRUCTION PROCESS

The reconstruction of the electromagnetic or acoustic phaso-gram to produce the optical image of the flaw requires a simple optical computer as shown in Figure 5. The laser provides the source coherent light to illuminate and reconstruct the phasogram for viewing. The spatial filter shapes or filters the beam to ensure the light source approaches a point source. The mechanical or electronically timed shutter provides the necessary light expo-sure when photographing the flaw images for permanent records. The adjustable mechanical aperture (i.e., rectangular) provides the required light over the entire area of the phasogram. The lens position is variable and moving the lens brings the true (real) image of the flaw into focus on the viewing screen as shown in Figure 5. Different lens positions correspond to different flaw depths in the test sample. The phasogram is inserted in a liquid gate containing a solution with an index of refraction approximating

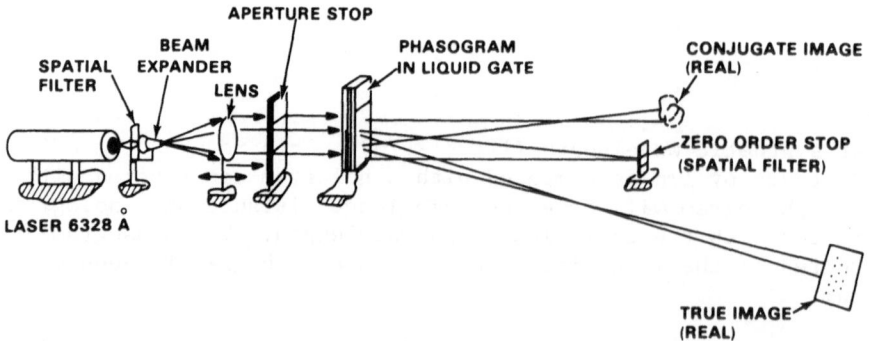

•Figure 5.   Schematic of the Optical Phasogram Reconstruction
            System

that of the phasogram film (i.e., Polaroid 46L). The liquid gate
essentially eliminates the undesired effects of film thickness
variations (i.e., phase errors). The solution surrounds the film
between two optical flats making the film appear as thick as the
width of the gate to the coherent light. The optical smooth
surfaces now represent the film surfaces, thus eliminating the
thickness variations.

The image screen (ground glass) is located usually at a
specified distance from the phasogram. The flaw images are then
viewed directly on a television monitor and permanent records
obtained merely by replacing the monitor screen with a camera.

PHASOGRAPHY EXPERIMENTAL RESULTS

The initial experiments consisted of verifying the various
basic image parameters using stationary source and scanned receiver
acoustic configuration. The acoustic system was used in the eddy
current research because the image parameters and reconstruction
techniques are well known and this reduced the number of unknown
variables.

The objective of the initial acoustic experiments was to
compare phasograms constructed at 2.5 MHz (i.e., phase multiplication factor of 2) with 5 MHz and 2.5 MHz holograms. Phasography
reconstruction theory predicts the phasogram and the 5 MHz hologram
should have the same reconstruction parameters and identical image
magnifications. This was the first step to verify or refute the
reconstruction theory.

Figure 6 shows the acoustic holograms, phasograms, and their
reconstructed images for comparison. The 5 MHz hologram and the
2.5 MHz phasogram (m = 2) have essentially the same fringe structure. They appear to be identical and easily differentiated from
the 2.5 MHz hologram. The 2.5 MHz hologram fringe structure (i.e.,
spacing, etc.) contains lower spatial frequencies as compared with
the phasogram constructed at the same frequency with a phase
multiplication factor of 2.

Figure 6 also shows the three reconstructed images of the
small point object (1.5 mm diameter). The 5 MHz hologram and
2.5 MHz phasogram images were reconstructed under identical
conditions, as if the phasogram was a synthetic 5 MHz hologram.
The effective reconstruction source distance was 5.1 meters from
the hologram. The true image-to-hologram distance was 6.1 meters
and the lateral magnification 0.005. The 2.5 MHz image was also
reconstructed at 6.1 meters with an effective source distance of
5.6 meters. The lateral magnification was exactly one-half the
5 MHz value (i.e., 0.0025).

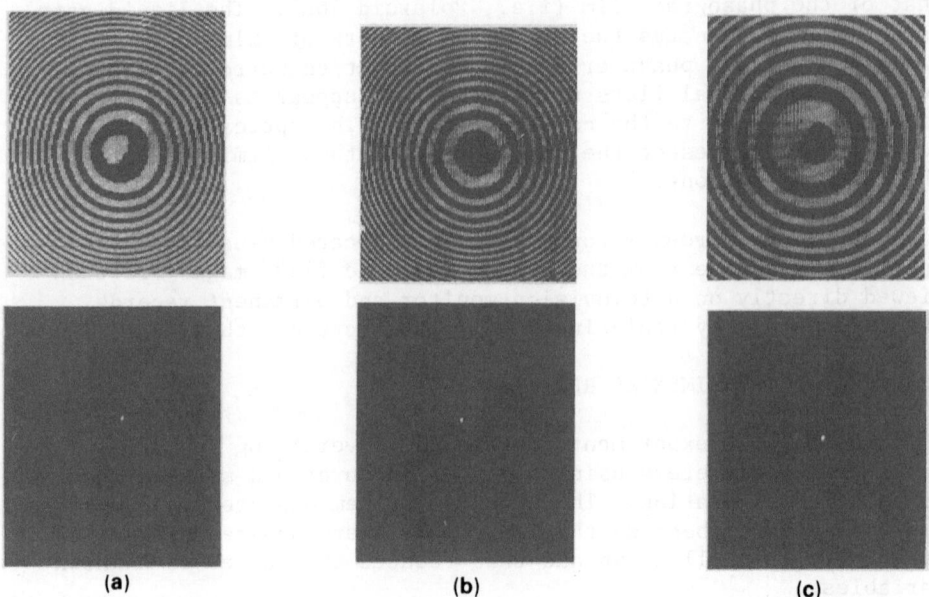

(a)   (b)   (c)

Figure 6.   Stationary Source-Scanned Receiver Holograms, Phaso-
            gram (m = 2) and the Reconstructions of a Point Object;
            (a) 2.5 MHz Phasogram Image, (b) 5 MHz Hologram/Image,
            and (c) 2.5 MHz Hologram Image.

     A large non-symmetrical metal letter "F" was imaged in the
next sequence of experiments to illustrate that the image magnifications are equal for the 5 MHz hologram and the 2.5 MHz phasogram
(m = 2). Figure 7 shows the phasogram construction geometry and
the large 12.7 cm letter "F". The phasograms and holograms were
constructed using the optimum scanning configuration (i.e., simultaneous source-receiver).

     Figure 8 shows the holograms, phasogram and their reconstructed
images of the letter "F". The fringe structure of the 5 MHz hologram and the 2.5 MHz phasogram illustrate greater spatial frequency
content as compared with the 2.5 MHz hologram. The hologram and
phasogram appear to be similar in their fringe densities.

     The reconstructed images of the letter "F" are shown below
their respective hologram or phasogram in Figure 8. The effective
reconstruction source distance was 4.3 meters from the hologram
or phasogram. The true image-to-phasogram distance was 6.1 meters
and the lateral magnification 0.3. The 2.5 MHz phasographic image
is exactly the same size (i.e., identical magnification) as the
5 MHz holographic image. The 2.5 MHz holographic image is shown
for comparison and its image lateral magnification of 0.07.

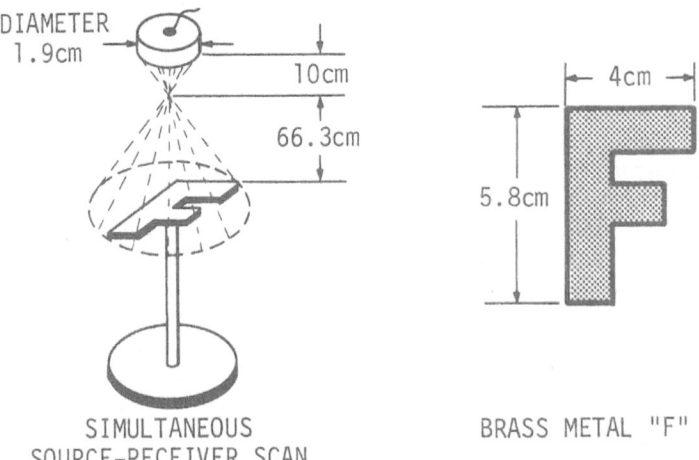

Figure 7. Phasogram Construction Geometry for the Letter "F"

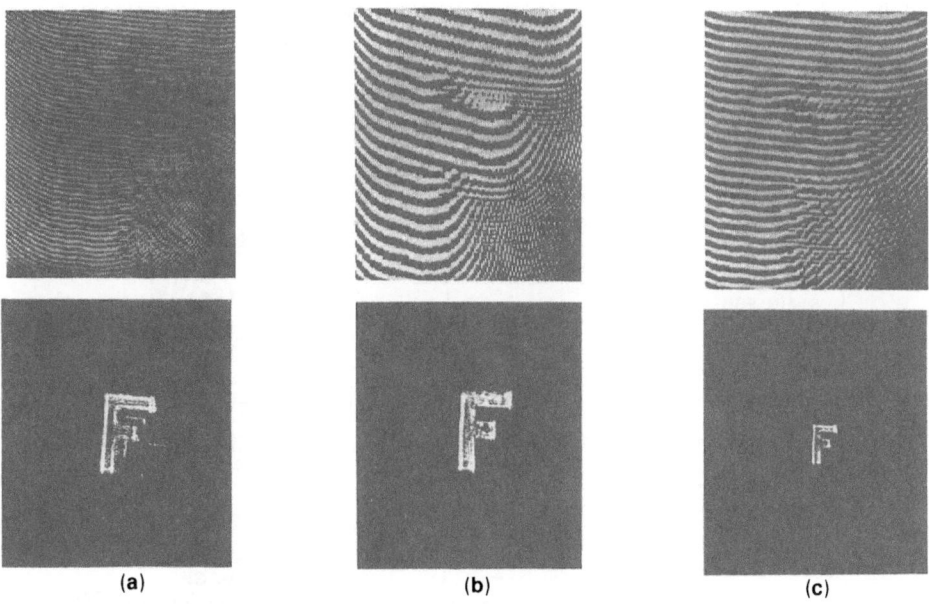

Figure 8. Simultaneous Source-Receiver Scanned Holograms, Phasogram and the Reconstructions of the Letter "F";
(a) 2.5 MHz Phasogram (m = 2), (b) 5 MHz Hologram, and
(c) 2.5 MHz Hologram/Image.

The preliminary results of the initial acoustic experiments verify the reconstruction parameters and image magnifications are identical between the phasogram and the equivalent frequency hologram. The phasographic process appears to synthetically reduce the construction wavelength by the phase multiplication factor.

Figure 9 shows the eddy current phasography construction geometry, phasogram and the reconstructed image of a circular defect. The illumination or test frequency was 500 kHz and the air gap approximately 0.1 mm. The electromagnetic probe scanned a 2.4 cm x 2.4 cm aperture above the stainless steel plate with the 6 mm circular defect. The defect was located 0.25 mm below the top surface of the plate. The eddy current defect phase signal was multiplied by forty. The phasogram fringe spacing appears to be very similar to a point object hologram. The inner four fringe spacings correlate very closely with the theoretical predicted spacings, but not the outer fringe. This outer fringe is the result of phase errors at the aperture extremes and is eliminated in the reconstruction process by aperture reduction techniques.

The optical reconstruction graphically illustrates the unique eddy current image of a flat top hole in stainless steel. Figure 10 shows the eddy current phasography construction geometry, phasogram and the reconstructed image of a simulated surface crack in stainless steel. The slot is 6.25 mm in length, 0.25 mm in width, and 0.5 mm in height. The phasogram was constructed at 500 kHz with a phase multiplication factor of forty. The general shape of

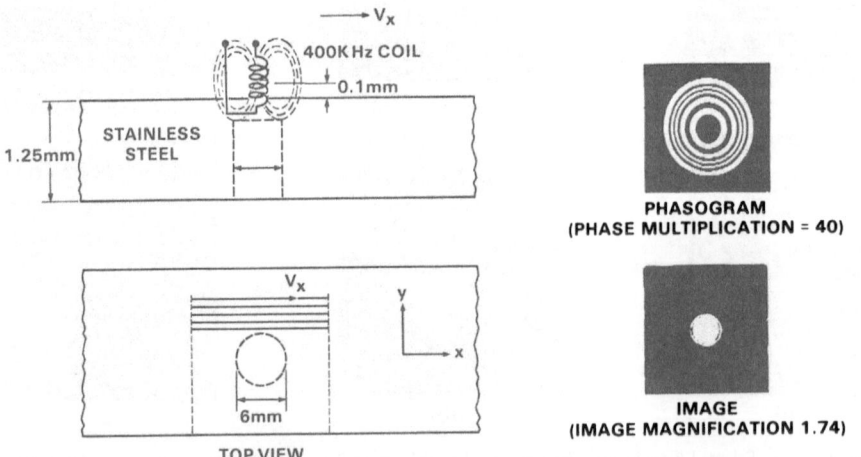

Figure 9. Eddy Current Phasography Point Object Construction Geometry, Phasogram and the Reconstructed Image

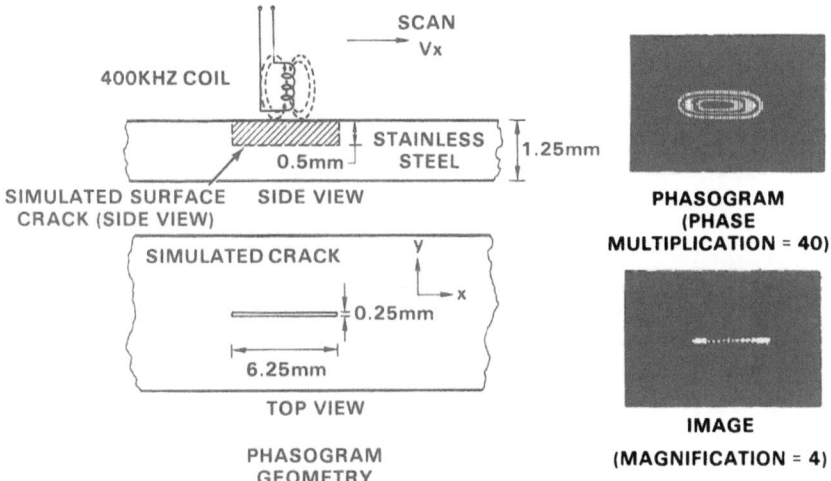

Figure 10. Eddy Current Phasography Simulated Surface Crack
            Construction Geometry, Phasogram and the Reconstructed
            Image.

the phasogram resembles a line object hologram. The outer fringe
spacing error is again the result of phase errors as the probe
reaches the aperture extremes. The reconstructed image shows the
top view of the surface slot. This is an exceptional image
considering the extremely low illumination frequency.

     Figure 11 shows the comparison of conventional eddy current
methods and phasography for the inspection of steam generator
tubing. A rotating probe is used to construct the two-dimensional
phasograms as shown in Figure 11. The conventional absolute
signals construct only one-dimensional profiles of the hole and
slot for analysis. Phasography provides two-dimensional images
that show the unique defect geometries. This should greatly
enhance the interpretation analysis and reduce the required
operator training.

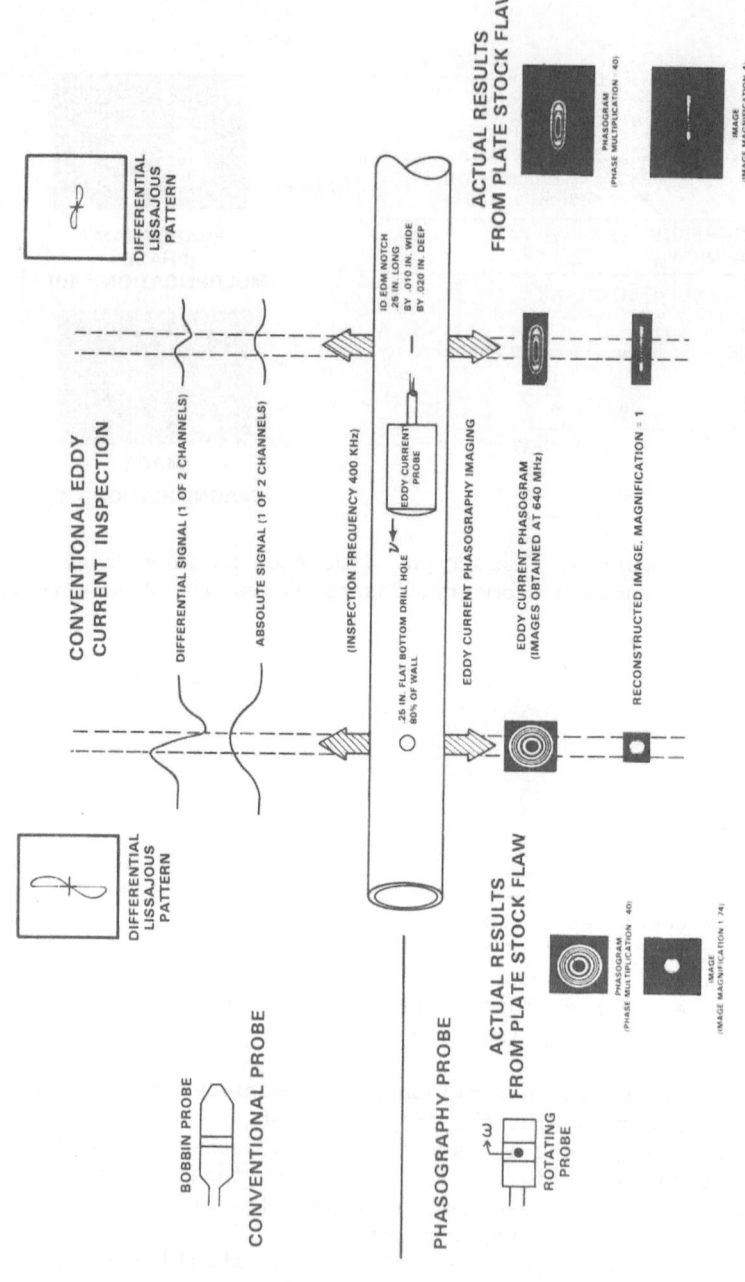

Figure 11.   Comparison of Conventional Eddy Current Techniques and Phasography

REFERENCES

1. B. P. Hildebrand and Kenneth Haines, Holography by Scanning,
    J. Opt. Soc. Am., Vol. 59, p. 1-19, 1969.

2. B. P. Hildebrand and B. B. Brenden, "Introduction to Acoustical
    Holography", Plenum Press, New York (1972).

3. Hugo L. Libby, "Introduction to Electromagnetic Nondestructive
    Test Methods", Wiley-Interscience, New York (1971).

4. Robert C. McMasters, "Nondestructive Testing Handbook",
    Ronald Press, New York (1963).

A 50 MHZ SYNTHETIC FOCUS SYSTEM

K. Liang, K. Peterson, S. Bennett, C. H. Chou,
B. T. Khuri-Yakub, and G. S. Kino

Ginzton Laboratory
Stanford University
Stanford, California 94305

INTRODUCTION

Last year at this conference we discussed a number of difficulties which can occur in synthetic aperture and other short pulse imaging systems. In this paper, we have carried out studies in both a linear and cylindrical format to understand some of these problems more clearly and to arrive at solutions for them. We have studied the effects of a finite number of digital phase samples on the sidelobe level, the effect of apodization on grating and sidelobe levels, the effect of a finite number of transducer elements, problems of imaging specular reflectors, and the use of selective back projection methods to improve images of specular reflectors. We show here that the use of this latter technique may be of great importance because it provides an alternative approach of imaging, which does not make use of the normal phase cancellation effects employed in conventional imaging systems.

We also describe the first experiments carried out with a 50 MHz synthetic aperture imaging system in a cylindrical format. These results show excellent imaging capabilities, both in metals and ceramics and make use of a technology which we believe can be extrapolated fairly easily up to frequencies as high as 300 MHz .

REVIEW OF SYNTHETIC APERTURE IMAGING PRINCIPLES

    We have described the basic principles of both linear array
and cylindrical synthetic aperture imaging systems in several
previous papers.[1-5] A brief review and the changes we have made
in the two systems will be given here.

    In the cylindrical imaging system, we employ a single
transducer immersed in water, which transmits a short acoustic
pulse impinging on the surface of a cylindrical rod. The rod is
rotated through 360^0 using as many as 256 aperture sampling
angles. A parallel beam, with a center frequency of 50 MHz ,
diverges as it enters the sample. The echoes are then received on
the same transducer and pass through a sampling oscilloscope and
into an A-to-D converter to the memory of the computer.
Therefore, it is as if there is a point source a short distance
from the surface of the sample and a point receiver at the same
point.

    A similar set-up is used at lower frequencies with a
32-element linear acoustic array. Now, instead of mechanical
scanning, each element of the array is addressed in turn through a
multiplexer. In both cases, the impulse response of the
transducer elements is a short rf pulse approximately two cycles
long between 1/e points.

    A focused image is formed by applying the appropriate time
delays to each of the digitized echoes and then summing the
contribution from each array element. The summed signal is
normally envelope-detected and an intensity modulation CRT image
or graphic amplitude display is generated. The time delay
information is usually stored as a look-up table; in some cases it
is calculated directly point by point.

    Our real-time linear array system is capable of focusing and
displaying images at a rate of approximately 30 frames per
second, using a digital clock rate of the order of 10 MHz .
Computer reconstruction of 256 data sets (256 transducer points)
can take as much as 15 minutes.

    During the last six months, we have constructed computer
routines to work with synthetic images that enable us to vary
various parameters of the imaging system, such as the number of
digital samples, apodization of the array, and the number of
elements employed. We are able to obtain contour plots of the
output as well as graphic plots through any cross-section of the
image and direct numerical output of the amplitude of each
point. This is possible with both simulated and
experimentally-derived data. We have been able to obtain

extremely accurate checks on our real-time digital imaging system which confirm that it is working as accurately as a complete computer-processed system.

IMAGE PROCESSING WITH A 32-ELEMENT LINEAR ARRAY SYSTEM

We have studied a 32-element synthetic aperture system using simulated data for a single line reflector. In all but one case, the width of the aperture is taken to be 16 mm (the grating lobe result is displayed for a 32 mm aperture). The wavelength at the center frequency is $\lambda = 0.45$ mm and the point reflector is taken to be at the center of the field a distance 80 mm or (178 λ) from the array. The reconstructed region is 32 mm x 7.5 mm (71 λ x 17 λ) .

Fig. 1 shows a reconstruction using rectangular apodization of the array and three digital samples per wavelength. The pulse is a Gaussian modulated cosine with the 1/e points 2 rf cycles apart. The figure shows contour plots around the image point. The first contour is at the -3 dB level. All the other contours are at 6 dB intervals (-6, -12, -18, ...). The contours are essentially elliptical in the neighborhood of the point and then become more erratic further away. The transverse definition at the 3 dB points is very close to the 1.1 mm (2.5 λ) predicted by simple paraxial theory.

In Fig. 2, a finely sampled cross-section of the undetected waveform along a line parallel to the array, through the center of the reflector, is shown. The sidelobe level is somewhat higher than we might have expected from simple theory. We would expect that for such a Gaussian pulse, the first sidelobe should have a level of -18.7 dB ,[1] and then the amplitude should drop off rather rapidly, eventually reaching a level of approximately -30 dB at the far-out sidelobes.

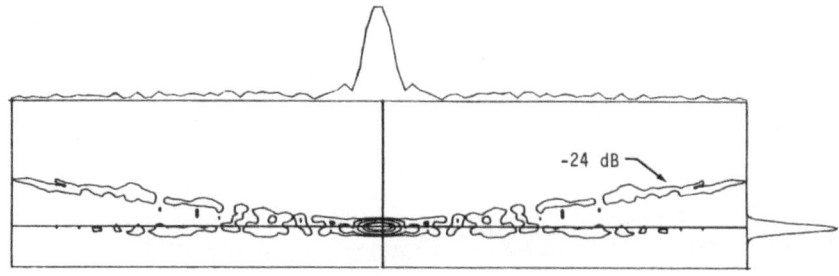

Fig. 1. Contour plot of a reconstructed point reflector using three digital samples per wavelength and uniform aperture apodization (aperture width = 16 mm, range = 80 mm).

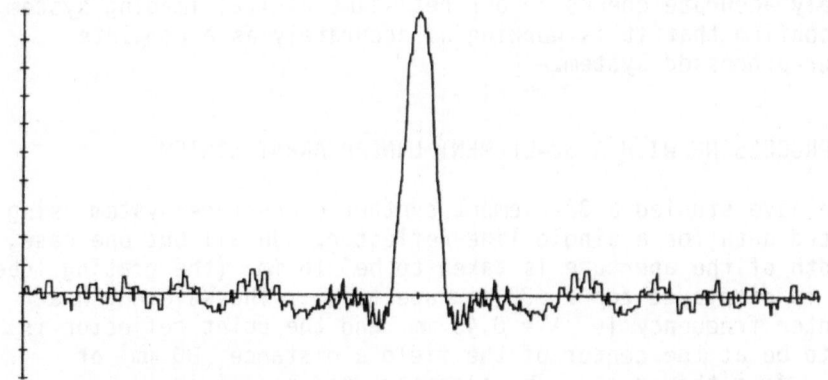

Fig. 2. Finely samples cross-section through the point reflector
        in Fig. 1, showing noise due to phase quantization
        errors.

The results obtained with an array apodized with Hanning
weighting does not show a great improvement. The main lobe
beamwidth at the 3 dB points increases to 3.87 λ , i.e., a
factor of 1.54 . This is slightly better than the theoretically
expected result of 1.62 for a cw waveform; the sidelobe level
remains about the same as for the unapodized array.

It is apparent that the sidelobes are associated with digital
phase errors. A similar set of results for a continuous system,
i.e., one with a very large number of digital samples, is shown in
Figs. 3, 4, and 5. The sidelobe levels now fall off rapidly with
distance, as would be expected for a short pulse imaging system.
Without apodization, the maximum value at the first sidelobe level
is -18.9 dB , as would be expected with this Gaussian pulse. The
sidelobe levels are virtually negligible beyond that point. With
apodization, the point spread function looks virtually Gaussian.

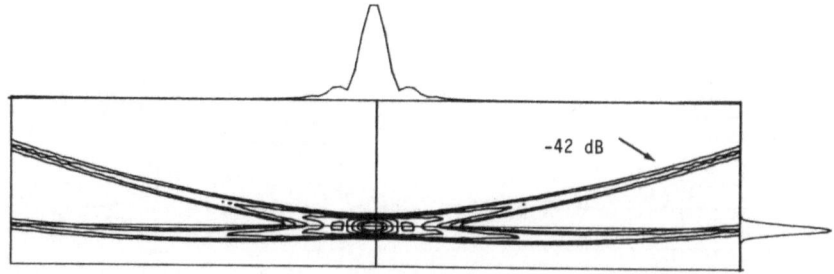

Fig. 3. Contour plot of a reconstructed point reflector using
        continuous sampling of phase and uniform aperture
        apodization.

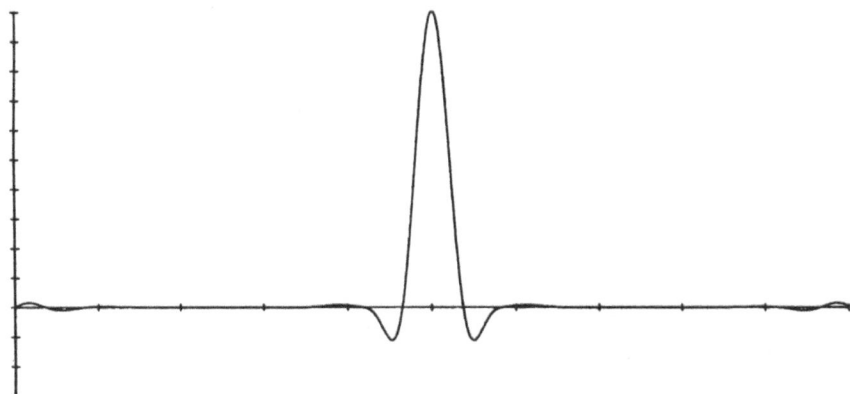

Fig. 4. Finely sampled cross-section through the point reflector
in Fig. 3, showing no phase quantization noise.

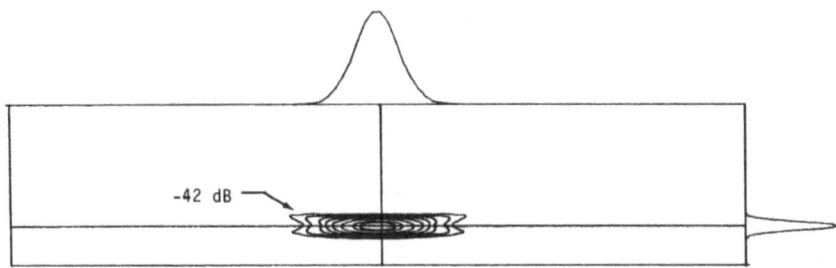

Fig. 5. Contour plot of reconstructed point reflector using
continuous sampling of phase and Hanning aperture
apodization.

It is interesting now to look at the region well away from
the main lobe. In a continuously sampled system, which is
unapodized, there tends to be phase cancellation of sidelobes some
distance from the point reflector, but there is a build-up at the
edges of the range response due to incomplete cancellation.
Apodization, as shown in Fig. 5, eliminates this problem and the
response is much cleaner. We might expect this for the usual
reasons that apodization eliminates edge errors.

The digital phase error formula we gave last year at this
conference predicts a maximum sidelobe level of

$$S_1 = \frac{1}{\mu - 1}\sqrt{\frac{z_0 \lambda}{D^2 \mu}} \tag{1}$$

where μ is the number of digital samples per cycle, z_0 is the range, and D is the width of the aperture. For a three digital phase system, this formula gives a sidelobe level of -19.3 dB , which is in excellent agreement with the computed sidelobe levels (-18.9 dB) that we have observed.

The behavior of the grating lobes is also of interest. We might expect from our earlier results that, for the Gaussian pulse used, the grating lobe level would be -25 dB .[3] The response near the first grating lobe is shown in Fig. 6 for an unapodized array. We observe that the maximum grating lobe level is -24 dB , approximately what would be expected from the theory. The grating lobe is spread out in the range direction approximately uniformly because the non-paraxiality of this system leads to spreading of the contributions from each element of the array over an approximately uniform region. In this case, aperture apodization tends to reproduce itself as a spread of the response in the range direction, which follows the amplitude variation of the response of the array. This is because only a few transducers at a time contribute to a point in the grating lobe.

We conclude that digitization can lead to higher sidelobe levels than are desirable, unless enough phase samples are used. The digital phase error formula that we have employed seems to give good predictions and the grating lobe levels are approximately what one might expect from simple formulae. Any technique that could eliminate problems due to digitization, i.e., filling in the phases or eliminating the sidelobe contributions due to these phase errors, would be desirable for improving the sidelobe level.

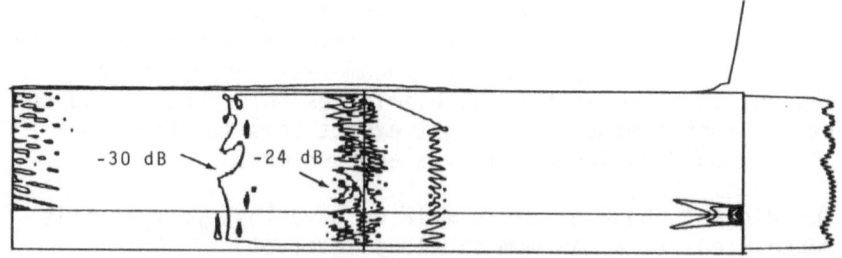

Fig. 6. Contour plot of first grating lobe using continuous
        sampling of phase and uniform aperture apodization
        (aperture width = 32 mm, range = 80 mm).

IMAGING IN THE CYLINDRICAL FORMAT

The linear array system has two main limitations due to its format: (1) sidelobes due to the use of a finite number of transducer array elements (32); and (2) the fact that we can only observe specular reflectors over a limited range of angles.

One approach to this problem is to scan the transducer array mechanically in a circular path centered on the center of the field of view. In this way, a composite picture can be obtained over a wide angular range. By adopting this procedure, one should obtain a definition at least as good as the original focused imaging system. The main lobes from any particular point should coincide, while the sidelobes will not coincide with each other. Consequently, the ratio of the main lobe to sidelobe level is increased and the results are very much like those to be expected from an incoherent imaging system. Furthermore, it is possible by this procedure to image specular reflectors over a wider range of angles.

A simple example of this procedure is shown in Fig. 7. The aim was to image a cylindrical hole, using surface acoustic waves at a center frequency of 3.3 MHz . A picture of this hole, taken from a single viewing angle, is shown in Fig. 8. Several pictures were taken from different angles, and by using fiduciary marks to obtain accurate locations, the pictures were superimposed upon each other photographically (see Fig. 9). The circular shape of the hole shows up clearly and the effective sidelobe levels are considerably decreased, thus forming a far better quality image. We suggest that this procedure could in fact be adopted in practice by supporting the array from a mechanical medical B-scan system of high accuracy. As is done with medical B-scan, a picture could be painted of the region of interest. In this case though, the individual pictures which contribute to the final image would themselves be dynamically focused images, thus improving the definition.

Using an unfocused system, Gebhardt et al[6] in fact demonstrated such a procedure with an NDT application with a rectilinear mechanically-scanned electronic radial sector scan system. The improvement here is that we use a focused beam.

Last year we described a prototype 50 MHz cylindrical imaging system and discussed the theory of this system.[2,4] Since that time we have carried out initial tests with the system at 50 MHz using an unfocused beam to image artificial bulk defects inside a cylindrical rod. As discussed earlier in this paper, this gives the effect of insonifying from a point just outside the rod, as shown in Fig. 10.

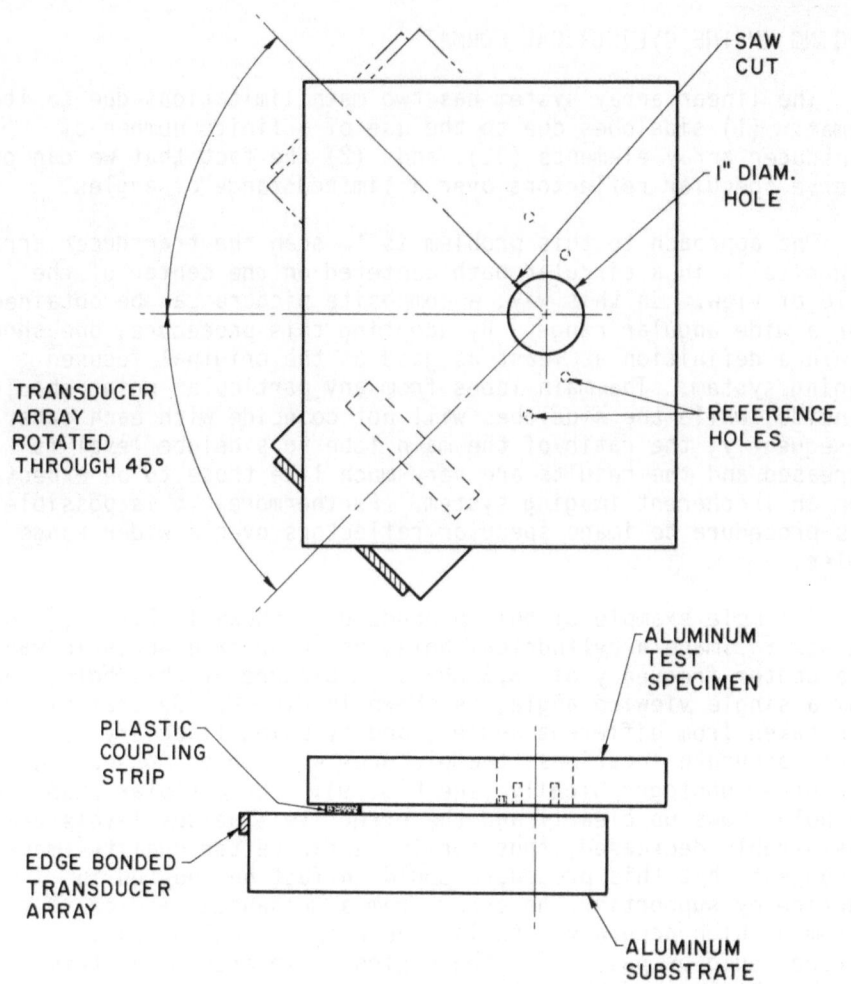

Fig. 7. Geometry of the composite image experiment. A focused
image is obtained from each of seven different viewing
angles. The seven viewing angles are spread over 90° at
15° intervals.

Two examples of imaging with this system are shown in
Figs. 11 and 12, respectively. In Fig. 11, an image of a 370 µm
hole in an aluminum rod is shown. The wavelength of the acoustic
waves at the center frequency is 126 µm , so the hole is
approximately 3 λ in diameter. Note that the hole is fairly
well defined in the image, although the exact location of the hole
boundary is ambiguous due to the presence of concentric rings
corresponding to the transducer impulse response.

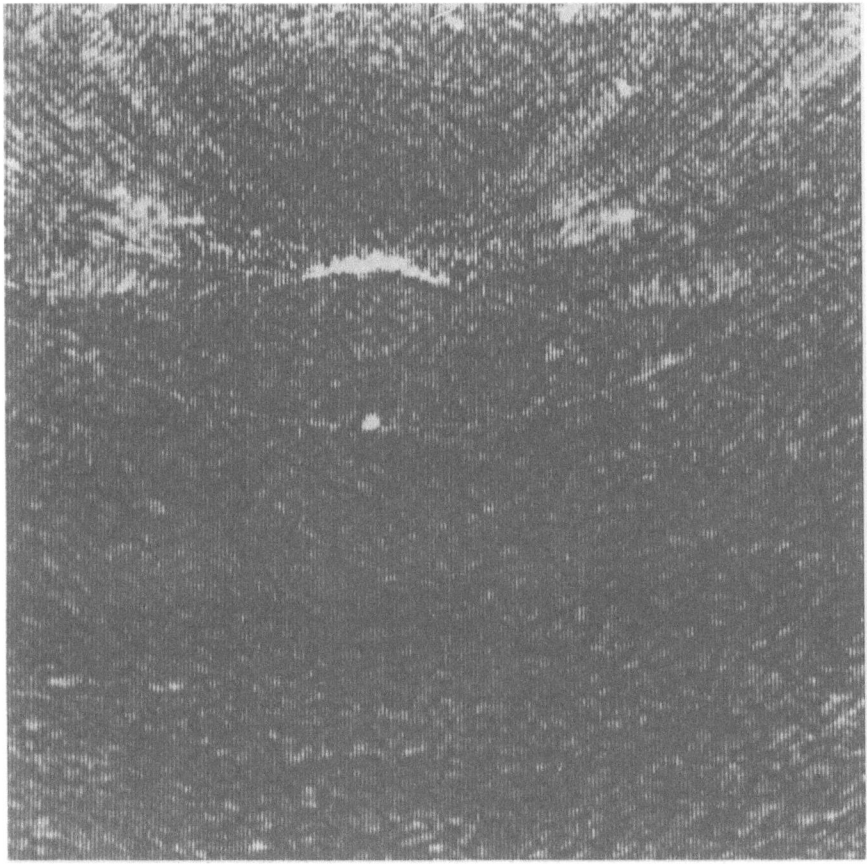

Fig. 8. A focused image of the hole and SAW cut from a single
       viewing angle.

    Figure 12 is a second picture of two holes in an aluminum
rod. A is a 370 μm hole located 0.25 mm off the center of
the rod, and B is a 600 μm hole located 1 mm off center.
The two holes are clearly delineated in the picture. In addition,
there are some extra sidelobes due to multiple echoes of waves
between the holes and mode-converted waves that travel along the
surfaces of the hole.

    These results were obtained by using a single transducer as
the source and receiver; the sample rod was rotated to synthesize
a 256-element circular array enclosing the rod. Last year, we
presented computer simulation results of the images of voids and
inclusions in ceramic material using only a synthesized 32-element

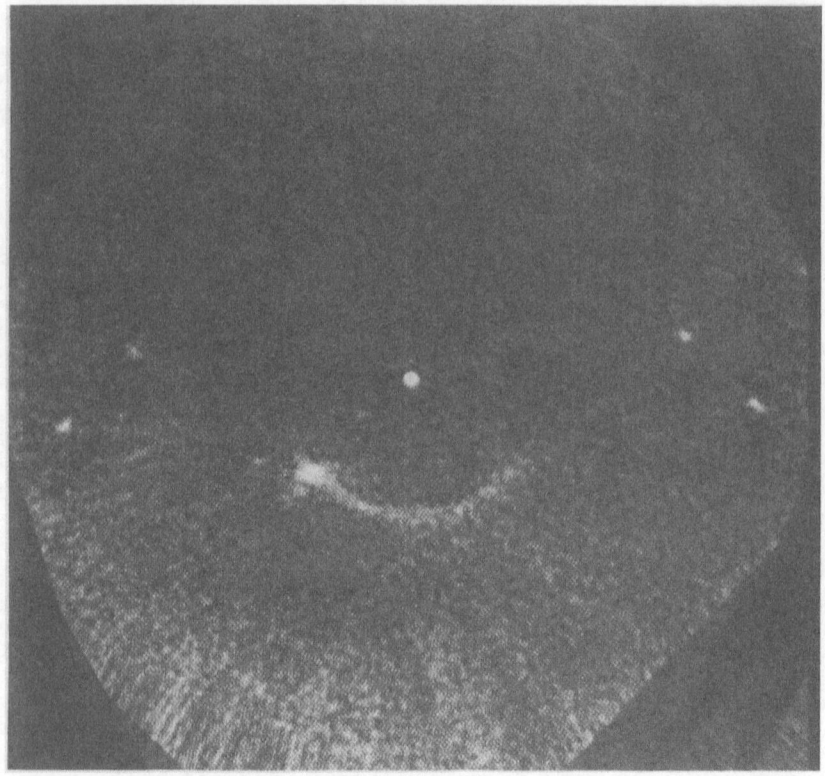

Fig. 9. The composite image obtained by photographically combining
the images from seven different angles.

circular array and linear back-projection as the reconstruction
algorithm.[2,4] Typically, because of the sparse sampling of the
aperture, that procedure introduced in the image the equivalent of
grating lobes and far higher sidelobe levels than those with a
densely-sampled aperture, so the quality of the images was
considerably deteriorated. Therefore, the question remains of how
best to either improve the quality of the images or at least not
to degrade it by using the fewest aperture samples possible.
Obviously, it is desirable to use as small a number of aperture
samples as possible in order to limit the data processing time and
storage required.

    We described last year, a new technique called selective
back-projection,[4] that is specially tailored for reconstructing
images of specular reflectors. The principles of this technique
can be summarized as follows:

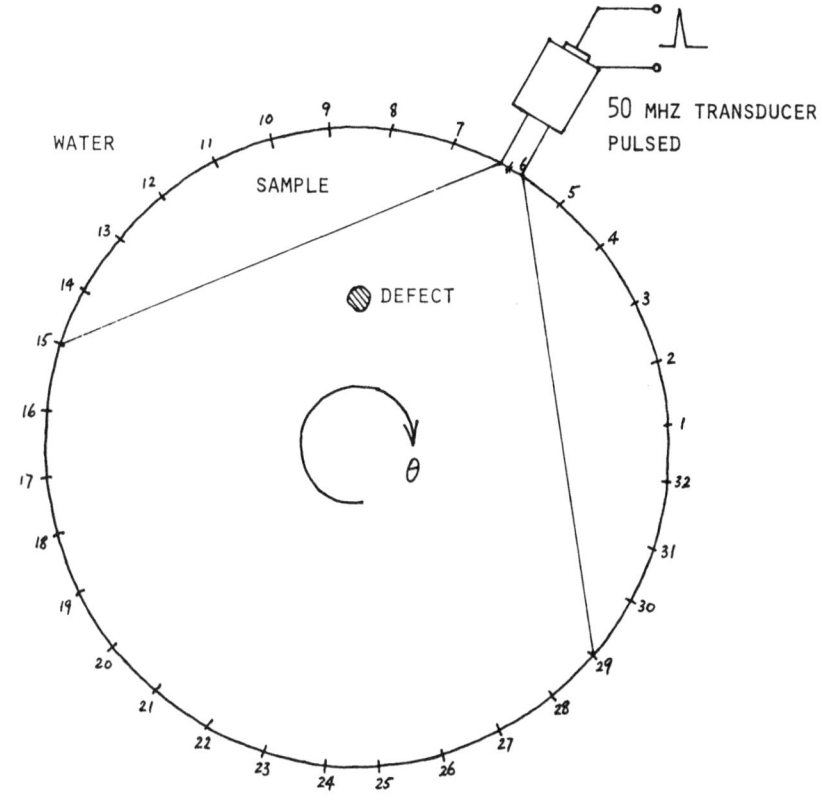

SAMPLE ROTATED TO SYNTHESIZE
A CYLINDRICAL APERTURE

Fig. 10. Synthetic cylindrical aperture.

(1) carry out imaging in the normal way, using time-delay
focusing; and

(2) examine the contributions from all transducer positions
to an image point. Any point which has fewer than M neighboring
transducer positions contributing to it is set to zero.
Otherwise, the image point value is obtained by summing over only
the M or more contributions.

The advantage of this technique is that it tends to eliminate
sidelobes, whose contributions are only from a limited number of
non-adjacent elements of the array. It also tends to produce a
far better quality image of a specular reflector, for which only a
limited number of array elements contribute to a particular image
point.

IMAGING A SINGLE HOLE IN AN
ALUMINUM ROD

GEOMETRY

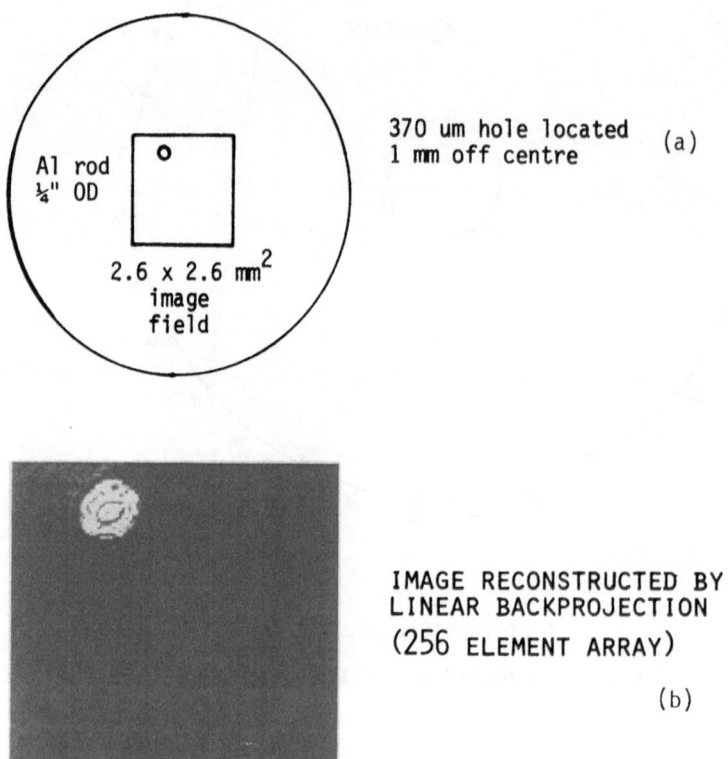

370 um hole located   (a)
1 mm off centre

IMAGE RECONSTRUCTED BY
LINEAR BACKPROJECTION
(256 ELEMENT ARRAY)

(b)

Fig. 11. A 50 MHz image of a 370 μm hole in an aluminum rod,
reconstructed using data from a synthetic 256-element
array.

We have applied this selective back-projection algorithm to
reconstruct the image of the two holes in an aluminum rod, shown
in Fig. 13. The reconstruction is done using data corresponding

IMAGING 2 HOLES IN AN
ALUMINUM ROD

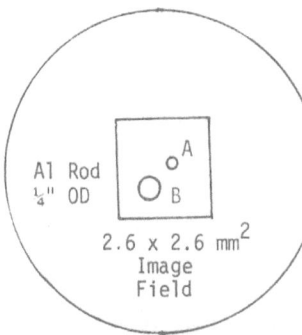

Hole A: 370 μm OD located
        250 μm off centre
                                (a)
Hole B: 600 μm OD located
        1 mm off centre

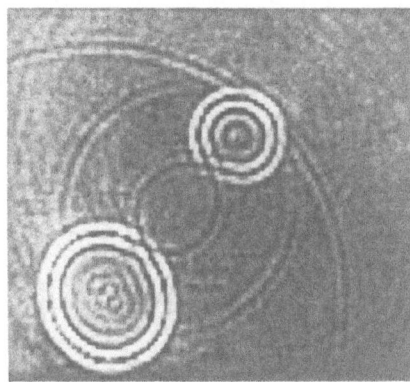

IMAGE RECONSTRUCTED BY

LINEAR BACKPROJECTION

(256-ELEMENT CIRCULAR ARRAY)

(b)

Fig. 12. A 50 MHz image of two holes in an aluminum rod
         reconstructed using data from a synthetic 256-element
         array.

to a 32-element circular array. For comparison, the image was
reconstructed first with linear back-projection, as shown in
Fig. 13a. It is evident that the sidelobe levels are
objectionably high. Figs. 13b to 13e are images obtained by
selective back-projection with a succession of selectivity
factors M . The sidelobe streaks are significantly suppressed

and there is also notable improvement in the definition of the hole boundaries.

   In simple reconstructions with a relatively small number of aperture elements, as we showed last year, the selective back-projection does improve the image. Following this line of reasoning, we realized that it was not even necessary to keep the rf phase information, for now the selective back-projection scheme eliminates most of the information from any point other than the center of the main lobe anyway. Consequently, we have conducted computer simulations on a simplified model in which a short pulse of only one sign is allowed to excite a specular reflector. Then the image is reconstructed by selective back-projection. This procedure is equivalent to using the detected output pulses in the reconstruction process rather than the rf output.

   Fig. 14 shows computer simulated synthetic images of a 500 μm hole in aluminum ceramic (λ = 130 μm) obtained by selective back-projection using a half-cycle 10 ns unipolar pulse. Again, the linearly back-projected image is shown for comparison in Fig. 14a. The resulting image has a pronounced star pattern artifact reminiscent of that in unfiltered tomographic reconstruction. Fig. 14b is a selectively back-projected image with a selectivity factor of 1 . Very good definition on the hole edge, equivalent to a fraction of a wavelength, is obtained with excellent sidelobe levels. Images obtained with M = 2 and 5 are shown in Figs. 14c and 14d. Finally, with M = 6 , no image is obtained as expected, because only a limited number of array elements contribute to any image point on the circular hole.

   The advantage of using a narrow unipolar pulse in selective back-projection can easily be realized in reconstructing a real image by preprocessing the experimental pulse-echo data. The time trace corresponding to a transducer element is first envelope-detected to remove the carrier frequency. Then each peak in the envelope is replaced by a half-cycle unipolar pulse with a height equal to the peak value, whereas all other data points are set to zero. This process is essentially equivalent to improving the time resolution and hence spatial definition of the echoes from the specular reflectors. The images are then reconstructed using this pulse sharpened data set and the selective back-projection algorithm. Fig. 15 shows the enormous improvement in the hole boundary definition achieved by applying this reconstruction technique instead of simple selective back-projection. Images for different selectivity factors M are shown to indicate the effect of varying M on edge-sharpness of the image.

SELECTIVE BACKPROJECTION

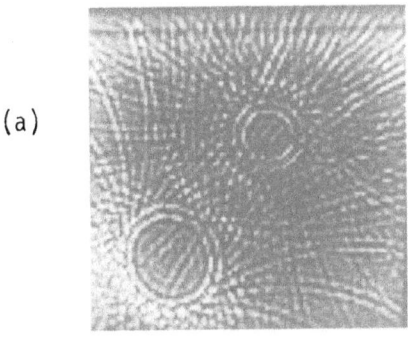

(a)

RECONSTRUCTED BY LINEAR
BACKPROJECTION
32-ELEMENT CIRCULAR ARRAY
(for comparison)

SELECTIVELY BACKPROJECTED IMAGES

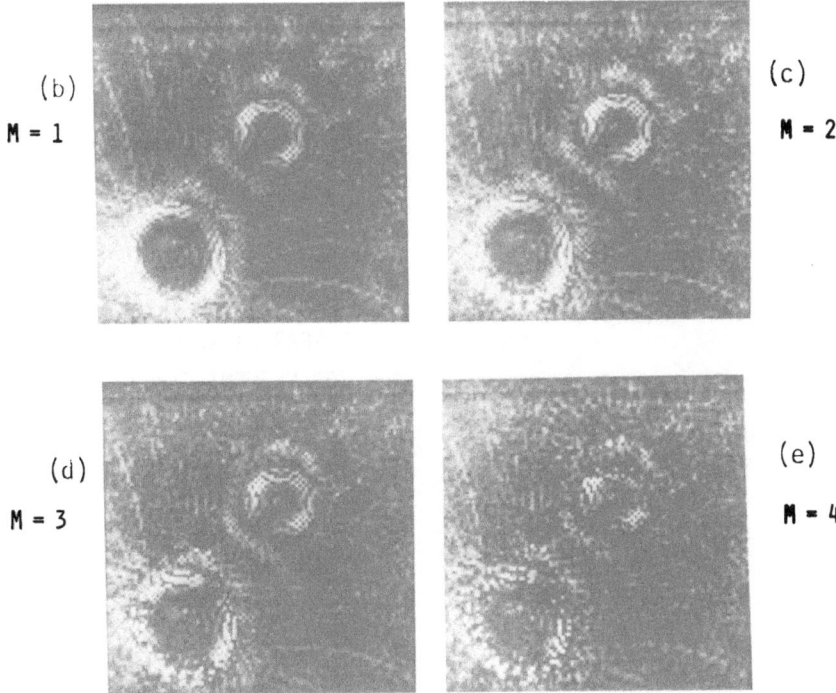

(b)
M = 1

(c)
M = 2

(d)
M = 3

(e)
M = 4

Fig. 13. Selectivity back-projected images of two holes in an
aluminum rod using different selectivity factors.

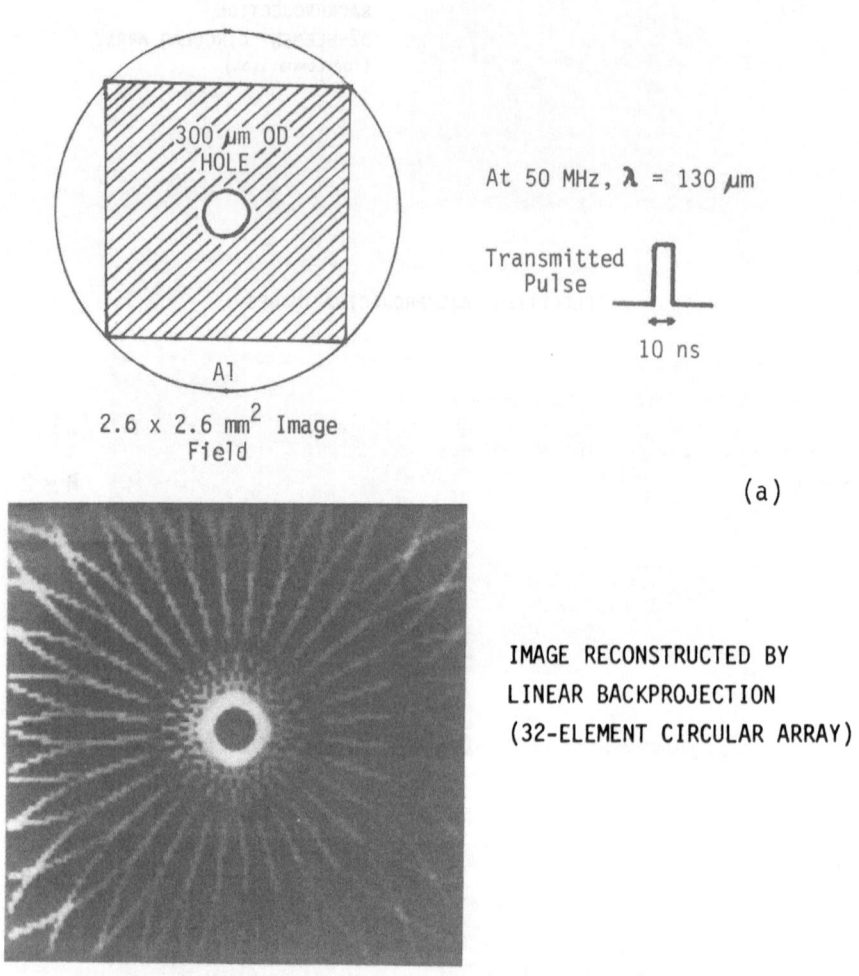

SIMULATED GEOMETRY

300 µm OD HOLE

Al

2.6 x 2.6 mm² Image Field

At 50 MHz, λ = 130 µm

Transmitted Pulse

10 ns

(a)

IMAGE RECONSTRUCTED BY
LINEAR BACKPROJECTION
(32-ELEMENT CIRCULAR ARRAY)

Fig. 14. Simulated images of a 300 µm hole in aluminum reconstructed by selective back-projection and using a half-cycle long transmit pulse.

SELECTIVE BACKPROJECTION

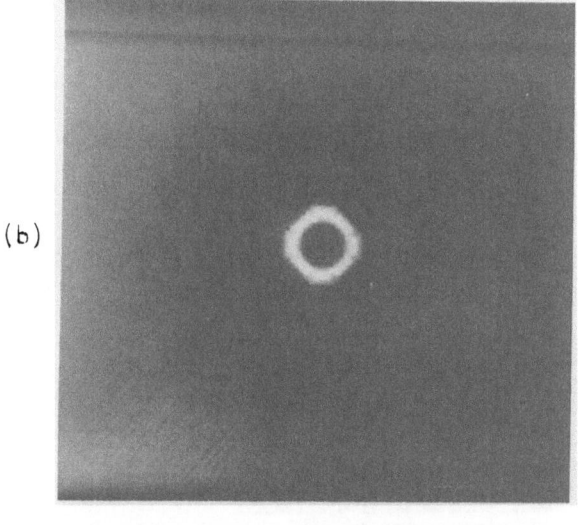

(b)   SELECTIVITY
                                                 FACTOR: 1

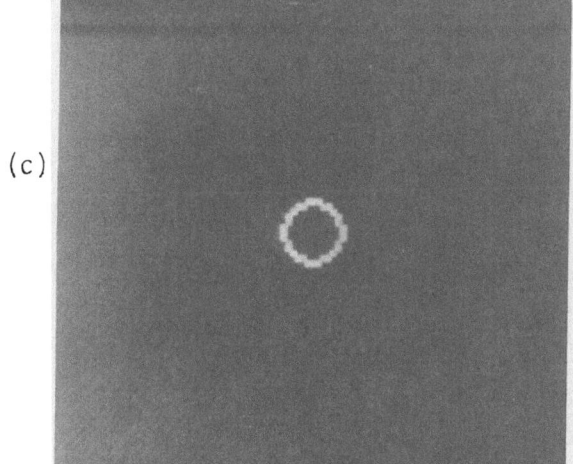

(c)   SELECTIVITY
                                                 FACTOR: 2

(Figure 14 continued)

SELECTIVE BACKPROJECTION

(d) 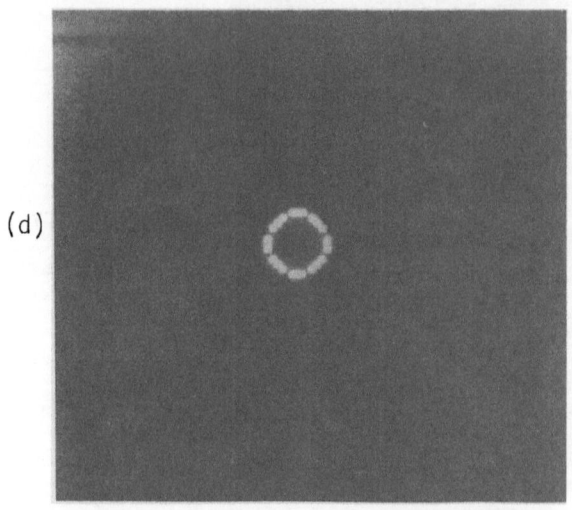   SELECTIVITY
                              FACTOR: 5

(e)    SELECTIVITY
                              FACTOR: 6

SELECTIVE BACKPROJECTION WITH
PULSE SHARPENING

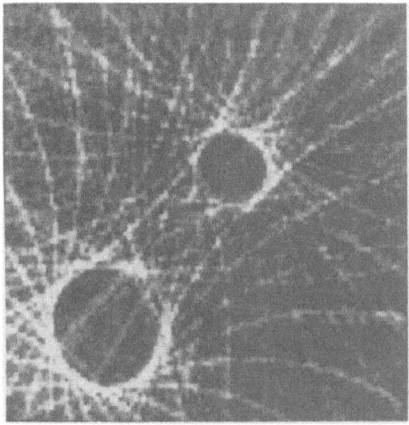

RECONSTRUCTED BY LINEAR
BACKPROJECTION
32-ELEMENT CIRCULAR ARRAY
(for comparison)

(a)

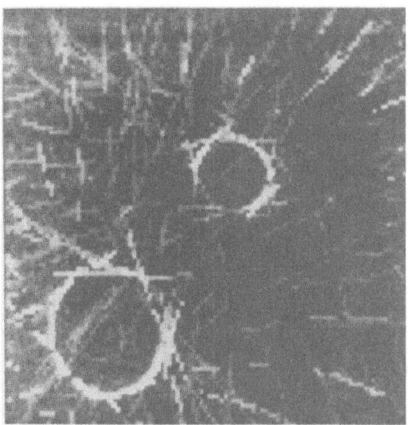

RECONSTRUCTED BY
SELECTIVE BACKPROJECTION
M = 1

(b)

(c) continued on next page

Fig. 15. Selectively back-projected images of two holes in an
aluminum rod using pulse-sharpening.

CONCLUSION

We have reviewed the synthetic aperture imaging technique and
a number of procedures for improving the quality of the image, in
particular the quality of the images of specular reflectors. We
have found that digital phase errors are important and could
increase sidelobe levels. Grating lobe levels are not as strongly
dependent on the digital sampling rate.

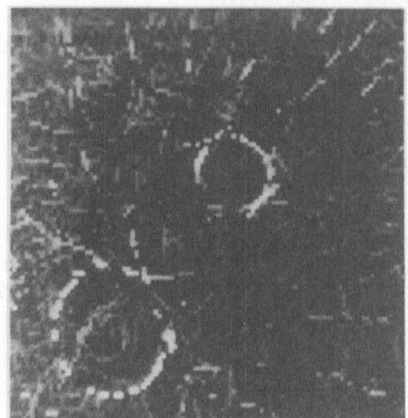

RECONSTRUCTED BY
SELECTIVE BACKPROJECTION
M = 2

(c)

Figure 15. (Continued)

A new reconstruction procedure of selective back-projection, using detected waveforms and pulse sharpening, has been demonstrated. Computer simulation and experimental results appear to be extremely promising. We believe that this technique is an entirely new and important way for imaging specular reflectors and for imaging in situations where the image field is relatively uncluttered and the aperture is sparsely sampled. Extrapolations of this technique, which can be used in situations where the image field is denser, such as body tissue, do appear to be possible.

We have also demonstrated that we can, for the first time, carry out synthetic aperture imaging at frequencies as high as 50 MHz . We are currently working on a contacting system to obtain images at freuqencies as high as 300 MHz . Ultimately, this should lead to the possibility of carrying out acoustic microscopy in regions inside a solid material without the normal limitations due to aberration of lenses.

ACKNOWLEDGMENT

This work was sponsored by the Ames Research Laboratory for the Defense Advanced Research Projects Agency and the Air Force Materials Laboratory under Contract No. SC-81-009 and the Air Force Office of Scientific Research under Contract No. F49620-79-C-0217.

REFERENCES

1. S. Bennett, D. K. Peterson, D. Corl, and G. S. Kino, "A Real-Time Synthetic Aperture Digital Acoustic Imaging System, Proc. Acoustic Imaging Conference, Cannes, France,1980.
2. K. Liang, B. T. Khuri-Yakub, C-H. Chou, G. S. Kino, "A Three-Dimensional Synthetic Focus System," IEEE Ultrasonics Symposium, Boston, Mass., November, 1980.
3. S. Bennett, D. K. Peterson, D. Corl, G. S. Kino, "Real-Time Synthetic Aperture Imaging System," IEEE Ultrasonics Symposium, Boston, Mass., November, 1980.
4. K. Liang, B. T. Khuri-Yakub, C. H. Chou, and G. S. Kino, "A Three-Dimensional Synthetic Focus System," Proc. Acoustic Imaging Conference, Cannes, France, 1980.
5. G. S. Kino, "Acoustic Imaging for Nondestructive Evaluation," Proc. IEEE, $\underline{67}$ (4), 510-522, April, 1979.
6. W. Gebhardt, F. Bonitz, H. Woll, and V. Schmitz, "Determination of Crack Characteristics, Size and Orientation of Defects by Phased Array Techniques in NDT," Third International Conference on Nondestructive Evaluation in the Nuclear Industry, Salt Lake City, Utah, February, 1980.

BIBLIOGRAPHIC INDEX

INDEX